Wagner · Hanke · Bode · Durst (Eds.)
High Performance Computing in Science and Engineering,
Munich 2002

Springer-Verlag Berlin Heidelberg GmbH

Siegfried Wagner · Werner Hanke
Arndt Bode · Franz Durst

Editors

High Performance Computing in Science and Engineering, Munich 2002

Transactions of the First Joint HLRB and KONWIHR
Status and Result Workshop, October 10-11, 2002,
Technical University of Munich, Germany

With 253 Figures, 79 in Color, and 50 Tables

 Springer

Editors

Siegfried Wagner

Institut für Aerodynamik
und Gasdynamik
Universität Stuttgart
Pfaffenwaldring 21
70550 Stuttgart, Germany
e-mail: wagner@iag.uni-stuttgart.de

Werner Hanke

Institut für Theoretische Physik
und Astrophysik
Universität Würzburg
Am Hubland
97074 Würzburg, Germany
e-mail: hanke@physik.uni-wuerzburg.de

Arndt Bode

Lehrstuhl für Rechnertechnik
und Rechnerorganisation
Institut für Informatik
Technische Universität München
Boltzmannstraße 3
85748 Garching, Germany
e-mail: bode@in.tum.de

Franz Durst

Lehrstuhl für Strömungsmechanik
Friedrich-Alexander-Universität
Erlangen-Nürnberg
Cauerstraße 4
91058 Erlangen, Germany
e-mail: durst@lstm.uni-erlangen.de

Cataloging-in-Publication Data applied for

Bibliographic information published by Die Deutsche Bibliothek

Die Deutsche Bibliothek lists this publication in the Deutsche Nationalbibliografie;
detailed bibliographic data is available in the Internet at <http://dnb.ddb.de>.

Mathematics Subject Classification (2000):
65Cxx, 65K05, 68M20, 70-08, 70Fxx, 74F10, 74F15, 74L99, 76-04, 76G25, 76Txx, 81-04, 81-08,
81V05, 81V10, 83-04, 85-08, 85A40, 86-04, 86A15, 92-04, 92Exx

ISBN 978-3-540-00474-5   ISBN 978-3-642-55526-8 (eBook)
DOI 10.1007/978-3-642-55526-8

http://www.springer.de
© Springer-Verlag Berlin Heidelberg 2003
Originally published by Springer-Verlag Berlin Heidelberg New York in 2003

The use of general descriptive names, registered names, trademarks, etc. in this publication does not
imply, even in the absence of a specific statement, that such names are exempt from the relevant pro-
tective laws and regulations and therefore free for general use.

Typeset by the authors using a Springer TeX macro package
Cover design: *design & production* GmbH, Heidelberg

Printed on acid-free paper   46/3142/LK - 5 4 3 2 1 0

Preface

High-Performance Computers (HPC) have initiated a revolutionary development in research and technology since many complex and challenging problems in this area can only be solved by HPC and a network in modeling, algorithms and software. In 1998 the Deutsche Forschungsgemeinschaft (German Research Association) recommended to install an additional Federal High-Performance Computer followed by the one in Stuttgart. In January 1999 the Wissenschaftsrat (German Science Council) decided that the Leibniz-Rechenzentrum (Computing Center) of the Bavarian Academy of Sciences in Munich should run the second Federal High-Performance Computer in Germany. The investment cost of this Höchstleistungsrechner in Bayern (HLRB) was borne by the Federal Government of Germany and the Free State of Bavaria whereas the operating cost was at the expense of the Bavarian Government only. The operation of the HLRB is organized in combination with the

- Leibniz-Rechenzentrum (LRZ) of the Bavarian Academy of Sciences as the operating authority of the HLRB
- Steering Committee of the HLRB
- Competence Network for Technical/Scientific High-Performance Computing in Bavaria (KONWIHR).

In 2000 a Hitachi SR8000-F1 was installed. It was the first Teraflops Computer in Germany and reached a peak performance of two Teraflops after an extension at the end of 2001. The goal of HLRB is to provide computer facilities necessary to solve challenging scientific and technological problems that cannot be solved on big servers but require large (storage) high-performance (very fast) computers and efficient software. Scientists from universities and research institutions of the Federal Republic have access to the HLRB.

Since its start of operation in April 2000, 95 projects were admitted to use the SR8000-F1 after careful evaluation. The Competence Network KONWIHR has so far supported 36 projects. All projects were evaluated by external reviewers and by the members of the steering committee representing the major disciplines. Only those projects were admitted that could only be solved on a HPC and not on workstations or other servers. In addition, the users have to prove their computing expertise. The research projects currently computed at HLRB mainly stem from solid-state physics, high-energy physics, geophysics, astrophysics, biophysics, cosmology, computational fluid dynamics (CFD), chemistry, applied mathematics, and informatics.

After more than two years of HLRB operation, the steering committee together with KONWIHR decided to carry out a joint HLRB and KONWIHR Result and Reviewing Workshop in October 2002. Since each project leader has to write a progress report every year anyway these manuscripts could be prepared in a form ready for publication. External reviewers and members of the steering committee evaluated these manuscripts and selected those to be presented during the workshop and to be included in the volume. Thus, 39 papers are contained in the present volume. They represent the main subjects that are already mentioned above. They also show the broad field of application of high-performance computing in science and engineering.

The scientific projects deal with numerical simulations of problems that come from physics (49,2 %), fluid dynamics (38,2 %), chemistry (5,9 %), biology (1,5 %), geophysics (0,9 %), biophysics (0,7 %) and others, like applied mathematics, informatics etc. The percentage is based on the total CPU hours from start in the year 2000 up to the third quarter of 2002.

The computing times were very long for many projects. For instance, during September 2002, almost 20 % of the total available computing capacity was occupied by one single, but very large project in solid-state physics. The research team investigated metal-insulator transitions and developed a new computation scheme that merges electronic band structure calculations and the dynamical mean field theory. A further large project in solid-state physics studied the microscopic mechanism behind high-temperature superconductivity. Both these solid-state projects aim at replacing the still mainly practiced empirical search for improved material properties (such as higher superconducting transition temperatures) in technologically very interesting substances by a guided search, based on a first-principle understanding. A very demanding problem in fundamental physics was the simulation of strongly interacting particles by quantum field theory, the quantum chromodynamics. Models of type Ia supernova explosions were used to study the expansion history of the universe. Also in computational fluid dynamics (CFD), direct numerical simulation (DNS) was applied to simulate boundary layer interaction along a curved wall with oscillating oncoming flow. DNS and LES (large eddy simulation) were applied to study passively controlled turbulent boundary layer flows over a sharp-edged and a smoothly contoured backward facing step. Up to 50 million grid points were needed to resolve the flow structure and its driving mechanisms within this project. In an additional project, the Lattice Boltzmann procedure was investigated regarding its applicability for direct numerical simulation of three-dimensional incompressible turbulence. LES and DES (detached eddy simulation) were the tools to simulate the flow around high-lift airfoils where RANS (Reynolds-averaged Navier-Stokes) methods usually fail when the boundary layer separates.

With this first proceedings volume of the HLRB/KONWIHR workshop the continuous financial supports by the Free State of Bavaria and the Deutsche Forschungsgemeinschaft are gratefully acknowledged. They not only provided the budget to establish and run the HLRB, but also supported many projects during the period covered by the workshop. The recommendations of the Wissenschaftsrat were necessary to establish the Federal High-Performance Computing Center in Munich and are appreciated by all users. Finally, we would like to thank the Springer-Verlag for publishing this volume.

München, October 2002

Siegfried Wagner
Werner Hanke
Arndt Bode
Franz Durst

Table of Contents

Part III Biosciences

Part IV Chemistry

Part VI Geophysics

Part VII Fundamental Physics

Part VIII Computer Science

Part I

High Performance Systems

Leibniz-Rechenzentrum

Bayerische Akademie der Wissenschaften
Barer Str. 21
80333 München

This introductory part describes the high performance hardware of the Hitachi SR8000-F1. All computations of projects described in this book were run on this platform.

TeraFlops Computing with the Hitachi SR8000-F1: From Vision to Reality

Reinhold Bader, Matthias Brehm, Ralf Ebner, Helmut Heller, Ludger Palm, and Frank Wagner

Leibniz–Rechenzentrum der Bayerischen Akademie der Wissenschaften
Barer Straße 21
80333 München, Germany
{*Bader, Brehm, Ebner, Heller, Palm, Wagner*}*@lrz-muenchen.de*

Abstract. These proceedings are concerned with results from scientific investigations obtained with calculations done on the Hitachi SR8000-F1 supercomputer at Leibniz Computing Center. The following contribution gives an overview of the innovative Hitachi architecture and the configuration of the machine at LRZ.

1 HPC services at the Leibniz Computing Center

Toward the end of the 1990's the German Scientific Council had decided on a policy concerning high performance computing resource development in Germany. This policy encompasses a promotion of selected computing centers to "centers for national high performance computing", which in intervals of around two years receive funding for the installation of state-of-the art high performance computers as well as for the competence infrastructure required to enable efficient usage of these systems by the scientific community. The Leibniz Computing Center (Leibniz–Rechenzentrum, LRZ) of the Bavarian Academy of Sciences in Munich was the first institution selected for promotion in 1999, following a competition between computing centers throughout Germany.

In the first quarter of 2000, a Hitachi SR8000-F1 intended to serve as German Federal Top-Level Compute Server in Bavaria was installed at LRZ; the machine was augmented to its final configuration by half its computing power in January 2002. With a peak performance of 2.02 Tera-Flop/second on its 168 compute nodes the SR8000-F1 even in 2002 is still the fastest computer in civilian use throughout Europe; it appears on the 14th position on the 19th TOP-500 list of supercomputers (June 2002) with a LINPACK performance of 1.65 Tera-Flop/second. The decision to purchase the Hitachi machine was based on the results of HPC application benchmarks which took into consideration that the required system should offer optimal performance for real-world applications. Access to the machine is open to all scientists throughout Germany whose compute requirements exceed the capacity of any other available computing platforms. A refereeing process taking around two months is required to decide whether a submitted project is suitable for the SR8000-F1.

2 Using the Hitachi SR8000-F1 at LRZ

With the SR8000-F1 the Leibniz Computing Center offers one of the most innovative scalable supercomputer architectures for problems with extreme requirements for memory and performance. Earlier supercomputers were based on one of the two design principles described in the following:

- Parallel vector systems – an example is furnished by the Fujitsu-Siemens VPP700/52 run by LRZ – offer extremely good memory bandwidth, resulting in a very efficient usage of the single vector node. However, the price/performance ratio nowadays is very bad, and scaling to very high performance additionally requires an expensive crossbar connection. For scientific applications, which for the most part are very memory intensive, this kind of system is very suitable.
- Massively parallel RISC systems use standard components from workstation and PC technology which are then networked and integrated to form a multi-computer. Due to the very good price/performance ratio it is possible to achieve very high peak performance for such a system. However for the algorithms typically used in technical and scientific computing only a very small fraction of the peak performance is actually observed. This is mainly due to the transaction-oriented cache-based memory hierarchy resulting in low effective memory bandwidth; furthermore, standard networking technology limits scalability of parallel programs due to latency and bandwidth issues.

Hitachi, beginning in the mid-90's, has pioneered the development of an integration of the above two design principles into a single architecture. Other hardware vendors (Cray, IBM) have followed up on this work. The Hitachi approach is based on the following premises:

- Use of a standard microprocessor (IBM PowerPC) instruction set, resulting in a good price/performance ratio. Furthermore, portability of standard applications is easily achieved.
- Introduction of a second path to memory circumventing the cache. Hence not only transaction-dominated applications will run well, but also memory-intensive ones. This enables codes developed for vector computers to perform adequately.
- Introduction of a hardware interface for synchronization of processes in a shared memory environment. This enables emulation of the multipipelining well-known from vector systems.

3 Capabilities of the SR8000-F1

3.1 Hardware and Operating System

The Hitachi SR8000-F1 at LRZ is a clustered SMP system with 112 nodes, each containing 9 CPUs driven at 375 MHz. Each CPU has 2 multiply-add floating-point units. Of the 9 CPUs, 8 are available for computational work while the ninth serves as system controller and initializer. The computation CPUs may be used in MPP mode or, by making use of the special hardware features described above, be operated as a unified node with 8 vector pipes. Compared to a classical vector system one still retains the advantage of high scalar performance. Table 1 provides an overview of the technical data of the LRZ machine. Note that the "Triad Performance" printed in bold in the table below in fact turns out to be a good indicator of the performance sustainable by the scientific applications which run on the machine.

Table 1. Overview of the SR8000-F1 at LRZ

Number of Nodes	168
Processors per Node	8 (+1 for administrative purposes)
Number of Processors	1344
Peak Performance per Node	12 GFlop/second
Peak Performance of system	2016 GFlop/second
LINPACK Performance	1645 GFlop/second
LRZ benchmark weighted performance	668 GFlop/s
Triad Performance from Memory	**246 GFlop/s**
Memory per Node	8 GByte (6 available in user space)
Memory in whole system	1376 GByte
L1 Data Cache	128 kilo-Byte
Aggregated disk storage	10 Tera-Byte
Disk space for HOME Directories	960 GByte
Disk space for temporary data	5900 GByte
Aggregated I/O bandwidth	2.7 GByte/s
MPI Latency	14 microseconds
MPI Bandwidth	950 Mega-Byte/second

108 of the nodes are equipped with 8 GBytes of memory, and for programs with exceptionally high memory requirements 4 nodes with 16 GBytes memory each are available. The operating environment is a OSF/Mach-based micro-kernel UNIX, HI-UX/MPP. There is only one file system tree across the whole machine: This single system image provides for efficient system administration. Since applications need to be able to address more than 2 GBytes of memory, the operating system (as well as the compilers, whose features are described in more detail below) provide full support of the 64 bit programming model. The provided memory bandwidth is nominally 4 GByte/s for each CPU, or 32 GByte/s per node, measurements show that

3.6 GByte/s can be reached by a single CPU in a node, and more than 23 GByte/s by the full 8-CPU node. For intermediate storage of large amounts of data, two kinds of globally available filesystems with scalability features are available:

- **file-striped file system:** performs very well when many small to medium-sized files are processed in parallel. One file system of size 1.2 Tera-Bytes is available for this purpose.
- **block-striped file system:** scales by subdivision of a presumably very large file into blocks of subfiles. The MPI-IO subsystem requires the use of this file system. Three file systems with a total size of 4.7 Tera-Bytes are operated in this mode.

3.2 Programming Paradigms

Corresponding to the hybrid architecture there is a rich choice of programming models:

PVP: As mentioned above, two paths to memory are available to each CPU. The first path is the standard one commonly implemented on RISC architectures, where data are transferred from memory to cache. The memory latency can be hidden by issuing *prefetch* instructions. The second path to memory offers a possibility to load single words from memory to register directly. The efficiency of this depends on having many registers available on the CPU, and

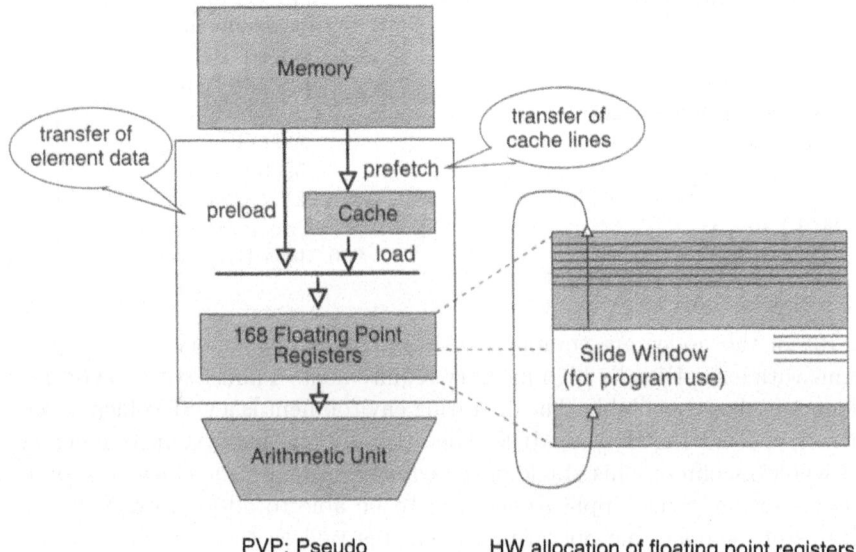

Fig. 1. Pseudo-Vector-Processing

again on being able to hide latency by issuing *preload* instructions. Of the total of 160 registers on the SR8000 CPU, 128 are available for preloads. Figure 1 illustrates these concepts, known under the name "pseudo-vectorization", or PVP. The prefetch and/or preload instructions are automatically inserted into the executable code in a suitably pipelined fashion by the compiler; for contiguous access to memory the compiler will usually issue prefetches, for strided access preloads are usually chosen. However, the user may override the automatic facility by inserting directives into his code. PVP should always be used on memory intensive codes, independently from the other models discussed below, since hiding the latencies can give up to nearly an order of magnitude performance boost.

COMPAS + MPI: By combining all the processors of a node to work on a single program in shared memory one obtains good performance for traditional vector codes. MPI is used only for communication of data between the nodes. Shared memory parallelization (COMPAS stands for "cooperative microprocessors in shared address space") may be either achieved via automatic compilation mechanisms assisted by user-inserted directives, or by usage of OpenMP directives which offer a portable way of programming

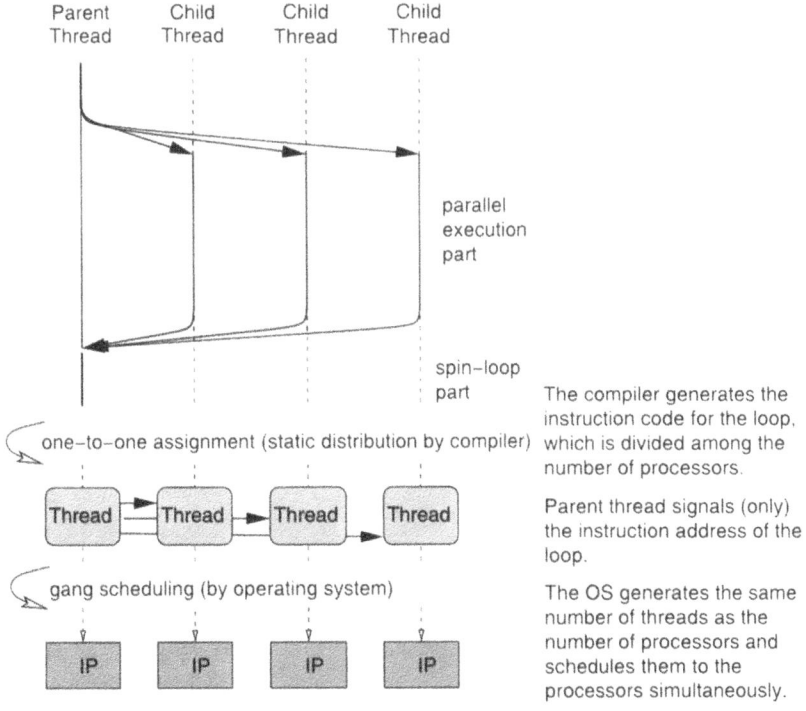

Fig. 2. Shared Memory Parallelization (COMPAS)

shared memory architectures. Figure 2 illustrates the method of distributing computational work among the processors of a node.

MPP: MPI is used for communication between instruction processors, be they within the same node or on two different nodes. This offers an easy migration path for codes traditionally run on a classic MPP system, e. g. the T3E.

The optimizing Compilers for Fortran 95, C, and C++ offered by Hitachi include support for PVP and COMPAS, version 2 of the OpenMP standard is supported within all programming languages; an implementation of the MPI-2 message passing standard is available for distributed-memory programming. Standard tools for debugging (Totalview version 5) as well as for MPI performance analysis and sub-routine tracing (VAMPIRtrace) have been ported to the SR8000-F1.

Part II

Computational Fluid Dynamics

Rolf Rannacher

Institut für Angewandte Mathematik
Universität Heidelberg
Im Neuenheimer Feld 293/294
69120 Heidelberg, Germany

Traditionally, Computational Fluid Dynamics (CFD) is one of the main consumers of super-computing power. The growing need for quantitatively accurate computation of complex viscous flow is extremely demanding and easily
reaches the limits of any existing computer. This is seen most prominently
in classical turbulence simulation and when the flow model is accompanied
by other physical mechanisms such as in fluid-structure interaction, in multiphase flow, or in flow underlying combustion and detonation processes. In
many of these cases successful and internationally competitive research relies
on the availability of powerful super-computers such as the Hitachi SR8000-
F1 at the LRZ and corresponding new software tools for exploiting their
parallel architecture. This requires a close cooperation of scientific users and
computer scientists. Such an interaction is realized for the group of projects
supported within the competence network KONWIHR in Bavaria.

This volume contains a representative collection of projects from CFD
which have been carried out on the SR8000-F1. Most of these projects are at
the forefront of computer-based physical and engineering research.

The first group of articles is concerned with the direct numerical simulation (DNS) of various kinds of turbulent flow which is particularly computation intensive. The group of F. Durst at Erlangen-Nürnberg (P. Lammers,
K. N. Beronov, G. Brenner and F. Durst) reports on results for the DNS of
channel flow by the so-called Lattice Boltzmann method. This new approach
to flow computation seems to develop into a competitor of the traditional grid
methods. The team of H. Wengle at Munich (J. Neumann and H. Wengle) re-
alizes the DNS of turbulent flow over a backward facing step, while W. Rodi
at Karlsruhe (J. Wissink and W. Rodi) uses DNS for turbulent boundary
layer separation along curved walls. Both applications carry the DNS closer

to technical applications for Reynolds numbers of moderate size. Hence, there is still a need for more accurate turbulence models for high-Reynolds number flow. This is the goal of the project of R. Friedrich at Munich (J. Kreuzinger and R. Friedrich) who develops a new four-equation turbulence model on the basis of detailed DNS results.

The second group of articles deals with the description of turbulent flow using Large-Eddy-Simulation (LES). This approach bridges the gap between the cost-intensive DNS and the Reynolds-averaged Navier-Stokes (RANS) model using turbulence models. Still, the computational costs are extremely high for achieving accuracy by the LES as required in technical applications. In the project of M. Breuer at Erlangen-Nürnberg (M. Breuer, N. Jovicic, and K. Mazaev) flow around a high-lift configuration is investigated, and F. Friedrich at Munich (F. Tremblay, M. Manhart and F. Friedrich) studies flow around a cylinder for higher Reynold numbers. Both simulations yield interesting new insight into the structure and the quantitative properties of boundary layer separation in turbulent flow.

The third group of articles concerns flows in combination with additional physical mechanisms. The collaborative project of E. Durst at Erlangen-Nünberg and E. Rank at Munich (M. Glück, M. Breuer, E. Durst, A. Halfmann and E. Rank) analyzes the fluid-structure interaction of thin wind-exposed structures. The group of R. Rank at Munich and M. Krafczyk at Braunschweig (M. Schulz, J. Tölke, M. Krafczyk, and E. Rank) uses the Lattice Boltzmann method for computing multiphase flow. Finally, the project of W. Hillebrand at the MPI Munich (J.C. Niemeyer, M. Reinecke, W. Hillebrand) simulates supernova explosions using a multidimensional model of subsonic propagation of thermonuclear flames. Such astrophysical computations are particularly demanding due to the extreme physical complexity of the underlying models.

The results reported in these articles demonstrate that the availability of high-performance computer power can not only lead to new insight into complex physical processes but also brings numerical simulations closer to answering critical questions in real-life applications.

Numerical Prediction of Deformations and Oscillations of Wind-Exposed Structures

Markus Glück[1], Michael Breuer[1], Franz Durst[1], Ansgar Halfmann[2] and Ernst Rank[2]

[1] Institute of Fluid Mechanics, University of Erlangen-Nürnberg
Cauerstraße 4, 91058 Erlangen, Germany
{glueck, breuer, durst} @lstm.uni-erlangen.de
[2] Institute of Computer Science in Civil Engineering, Technical University of Munich
Arcisstraße 21, 80290 München, Germany
{halfmann, rank} @bv.tum.de

Abstract. In this contribution a partitioned, but fully implicit coupling algorithm was applied to the numerical prediction of fluid-structure interactions occuring at lightweight membranes in civil engineering. The first test case was the steady deformation of a mobile umbrella. The second application was concerned with an unsteady wind gust exciting a membranous roof to perform structural oscillations.[*]

1 Introduction

Interaction phenomena between fluids and structures can be found in many engineering and also medical disciplines such as civil, mechanical and medical engineering, shipbuilding, and bio-medicine. Although the simulation tool presented in this paper was designed for *civil engineering* applications, it could also be applied to other fields. The partitioned *coupling approach* for time-dependent fluid-structure interactions, which is described in more detail in *Glück et al.* [5] can be applied to thin shells and membranous structures with large displacements. The frame algorithm connects a three-dimensional, finite-volume based multi-block flow solver for incompressible fluids [3] with a finite-element code for geometrically non-linear structural problems [7] using a commercial coupling interface (MpCCI [1]). Thus a high modularity is achieved and the whole range of opportunities with these two powerful codes – each of them highly adapted to its specific field of application – can be used also for coupled simulations.

---

[*] Results presented at the GAMM Conference, Augsburg, Germany, March 25-28, 2002.

2 Application to a mobile umbrella

In a joint project with the Institute of Aerodynamics and Gasdynamics (IAG) at the University of Stuttgart the *wind tunnel model* of a mobile umbrella was examined (see Fig. 1). In reality, this umbrella is used as temporary roofage for parties, concerts etc. In the experiment, a caoutchouc membrane with a diameter of 600 mm was pre-strained by 32 radial cables. In the middle, the roof was mounted at a vertical pole. The elevation of the deformed umbrella was optically measured by two laser triangulators along several radii. For a detailed description of the *wind tunnel tests*, see *Wagner et al.* [8] and concerning the *numerical simulation*, see *Halfmann* [6].

The two different *CFD grids* contained 153,600 and 1,228,800 control volumes, respectively, with 1920 and 3072 of them flanking the roof. One part of the finer one is depicted in Fig. 2. Fig. 3 shows the whole CSD model consisting of the membran, the pillars and the cables in the initial state. The *CSD mesh* of the membran consisted of 2752 quadrilateral elements. Altogether, three nodes of the *HITACHI SR8000-F1* were used (two for CFD, one for CSD). A maximum memory of 880 MB per node was required for the coupled simulation on the finest CFD grid.

The umbrella was attacked by a *time-independent* flow in positive x direction parallel to the ground. The CFD calculation was carried out in a stationary way by means of the standard k-ε turbulence model, because a *steady deformation state* was found in the experiment. For the velocity at the inlet to the fluid domain a power law was assumed in accordance with the wind tunnel measurements:

Fig. 1. Wind tunnel model [8] (deformed state, flow from the left)

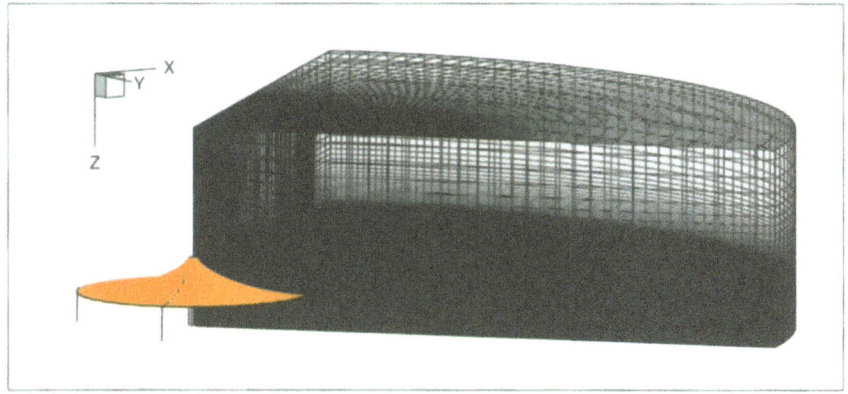

Fig. 2. Part of the CFD grid around the mobile umbrella

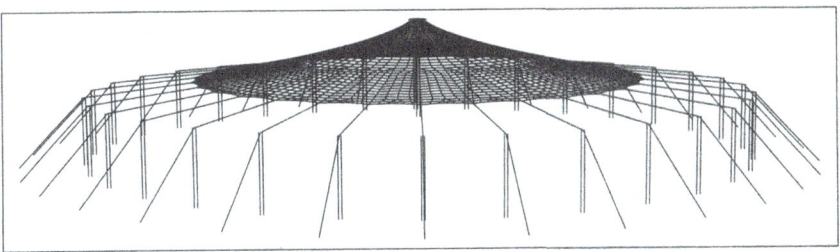

Fig. 3. Total view of the CSD model (initial state)

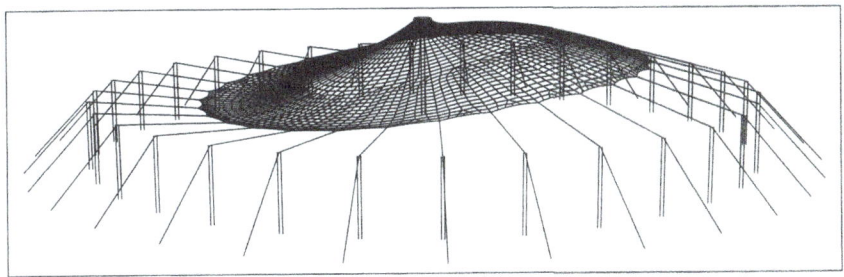

Fig. 4. Total view of the CSD model (deformed state, flow from the left)

$$\frac{U(z)}{U_{\text{ref}}} = \left(\frac{z}{z_{\text{ref}}}\right)^{\alpha} , \qquad \alpha = 0.102 , \qquad z_{\text{ref}} = 0.4\,\text{m} , \qquad U_{\text{ref}} = 12.4\,\text{m/s} . \quad (1)$$

The Reynolds number, based on U_{ref} and the difference in the height between the midpoint and the edge of the membranous umbrella, was $Re \approx 57{,}000$. A *turbulence intensity* of $Tu = 5\,\%$ has been measured in the wind tunnel in the oncoming flow. The turbulent kinetic energy $k(z)$ and the dissipation rate $\varepsilon(z)$ at the inlet were specified as follows:

$$k(z) = \frac{3}{2}\,Tu^2\,U(z)^2 , \qquad \varepsilon(z) = \frac{k(z)^{3/2}}{L_{\text{t}}(z)} , \qquad L_{\text{t}}(z) = 5.138\,z . \quad (2)$$

Fig. 4 shows the result of the coupled fluid-structure simulation. The diagrams in Fig. 5 represent the displacement of the membran along two chosen radii. The two different CFD grids produced almost the same results. That means, *grid independence* concerning the flow calculation has already been reached. In general, the numerical results coincide satisfactorily with the experimental data [8]. The correspondence is very good in the rear part of the umbrella. For example, the local minimum near the trailing edge at the outer radius $R_2 = 270$ mm at $\phi = 180°$ is represented in the numerical simulations, too. The offset of the front part of the membran could be due to several influences: the weakness of the k-ε model for turbulent flows around bodies, the diameter of the circumferential cable, which was neglected in the CFD calculation, or uncertainties in the modeling of the cables within the CSD calculation, because the materials used in the experiment did not fully satisfy the linear Hooke's law.

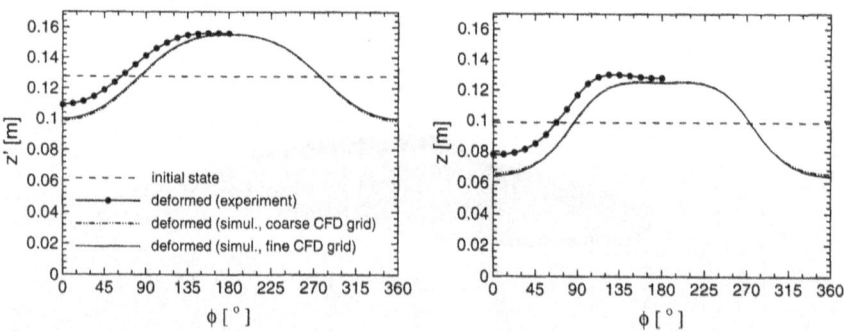

Fig. 5. Height z above ground of the mobile umbrella at $R_1 = 120$ mm (left) and $R_2 = 270$ mm (right)

3 Application to a membranous roof

The membrane roof in Fig. 6 was built in front of the entrance to an office building in Dresden (Germany) in 2000. This building acted in a slightly modified shape as a model for the present test case. Between the pylons a thin glass-fiber membrane is tautened by circumferential cables.

Fig. 6. Membranous roof in front of an office building in Dresden (Germany)

The superposition of a *constant basic wind flow* and a *time-dependent wind gust* was taken into consideration to realize an *unsteady* fluid-structure interaction. In a previous investigation a constant wind flow did not lead to a dynamic response (see *Glück et al.* [4]). In this first test case for a *dynamically* coupled calculation of the flow around a *membrane structure*, importance was attached first to the appropriate simulation of the structural response, whereas especially the structural oscillations at the end of the wind gust were to be examined. For that reason and because of the known weakness of the k-ε model for unsteady turbulent flows around bluff bodies, a simulation of the turbulent wind field was not taken into account. Instead, a laminar flow was assumed with a Reynolds number of $Re = 120$. The *basic flow* was assumed to be parallel to the ground in the positive x direction with $U_\infty = 10$ m/s at a height of 10 m above the ground. A *wind gust* was superimposed, following a Gaussian curve in both time and space:

$$W(x,t) = W_{max}\, e^{-0.5(t-t_m)^2}\, e^{-0.005(x-x_m)^2}\,, \tag{3}$$

with $x_m = -12$ m and $t_m = 3$ s. The vector field of the inflow velocity is depicted in Fig. 7. It is given by:

$$\mathbf{U}_{inflow}(x,z,t) = \begin{pmatrix} U(z) \\ 0 \\ W(x,t) \end{pmatrix}. \tag{4}$$

Both the CFD and CSD simulations were based on a time step $\Delta t = 0.4$ s. The *CSD mesh* consists of 1409 nodes and 1311 finite elements, whereas the *CFD mesh* includes 1,024,000 finite volumes, of which 3072 contact the membrane surface. Altogether, five nodes of the *HITACHI SR8000-F1* were used (four for CFD, one for CSD). A maximum memory of 390 MB per node was required for the coupled simulation. The calculation time (wall-clock time) amounted to 180 hours (approximately 50 minutes per time step), whereas 62.6 % accounted for the CFD simulation, 37.2 % for the CSD simulation, and 0.2 % for the adaption of the fluid mesh to the prevailing structural shape. In consideration of the fact that the CSD mesh is much coarser than the CFD grid, it becomes obvious that there is the necessity to improve the performance of the CSD code on the *HITACHI SR8000-F1*.

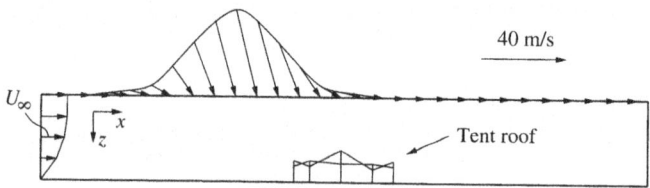

Fig. 7. Illustration of the inflow velocity $\mathbf{U}_{inflow}(x,z)$ at time $t = t_m = 3$ s with velocity reference vector (to scale)

Fig. 8 shows a sample of four states of the flow field and the structural deformations at various time instants. The grey scales in the fluid represent the *velocity magnitude* near the walls, while the shading of the membrane embodies its *deformation*. Some areas of the membrane were moving upwards and others downwards during the impact of the gust. A few parts of them showed a noticeable overshoot and subsequent small oscillations, before they reached the steady deformation state – which corresponds to the constant basic flow without superimposed wind gust – again.

The fluid loads at the CSD nodes as well as the structural displacements at a certain time are depicted in Fig. 9. An evaluation of the forces that occurred within the material in both the x and y directions – which nearly coincide with the weft and warp directions of the fabric – yielded the *maximum membrane forces* $\sigma_x = 29.1$ kN/m and $\sigma_y = 73.7$ kN/m at $t = 3.16$ s. Both values were lower than the given *tensile strengths* of the material. Hence, the stability of the structure is assured concerning the above-mentioned wind load.

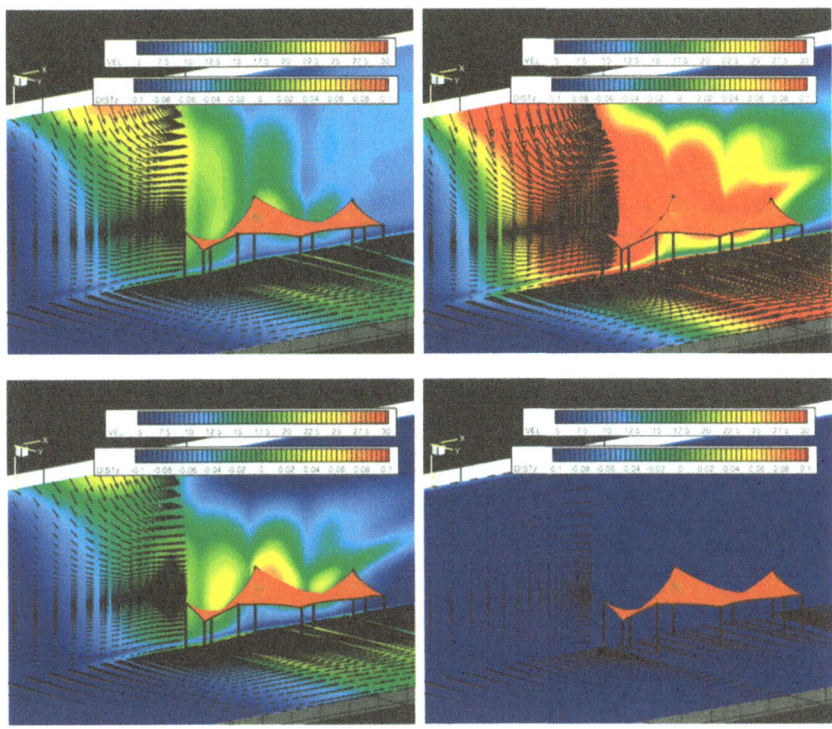

Fig. 8. Velocity distribution (horizontal and vertical cutting plane) and deformation of the membrane at four different times: $t_1 = 2.0$ s (top left), $t_2 = 3.0$ s (top right), $t_3 = 4.2$ s (bottom left), $t_4 = 6.4$ s (bottom right)

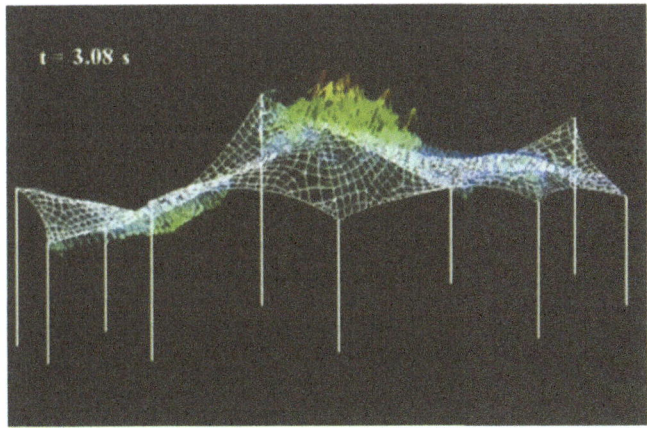

Fig. 9. Fluid loads at the structural nodes at $t = 3.08$ s (displacements exaggerated by factor six)

4 Performance of the CFD code

The coupled computations run on the SMP cluster Hitachi SR8000-F1 at the HLRB Munich (Germany). The fluid code – which consumes most of the total CPU time – has been optimized for running on this hardware. The architecture of the Hitachi SR8000-F1 allows the use of three different levels of parallelization (see *Brehm et al.* [2]), which are all taken into account to achieve a high performance of the CFD code:

- *pseudo-vector processing,*
- *intra-node auto-parallelization* over all processors of one SMP node (via *COMPAS*),
- *inter-node parallelization* using several SMP nodes and a communication library for the data exchange between the nodes such as MPI.

A very fruitful and still ongoing co-operation with the Computing Center in Erlangen (Germany) yielded a performance increase from 1342 MFlops to 1794 MFlops for a single SMP node.

5 Conclusions

A coupled algorithm for the numerical simulation of fluid-structure interactions was presented. Both disciplines employ separate highly adapted codes being coupled by a neutral coupling interface with a partitioned but fully implicit algorithm. Two real-life examples from civil engineering have been introduced. For one of them the numerical results were validated successfully at experimental data from a wind tunnel test.

Acknowledgement. The financial support of FLUSIB by the Bavarian State Ministry for Science, Research and the Arts in the Competence Network KONWIHR is gratefully acknowledged. The authors also want to thank Dipl.-Ing. U. Kaiser and Prof. Dr.-Ing. S. Wagner from the IAG of the University of Stuttgart (Germany) for providing the results of several wind tunnel tests. The support of Dr.-Ing. J. Bellmann from SOFiSTiK AG, Oberschleißheim (Germany) and the HLRB Munich (Germany) providing the SMP cluster Hitachi SR8000-F1 used to perform the numerical simulations is gratefully acknowledged. The authors also want to thank Dr. rer. nat. G. Wellein and Dipl.-Ing. F. Deserno from the Computing Center in Erlangen (Germany) for the optimization of the fluid code for the above mentioned SMP cluster.

References

1. Ahrem, R., Hackenberg, M.G., Post, P., Redler, R., Roggenbuck, J. (2000): MpCCI – Mesh Based Parallel Code Coupling Interface. Institute for Algorithms and Scientific Computing (SCAI), GMD, http://www.mpcci.org/.

2. Brehm, M., Bader, R., Ebner, R. (2001): Höchstleistungsrechner in Bayern (HLRB): The Hitachi SR8000-F1. http://www.lrz-muenchen.de/services/compute/hlrb/.

3. Durst, F., Schäfer, M. (1996): *A Parallel Block-Structured Multigrid Method for the Prediction of Incompressible Flows*. Int. J. Num. Methods Fluids, vol. 22, pp. 549-565.

4. Glück, M., Breuer, M., Durst, F., Halfmann, A., Rank, E. (2001): Computation of Fluid-Structure Interaction on Lightweight Structures. J. Wind Eng. Ind. Aerodyn. **89/14-15**, 1351-1368.

5. Glück, M., Breuer, M., Durst, F., Halfmann, A., Rank, E. (2002): Computation of Wind-Induced Vibrations of Flexible Shells and Membranous Structures. J. of Fluids and Structures, in press.

6. Halfmann, A. (2002): Ein geometrisches Modell zur numerischen Simulation der Fluid-Struktur-Interaktion windbelasteter, leichter Flächentragwerke. Dissertation, Lehrstuhl für Bauinformatik, Technische Universität München.

7. Katz, C., Bellmann, J. (1988-1998): ASE-Handbuch. SOFiSTiK GmbH, Katz+Bellmann, Oberschleißheim.

8. Wagner, S.N., Bergmann, D., Kaiser, U. (2001): Membranschirm im Original unter Windeinfluss. Institute report (unpublished), Institute of Aerodynamics and Gasdynamics (IAG), University of Stuttgart.

Large-Eddy and Detached-Eddy Simulation of the Flow Around High-Lift Configurations

Michael Breuer[1], Nikola Jovičić[1], and K. Mazaev[2]

[1] Institute of Fluid Mechanics, University of Erlangen-Nürnberg
   Cauerstr. 4, 91058 Erlangen, Germany
   breuer/njovicic@lstm.uni-erlangen.de
[2] Saint Petersburg State Marine Technical Univ.,
   Lotsmanskaja Street 3, 190008 Saint Petersburg, Russia
   kmazaev@softimpact.ru

Abstract. The paper is concerned with separated turbulent flows past a flat plate and an unswept wing (NACA–4415 airfoil) at high angles of attack. For both configurations large-eddy simulations (LES) were carried out. Additionally, the flow past an infinitely thin plate was simulated by detached-eddy simulation (DES). First a relatively low chord Reynolds number $Re_c = 20,000$ was chosen for the airfoil case yielding a leading-edge stall for both angles of attack ($\alpha = 12°$ and $18°$) investigated. The flow field in this range of Re_c and α is dominated by asymmetric vortex shedding predefined by strong vortices developing almost periodically in the vicinity of the trailing edge. The life cycle of these vortical structures is controlling the entire flow field. At the leeward side of the airfoil, a large clockwise rotating recirculation region of nearly constant pressure exists. For $\alpha = 18°$ the features of the flow around the flat plate were found to be very similar to the airfoil case, because both flows separate at comparable positions and are dominantly controlled by the shedding motion of the trailing-edge vortex. The DES predictions of the flat plate lead to reasonable agreement with the LES results concerning the mean flow field. However, comparing DES and LES predictions carried out on the same grid yielded remarkable deviations especially in the representation of the free shear layer and the higher-order statistics. Hence the standard DES formulation needs some adjustments. Finally, some first LES results for the airfoil flow at a higher Reynolds number $Re = 100,000$ are presented showing a completely different flow behavior than the low-Re case. This includes a laminar separation bubble, transition behind the bubble, and separation of the turbulent boundary layer at about 70% chord length.*

Key words: Large-Eddy Simulation, Detached-Eddy Simulation, Separated Turbulent Flow, High-Lift Configuration, High-Performance Computing

---

* This contribution will be presented at the Fifth World Congress on Computational Mechanics, held in Vienna, Austria from July 7 to 12, 2002.

1 Introduction

The large-eddy simulation (LES) technique is a promising tool for highly unsteady turbulent flows which are dominated by large separation and recirculation regions, vortex formation and shedding, or shear layer instabilities and transition. Such phenomena typically occur in bluff-body flows but are also present in flows past streamlined geometries exposed to extreme operating conditions. A typical aerodynamic application is the flow around airfoils at maximum lift and beyond, where a leading-edge or trailing-edge stall is observed. A flow of this kind was experimentally investigated within the COSTWING experiment [1]. The adequate numerical simulation of such flows and the detailed analysis of the flow characteristics are the main objectives of the present study. The experimental setup consists of a nominally 2D airfoil based on a NACA-4415 profile which is mounted inside a channel (see Fig. 1 (a)). Compared with other activities such as the European project LESFOIL [2], which aims at the flow around an Aerospatiale A-airfoil at a chord Reynolds number $Re_c = u_\infty c/\nu = 2.1 \times 10^6$ and an angle of attack $\alpha = 13.3°$, the Re_c-range of COSTWING is between one or two orders of magnitude smaller ($Re_c = 8 \times 10^4$ to 8×10^5). This circumstance simplifies the predictions and avoids at least some of the difficulties encountered in LESFOIL [2]. The objective of COSTWING is twofold: on the one hand the physics of separated turbulent flows should be studied in detail including detection of coherent structures; on the other hand the experiment should serve as a database for the validation of numerical simulations. In the latter sense it is used here. The main goals are to study the physics of stalled airfoil flows in detail and to evaluate the usefulness of LES [3, 4] and DES [5] for this kind of flows.

LES predictions of the flow past the airfoil were carried out for different conditions ($Re_c = 20,000$ and $100,000$; $\alpha = 12°$ and $18°$). Additionally, detailed investigations were directed towards the simplest high-lift configuration, namely an inclined flat plate ($Re_c = 20,000$ and $\alpha = 18°$ [4]). This configuration was predicted based on LES as well as using the detached-eddy simulation (DES) technique proposed by Spalart et al. [6]. DES is a *hybrid non-zonal approach* combining Reynolds-Averaged Navier-Stokes turbulence modeling (RANS) and LES. The main objective is to reduce the high computational resources required for pure LES, which seems to be necessary according to the analysis by Spalart et al. [6]. Otherwise, prediction of realistic applications will be out of reach during the next couple of decades, even if the assumptions of Spalart may be too pessimistic. This matter of fact is the main motivation of several initiatives (see, e.g., [7,8,9,10,11]) aiming at the development of *hybrid* methods combining the advantages of both LES and RANS.

The basic principles of DES are described in [6] and first results are presented in [12]. The main difference compared with other attempts of *coupling* RANS and LES is the fact, that the flow is not explicitly subdivided into two zones, and, as a consequence, there is no problem of explicit boundary conditions between RANS and LES regions. The main disadvantage of the pure RANS approach is clear: RANS describes flows in a statistical sense typically leading to time-averaged pressure and velocity fields. Generally this approach is not able to distinguish between quasi-periodic large-scale and turbulent chaotic small-scale features of the flow. This leads to huge troubles when the flow field is governed by both phenomena. A typical representative is a bluff-body flow. RANS is not able to reproduce the unsteady characteristics reasonably resulting in a non-adequate description of unsteady phenomena, such as vortex formation and shedding. LES, on the other hand, operates with unsteady fields of physical values; the governing equations are the Navier-Stokes equations, but, unlike the RANS approach, spatial filtering is applied instead of averaging in time, and turbulent stresses are divided into resolved and modeled ones. The latter can be found from very simple (in comparison with *standard* turbulence models) subgrid scale models. As mentioned above, the main disadvantage of LES is the high computational costs resulting from extremely fine grids used for the direct prediction of the *non-modeled* vortical structures, and additionally very fine time steps required for resolving turbulent time-scales.

The hybrid DES approach combines RANS and LES in a natural way. Near solid boundaries the governing equations *work* in RANS-mode, i.e., all turbulent stresses are modeled with the help of a statistical turbulence model. Furthermore, pressure and velocity fields are time-averaged and unsteady attached vortical structures are not resolved directly. Far from solid boundaries, the method *switches* to the LES-mode. From the physical point of view it means resolving of all large-scale vortical structures and modeling of small eddies based on a subgrid scale model. From the numerical point of view one has to deal with unsteady fields of pressure and velocity in this mode. However, the computational domain contains no explicit boundaries or interfaces between both modes of operation. It is quite natural, that RANS *works* within the attached boundary layer region, where statistical turbulence models in general perform properly, and LES is applied far away from the boundaries in the detached flow region, where large unsteady vortices, which should be resolved directly, are present. Exactly this kind of division allows to use a rather coarse highly stretched grid in the wall-normal direction, because for RANS it is not necessary to apply extremely fine resolutions. Furthermore, unlike in the LES case which requires a fine resolution in all three spatial directions, the spanwise and streamwise resolution does not have to be very high for RANS [13]. Both factors raise hope that DES can be used with acceptable effort even for high-Re flows, typically encountered in technical applications.

Due to the modeling approach described above, the main application area of DES should be in accordance with the DES concept itself: flows, for which RANS prediction do not work properly, namely unsteady turbulent flows with large separation regions. The flow around high-lift airfoils with massive separation is a typical representative [12]. Other applications leading to a mixed idea about the performance of this approach can be found in [6, 7, 8, 13, 14, 15, 16, 17]. The test case for DES chosen in the present investigation, is in favor of the general concept. It allows a detailed investigation especially of the LES-mode of the DES predictions. Owing to the fixed separation at the sharp leading edge of the plate, the RANS region is restricted to a narrow gap near the wall. Hence, the DES approach has to work nearly completely in LES-mode which allows to evaluate the performance of the S-A model as a subgrid scale model and additionally the coupling with the RANS region.

The paper is organized as follows. First, the governing equations are given for LES and DES. Second, in Sect. 3 both modeling approaches are briefly described. Section 4 gives an overview of the numerical solution method applied. In the subsequent section the geometry and the details of the flow configurations are defined. Finally, in Sect. 6 the LES results for the airfoil flow at different conditions (Re_c, α) and the comparison of the LES and DES predictions for the flat plate flow are presented.

2 Governing Equations

In the present study an incompressible fluid with constant fluid properties is considered. The governing equations are the Navier-Stokes equations describing the conservation of mass and momentum. In the case of LES, these equations are filtered in space with the filter width Δ in order to separate large- and small-scale motions, which leads to the so-called *filtered* Navier-Stokes equations. In the RANS (DES) approach, the same equations are time-averaged (or ensemble-averaged) leading to the well known RANS equations. Both kinds of equations can be written in a common (dimensionless) form:

$$\frac{\partial \overline{u}_j}{\partial x_j} = 0 \, , \tag{1}$$

$$\frac{\partial \overline{u}_i}{\partial t} + \frac{\partial (\overline{u}_i \overline{u}_j)}{\partial x_j} = -\frac{\partial \overline{p}}{\partial x_i} - \frac{1}{Re} \frac{\partial \tau_{ij}^{mol}}{\partial x_j} - \frac{\partial \tau_{ij}^{turb}}{\partial x_j} \, , \tag{2}$$

$$\tau_{ij}^{mol} = -2 \, \nu \, \overline{S}_{ij} \quad \text{and} \quad \overline{S}_{ij} = \frac{1}{2} \left(\frac{\partial \overline{u}_i}{\partial x_j} + \frac{\partial \overline{u}_j}{\partial x_i} \right), \tag{3}$$

where u_i, p and $\nu = \mu$ ($\rho = 1$) are the velocity components, the pressure and the viscosity, respectively. In LES the overbar ($\overline{\cdot}$) defines the resolved scales, whereas in RANS it denotes the time-averaged components. $Re = Re_c$ is the characteristic Reynolds number defined above. The term τ_{ij}^{mol} describes the momentum transport due to molecular motion, which for a Newtonian fluid is given by (3). Owing to filtering (LES) or time-averaging (RANS) of the non-linear convective term in the momentum equation (2), the additional term

τ_{ij}^{turb} arises which has to mimic the momentum transport due to turbulence motion. In the case of LES, τ_{ij}^{turb} is denoted as the subgrid scale (SGS) stress tensor and restricted to the influence of the non-resolved small-scale structures on the resolved large eddies. However, in RANS τ_{ij}^{turb} signifies the Reynolds stress tensor describing the influence of the entire spectrum of all length scales on the averaged flow field. In both cases this unknown has to be modeled. The modeling approach for DES is described in detail in the subsequent section. For LES only a brief description is given. More details on this topic can be found in [18, 19, 20].

3 Modeling Approaches

3.1 Detached-Eddy Simulation

As mentioned above, the DES approach is based on a non-explicit splitting of the computational domain into two zones. In the first region near solid walls, the *conventional* RANS equations have to be solved. Within the second region, the governing equations are the filtered Navier-Stokes equations of the LES approach. In principle the DES concept allows to apply totally different models for the two zones, which would be a natural choice since the models have clearly distinguishable tasks. However, following the concept of Spalart et al. [6, 7, 12, 14], in both cases the one-equation Spalart-Allmaras (S-A) turbulence model [21] is used either as a *conventional* turbulence model or as a SGS model, respectively. The model and the mechanism, which allows to apply it both in RANS and in LES mode, are described briefly. The S-A model is based on the Boussinesq's approximation which describes the stress tensor τ_{ij}^{turb} as the product of the strain rate tensor $\overline{S}_{ij}$ (see (3)) and an eddy viscosity ν_T:

$$\tau_{ij}^{turb,a} \;=\; \tau_{ij}^{turb} - \delta_{ij}\,\tau_{kk}^{turb}/3 \;=\; -2\,\nu_T\,\overline{S}_{ij} \qquad (4)$$

where $\tau_{ij}^{turb,a}$ is the anisotropic (traceless) part of the stress tensor τ_{ij}^{turb} and δ_{ij} is the Kronecker delta. The trace of the stress tensor is added to the pressure resulting in the new pressure $P = \overline{p} + \tau_{kk}^{turb}/3$. The determination of the eddy viscosity ν_T is based on the solution of an additional transport equation. For the low-Re variant of the S-A model, the new eddy viscosity variable $\tilde{\nu}$ is introduced leading to the following governing equation:

$$\underbrace{\frac{\partial \tilde{\nu}}{\partial t}}_{\text{local change}} + \underbrace{u_i \frac{\partial \tilde{\nu}}{\partial x_i}}_{\text{convection}} \;=\; \underbrace{c_{b1}\,\tilde{S}_\nu\,\tilde{\nu}}_{\text{production}} + \underbrace{\frac{1}{\sigma}\left\{ \nabla \cdot \left[(\nu + \tilde{\nu})\,\nabla\tilde{\nu}\right] + c_{b2}\,(\nabla\tilde{\nu})^2 \right\}}_{\text{diffusion}}$$

$$- \underbrace{c_{w1} f_w \left[\frac{\tilde{\nu}}{d}\right]^2}_{\text{destruction}}. \qquad (5)$$

The left-hand side of (5) represents the local and convective changes of the transport variable $\tilde{\nu}$; the right-hand side includes the production term, the diffusion term and finally the destruction term for the reduction of the stresses in the vicinity of solid walls. It represents the dissipation of the turbulent kinetic energy in the near-wall region. Due to its physical nature, in RANS-mode the destruction term is directly related to the reciprocal of the minimal wall distance d. Hence the boundary condition for the intermediate variable $\tilde{\nu}$ at solid walls is $\tilde{\nu} = 0$. The production term includes the scalar quantity $\tilde{S}_\nu$ which is expressed by the magnitude of the vorticity $S_\nu = |\omega|$ plus a near-wall correction. The derivation of the missing relations (ν_T, χ, f_{v1}, f_{v2}, f_w, g, r) and the values of the model parameters (κ, σ, c_{b1}, c_{b2}, c_{v1}, c_{w1}, c_{w2}, c_{w3}) can be found in [21].

This is the standard RANS formulation of the S-A model. The modification for the DES approach is based on the following idea [6,7,12,14]: under the condition of local equilibrium, the production term ($\sim \tilde{S}_\nu \tilde{\nu}$) in (5) is balanced by the destruction term ($\sim (\tilde{\nu}/d)^2$). This local balance leads to the relation $\tilde{\nu} \sim \tilde{S}_\nu d^2$, which is very similar to the relation given by the Smagorinsky SGS model [22] used for LES (see Sect. 3.2) if the wall distance d is replaced by the filter width Δ. This analogy allows in principle to apply the S-A model as a SGS model if the wall distance d in all relations of the S-A-model is substituted by a characteristic length scale proportional to Δ. Replacing d by the new variable $\tilde{d}$ defined by $\tilde{d} \equiv \min(d, C_{DES} \times \Delta)$ with $\Delta \equiv \max(\Delta x, \Delta y, \Delta z)$ and $C_{DES} = 0.65$ leads to a uniform model for the RANS and the LES-mode of the DES approach. If the grid is fine enough to resolve vortical structures far away from the wall or within separated flow regions and is almost isotropic such as for a conventional LES, the relation $d > \Delta$ will guarantee that (5) is used as a SGS model. Otherwise, because of extremely fine grids required for LES computations of the near-wall flow (especially at high Re), it is not desirable to use this technique in this specific region. Therefore, the RANS approach should be applied for $d < \Delta$ and $\tilde{d}$ should be defined in the original manner ($\tilde{d} = d$). Within thin boundary layers near walls, RANS also requires refined grids. However, in contrast to LES, an appropriate resolution of the wall normal direction (e.g. Δy) is often sufficient. This leads to highly stretched grids in the vicinity of walls, where only the Δy values are small in comparison with the wall distance d, but Δx and/or Δz are much larger. Hence it should be mentioned that the definition of the filter size Δ in the LES-mode differs from that of conventional SGS models which typically scale the filter width with the size of the control volume, i.e., $\Delta = (\Delta x \times \Delta y \times \Delta z)^{1/3}$. However, the modified definition of Δ guarantees the desired behavior of the model including the switch between both modes. The region corresponding to $d \sim \Delta$, is called *grey area* [6,7,12,14], because it is not clear, what exactly happens in this region and how significant the role of modeled and resolved vortical structures existing between the RANS and LES zones are.

Unlike in LES, due to the modified definition of Δ not only the small-scale eddies which are smaller than the grid cells will be filtered, but the eddies scaled with the grid cells in a boundary layer will also not be resolved, because they are in the region modeled by RANS. Therefore, it is clear, that DES should be most effective for flows consisting of regions with thin attached boundary layers (I) and separated regions which are controlled by large unsteady vortical structures (II). Flows of kind (I) can be predicted reliably and efficiently by RANS, whereas LES is also appropriate but extremely expensive. On the other hand, flow phenomena of kind (II) cannot be computed reliably by RANS. For this purpose LES is the appropriate tool. Furthermore, such flows with attached and separated regions are the most natural choice for DES in order to guarantee proper coupling of RANS and LES zones. In conclusion, DES is expected to work most efficiently for this special class of flows. The expectations are to save computational resources compared with pure LES and to predict separated flows more reliable than with pure RANS.

3.2 Large-Eddy Simulation

For the LES approach, the term τ_{ij}^{turb} in (2) represents the subgrid scale stress tensor describing the influence of the non-resolved subgrid scales on the resolved part of the turbulent flow. For the description of this effect, it is typically assumed that the non-resolvable small-scales in an LES are much less problem-dependent than the large-scale turbulence so that the subgrid scale turbulence can be represented by relatively simple models, e.g., zero-equation eddy-viscosity models. The well known and most often used Smagorinsky model [22] is also based on the Boussinesq's approximation (4). The eddy viscosity ν_T itself is a function of the strain rate tensor $\overline{S}_{ij}$ and the subgrid length $l = C_s \, \Delta \, \mathcal{D}$, where the filter width is defined by $\Delta = (\Delta x \times \Delta y \times \Delta z)^{1/3}$ and $\mathcal{D}$ is the Van Driest damping function required in order to take the reduction of the subgrid length near solid walls into account. The parameter C_s is the well known Smagorinsky constant which has to be prescribed as a fixed value in the entire integration domain or can be determined as a function of time and space by the dynamic procedure originally proposed by Germano et al. [23] and later improved by several authors, e.g., Lilly [24]. In the present investigation, the fixed parameter version of the Smagorinsky model with a standard constant $C_s = 0.1$ was applied in all simulations. The application of the dynamic version of the model is planned for future investigations.

4 Numerical Methodology

The same computer code $\mathcal{LESOCC}$ [18, 19] is used for both approaches to reduce the influence of different flow solvers. The code is based on a 3-D finite-volume method for arbitrary non-orthogonal and block-structured grids. All viscous fluxes are approximated by central differences of second-order accuracy, which fits the elliptic nature of the viscous effects. As shown in [19] and by other authors, e.g., Kravchenko and Moin [25], the quality of LES predictions is strongly depending on low-diffusive discretization schemes for the non-linear convective fluxes in the momentum equation (2). Although several schemes are implemented in the code applied, the central scheme of second-order accuracy is preferred for the LES and the DES predictions in the present work.

Time advancement is performed by a predictor-corrector scheme. A low-storage multi-stage Runge-Kutta method (three sub-steps, second-order accuracy) is applied for integrating the momentum equations in the predictor step. Within the corrector step the Poisson equation for the pressure correction is solved implicitly by the incomplete LU decomposition method. Explicit time marching works well for LES/DES with small time steps which are necessary to resolve turbulence motion in time. The pressure and velocity fields on the non-staggered grid are coupled by the momentum interpolation technique of Rhie and Chow [26]. The S-A model for the DES computations requires the solution of the additional scalar transport equation (5), which is discretized in the same way as the momentum equations (2). The code is highly vectorized and additionally parallelized by domain decomposition with explicit message-passing based on MPI allowing efficient computations especially on vector-parallel machines and SMP clusters. The simulations were partially carried out on the Fujitsu VPP 300/700 applying 4 processors and partially on the Hitachi SR 8000-F1 applying 8–16 nodes (64–128 processors). A variety of different test cases (see, e.g., [18, 19, 20, 4, 3, 27]) served for the purpose of code validation.

5 Description of the Flow Configurations

Because the COSTWING experiment [1] was especially designed as a validation test case for numerical simulations, large emphasis was put on a detailed definition of the corresponding boundary conditions. Figure 1 (a) shows a two-dimensional sketch of the configuration. The NACA-4415 profile is mounted inside a plane channel of height $3\,c$, where c describes the chord length of the profile. Upstream of the profile the channel has a length of $2\,c$, whereas downstream a length of $3\,c$ is assumed. At the Re_c chosen, no-slip boundary conditions can be applied at the surface of the profile. In order to save grid points the boundary layers of the channel walls (see Fig. 1) are not resolved and approximated by slip-conditions $(\partial u/\partial y = v = \partial w/\partial y = 0)$. Owing to this channel configuration a lot of grid points can be saved because the far-field does not have to be resolved. The experiment was especially designed in such a way that either statistically two-dimensional or span-wise periodical flow structures can be expected. Therefore, periodicity in the spanwise direction is assumed and a spanwise computational domain of depth $z_{max} = 1.0 \times c$ is chosen. This choice is based on a detailed investigation [4] for the flow around the inclined flat plate mounted in the vertical middle of the same channel configuration and operated at the same flow conditions ($Re_c = 20,000$, $\alpha = 18°$). It showed that this spanwise extension is on the one hand necessary to assure reliable results and on the other hand represents a well-balanced compromise between spanwise extension and spanwise resolution. At the inlet section a constant velocity u_∞ is prescribed, whereas at the outlet a convective boundary condition assures that vortices can pass through the outflow boundary [19].

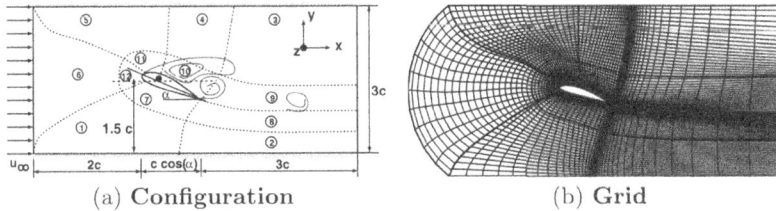

(a) **Configuration**   (b) **Grid**

Fig. 1. (a) Two-dimensional sketch of the geometric configuration including block boundaries; (b) x-y-plane of the grid used for $Re_c = 20,000$ and $\alpha = 18°$ (only every fifth grid line is shown)

Curvilinear block-structured grids consisting of up to 16 blocks and a maximum of ≈ 16.23 million control volumes (up to 100 CVs in spanwise direction) are applied. One typical example is shown in Fig. 1 (b). The grid points are clustered in the vicinity of the airfoil/plate and at the leading and trailing edge. For both cases a variety of different computations have been carried out using different grids and different spanwise extensions of the domain. The details are documented in [3] for the airfoil case and in [4,5] for the flat plate configuration. Additionally, different flow parameters (Re_c, α) were investigated, partially by applying both simulation techniques (LES, DES). Furthermore, in [5] the results are also compared with pure RANS predictions.

The present paper is restricted to the most interesting cases and some recent results. Table 1 gives an overview of the cases considered. First the airfoil flow is analyzed at $Re_c = 20,000$ and two different angles of attack ($\alpha = 12°$ and $18°$). Then the practically more interesting case of $Re_c = 100,000$ ($\alpha = 18°$) is considered. All these predictions are based on LES. Finally, the flow around the flat plate is analyzed at $Re_c = 20,000$ and $\alpha = 18°$. Here the main objective is to compare the performance of DES and LES. Therefore, simulations based on the same relatively coarse grid are compared with a reference solution obtained by LES on a much finer grid.

Table 1. Overview of some simulation parameters and grid properties

Case	Re_c	α [°]	Type	No. of CVs	CVs in z-direct.	Remark
			NACA-4415 Airfoil			
A-20-12	20,000	12	LES	8.34×10^6	76	low-Re, low α
A-20-18	20,000	18	LES	8.34×10^6	76	low-Re, high α
A-100-18	100,000	18	LES	16.23×10^6	100	high-Re, high α
			Inclined Flat Plate			
P-DES-C	20,000	18	DES	0.99×10^6	36	coarse DES
P-LES-C	20,000	18	LES	0.99×10^6	36	coarse LES
P-LES-F	20,000	18	LES	8.97×10^6	76	fine LES, ref. sol.

6 Results and Discussion

6.1 Flow Around the Airfoil at Varying Re_c and α

Low Reynolds Number $Re_c = 20,000$ For the low-Re case two different angles of attack were investigated. Figure 2 depicts the instantaneous flow fields by contours of the vorticity component ω_z for $\alpha = 12°$ and $18°$ at arbitrarily chosen time instants and z-planes. As expected, the flow at the lower side of the profile is attached and laminar. At the upper side the flow separates shortly after the leading edge induced by a strong positive pressure gradient in this region (leading-edge stall). It forms a large clockwise rotating recirculation region on the leeward side of the airfoil with a nearly constant pressure. Behind the trailing edge strong counterclockwise rotating vortex structures are clearly visible which play an important role for the dynamics of the entire flow. This characteristic flow feature was already analyzed in [3] for $\alpha = 18°$. Therefore, the main results will be discussed here only briefly.

The flow around the airfoil at $Re_c = 20,000$ and $\alpha = 18°$ is strongly dominated by an asymmetric vortex shedding cycle with a Strouhal number $St' = fc'/u_\infty = 0.197 \approx 0.2$ scaled with the windward width $c' = c \times \sin \alpha$. The development and shedding behavior of the trailing-edge vortex controls the entire flow field. It develops almost periodically in the vicinity of the trailing edge. The life cycle of this strong vortical structure also determines the structure and the size of the large clockwise rotating recirculation region on the leeward side of the airfoil which originates from the separation at the leading edge. In the separated shear layer a Kelvin-Helmholtz instability indicating the location where transition takes place, is observed.

Additional to Fig. 2 which compares the unsteady flow fields for both angles of attack, Fig. 3 shows the corresponding streamlines of the spanwise and time-averaged flow ($\Delta T_{avg} \approx 80$). These figures clearly demonstrate that no fundamental differences in the flow structures exist between $\alpha = 12°$ and $18°$. In both cases the flow separates shortly after the leading edge forming

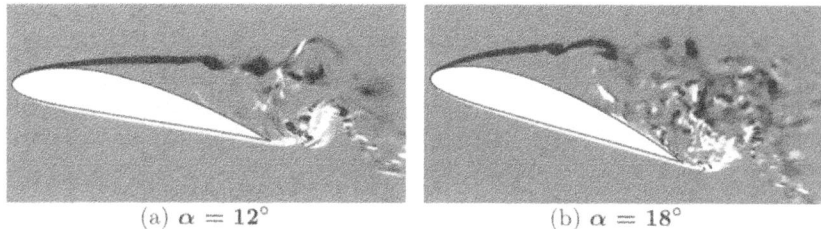

(a) $\alpha = 12°$ (b) $\alpha = 18°$

Fig. 2. Instantaneous flow field past the inclined airfoil at $Re_c = 20,000$, contours of vorticity ω_z; (a) $\alpha = 12°$, (b) $\alpha = 18°$

a large clockwise rotating recirculation region which is smaller for $\alpha = 12°$ than for $18°$. Furthermore, also in the averaged flow of both cases the strong counterclockwise rotating vortex structure behind the trailing edge is clearly visible. As explained above, this plays an important role for the dynamics of the entire flow development. Beside the smaller extension of the recirculation region, further deviations are given by the delayed occurrence of the Kelvin-Helmholtz instability including the subsequent transition and the reduced strength of the trailing-edge vortex. However, no principal deviations between both cases are found.

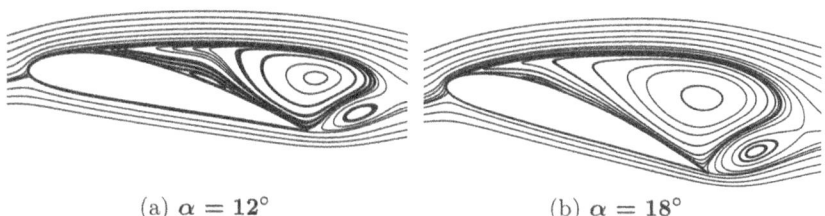

(a) $\alpha = 12°$ (b) $\alpha = 18°$

Fig. 3. Streamlines of the spanwise and time-averaged flow past the inclined airfoil at $Re_c = 20,000$; (a) $\alpha = 12°$, (b) $\alpha = 18°$

The averaged pressure and shear stress distribution at the surface (index s) of the airfoil are depicted in Fig. 4 by the dimensionless quantities $C_p = (p_s - p_\infty)/(0.5\rho_\infty u_\infty^2)$ and $C_f = \tau_s/(0.5\rho_\infty u_\infty^2)$. As expected, the higher angle of attack leads to a stronger suction peak at the leading edge. Correspondingly, the flow separates earlier, i.e., at $x/c \approx 0.03$ for $\alpha = 18°$ and $x/c \approx 0.148$ for $\alpha = 12°$, respectively. A region of positive wall shear stress is found at about $0.50 \lesssim x/c \lesssim 0.72$ for $\alpha = 12°$ and $0.23 \lesssim x/c \lesssim 0.60$ for $\alpha = 18°$, respectively. However, this flow feature is extremely thin and restricted to the direct vicinity of the surface (see Fig. 3). Owing to the stronger circulation of the recirculation region at $\alpha = 18°$, higher values of negative wall shear stress are observed near the trailing edge. In both cases nearly constant pressure and shear stress distributions are detectable in the intermediate section of the airfoil where the flow is completely separated.

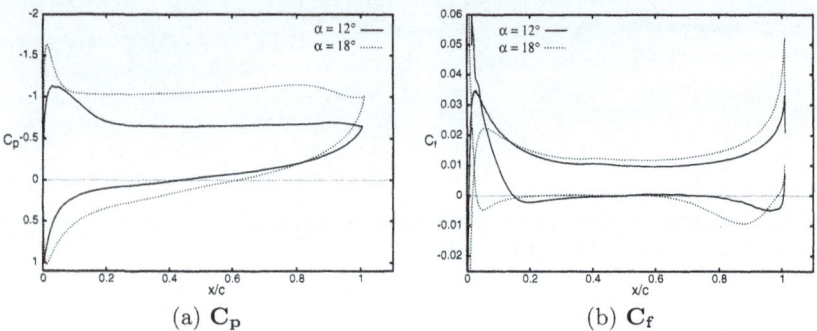

(a) $\mathbf{C_p}$ (b) $\mathbf{C_f}$

Fig. 4. Surface pressure and shear stress distribution of the spanwise and time-averaged flow past the inclined airfoil at $Re_c = 20,000$ and $\alpha = 12°$ and $18°$; (**a**) C_p, (**b**) C_f

Higher-order statistics of the LES predictions can be evaluated looking at each component of the Reynolds stress tensor separately, see [3,5]. However, only the contours of the turbulent kinetic energy $k = \frac{1}{2}\left(\overline{u'u'} + \overline{v'v'} + \overline{w'w'}\right)$ are shown here in Fig. 5. It has to be mentioned that the components of the tensor include the periodic and the resolved turbulent fluctuations. No attempt was made to separate the periodic and turbulent components. For both cases the largest values of k can be found in the vicinity of the trailing edge, mainly caused by the quasi-periodic shedding motion of the trailing-edge vortex. However, the maxima are quite different, i.e., $k_{max} \approx 0.22$ for $\alpha = 12°$ and $k_{max} \approx 0.34$ for $\alpha = 18°$. Further deviations are clearly visible owing to the delayed shear layer instability for $\alpha = 12°$ discussed above. For the high-α case a local maximum is found in the region of the Kelvin-Helmholtz instability where transition of the flow occurs. Such a characteristic feature is not visible for the low-α case because the transition region is shifted downstream and merges with the trailing-edge processes.

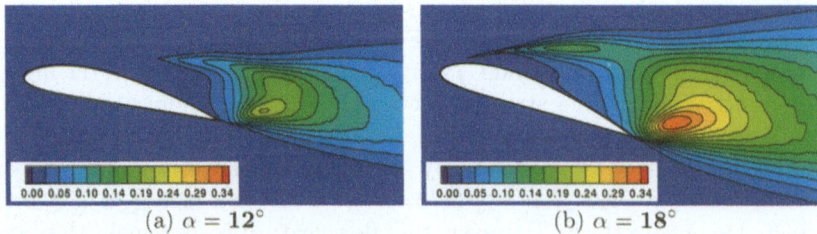

(a) $\alpha = 12°$ (b) $\alpha = 18°$

Fig. 5. Distribution of the turbulent kinetic energy k of the flow past the inclined airfoil at $Re_c = 20,000$; (**a**) $\alpha = 12°$, (**b**) $\alpha = 18°$

Table 2 summarizes some important integral parameters for both cases. Additionally, results for the flat plate at the same Reynolds number and an angle of attack $\alpha = 18°$ are provided. Almost identical lift and drag coefficients $(\overline{C}_l, \overline{C}_d)$ are found for the time-averaged flow around the airfoil and plate at the same angle of attack. Concerning the fluctuations of the integral values given by standard deviations $(\sigma_{C_l}, \sigma_{C_d})$, slightly larger values are observed for the airfoil. The Strouhal number of the airfoil is only marginally smaller than for the flat plate $(St' \approx 0.2)$. This value is consistent with experimental findings for inclined thin plates which observed a nearly constant $St' \approx 0.15$ for $20° \leq \alpha \leq 90°$ and a strong increase of $St' \to 0.2$ for $\alpha \leq 20°$ (see, e.g., Knisely [28] and the cited refs.). For $\alpha = 12°$ the Strouhal number for the airfoil is still comparable whereas the mean values as well as the standard deviations of the lift and drag coefficients strongly decrease.

Finally, it should be mentioned that the case of $Re_c = 20,000$ and $\alpha = 18°$ was compared with similar LES predictions carried out by Evans and Friedrich [29] and showed good agreement concerning all characteristic features discussed above.

Table 2. Comparison of important parameters for the flow past the airfoil and the flat plate at different Re_c and α; $St = fc/u_\infty$, $St' = St \times \sin\alpha$

Configuration	Re_c	α [°]	$\overline{C}_l$	$\overline{C}_d$	σ_{C_l}	σ_{C_d}	St	St'
NACA-Airfoil	20,000	12	0.749	0.186	4.1×10^{-2}	6.9×10^{-3}	0.90	0.187
NACA-Airfoil	20,000	18	1.118	0.391	8.9×10^{-2}	2.4×10^{-2}	0.63	0.195
Flat Plate	20,000	18	1.123	0.379	6.9×10^{-2}	2.2×10^{-2}	0.66	0.204
NACA-Airfoil	100,000	18	1.529	0.128	2.0×10^{-2}	4.0×10^{-3}	–	–

High Reynolds Number $Re_c = 100,000$ In accordance to Fig. 2 (b) for the low-Re case, Fig. 6 depicts the instantaneous flow field by contours of the vorticity component ω_z for the high-Re case at the same angle of attack $\alpha = 18°$. The corresponding streamlines of the spanwise and time-averaged flow are shown in Fig. 8. A totally different flow structure than for $Re_c = 20,000$ described above is observed for $Re_c = 100,000$. The flow separates shortly behind the nose $(x/c \approx 0.02)$ but reattaches at about $x/c = 0.12$ forming a closed *laminar* separation bubble which is extremely thin. In a zoomed animated sequence of the nose region, one can observe that the size and the location of the bubble is slightly varying in time. In the rear part of the bubble the highest value of the turbulent kinetic energy is found (see Fig. 9) indicating that transition takes place at this position. Downstream of the bubble a turbulent boundary layer is visible at the leeward side of the airfoil (Figs. 6 and 9 (a)).

Owing to transition to turbulence the boundary layer remains attached up to $x/c \approx 0.7$, although it is exposed to an adverse pressure gradient as shown in Fig. 7. In the vicinity of the trailing edge a closed recirculation region is

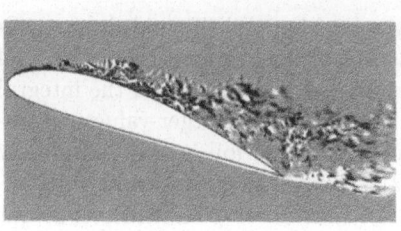

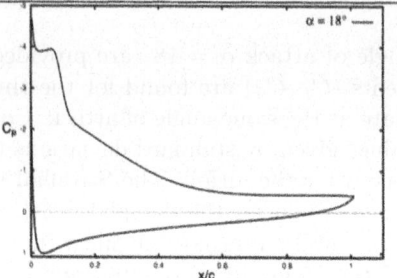

Fig. 6. Instantaneous flow field past the inclined airfoil at $Re_c = 100,000$ and $\alpha = 18°$, contours of vorticity ω_z

Fig. 7. Surface pressure distribution C_p of the spanwise and time-averaged flow past the inclined airfoil at $Re_c = 100,000$ and $\alpha = 18°$

observed which is much smaller than for the low-Re case. Furthermore, no trailing-edge vortex playing a dominant role and leading to the asymmetric vortex street past the airfoil at $Re_c = 20,000$ is present at $Re_c = 100,000$. Hence the frequency spectra of the lift and drag coefficients (not given here) do not show a strong peak which can be identified as the corresponding Strouhal number. In Table 2 the remaining integral parameters are summarized. Compared with the low-Re case the mean lift coefficient increases about 36%, whereas the drag coefficient drastically decreases to about one third yielding a four times better glide ratio (lift to drag). In conclusion, the high-Re case shows an instantaneous and time-averaged flow structure which completely deviates from the low-Re counterpart. In contrast to the *leading-edge stall* at $Re_c = 20,000$, the LES predictions for the airfoil flow at $Re_c = 100,000$ show a typical *trailing-edge stall* with a laminar separation bubble, transition to turbulence, and reattachment of the turbulent flow.

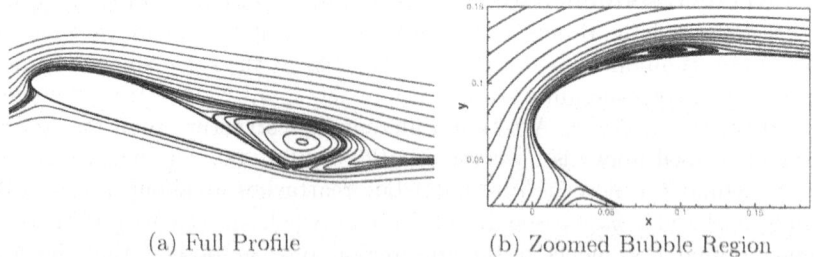

(a) Full Profile   (b) Zoomed Bubble Region

Fig. 8. Streamlines of the averaged flow past the inclined airfoil at $Re_c = 100,000$ and $\alpha = 18°$

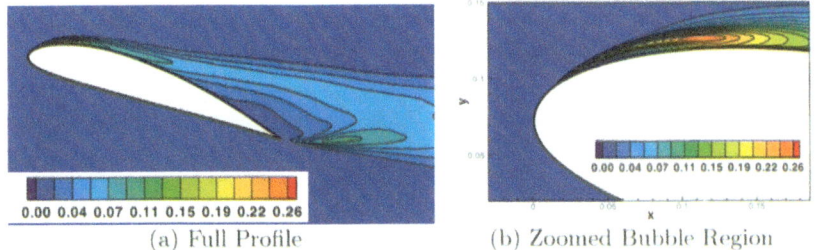

(a) Full Profile   (b) Zoomed Bubble Region

Fig. 9. Distribution of the turbulent kinetic energy k, airfoil flow at $Re_c = 100,000$ and $\alpha = 18°$

6.2 Flow Around the Flat Plate – Comparison of DES and LES Results

Based on the results shown in the previous section, it is on the one hand obvious that LES predictions can provide detailed and useful information about complex turbulent flows. How reliable these computations are, will be proved further in the near future, when the experimental COSTWING results [1] will be available. On the other hand, it is also obvious, that these computations require highly resolved grids with several million grid points for $\mathcal{O}(Re = 10^5)$. Hence for practical applications with Reynolds numbers which are one to three orders of magnitude higher, LES leads to tremendous large number of grid points as estimated by Spalart et al. [6].

Hence the motivation for the search for less expensive alternatives such as DES is evident. Here this new approach is applied to the simpler test configuration of an inclined flat plate at $Re_c = 20,000$ and $\alpha = 18°$ which yields a similar instantaneous and time-averaged flow structure as the airfoil at the same flow conditions [4] discussed above. Figure 10 depicts the contours of the vorticity component ω_z taken from the instantaneous flow at an arbitrary time instant. The three sub-figures show the results of the DES prediction on the coarse grid (denoted as P-DES-C, see Table 1), the corresponding LES results on the same grid (P-LES-C), and the reference LES solution on a much finer grid (P-LES-F), respectively. Compared with P-LES-F the contours obtained by the other simulations are visibly smoother and do not show a large variety of resolved small-scale structures. This is particularly noticeable in the region of the leading-edge shear layer. Contrarily the P-LES-F prediction reproduces this flow feature sharply, thus allowing to resolve in detail the motion of single eddies in this region. Consequently, for the P-LES-F case and at least allusively also for P-LES-C, it is possible to detect a Kelvin-Helmholtz instability occurring in the shear layer and leading to transition. Such detailed flow structures are not resolved by P-DES-C. Therefore, DES fails to reflect the flow in this region properly. This also becomes evident looking at the higher-order statistics below. Although the same grid is applied in the P-DES-C and the P-LES-C case, the latter tends more clearly towards the reference case P-LES-F. Hence, the grid resolution is not solely responsible for the quality of the results. Contrarily the modeling aspect plays a dominant role for this behavior.

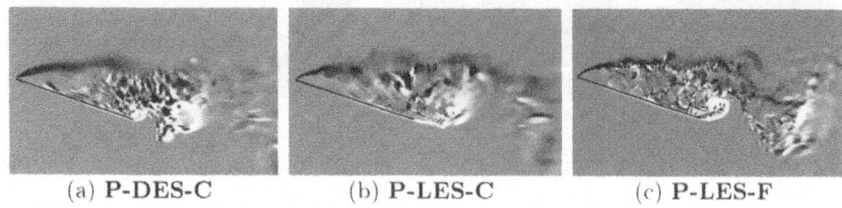

(a) **P-DES-C**   (b) **P-LES-C**   (c) **P-LES-F**

Fig. 10. Instantaneous flow field past the inclined plate at $Re_c = 20,000$ and $\alpha = 18°$, contours of vorticity ω_z; (**a**) P-DES-C, (**b**) P-LES-C, (**c**) P-LES-F (see Table 1)

In Fig. 11 the averaged flow field of all simulations is given by streamlines. The oncoming flow separates at the leading edge and forms a large recirculation region rotating in clockwise direction on the leeward side of the plate. Similar to the airfoil flow, a second vortex originates at the trailing edge due to the roll-up of the shear layer there. This vortex is rotating counterclockwise and is smaller in size. All performed simulations are capable of predicting these overall flow features but at a closer look, there are deviations between them. One difference appears in the location of the centers of the recirculation region and the trailing-edge vortex. The DES result on the coarse grid (P-DES-C) exhibit for both structures an offset in upstream direction of about $0.068\,c - 0.078\,c$, whereas the centers of the P-LES-C prediction nearly coincide with that of P-LES-F. As visible in Fig. 11, this observation corresponds to different locations of the trailing-edge vortex. In the DES prediction this flow structure resides partly upon the end of the plate, whereas in both LES computations the trailing-edge vortex is shifted completely behind the plate.

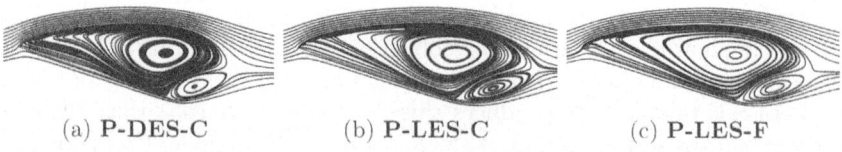

(a) **P-DES-C**   (b) **P-LES-C**   (c) **P-LES-F**

Fig. 11. Streamlines of the spanwise and time-averaged flow past the inclined plate at $Re_c = 20,000$ and $\alpha = 18°$; (**a**) P-DES-C, (**b**) P-LES-C, (**c**) P-LES-F

Higher-order statistics of the DES and LES predictions are shown in Fig. 12 by contours of the turbulent kinetic energy k. Similar to the airfoil flow the largest values of k can be found in the vicinity of the trailing edge, mainly caused by the quasi-periodic shedding motion of the trailing-edge vortex. However, the shear layer originating at the leading edge of the plate is not equally well captured by DES and LES. In the P-LES-F case a local maximum in the k-distribution is located somewhat downstream of the leading edge. It reflects the Kelvin-Helmholtz instability detected in the free shear layer of the flat plate similar as in the airfoil case. This local maximum of k is not observed in the other simulations depicted. Whereas the LES on the coarse grid (P-LES-C) tends to reproduce the shear layer similar to P-LES-F, P-DES-C is not capable of doing so and predicts almost vanishing turbulent kinetic energy in the shear layer region. This remarkable deviation between the DES and LES results coincides with the findings from the distribution of the unsteady vorticity component ω_z which already demonstrated that the DES prediction is not able to resolve the Kelvin-Helmholtz instability in the free shear layer.

Additionally, strong deviations exist for the maximum values of k found in the center of the trailing-edge vortex. The DES prediction ($k_{max} \approx 0.43$) overestimates the maxima of the resolved Reynolds stresses to a great extent, i.e. 39% for P-DES-C according to the reference value from P-LES-F ($k_{max} \approx 0.31$). Contrarily for the LES on the same grid as used for the DES, the resolved Reynolds stresses and k are slightly underpredicted (P-LES-C: $k_{max} \approx 0.29$) and therefore show the expected behavior for a coarser resolution with respect to the fine one applied in the reference simulation.

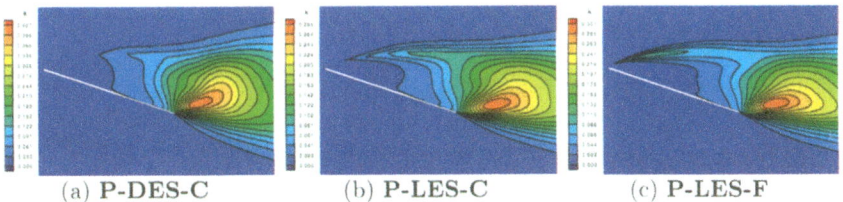

(a) **P-DES-C**   (b) **P-LES-C**   (c) **P-LES-F**

Fig. 12. Resolved Reynolds stresses in terms of the turbulent kinetic energy k, flow past the inclined plate at $Re_c = 20,000$ and $\alpha = 18°$; (a) P-DES-C, (b) P-LES-C, (c) P-LES-F

Further interesting differences are found in the distribution of the eddy viscosity with respect to the molecular viscosity not shown here. Starting with the P-LES-F case, increased values of ν_T/ν can be observed in the shear layers and in regions of large vortical structures in the wake. The shear layer originating from the leading edge is reproduced particularly detailed. In comparison with P-LES-F ($\nu_T/\nu|_{max} \approx 2$), P-LES-C yields expectedly to higher values of $\nu_T/\nu|_{max} \approx 6$. This is quite natural, since on a coarser grid the filter size is larger and hence the eddy viscosity inherits a greater part in describing the turbulent motion. A strongly deviating distribution of ν_T/ν is found for P-DES-C exhibiting very high eddy viscosities in the leading-edge shear layer ($\nu_T/\nu \approx 20$). A detailed analysis has shown that the modified S-A model suggested by Spalart et al. [6, 7, 12, 14] is not properly adjusted to the modeling requirements in the LES-mode of the DES approach and hence need some modifications. For a complete analysis and a description concerning these adjustments, we refer to [5], where also a comparison with pure RANS computations can be found.

7 Performance of the Code

The LES computations of the flow past the flat plate and the wing were carried out on the SMP cluster Hitachi SR 8000–F1 [30]. The architecture of this machine allows the use of three hierarchical levels of parallelization, which were all taken into account for $\mathcal{LESOCC}$:

– *lowest level:* pseudo-vector processing on each RISC-processor of a SMP node via the pre-load mechanism from the main memory or the pre-fetch mechanism from the cache.
– *intermediate level:* intra-node auto-parallelization over all processors of one SMP node supported by the compiler applying COMPAS (Co-Operative Micro Processors in single Address Space).
– *highest level:* inter-node parallelization applying the domain decomposition technique with explicit message passing on several SMP nodes with the help of a communication library (MPI) for the data exchange between the nodes.

The optimization of the code for this special architecture was carried out with the support of RRZE Erlangen (Germany). Using a single SMP node and applying the first two parallelization techniques described above, typically lead to a measured performance of ≈ 3 GFlops. For the production runs 8–16 nodes of the cluster were used. The calculation denoted as **A-20-12** (see Table 1) was carried out e.g. on 12 nodes taking all levels of parallelization into account. A total sustained performance of about 32.0 GFlops equivalent to an averaged performance of about 2.67 GFlops per SMP node was achieved. Hence small losses acceptable for practical applications occur due to a non-optimal load-balancing, local and global communication and additional operations.

8 Conclusions

The outcome of the present investigation is twofold. In the first part detailed LES predictions of the flow around an unswept wing were carried out to study the physics of stalled airfoil flows. It was found that at a quite low Reynolds number $Re_c = 20,000$ a leading-edge stall occurs at both angles of attack ($\alpha = 12°$ and $18°$) investigated. The flow separates shortly after the leading edge because it is exposed to a strong adverse pressure gradient. The separated flow forms a free shear layer, where a Kelvin-Helmholtz instability is observed and transition to turbulence takes place. At the leeward side of the airfoil a large clockwise rotating recirculation region with a nearly constant pressure level is found. Furthermore, a quasi-periodic formation and shedding motion ($St' \approx 0.2$) of a strong vortical structure in the vicinity of the trailing edge is detected leading to an asymmetric vortex street in the wake. These flow features are present for both angles of attack. A completely different flow structure was observed when the Reynolds number is increased to $Re_c = 100,000$. Now the flow also separates shortly after the nose but reattaches quickly and forms a closed *laminar* separation bubble. In the rear part of the bubble transition to turbulence takes places leading to an attached boundary layer which finally separates at $x/c \approx 0.7$ typical for a trailing-edge stall. In the second part of the paper the standard DES concept was evaluated based on the flow around an inclined flat plate, which shows similar flow characteristics as the airfoil case. The DES results were compared with LES data achieved on the same coarse grid and with a reference LES solution on a much finer grid. This comparison clearly showed some deficiencies of the DES approach which have to be attributed to the modeling strategy applied in the LES-mode of the DES concept. Appropriate improvements are subject of ongoing activities.

Acknowledgement. This work was supported by a fellowship from the German Academic Exchange Program (DAAD) to K. Mazaev. Additionally, the LES contributions were supported by the *Deutsche Forschungsgemeinschaft* under contract number BR 1847/2. The computations were mainly carried out on the German Federal Top-Level Computer Hitachi SR 8000-F1 at HLRB. The assistance of Dr. G. Wellein of the RRZE Erlangen (Germany) in optimizing the code $\mathcal{LESOCC}$ used for the simulations and leading to a noticeable gain in performance is greatly appreciated. All kinds of support are gratefully acknowledged.

References

1. Lerche, Th., Dallmann, U.Ch. (1999): Das Prinzipexperiment COSTWING I: Dokumentation der Aufbauphase. Institut für Strömungsmechanik, DLR Göttingen, IB 223–99 A04.
2. Davidson, L. (2000): LESFOIL: An european project on large-eddy simulations around a high-lift airfoil at high Reynolds number. In: ECCOMAS 2000, European Cong. on Comp. Meth. in Appl. Sci. and Eng., Barcelona, Spain, 11–14 Sept.
3. Breuer, M., Jovičić, N. (2001): An LES investigation of the separated flow past an airfoil at high angle of attack. 4th Workshop on DNS & LES, Enschede (NL), July 18–20, 2001, ERCOFTAC Series, vol. 8, pp. 165–172, DNS and LES IV, eds. B.J. Geurts et al., Kluwer Acad. Publ., Dordrecht.

4. Breuer, M., Jovičić, N. (2001): Separated flow around a flat plate at high incidence: an LES investigation. J. of Turbulence (http://iop.jot.org), **2**, 1–15.
5. Breuer, M., Jovičić, N., Mazaev, K. : Comparison of DES, RANS, and LES for the separated flow around a flat plate at high incidence, submitted for publication.
6. Spalart, P.R., Jou, W.-H., Strelets, M., Allmaras, S.R. (1997): Comments on the feasibility of LES for wings, and on a hybrid RANS/LES approach. First AFOSR Int. Conf. on DNS/LES, Ruston, LA, Advances in DNS/LES, C. Liu & Z. Liu (eds.), Greyden, Columbus, OH.
7. Spalart, P.R. (2000): Trends in turbulence treatments. AIAA Paper 2000–2306, FLUIDS 2000, Computational Fluid Dynamics Symp., Denver, Colorado, USA, June 19–22.
8. Spalart, P.R. (2000): Strategies for turbulence modelling and simulations. Int. J. of Heat and Fluid Flow, **21**, 252–263.
9. Speziale, C.G. (1996): Turbulence modeling for time-dependent RANS and VLES: a review. AIAA Journal, **36**(2), 173–184.
10. Speziale, C.G. (1998): A combined large-eddy simulation and time-dependent RANS capability for high-speed compressible flows. J. of Sci. Comput., **13**, 253–274.
11. Germano, M. (1999): From RANS to DNS: towards a bridging model. 3rd Workshop on DNS & LES, Cambridge (U.K.), May 12–14, 1999, ERCOFTAC Series, **7**, 225–236, DNS and LES III, eds. P.R. Voke et al., Kluwer Acad. Publ., Dordrecht.
12. Shur, M., Spalart, P.R., Strelets, M., Travin, A. (1999): Detached-eddy simulation of an airfoil at high angle of attack. 4th Int. Symp. on Eng. Turb. Mod. & Meas., Corsica, France, May 24–26, 1999, Eng. Turb. Mod. & Exp. 4, 669–678, eds. W. Rodi et al., Elsevier Sci. B.V., Amsterdam.
13. Nikitin, N.V., Nicoud, F., Wasistho, B., Squires, K.D., Spalart, P.R. (2000): An approach to wall modeling in large-eddy simulations. Phys. of Fluids, **12**(7), 1629–1632.
14. Travin, A., Shur, M., Strelets, M., Spalart, P.R. (2000): Detached-eddy simulations past a circular cylinder. J. of Flow, Turbulence and Combustion, **63**(1/4), 293–313.
15. Forsythe, J.R., Hoffmann, K.A., Dietiker, J.F. (2000): Detached-eddy simulation of a supersonic axisymmetric base flow with an unstructured solver. AIAA Paper 2000–2410, FLUIDS 2000, Computational Fluid Dynamics Symp., Denver, Colorado, USA, June 19–22.
16. Strelets, M. (2000): Detached-eddy simulation of massively separated flows. AIAA Paper 2001-0879.
17. Squires, K.D., Forsythe, J.R., Spalart, P.R. (2001): Detached-eddy simulation of the separated flow around a forebody cross-section. 4th Workshop on DNS & LES, Enschede (NL), July 18–20, 2001, ERCOFTAC Series, **8**, 481–500, DNS and LES IV, eds. B.J. Geurts et al., Kluwer Acad. Publ., Dordrecht.
18. Breuer, M., Rodi, W. (1996): Large-eddy simulation of complex turbulent flows of practical interest. In: Flow Simulation with High-Performance Computers II, ed. E.H. Hirschel, Notes on Numerical Fluid Mechanics, **52**, 258–274, Vieweg Verlag, Braunschweig.
19. Breuer, M. (1998): Large-eddy simulation of the sub-critical flow past a circular cylinder: numerical and modeling aspects. Int. J. for Numerical Methods in Fluids, **28**, 1281–1302.

20. Breuer, M. (2000): A challenging test case for large-eddy simulation: high Reynolds number circular cylinder flow. Int. J. of Heat and Fluid Flow, **21**(5), 648–654.

21. Spalart, P.R., Allmaras, S.R. (1994): A one-equation turbulence model for aerodynamic flows. La Recherche Aérospatiale, **1**, 5–21.

22. Smagorinsky, J. (1963): General circulation experiments with the primitive equations, I, the basic experiment. Mon. Weather Rev., **91**, 99–165.

23. Germano, M., Piomelli, U., Moin, P., Cabot, W.H. (1991): A dynamic subgrid scale eddy viscosity model. Phys. of Fluids A, **3**(7), 1760–1765.

24. Lilly, D.K. (1992): A proposed modification of the Germano subgrid scale closure method. Phys. of Fluids A, **4**(3), 633–635.

25. Kravchenko, A.G., Moin, P. (1997): On the effect of numerical errors in large-eddy simulation of turbulent flows. J. of Comput. Physics, **131**, 310–322.

26. Rhie, C.M., Chow, W.L. (1983): A numerical study of the turbulent flow past an isolated airfoil with trailing-edge separation. AIAA Journal, **21**, 1525–1532.

27. Rodi, W., Ferziger, J.H., Breuer, M., Pourquié, M. (1997): Status of large eddy simulation: results of a workshop. Workshop on LES of Flows Past Bluff Bodies, Rottach-Egern, Tegernsee, Germany, June 26–28, 1995, J. of Fluids Engineering, **119**(2), 248–262.

28. Knisely, C.W. (1990): Strouhal numbers of rectangular cylinders at incidence: a review and new data. J. of Fluids and Structures, **4**, 371–393.

29. Evans, G.S., Friedrich, R. (2002): LES of flow around a NACA 4415 airfoil at high angle of attack. Personal communication.

30. Brehm, M., Bader, R., Ebner, R. (2001): High-performance computers in Bavaria (HLRB): The Hitachi SR 8000-F1 (in German). http://www.lrz-muenchen.de/services/ compute/hlrb/ .

Direct Simulation with the Lattice Boltzmann Code BEST of Developed Turbulence in Channel Flows

Peter Lammers, Kamen N. Beronov, Gunther Brenner, and Franz Durst

Institute of Fluid Mechanics
Friedrich-Alexander University Erlangen–Nuremberg
Cauerstrasse 4
91058 Erlangen, Germany
{*plammers,kberonov,brenner,durst*} *@lstm.uni−erlangen.de*

1 Introduction

In spite of the dramatic increase of the performance of recent supercomputers the direct numerical simulation (DNS) of turbulent flows is still an expensive venture in view of the high memory and CPU time requirements. Today, the DNS is limited to very low Reynolds number and simple flow geometries which are far away from technical relevance. Therefore, there is an increasing demand in the development of numerical schemes to simulate fluid flows resolving the turbulent scales in order to exploit existing and future supercomputers more efficiently. In that context, the lattice Boltzmann method have challenged the traditionally used DNS method based on FV or pseudospektral . The potential of these LBM have been clearly shown in various publications [4]. The goal of the present paper is to specify the advantages of the LBM to DNS more quantitatively and to make use of the LBM to investigate new phenomena related to wall bounded turbulence.

This report summarises activities at the LSTM-Erlangen that have led to the development of a 3D lattice Boltzmann solver (BEST) for the simulation of time dependent, incompressible flows on high performance parallel computers as the main target computer architecture.

To show the potential for DNS applications, BEST is employed to investigate 3D turbulence in a quasi 2 dimensional channel flow. The simulations were run up to large physical times, well into the statistically steady–state regime, which is referred to as developed turbulence here. Statistics were assembled in that regime and compared to corresponding data obtained from DNS with a classical Chebyshev-pseudospectral code (CPS). The overall agreement, presented in this paper, is very good.

Beside this, the present computations with BEST served another practical purpose — to estimate the minimal spatial resolution required for an adequate computation of various turbulence statistics in the channel. The efficiency of BEST made it possible to carry out the first systematic investigation of the effect of the streamwise length of the computational box on the computed Reynolds stresses and higher moments. Theoretical arguments and experimental observations have suggested that only long domains (in terms of grids accessible to numerical simulation) can yield reliable statistics. But a quantification of the required length through systematic parameter studies using DNS had remained too expensive.

Summarising the presented results the main conclusion of this paper might be formulated this way: Incompressible turbulent flows within simple geometries, such as in a 2D channel, have been computed for several decades now, using classical DNS methods, in particular the highly precise pseudospectral method. It may appear even surprising that the "simple" LB scheme produces results of comparable quality at considerably lower cost, even in such idealised flows. We belief that in more complicated geometries the advantage of LB methods grows to the extent of a qualitative difference.

The paper is organised as follows: The relevant physical aspects of the flow under investigation are presented in Sect. 2. Section 3 gives a brief introduction of the lattice Boltzmann method including remarks on the computational cost of CPS versus LBM (Sect. 3.1) and a short description of the code BEST (Sect. 3.3) together with its performance characteristics on the Hitachi SR8000–F1 at the Leibniz Computer Center (LRZ) in Munich. The validation of BEST against a pseudospectral code and the comparison of the results is discussed in Sect. 4.1. Results of the investigation of the influence of the computational box on the statistics are presented in Sect. 4.2. Computations up to $4096{:}256^2$ were required.

In Sect. 5 the main aspects of this advantage are discussed and a list of examples is given, of turbulent flows for which the LB method is the DNS method of choice. A summary and discussion of the presented results is also placed in Sect. 5. The promising potential for applications of the BEST code in its current version and after further development are outlined in this context.

2 Physical problem: channel turbulence

The test problem selected to compare the performance of the present LB code for the simulation of turbulence in a wall bounded shear flow is the well known minimal channel defined by [10]. Simulated Reynolds numbers are moderate, to allow for multiple and sufficiently long runs. While away from the wall a kind of self–similar cascade dominated by inertial effects can be discerned [14], most of the interesting physics, including the process of generation of turbulence [9], is taking place near the wall, where all involved forces — pressure, viscosity, and inertia — are important. Figure 1 shows the developed turbulence instantaneous vorticity field $\boldsymbol{\omega} = \nabla \times \boldsymbol{v}$ in the vicinity of a wall. The velocity field $\boldsymbol{v}$ is nearly parallel to the wall in that region, with very long–range correlation in the streamwise direction x. In the simulated flow v is assumed periodic in x and in the spanwise direction z, but with large periods. The non–slip condition is imposed at $y = \pm H$, H being the channel half–width.

In the present context, resolution is characterized by two measures: (i) the overall dimensions of the computational box and (ii) the stepsize (equal in all three coordinate directions) of a uniform LB grid spanning that box. Lengths are measured in three alternative units: (i) the grid stepsize Δ, (ii) the channel half–width H, (iii) the wall unit $\delta^+ = \nu/u_\tau$, where ν is the kinematic viscosity and u_τ the skin friction velocity, defined through $u_\tau^2 = (-dP/dx)\rho H$. Here $\rho = 1$ is the density and dP/dx is the mean pressure gradient driving the flow.

The Reynolds number Re is defined with a suitable length scale (H or δ^+) and velocity unit (u_τ, average, or centerline velocity). Here only the definition $Re_\tau = Hu_\tau/\nu$ will be used, implying $Re_\tau = H/\delta^+ = H^+$. Several other Reynolds numbers are in use. One is based on U_c, the average centerline velocity: $Re_c = HU_c/\nu$. Another one comparable in magnitude with the one before mentioned is based on U_m, the mean velocity: $Re_m = 2HU_m/\nu$. In a turbulent channel flow $Re_m > Re_c \gg Re_\tau$. It is natural to first prescribe Re_τ, which alone defines the turbulence physically, and then choose u_τ and Δ^+ from numerical resolution and stability consideration. Then dP/dx and ν are not physical input data and have instead to be determined as follows:

$$\nu = u_\tau/\Delta^+, \qquad |dP/dx| = Re_\tau \Delta^+ (\nu/H)^2 = u_\tau^2 \Delta^+/Re_\tau. \quad (1)$$

It has been documented throughout the literature on DNS of turbulence that a step size of $\Delta = 1.5\eta - 2\eta$, where η is the dissipative (Kolmogorov) lengthscale, is an upper limit, above which the fine structure of turbulence is not resolved, while $\Delta = \eta$ guarantees full resolution. For the present 2D–channel turbulence, it is estimated [15, exercise 7.8] that $\eta^+ \approx 1.5$ at the wall and η increases inward. A uniform grid step size $\Delta^+ \approx 1.5$ in wall units would therefore guarantee a fully resolved DNS and $\Delta^+ \approx 2.3$ would still be stable and give satisfactory resolution over the whole flow domain. A prerequisite for the LB scheme to approximate the incompressible Navier–Stokes dynamics is that the flow velocity v be uniformly small compared to the "sound speed" Δ_x/Δ_t which equals 1 in lattice units. The relative error due to numerical compressibility scales with v^2. To keep it below 1% even at the centerline, where velocities are in the range $10 - 20\,u_\tau$ for $Re_\tau = O(10^2)$, one may specify $u_\tau = 0.005$ in lattice units.

In order to obtain converged statistics, the initial flow state has to be first allowed to evolve towards a statistically steady state, usually referred to as *developed turbulence*. Then mean profiles (e.g. of velocity $\langle v_x \rangle (y)$ or pressure $\langle P \rangle (y)$ — quite separate from the driving mean pressure gradient) are accumulated, and only in a third phase the single–point moments of turbulent fluctuations (e.g. Reynolds stresses $\langle v_i' v_j' \rangle$, skewness $S[v_j'] = \left\langle v_j'^3 \right\rangle / (\langle v_j'^2 \rangle)^{3/2}$ and flatness $F[v_j'] = \left\langle v_j'^4 \right\rangle / (\langle v_j'^2 \rangle)^2$ for the turbulent velocity fluctuations

Fig. 1. Snapshot of an isosurface of vorticity amplitude.

v_j) can be computed. The convergence of two–point as well as higher–order one–point statistics requires even longer integration time. The times required for the accumulation of statistics is measured in the physical time unit $\tau = H/u_\tau$, independent of the numerical method and resolution used. Depending on initial condition, flow domain size (in H units) and Re_τ, as well as on the criterion chosen to indicate the statistically steady state, the initial transient lasts typically 10–30 τ. Mean profiles converge faster, but statistics of fluctuations take longer. Higher–order statistics and longer (in H units) domains imply longer (in τ units) times needed for statistics to converge.

3 Numerical method: lattice Boltzmann DNS

While lattice Boltzmann methods (LBM) are now well developed and widely used for the simulation of Navier–Stokes [4] as well as more complex fluid dynamics, direct numerical simulations of incompressible flows at high Reynolds numbers using LBM have not yet obtained popularity. Most simulations of that kind are of instabilities or can be classified into the large–eddy simulation (LES) or Reynolds–averaged Navier-Stokes variety with k–ϵ modelling. Direct numerical simulations (DNS) of three–dimensional (2D) Navier–Stokes turbulence in flows of engineering interest with the use of LBM have been made relatively early (e.g. [6,2]), while the research community in turbulence modeling remains true to traditionally used high–resolution (pseudospectral or compact finite difference) methods. For the case of 2D channel turbulence in particular, the reference method is the Kim,Moin and Moser pseudospectral code using Chebyshev polynomials in the wall–normal direction [10].

3.1 Lattice Boltzmann vs. pseudospectral methods

Pseudospectral methods (PS) have the maximal accuracy as compared to other methods for fixed number N of degrees of freedom in the discretization.

This is true both for amplitude and phase, for all variables and their derivatives, but only under the condition that the computational grid accomodating the (fixed number of) degrees of freedom is *sufficiently fine* to resolve the fine flow structure. The cost of PS algorithms based on Fast Fourier transforms is of order $O(N \log N)$.

The nonlocal character of the involved Fourier decompositions and the specific fast algorithms for the spectral transforms prevent local refinement of resolution: In order to improve resolution at the channel centerline, one is forced to compute what turns out by comparison with other discretizations to be an unnecessarily large number of points near the wall. In the homogeneous directions, the same resolution must be used at all distances from the wall.

LB schemes are free from this restriction and allow flexible grid refinement. In contrary to the nonlocal FFT operations in PS codes, LB algorithms are strictly local, which reduces the inter–processor communication cost dramatically: from $O(N^{4/3} \log N)$ to $O(N^{2/3})$ in 3D and from $O(N^{3/2} \log N)$ to $O(N^{1/2})$ in 2D. Like all other explicit methods, LB methods use smaller time steps when spatial resolution is increased. As for LB methods this step is proportional to the spatial grid step, in 3D the overall *computation cost* is $O(N^{4/3})$ and the corresponding *communication cost* is $O(N)$.

While these asymptotic estimates suggest that LB codes are overall more efficient than PS codes, it remains to verify (i) if the predicted difference in speed is observed already at practically relevant resolutions N, and (ii) if this does not come at the cost of an inacceptable deterioration of the numerical solution quality. In Sect. 3.3 and Sect. 4.1 respectively, it is shown that in both aspects the BEST code presents a viable alternative to a Chebyshev–PS code for 2D channel turbulence.

3.2 The 3D lattice Boltzmann model

The first step to specify a LB method is the choose of the lattice model. This includes (a) a finite velocity model which discretizes the virtual velocity space, and (b) a corresponding equilibrium distribution of probabilities for the discrete velocities. Having chosen the lattice, one then has to select a relaxation mechanism which assures that these probabilities remain close to their equilibrium values. Such proximity is a precondition for the validity of the Chapmann–Enskog expansion, the procedure which gives the relation between LB and Navier–Stokes dynamics.

The 3D lattice model used here is the 19-velocity D3Q19 model [16], obtained as projection from 4D onto 3D of the face–centered hypercubic (FCHC) lattice [3] which guarantees sufficient symmetry to reproduce precisely the Navier–Stokes hydrodynamics upon Chapman–Enskog expansion. The modification proposed in [7] for computations of (low–Reynolds–number) incompressible flows is adapted for this lattice. The resulting lattice Boltzmann model is referred to as D3Q19i.

While multiple–relaxation–time (MRT) schemes (also called moment models) of relaxation toward equilibrium have been developed and demonstrated to have advantages with respect to the quality and stability of the numerically computed flows [5] in different applications, mostly at low to moderate Reynolds numbers, we have preferred to use the single–relaxation–time (SRT) relaxation scheme (also called BGK method), for several reasons: (i) It is the simplest scheme to understand and to implement. (ii) The advantages of MRT over SRT were expected to be even more pronounced in turbulent flows as compared to the laminar situations in which they have already been demonstrated. But preliminary experiments with the MRT method obtained for D3Q19i following [11] have suggested that the advantage is not as large as, and perhaps even less than expected. The overall cost of the MRT scheme appeared noticeably higher than that of BGK relaxation. (iii) The theoretical links not only to gas–kinetic theory [8] but also to the kinetic theory of turbulence [12] are well understood for the BGK model.

While standard LB schemes advance in effect the fluid density and velocity $v(x,t)$, their "incompressible" counterparts assume a constant density and advance a "pressure" $p(x,t)$. For the D3Q19i model, this proceeds according to the scheme

$$p(x,t) = q\,, \quad q(x,t) \;=\; \sum_{j=0}^{18} q_j(x,t)\,, \quad v(x,t) \;=\; \sum_{j=0}^{18} q_j(x,t)\,\xi_j\,, \quad (2)$$

$$q_j(x+\xi_j, t+\Delta t) \;=\; q_j(x,t) - \frac{\Delta t}{3\nu + 1/2}\Big(q_j(x,t) - \bar{q}_j(x,t)\Big)\,, \quad (3)$$

$$\bar{q}_j(x,t) \;=\; w(|\xi_j|^2)\Big(q + 3v_j + \frac{9}{2}v_j^2 - \frac{3}{2}|v|^2\Big)\,, \quad (4)$$

$$v_j = v(x,t)\cdot\xi_j\,, \qquad w(0) = \frac{4}{9}\,, \quad w(1) = \frac{w(0)}{4}\,, \quad w(2) = \frac{w(1)}{4}\,,$$

$$\xi_1 = (+,0,0)\,, \quad \xi_2 = (+,+,0)\,, \quad \xi_3 = (0,+,0)\,, \quad \xi_4 = (-,+,0)\,,$$

$$\xi_5 = (0,-,0)\,, \quad \xi_6 = (-,-,0)\,, \quad \xi_7 = (-,0,0)\,, \quad \xi_8 = (+,-,0)\,,$$

$$\xi_9 = (0,0,+)\,, \quad \xi_{10} = (+,0,+)\,, \quad \xi_{11} = (0,+,+)\,, \quad \xi_{12} = (-,0,+)\,,$$

$$\xi_{14} = (0,0,-)\,, \quad \xi_{15} = (+,0,-)\,, \quad \xi_{16} = (0,+,-)\,, \quad \xi_{17} = (-,0,-)\,,$$

$$\xi_{18} = (0,-,-)\,, \quad \xi_{13} = (0,-,+)\,, \quad \xi_0 = (0,0,0)\,,$$

where e.g. $(+,0,-)$ stands for $(\Delta x, 0, -\Delta z)$ and fields are given at the mesh points $x = (x,y,z)$ of a regular cartesian grid. This is a characteristic method for the BGK equation

$$(\partial_t + \xi\cdot\nabla_x)\, q(\xi, x, t) = -\omega\,(q - \bar{q})\,, \qquad\qquad \omega = \nu/3\,.$$

It appears as an explicit first–order scheme but is in fact second–order in time. It approximates the incompressible Navier–Stokes equations [7] up to deviations of order $O\Big(|v|^3 + |\nabla q|^2/(1+q^2)\Big)$.

3.3 The BEST code: implementation and performance

The lattice Boltzmann code BEST (Boltzmann equation solver tool) developed at LSTM, implements the most popular LB models for 2D and 3D Navier–Stokes flows, in both their the original and "incompressible" versions. It is ported to a broad range of computer architecture including Hitachi SR, purely vector computers like Fujitsu VPP and NEC SX, massively parallel systems like Cray T3E, and RICS architecture like SGI Origin. BEST is parallized for distributed memory architectures using MPI and for shared memory architectures using OpenMP.

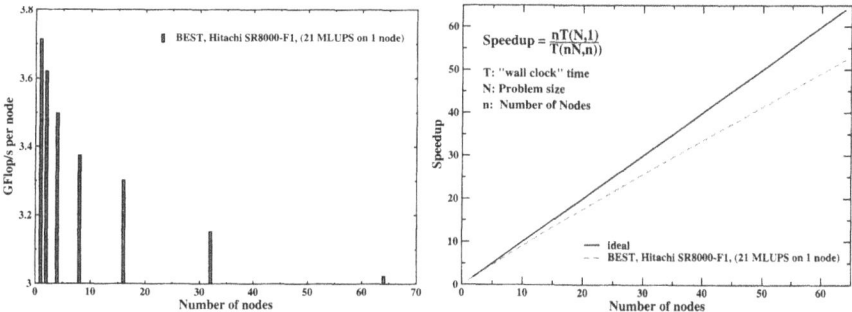

Fig. 2. Speedup of BEST on the SR8000–F1

The vectorisation on vector computers is over 90%, which is equivalent to 50% of the peak performance on a single vector node. For better performance on the Hitachi SR8000–F1, an inner Node parallelisation technique (COMPAS) developed by Hitachi is used. This assures some 30% of the theoretical peak performance of the SR8000–F1, or 21 Mlups (lups = lattice–site updates per second) on one node. The parallel speedup with increasing number of nodes used in a test computation with BEST and the absolute performance in Gflops (flops = floating–point operations per second) can be seen in Fig. 2.

A comparison between LB and Chebyshev–pseudospectral (CPS) methods was taken up in Sect. 3.1 in terms of asymptotic estimates of computational cost. It is of interest to compare the actual cost, at moderate resolutions, for the time of the LB code BEST and a standard CPS code take to advance a flow at $Re_\tau \approx 180$, defined on a rectangular 128^3 grid over one physical time unit τ (its definition was given in Sect. 2). The LB code uses fixed time–step and identical·cubic grid cells. The CPS code uses adaptive (CFL–dependent) time–step and its grid is nonuniform in the wall–normal direction: the y–step between neighbouring grid points varies from $\Delta^+ = 0.16$ between the two points closest to the wall to $\Delta^+ = 4.34$ at the centre of the channel. The resolution of the LB code is $\Delta^+ = 2.8$, uniform in the three coordinate directions. Step sizes Δ^+ are measured in wall units δ. Thus, LB resolution in y is finer than the CPS one at $y^+ > 10$. Running on a single node of the SR8000–F1 with these specifications, BEST needed 32 min, whereas the CPS code needed 158 min.

4 Results

Two series of LB simulations were carried out, one with $\Delta^+ \approx 2.36$ and $Re_\tau \approx 150$ (so $\nu \approx 0.02$ and $H = Re_\tau/\Delta^+ \approx 63.5$ in lattice units, and the resolution in the wall–normal direction y is 128 grid steps), and a second with $\Delta^+ \approx 1.41$ and $Re_\tau \approx 180$ (so $H \approx 127.5$ and wall–normal grid size is 256). The discussion in Sect. 2 on the smallest scales that have to be resolved in a turbulence DNS implies that the first series is close to the crudest resolution for which turbulence structure can be properly represented and the DNS be stable, while the second run is so well resolved, that there is no point in further increasing the resolution. The comparison in Sect. 4.1 of statistics obtained from that run and corresponding CPS data, which include much more grid points than the LB code in the near wall region, shows among other things, that there is no point indeed in paying the price for a grid step size smaller than $\Delta^+ \approx 1.5$ in a DNS with a LB method.

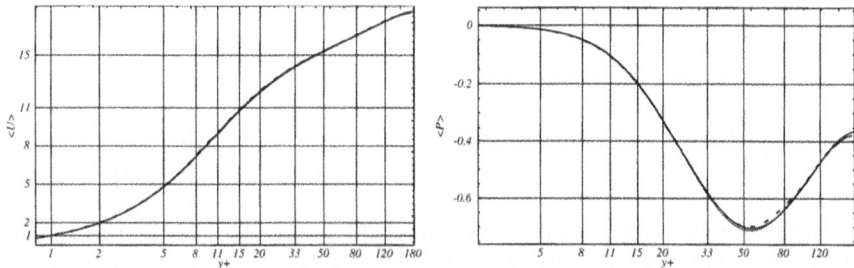

Fig. 3. Comparison of LB and CPS results for *mean profiles* — left: streamwise velocity, right: pressure. Line styles — thin solid: BEST , bold dashed: CPS data from [13], thin dotted: CPS data from in–house DNS [17].

In each of the two series of runs, the computational domain had a square cross–section and its length (necessarily proportional, for the usual cubic LB grid cell geometry, to the number of grid points in streamwise direction x) was varied. Detailed results will are presented below only for $Re_\tau = 180$ which corresponds to a standard DNS data set [10, 13].

For the comparisons in Sect. 4.1 only well converged LB data are used, meaning a long–enough averaging time of over 25τ after allowing for an initial transient of about 25τ, as well as a sufficiently large computational domain. The establishment of a sufficient streamwise length is the subject of Sect. 4.2 while the estimation of sufficient width will not be discussed in this article. For the statistics shown, a domain of streamwise length of $32H$ and spanwise

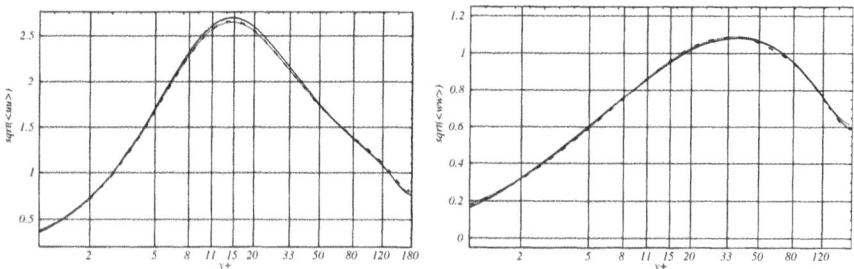

Fig. 4. Comparison of LB and CPS results for the turbulence velocity *intensities* — left: streamwise velocity, right: spanwise velocity. Line styles — as in Fig. 3.

width of $2H$ were taken and verified to be sufficient. While its width is notably smaller than the physical–space width covered by the reference simulations [13], its length is larger than in the reference DNS. Resolution in x and z, and in the core flow also in y, is several times finer in the LB than in the CPS simulation.

4.1 Validation: LB vs. pseudospectral DNS

The computation on correct mean profiles is the easiest test, passed by many different numerical methods. As seen in Fig. 3, which compares the mean profiles of (streamwise) velocity and pressure, the comparison with the reference data base [13] is excellent, so BEST is no bad exception to the rule. Normalization is by u_τ and u_τ^2 respectively, as $\rho = 1$. It may be noted in passing that LB methods are weakly compressible by construction and for the present incompressible computations the implemented D3Q19i method is in effect an artificial compressibility scheme. This may affect the fluctuating pressure, but not the mean turbulent pressure distribution, which in the present case can be related to a Reynolds stress component: $\langle P \rangle (y) + \left\langle v_y'^2 \right\rangle (y) = \text{const.}$ Data agree with this analytic constraint.

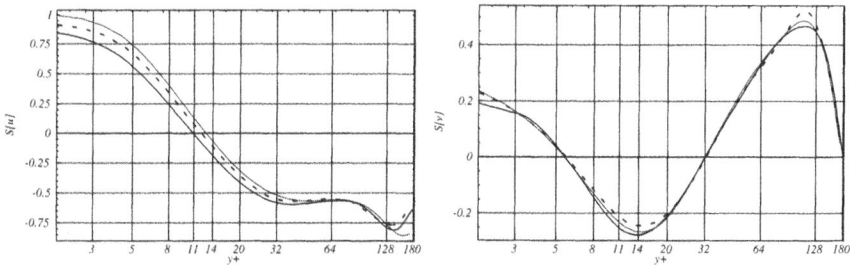

Fig. 5. Comparison of LB and CPS results for the *skewness* of velocity components — left: streamwise velocity, right: wall–normal velocity. Line styles — as in Fig. 3.

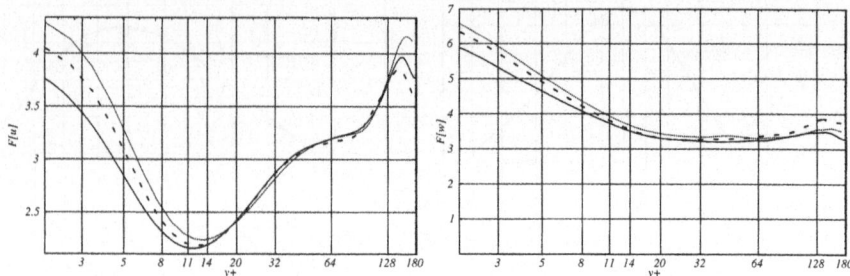

Fig. 6. Comparison of LB and CPS results for the *flatness* of velocity components — left: streamwise velocity, right: spanwise velocity. Line styles — as in Fig. 3.

The next test is the correct prediction of the y–distribution of Reynolds stresses. Those for which the largest relative deviation between LB and CPS results was observed, namely $\left\langle v_x'^2 \right\rangle$ and $\left\langle v_z'^2 \right\rangle$, are represented in Fig. 4 normalized respectively to $\left\langle v_x'^2 u_\tau^{-2} \right\rangle^{1/2}$ and $\left\langle v_z'^2 u_\tau^{-2} \right\rangle^{1/2}$. There is an excellent quantitative agreement, for these and of course for the other Reynolds stresses not shown here. The largest difference, at the peak of $\left\langle v_x'^2 \right\rangle$, can be attributed to the difference in flow domain length.

The last and most stringent test presented here is a comparison of higher–order one–point statistics, namely y–distributions of skewness and flatness of fluctuating velocity components. Skenesses are shown in Fig. 5 and flatnesses in Fig. 6. It is striking to observe that the difference between BEST results and the reference database is comparable to and often smaller than that between two runs of the *same CPS code*. (R. Volkert at LSTM has done the in–house repetition of the computations at $Re_\tau \approx 180$ reported in [13] and its references. The differences between his results and the [13] data can be due to only two factors — a difference in the time span of averaging or a difference in initial conditions. Both of these are due to the minor incompletenesses in the run specifications that can be extracted from [13] and its references. For further details see [17]) It may be further noted that the two CPS simulations are driven by enforcing a constant flux, whereas the BEST simulation is forced by a constant pressure drop, and the initial flow fields used by the two different codes are entirely different.

4.2 New results: toward a "minimal channel" specification

A growing list of journal publications report evaluations of sensitive higher–order statistics in uniform–shear homogeneous turbulence and in 2D channel flow on the basis of DNS with a marginal resolution (necessarily so when turbulence is simulated on a 128^3 grid, say) and, more importantly, using flow domains of very limited extent in the streamwise direction. Such use of relatively low–Re numerical results without proper validation in support of statements about fundamental physical issues like the effect of shear on the scaling of velocity structure functions, the (possible lack of) isotropy at the smallest scales of turbulence, or the change in anomalous scaling in the near–wall region, have recently invoked some informal cautioning comments. One of the main questions that have been included in this relation into the current agenda of the turbulence computing community is the quantification of the admittedly large minimum length, in physical units, of the computational used for shear–flow DNS.

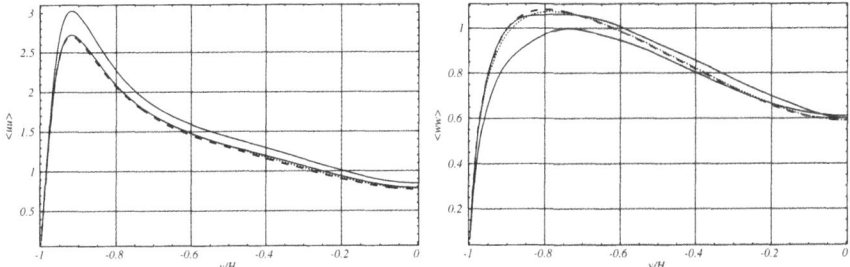

Fig. 7. Convergence of turbulence *intensities* obtained using BEST with stream–wise length of computational domain — left: streamwise velocity, right: spanwise velocity. Lines — solid: $L=2H$ and $L=8H$, , dashed: $L=32H$.

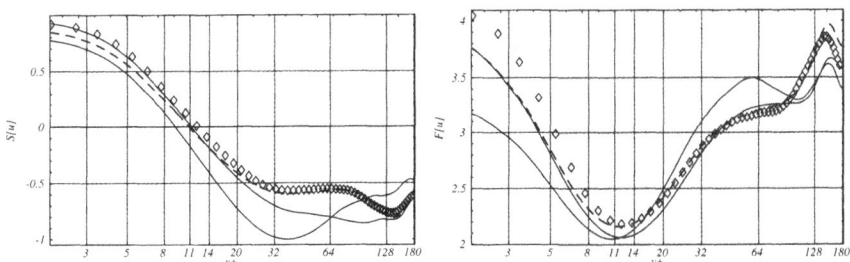

Fig. 8. Convergence of streamwise turbulence velocity statistics with streamwise length of computational domain — left: skewness, right: flatness. Symbols: CPS data from [13], lines — BEST data marked as in Fig. 7.

In the present context, this issue is related to physical questions like those about the size of the "minimal unit of wall turbulence" (see e.g. [9] and its references), about the possibility for very long–range correlations near the channel centerline, the scaling of streamwise correlation lengths with Re_τ, etc. What can feasibly established on the basis of DNS, however, even with signifficant computational resources, is only an *a posteriori* check for the error due to finite streamwise length of the (periodic) computational domain. Having fixed $Re_\tau = 180$, the domain cross–section and the small–scale resolution, only the streamwise length is increased, until the required statistics converge. For each given length a convergence in time is also presumed, of course.

An rapid-distortion theory (RDT) analysis of simple shear flow may be used to estimate that, in order to obtain all Reynolds stresses and in particular $\left\langle v_x'^2 \right\rangle$ correct to within 1%, it is necessary to assure a streamwise low–wavenumber resolution corresponding to 2^{11}–2^{12} wall units. Independently from that, observations of the near–wall vorticity structure in experiments on boundary–layer and channel turbulence have suggested that the minimum length of dominant vortex structures in that region is 1500–3000 wall units. In agreement with both estimates, Fig. 7 shows that, in order to compute credible $\left\langle v_x'^2 \right\rangle$ statistics, a domain length of 1.5–$6.0 \cdot 10^3$ wall units is required. The reference DNS [13] meet this requirement.

It can be expected that for higher–order statistics involving v_x the requirements will rise. And indeed, the results for its skewness and flatness shown in Fig. 8 suggest that, while the turbulence structure near the wall is covered well within 1500 wall units streamwise, an even longer domain would be required for convergence of the statistics in the core flow. It thus supports simultaneously two experimental observations — about the average size of near–wall vortices and about the possibility for very long–range correlations near the channel centerline.

5 Discussion and outlook

The advantages and broad potential of the LB approach toward large–Reynolds–number turbulence computations and the presented successful validation of the BEST code open the way for a new, powerful tool for turbulence computations, which features several distinctive advantages. It is ready for immediate application in new kinds of DNS expected to have significant and direct effect on the advance of turbulence modeling and theory. Later in this section, applications of BEST to LES and its relation to up–to–date subgrid–scale models (SGSM) are also discussed, including the necessity to go beyond the present status of SGSM for LB methods and an outline of a promising way to go.

5.1 Summary of results

The efficiency and reliability of the single–relaxation–time LB scheme and its implementation in the BEST code for DNS of 3D incompressible turbulence was verified by a comparison with DNS results for 2D channel flow turbulence, obtained by a classical Chebyshev–pseudospectral (CPS) method. As far as mean flow and Reynolds stresses are concerned, resolved LB simulations are in excellent agreement with results by the CPS method, available in the literature as well as from recent computations at LSTM. Even more impressive is the agreement for quantities like the pressure, where higher velocity moments and nonlocality are involved, or dissipation, where the resolution of derivatives of the computed fields and of small–scale flow structures are decisive. Agreement for velocity skewness and flatness are rather good over most of the flow domain, with noticeable errors only close to the wall and to the channel centerline — regions where certain velocity components have peculiar behaviour. In general, the scatter between data from LB and CPS simulations is comparable, even in such special regions, to the scatter between different simulations with the same CPS code.

For moderate– to low–Reynolds number bounded flows, which are thus at least approximately steady at large times, the "incompressible" LB model reproduced in Sect. 3.2 has been verified [18] to produce a correct average pressure field. Whether this will also hold for the *average pressure* in time–dependent, high–Reynolds–number flows, a regime outside the scope of rigorous applicability of the "incompressible" model, remained to be verified. A related open is the question to what extent possible discrepancies in the *fluctuating pressure* field will contaminate the velocity field and particularly its lower–order statistics in turbulent flows. If, for instance, turbulence is dominated by mean–flow gradients, nonlinearity and the "slow pressure" effects it induces are of lesser importance and turbulent velocities may be well predicted by the LBM. The results in Sect. 4.1 for both pressure and velocity statistics strongly suggest positive answers to both questions.

The regular grids required for fully resolved DNS even at moderate Reynolds numbers ($130 \leq Re_\tau \leq 200$ requires the large grids described in previous sections. This is despite an already numerous list of publications reporting without proper validation on DNS with marginal resolution at such moderate Reynolds numbers. Our systematic convergence study showed that even at these Re_τ, reliable higher–order and two–point statistics necessitate a grid size of at least 4096:512:256 in the streamwise:spanwise:wall–normal direction of a pressure–driven turbulent flow between two parallel plates (2D channel), irrespective of the numerical method and due to the large size of flow structures dominating such turbulence. Such DNS are expensive, so a systematic study of grid size on channel flows has been lacking, despite its considerable interest to the international turbulence research community. The high efficiency of the BEST code and the computing resources provided by HLRB have now made it possible.

5.2 BEST for turbulence DNS and LES

It was argued in Sect. 3.1 that, for DNS of *high–Reynolds–number channel* flows at least, LB codes have practical advantages over pseudospectral (PS) as well as over alternative low–precision codes. This was demonstrated in the sequel, comparing BEST to a PS code. In flows with complicated cross–sectional geometry, PS methods are impractical while codes of other types, e.g. finite–volume based, become too expensive if they are to yield high precision. On the other hand, the need for new turbulence DNS, in new geometries and at higher Re, has been pointed out in [1] as a condition for advance in the development of SGSM. Examples of large–scale DNS of fundamental interest, which are made possible by the availability of the BEST code on machines with performance comparable to the SR8000–F1, can readily be given. One such example is that the "numerical wind tunnel" announced some 10 years ago in Japan, can now finally be realized, e.g. by a model simulation of the (axisymmetric but not isotropic) turbulence behind a grid in a wind tunnel — at moderate Re but over $O(10^2)$ grid units distance behind the grid. Such a DNS would produce a data base of unprecedented spatial range and relevance to basic questions in turbulence modeling such as the return to isotropy rate or the dynamical relations between dimensional and componental anisotropy.

Even for 2D channel turbulence, a DNS at *very large Reynolds* numbers is more efficient using LB than PS methods, due to (i) smaller computational cost for fixed data size, (ii) smaller communication costs, and (iii) the possibility for local grid refinement. It can be estimated [15] how the smallest scales which must be resolved in a 2D channel DNS coarsen with increasing distance from the wall. Using a grid which far from the wall becomes coarser in all coordinate directions, and taking advantage of the experience with channel turbulence DNS with BEST, it can be estimated that, using the full resourses of the Hitachi SR8000–F1 machine, this code could achieve a record–high $Re_\tau \approx 1600$, while resolving better than any PS feasible simulation the longitudinal and transverse structure of turbulence, both close to the wall and in the more homogeneous core flow. The results from such a simulation are expected to bring about decisive advance in the fundamental understanding of wall–bounded turbulence, both in the near–wall region and in the log–layer.

A fundamental aspect of our present work is the development of large eddy simulation (LES) features of the BEST code. Smagorinsky–type SGSM are being validated against DNS in 2D channel and square duct turbulence. Both the standard and the Van Driest wall–corrected variants tend to be overdissipative, which can be improved by employing, within BEST, a more flexible model, e.g. dynamical Smagorinsky with averaging or a k–ϵ models. But even with such modifications, all LES based on LB methods that are known to us have used only eddy–viscosity type of modeling. Its weakness is the inability to account properly for the anisotropy of turbulence, a very pronounced and most important feature of wall–bounded flows. This can be improved by employing, again within BEST but only upon a required theoretical development, of tensor–type models such as the mixed–self–similarity model of invariant-theory-based models. The relation of LB methods to the

PDF methods which are today state of the art in reactive turbulent flow computations could then be exploited.

Acknowledgement. The presented work has been funded by KONWIHR, through the BESTWIHR project and through a grant by the Deutsche Forschungsgemeinschaft. The large–scale computations were carried out at the LRZ. But also the support by the RRZE at the University Erlangen–Nuremberg, the John von Neumann-Institut for Computing (NIC) in Jülich and the Computing-Center at the University Bayreuth are gratefully acknowledged. R. Volkert and M. Breuer at LSTM have kindly made available some of their data from pseudospectral simulations of channel turbulence at the same Re_τ as for the reported BEST runs.

References

1. R. J. Adrian, C. Meneveau, R. D. Moser, and J. Riley. Turbulence measurements for les.
   http://www.me.washington.edu/~riley/les/workshop/report/workshop7.pdf, Okt. 1999.
2. G. Amati, S. Succi, and R. Piva. Massively parallel lattice-Boltzmann simulation of turbulent channel flow. *Int. J. Mod. Phys. C*, 8(4):869–878, 1997.
3. R. Benzi, S. Succi, and M. Vergassola. The lattice Boltzman equation: Theory and applications. *Physics Reports (Review Section of Physics Letters)*, 222(3):145–197, 1992.
4. S. Chen and G. D. Doolen. Lattice Boltzmann method for fluid flows. *Annu. Rev. Fluid Mech.*, 30:329–364, 1998.
5. D. d'Humières, I. Ginzburg, M. Krafczyk, P. Lallemand, and L.-S. Luo. Multiple-relaxation-time lattice Boltzmann models in three dimensions. *Phil. Trans. R. Soc. Lond. A*, 360(1792):437–452, 2002.
6. J. G. M. Eggels. Direct and large-eddy simulation of turbulent fluid flow using the lattice-Boltzmann scheme. *Int. J. Heat and Fluid Flow*, 17:307–323, 1996.
7. X. He and L.-S. Luo. Lattice Boltzmann model for the incompressible Navier-Stokes equation. *J. Stat. Phys.*, 88(3/4):927–944, 1997.
8. X. He and L.-S. Luo. Theory of the lattice Boltzmann method: From the Boltzmann equation to the lattice Boltzmann equation. *Phys. Rev. E*, 56(6):6811–6817, Dec. 1997.
9. J. Jiménez and A. Pinelli. The autonomous cycle of near wall turbulence. *J. Fluid Mech.*, 389:335–359, 1999.
10. J. Kim, P. Moin, and R. Moser. Turbulence statistics in fully developed channel flow at low reynolds number. *J. Fluid Mech.*, 177, 1987.
11. P. Lallemand and L.-S. Luo. Theory of the lattice Boltzmann method: Dispersion, dissipation, isotropy, galilean invariance, and stability. *Phys. Rev. E*, 61(6):6546–6562, 2000.
12. T. S. Lundgren. Model equation for nonhomogeneous turbulence. *Phys. Fluids*, 12:485–497, 1969.
13. R. Moser, J. Kim, and N. Mansour. Direct numerical simulation of turbulent channel flow up to $Re_\tau = 560$. *Phys. Fluids*, 11, 1999.

14. A. E. Perry and M. S. Chong. On the mechanism of wall turbulence. *J. Fluid Mech.*, 119:173–217, 1982.

15. S. B. Pope. *Turbulent Flows*. Cambridge Univ. Press., 2000.

16. Y. H. Qian, D. d'Humières, and P. Lallemand. Lattice BGK models for Navier-Stokes equation. *Europhys. Lett.*, 17(6):479–484, Jan. 1992.

17. R. Volkert, M. Breuer, and F. Durst. Enhanced direct numerical simulations of the plane channel flow based on a spectral method. Technical report, Lehrstuhl für Strömungsmechanik, Universität Erlangen-Nürnberg, 2002.

18. T. Zeiser, Y.-W. Li, H. Freund, P. Lammers, J. Bernsdorf, G. Brenner, E. Klemm, G. Emig, and F. Durst. Flow field, mass transport and selectivity of chemical reactions in sphere-packed fixed-bed reactors. *9th International Conference on Discrete Simulation of Fluid Dynamics, Santa Fe, New Mexico USA*, 2000.

DNS of Homogeneous Shear Flow and Data Analysis for the Development of a Four-Equation Turbulence Model

Johannes Kreuzinger and Rainer Friedrich

Fachgebiet Strömungsmechanik
Technische Universität München
Boltzmannstr. 15
85748 Garching, Germany
johannes@flm.mw.tu-muenchen.de

1 Introduction

More powerful turbulence models than algebraic and two-equation models are of interest for the aircraft, automobile and chemical industries, to give only a few examples. The development of better models needs new concepts and the validation by complete and reliable data sets. Such a new concept is currently developed by Dr. Jovanović at TU Erlangen. His model is based on two transport equations for the second and third invariants II and III of the Reynolds stress anisotropy tensor in addition to the equations for k and ϵ. This forms an important step foreward: The influence of anisotropy can be treated like in full Reynolds stress models, but the effort will be much lower.

For model validation, data sets of all terms in the balance equations for k and ϵ are needed for various anisotropic flows. Especially for the ϵ-balance, only data for a channel at very low Reynolds number is available (Mansour [1]). The calculation of this balance demands high accuracy and resolution, because the dissipation rate is mostly determined by the small scales.

In the next two years extensive simulations of three different flows are planed: Homogeneous shear turbulence, channel flow and turbulent boundary layer flow with and without adverse pressure gradient. At first, only the incompressible case is of interest. In a second step, the influence of compressibility will be evaluated. Therefore all simulations are performed with a compressible Navier-Stokes solver.

This paper presents first results achieved for homogeneous shear turbulence on relatively coarse numerical grids of about $1.5 \cdot 10^6$ points.

2 Homogeneous shear turbulence

2.1 Description of the flow case

In homogeneous shear turbulence the mean flow is only one-dimensional and has a constant transverse gradient: $\bar{u} = (S\,x_3, 0, 0)$, $S = \frac{\partial \bar{u_1}}{\partial x_3}$ denoting the shear. Statistical properties are independent of position in space, therefore diffusion disappears. The turbulent kinetic energy is governed by production and dissipation, and production is generally higher than dissipation. Therefore the turbulent kinetic energy increases. Also the integral length scales increase with time. After a transient phase the flow becomes self-similar, values normalized by S and k do not change any more (Pope [2], Tavoularis [3]). Compressibility reduces the growth of kinetic energy by modifying the structure of turbulence: fluctuations transverse to the mean flow are damped. This causes a reduced production rate (Sarkar [4]).

2.2 Balance equations in incompressible homogeneous shear turbulence

Balance equation for the Reynolds stress tensor $\rho\overline{u_i' u_j'}$

$$\rho\frac{\partial \overline{u_i' u_j'}}{\partial t} = P_{ij} + \Pi_{ij} + \epsilon_{ij}^s \tag{1}$$

The terms on the right hand side (RHS) denote production, pressure strain correlation and solenoidal dissipation rate. The balance equation for the turbulent kinetic energy k is the trace of this equation. In the incompressible case the trace of the pressure strain correlation Π_{ii} is zero.

Balance equation for the solenoidal dissipation rate ϵ^s

$$\frac{\partial}{\partial t}\overline{\frac{\partial u_i'}{\partial x_k}\frac{\partial u_i'}{\partial x_k}} = -2\overline{\frac{\partial u_i'}{\partial x_k}\frac{\partial u_j'}{\partial x_k}\frac{\partial \overline{u_i}}{\partial x_j}} - 2\overline{\frac{\partial u_i'}{\partial x_k}\frac{\partial u_i'}{\partial x_j}\frac{\partial \overline{u_j}}{\partial x_k}}$$
$$- 2\overline{\frac{\partial u_i'}{\partial x_k}\frac{\partial u_j'}{\partial x_k}\frac{\partial u_i'}{\partial x_j}} - 2\frac{\mu}{\rho}\overline{\frac{\partial^2 u_i'}{\partial x_k \partial x_j}\frac{\partial^2 u_i'}{\partial x_k \partial x_j}} \tag{2}$$

The terms on the right hand side are (Mansour [1]) mixed production P_ϵ^1, production by the mean velocity gradient P_ϵ^2, turbulent production P_ϵ^4 and dissipation Y.

3 DNS of compressible homogeneous shear turbulence

3.1 Numerical algorithm

We use a code that solves a characteristic-type formulation of the compressible Navier-Stokes equations for the variables pressure p, velocity u_i and entropy s (Sesterhenn [16]). The discretisation in space uses fifth order compact upwind schemes (Adams et al. [5]) and sixth order compact central schemes (Lele [6]) for the viscous terms. Integration in time is done by a Runge-Kutta scheme of third order.

A numerical algorithm for the simulation of homogeneous shear turbulence has to satisfy the following shear periodic boundary conditions:

$$\phi(t, x + m_1 L_x, y + m_2 L_y, z + m_3 L_z) = \phi(t, x - Sm_3 L_z t, y, z). \tag{3}$$

L_x, L_y, L_z denote the dimensions of the box in the three spatial directions, m_1, m_2, m_3 are arbitrary integers.

These boundary conditions can be satisfied by a coordinate system moving locally with the mean flow. Corresponding coordinate transformations have first been presented by Rogallo [7] and used by many others (e.g. Feiereisen [14], Lee [8], Blaisdell [9], Sarkar [15]). To avoid extremely skewed grids it is necessary to interpolate the data to a new grid at several times. The corresponding need for dealiasing causes an abrupt loss of energy at high wavenumbers (Blaisdell [9]). In our case we are particulary interested in these wavenumbers and their evolution in time. Therefore this algorithm cannot be used.

The algorithm used in the present simulation implements the shear periodic boundary conditions for a non moving grid. It was first proposed by Baron [10] and later used by Schumann [11] and Gerz [12]. Its implementation in the present code is sketched in Fig. 1. As with usual periodic boundary conditions one uses for the calculation of the derivatives at the boundary points stencils that include points of the opposite boundary, e.g. $\phi_{W'} = \phi_W$. In the case of the derivative in z-direction, this point from the opposite boundary N is taken from a position moved in x-direction by $-SL_z t$. In general this point does not match a grid point. Therefore interpolation in the periodic x-direction is necessary to get the value of ϕ_N.

In contrast to usual periodic boundary conditions a compact scheme has to include special boundary stencils. These boundary schemes are not perfect.

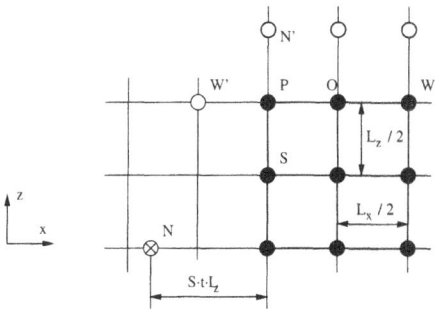

Fig. 1. Shear-periodic boundary condition

In Fig. 2 the transfer function of the compact upwind scheme proposed by Adams [5] is shown for the first five boundary points and a central point. There are remarkable errors and deviations from the central scheme in both the dispersion and dissipation even at moderate wavenumbers of $k\Delta x = 1.5$ up to the fifth point from the boundary. This point contributes to the calculation in case one already interpolates four layers of points. To avoid inhomogeneities in z-direction additional to the interpolation procedure an explicit derivation scheme is necessary.

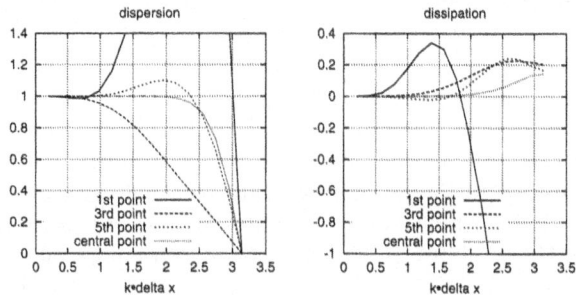

Fig. 2. Transfer function of the compact upwind derivation scheme at the 1st, 3rd and 5th point from the boundary and in the center of the domain

In this case we used a fifth order upwind scheme presented by Rai [13] for the convective terms and the usual central scheme of sixth order for the viscous terms. Because of the lower spectral resolution of these schemes compared to the compact ones, a finer grid has to be used. For a grid refined by a factor of $\frac{\Delta x}{\Delta z} = 1.75$ the transfer function of the explicit scheme resembles the one of the compact scheme up to wavenumbers of $k\Delta x \approx 2.3$. Similar behavior of the difference schemes in all directions is expected because the higher wavenumbers that are possible on the finer grid have considerable numerical dissipation.

3.2 Validation and first results

For validation purposes an unpublished case simulated by S. Sarkar was redone using the same Mach and Reynolds numbers, shear rate, initial spectra and spatial resolution:

$$Re_{\lambda 0} = \frac{q^2}{\nu \cdot \sqrt{\omega_i \omega_i}} \qquad 21.7$$

$$M_{t0} = q/c \qquad 0.1$$

$$S_0^* = \frac{S \cdot k}{\epsilon} \qquad 6.85$$

initial spectra $\qquad E(k) \sim k^4 e^{-2k^2/k_0^2}; \ k_0 = 12\frac{2\pi}{L}$

resolution $\qquad 96 \cdot 96 \cdot (96 \cdot 1.75)$ points

This case can just be calculated on a PC, it needs 1.2 GB of RAM and a calculation time of 60 h on a Pentium 1.9 GHz. Figure 3 shows good agreement of the behavior of k and ϵ^s compared to the values of Sarkar, even though this is not a fully resolved case (see below) and the spectral code of Sarkar has a higher effective spatial resolution than the finite difference discretisation used here.

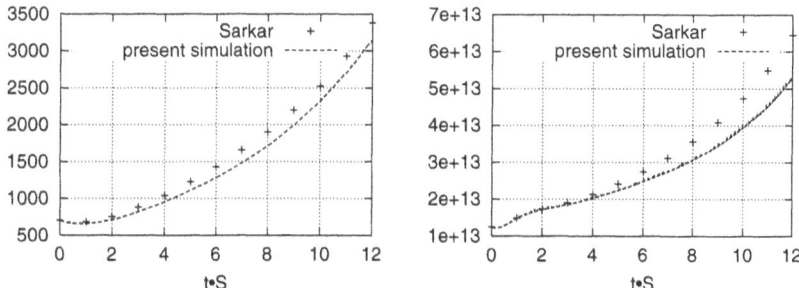

Fig. 3. Turbulent kinetic energy $k = \frac{1}{2}\overline{\rho u_i u_i}$ and solenoidal dissipation rate $\epsilon_s = \overline{\omega_i \omega_i}$

Both k and ϵ^s show a slightly too slow increase in time. This is an effect of the numerical scheme. The upwind scheme is dissipative for high wavenumbers. This results in a too low physical dissipation rate. Nevertheless the growth of k is too small, because the numerical dissipation is more than compensating the lacking physical dissipation [1]. The dilatational dissipation ϵ^d (not shown) evolves as in the case simulated by Sarkar. It is nearly zero in the beginning. At time $tS = 6$ it starts rising, but at the end it does not reach more than 1% of ϵ^s. The flow fields are practically incompressible.

The Reynolds stresses, normalized by $2k$ remain nearly constant after a transient phase ending at $tS = 4$, as expected. The values agree well with the experimental ones given by Tavoularis [3], see table 1.

Table 1. Normalized Reynolds stresses

	present simulation	Tavoularis' data [3]
b_{11}	0.53 ... 0.57	0.51 ± 0.04
b_{22}	0.28 ... 0.30	0.27 ± 0.03
b_{33}	0.14 ... 0.18	0.22 ± 0.02
b_{13}	-0.18 ... -0.16	-0.16 ± 0.01

---

[1] Sarkar was periodically filtering out high wavenumbers to get a stable numerical scheme, but obviously the effects of his filtering are smaller then the effect of the present numerical dissipation.

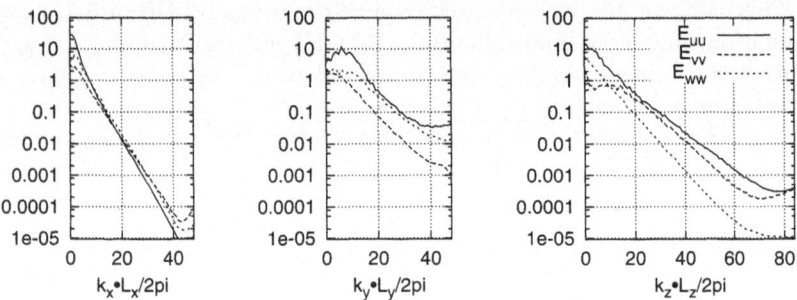

Fig. 4. One-dimensional enery spectra of velocity components

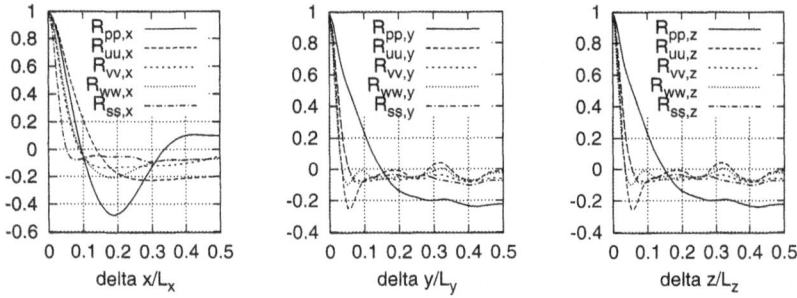

Fig. 5. Two point correlations in x-, y-, and z-direction

The spatial resolution of the simulation is checked at the time $tS = 4$, after the end of the transient phase in the Reynolds stresses. The power spectra of the velocities, see Fig. 4, show an energy pile-up in some cases. Also the spectra in y-direction show a low decay of only 2.5 orders of magnitude. This indicates insufficient resolution of the small scales. The ratio of grid spacing to Kolmogorov length scale reaches from 4.17 at the beginning of the simulation to 6.25 at the end of it. For complete resolution of the dissipation range Pope [2] gives a value of 2.1 in the case of a spectral scheme, which implies the need for grid refinement, resulting in a grid of $40 \cdot 10^6$ points. This underlines the necessity of high performance computing for this case. But even with the present resolution it is possible to get reasonable values for the transport terms of the k-balance and good qualitative trends for the terms of the ϵ^s-balance (see section 4.3).

The decay of two point correlations is a measure of sufficient resolution of the large scales, as to say sufficient size of the numerical box. As Fig. 5 shows for the pressure, the two point correlations do not decay. This is consistent with the instantaneous flow field (Fig. 6), that shows a large wave-like phenomenon in x-direction in the pressure. This phenomenon is not explained at the moment. The other values show a sufficient decay, only $R_{uu,x}$ is remaining at a value of -0.2. The instantaneous flow field (Fig. 6) shows in x-direction elongated structures. These structures were also seen by Lee [8], who compared them to streaky structures near walls. Because of these elongated structures a longer box is needed.

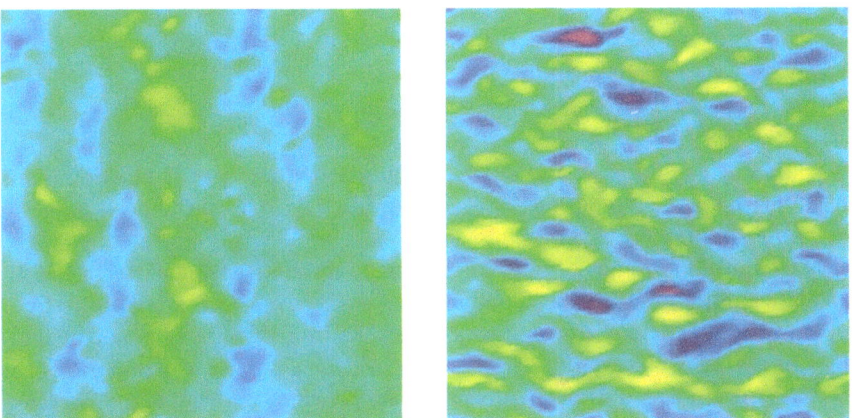

Fig. 6. Pressure p and fluctuating velocity component u'_1, x-y-plane, $tS = 4$, high values in blue, low ones in yellow

4 Analysis of the terms in the balance equations

4.1 Calculation of the terms

The terms of the right hand side of the balance equations (1,2) are calculated by averaging over all gridpoints at times where tS has integer values. k and ϵ^s are calculated every five numerical timesteps. The left hand sides of the balances are calculated from this time series by a second order central scheme.

By constructing the compressible balance equation for ϵ^s terms including high derivatives occur. Evaluating these terms numerically can lead to wrong results. This will be shown using the example of the term involving pressure fluctuations Π_ϵ.

This term in his original form for homogeneous turbulence can be written in different ways, that are numerically different:

$$\Pi_\epsilon = -2\overline{\frac{\partial u_i'}{\partial x_k}\frac{\partial}{\partial x_k}\left(\frac{1}{\rho}\frac{\partial p'}{\partial x_i}\right)}$$

$$= -2\overline{\frac{1}{\rho}\frac{\partial u_i'}{\partial x_k}\frac{\partial^2 p'}{\partial x_i \partial x_k}} - 2\overline{\frac{\partial u_i'}{\partial x_k}\frac{\partial p'}{\partial x_i}\frac{\partial(1/\rho)}{\partial x_k}} \quad (= 4.49 \cdot 10^{18}) \tag{4}$$

$$= +2\overline{\frac{1}{\rho}\frac{\partial^2 u_i'}{\partial x_k \partial x_i}\frac{\partial p'}{\partial x_k}} - 2\overline{\frac{1}{\rho}\frac{\partial}{\partial x_i}\left(\frac{\partial u_i'}{\partial x_k}\frac{\partial p'}{\partial x_k}\right)} - 2\overline{\frac{\partial u_i'}{\partial x_k}\frac{\partial p'}{\partial x_i}\frac{\partial(1/\rho)}{\partial x_k}}$$

$$(= 5.76 \cdot 10^{17}) \tag{5}$$

$$= +2\overline{\frac{1}{\rho}\frac{\partial^2 u_i'}{\partial x_k \partial x_k}\frac{\partial p'}{\partial x_i}} \quad (= -3.26 \cdot 10^{17}) \tag{6}$$

The numerical values in brackets are calculated at time $tS = 4$. The dominant terms on the right hand sides of each line are the first ones, the others are negligibly small.

Formulation (5) shows, that in the incompressible case this term has to vanish. Formulation (4), that includes a higher derivative of the pressure than appears in the Navier-Stokes equations results in a value that is not negligible, see section 4.2. Assuming very small variations of ρ, the formulations (5) and (6) contain correlations of a second derivative of velocity and a pressure derivative. Two variables a and b are correlated, if $\overline{ab}/(a_{rms}b_{rms}) = 1$ or at least non-zero. In formulation (6) the correlation of $\frac{\partial p'}{\partial x_i}$ and $\frac{\partial^2 u_i'}{\partial x_k \partial x_k}$ has a value of $-2.03 \cdot 10^{-3}$. One can regard the variables as uncorrelated. The resulting numerical values of (5) and (6) are most probably due to statistical uncertainty because of the limited number of statistically independent samples.

4.2 Evaluation of compressible terms

The calculation of all terms in the compressible balance equations of k and ϵ^s shows, that in the present case of weak compressibility the compressible terms are two orders of magnitude smaller than the incompressible ones, therefore they can be neglected. By transferring this result to the balances of Reynolds stresses, one has to take into account that the trace of the pressure strain term Π_{ii} is an explicitly compressible term, while Π_{ij} is not. The latter plays an important role in the distribution of fluctuating energy among the three spatial directions.

4.3 Balances for k, $\rho\overline{u_i'u_j'}$ and ϵ

Figure 7 demonstrates the evolution of the balances of k and ϵ in time. The correspondence or deviation of left hand side (LHS) and right hand side (RHS) is a measure of the quality of the calculated terms.

For the k-balance, the LHS is only slightly higher than the RHS. This is due to numerical dissipation, see section 3.2. The mismatch is at every

Table 2. Terms in the balance equations

Balance of k, $tS = 4$		Balance of ϵ^s, $tS = 4$	
RHS:		RHS:	
P	$7.52 \cdot 10^8$	P_ϵ^1	$8.42 \cdot 10^{18}$
$-\epsilon^s$	$-3.67 \cdot 10^8$	P_ϵ^2	$1.18 \cdot 10^{19}$
compressible terms	$6.03 \cdot 10^6$	P_ϵ^4	$1.57 \cdot 10^{19}$
ΣRHS	$3.91 \cdot 10^8$	Y	$-2.41 \cdot 10^{19}$
		compressible terms	$-2.33 \cdot 10^{17}$
		ΣRHS	$1.16 \cdot 10^{19}$

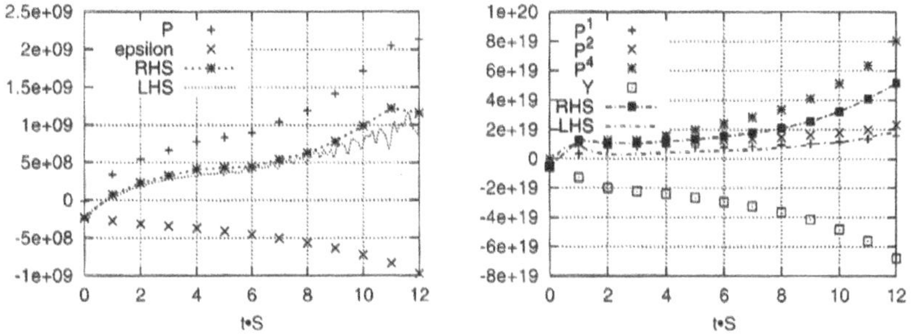

Fig. 7. Balance of turbulent kinetic energy k and dissipation rate $\epsilon^s = \overline{\dfrac{\partial u_i'}{\partial x_k} \dfrac{\partial u_i'}{\partial x_k}}$

time smaller then 10% of P, the biggest term of the RHS. The oscillation of the LHS beginning at $tS = 5$ may be explained by the formation of large structures that reach the box size. If this is the case, the box is too small to obtain good statistics.

LHS and RHS of the Reynolds stresses (not shown) reveal a good agreement too.

The dissipation balance shows a low discrepancy between LHS and RHS of 5% relative to the dissipation term Y at the beginning. The error increases to a value of 30 to 40 % in the time range of $1 \leq tS \leq 9$ and then increases further. This may be explained but the insufficient resolution of small scales. But the qualitative behavior of the LHS is matched by the RHS, even though the RHS is a difference of big terms, which makes it sensitive to relative errors in these terms.

5 Conclusions and outlook

Homogeneous shear turbulence has been simulated directly for a Taylor microscale Reynolds number of 21.7 and very low tubulence Mach number in order to obtain data for improved modelling of the Reynolds stress and the dissipation rate balance equations. A compressible Navier-Stokes solver has been used which is of high order spatial and temporal accuracy. Shear-periodic boundary conditions have been implemented in the code rather than coordinate transformations which suffer from some deficiencies. The present finite difference code has been validated against the spectral code of Sarkar and has shown good performance using even marginal spatial resolution.

The computed DNS data were used to analyze all terms in the balance equations for the Reynolds stresses and the k-dissipation rate. Encouraging results were obtained on a PC in a first step, which indicate that high data accuracy is required especially for the dissipation rate equation, if reliable model improvements are to be achieved. Additionally, the terms have to be computed not only from analytically correct formulations, but also from numerically suitable ones, otherwise errors can occur which are of the same order of magnitude as other calculated terms in the equation.

The results also show that higher resolution is needed in the near future which means that computations will shortly follow on the high-performance machines of the LRZ. The code has already been parallelized and tested on the Hitachi SR 8000 in the recent past. Hence, highly resolved computations of homogeneous shear turbulence and of turbulent channel flow will be started this fall. These simulations will provide an extensive and reliable data base for further development and improvement of a four-equation turbulence model designed by J. Jovanović, Erlangen.

References

1. Mansour, N.N., Kim, J., Moin, P. (1988): Reynolds-stress and dissipation-rate budgets in a turbulent channel flow. J. Fluid Mech., **194**, 15–44
2. Pope, S.B. (2000): Turbulent flows. Cambridge University Press, Cambridge
3. Tavoularis, S., Karnik, U. (1989): Further experiments on the evolution of turbulent streses and scales in uniformly sheared turbulence. J. Fluid Mech., **204**, 457–478
4. Sarkar, S. (1995): The stabilizing effect of compressibility in turbulent shear flow. J. Fluid Mech., **282**, 163–186
5. Adams, N.A., Shariff, K. (1996): A high-resolution hybrid compact-ENO scheme for shock-turbulence interaction problems. J. Comp. Phys., **127**, 27–51
6. Lele, S.K. (1992): Compact finite difference schemes with spectral-like resolution. J. Comp. Phys., **103**, 16–42
7. Rogallo. R.S. (1981): Numerical experiments in homogeneous turbulence. NASA Technical Memorandum 81315
8. Lee, M.J., Kim, J., Moin, P. (1990): Structure of turbulence at high shear rate. J. Fluid Mech., **216**, 561–583
9. Blaisdell, G.A., Mansour, N.N., Reynolds, W.C. (1991): Numerical simulations of compressible homogeneous turbulence. Stanford Report No. TF-50
10. Baron, F. (1982): Macro-simulation tridimensionelle d'écoulements turbulents cisaillés. Ph.D. thesis, Université Pierre et Marie Curie, Paris
11. Schumann, U. (1985): Algorithms for direct numerical simulation of shear-periodic turbulence. Lecture Notes in Physics, Springer, Heidelberg

12. Gerz, T. and Schumann, U., and Elghobashi, S.E. (1989): Direct numerical simulation of stratified homogeneous turbulent shear flows. J. Fluid Mech., **200**, 563–594

13. Rai, M.M., Moin, P. (1991): Direct simulation of turbulent flows using finite-difference schemes. J. Comp. Phys., **96**, 15-33

14. Feiereisen, W.J., Shirani, E., Ferziger, J.H., Reynolds, W.C. (1982): Direct numerical simulation of homogeneous turbulent shear flows on the Illiac IV computer: applications to compressible and incompressible modelling. In: Bradbury et al (ed.) Turbulent Shear Flows 3, 309–319

15. Sarkar, S., Erlebacher, G., Hussaini, M.Y. (1991): Direct simulation of compressible turbulence in a shear flow. Theor. Comput. Fluid Dyn., **2**, 291–305

16. Sesterhenn, J. (2001): A characteristic-type formulation of the equations for high order upwind schemes. Computers & Fluids, **30**, 37–67

Large-Eddy Simulations of High Reynolds Number Flow Around a Circular Cylinder

Frédéric Tremblay, Michael Manhart, and Rainer Friedrich

Fachgebiet Strömungsmechanik
Technische Universität München
Boltzmannstr. 15
85748 Garching, Germany
michael@flm.mw.tum.de

Abstract. Large-eddy simulations (LES) of the transitional and turbulent flow around a circular cylinder at a subcritical Reynolds number of 140000 are performed with a novel technique using staggered Cartesian grids. This technique is implemented in the code MGLET, a parallel Cartesian Navier-Stokes solver for DNS and LES which uses second-order central space and time discretizations and fractional time-stepping together with an iterative solver for the pressure-Poisson equation. The simulations were performed on the Hitachi SR8000-F1 with maximally 16 nodes. They demonstrate the effect of the grid resolution and the need to properly resolve the boundary and shear layers in order to predict the near wake flow reliably. A still unresolved computational issue is, how the necessary size of the computational domain in spanwise direction, namely several cylinder diameters, can be practically achieved with the present days HP computer capacities.

1 Introduction

The flow across a circular cylinder is one of the classical flow problems which are not fully understood. At subcritical Reynolds numbers based on freestream and cylinder diameter less than $2 \cdot 10^5$ and unperturbed incoming flow, the boundary layers are in a stage of laminar flow even after separation. The transition to turbulence occurs in the free shear layer leaving the separation line. Only at supercritical Reynolds numbers does transition occur within the boundary layer leading to its turbulent separation in the rear part of the cylinder. It is clear that an accurate prediction of transition processes needs adequate spatial and temporal resolution of the flow field. The resolution requirements for LES scale roughly with a power of the Reynolds number between two and three. If these requirements are not met, the near wake flow will differ from experimental observations in the sense that the computed recirculation zone will be too large. Measured and computed velocity profiles at equal positions will not match in this case.

The present investigation provides LES data of the flow around a circular cylinder at the high subcritical Reynolds number of 140000 and compares these data with results using DES [5] and LES performed on body-fitted co-ordinates [1]. These comparisons raised new questions about the effects of grid refinement. In the LES studies of [1], it was found that "grid refinement did not automatically lead to improved results for all quantities, where improvement is defined in this context in the sense of a better agreement with experiments". On coarse grids, typical for DES [5], an important overprediction of the mean recirculating region is noted. We use an immersed boundary technique to represent the cylinder within a cartesian grid owing to its high computational efficiency achieved in this flow configuration. Following thorough validation tests, the method was successfully applied for the DNS and LES of the flow past a circular cylinder at $Re = 3900$, see [6] and [7]. We have thus gained confidence in the reliability of results obtained with this technique and hence could proceed to flow at a higher Reynolds number.

2 Numerical method and computational details

The code MGLET, used for all our computations, is a parallel finite volume solver for the incompressible Navier-Stokes equations on staggered cartesian non-equidistant grids. It uses second-order central schemes in space and time and has a long tradition in Large-Eddy-Simulation (LES) and Direct Numerical Simulation (DNS) (see Werner and Wengle [8], Manhart and Wengle [4] and Manhart et al. [3]). A method to represent arbitrarily shaped bodies within cartesian grids which preserves second-order accuracy has been developed and implemented in the code MGLET. In this approach we apply Dirichlet velocity boundary conditions on cell faces located in the immediate vicinity of the body surface. The boundary conditions are applied in such a way that the physical location of the surface and its velocity are best represented. The cells inside the body are excluded from the computation by using a masking array. The discretization remains the same for all cells. A detailed description of the immersed boundary technique can be found in [6].

We performed three LES using the Smagorinsky SGS model without wall damping, the model coefficient C_s is set to 0.1. The first two computations use grids which do not properly resolve the boundary layer along the surface of the cylinder and the detached shear layers, LESH1 and LESH2 cases. The third and last computation, LESH3, is performed on a grid which resolves the boundary layer at the surface of the cylinder and has a grid spacing in the spanwise direction of $\Delta z/D = 1/128$ which is by a factor 2 smaller than that of Breuer [1].

In Tab. 1, the details of the grids used are presented. N_x, N_y and N_z are the number of grid points in the streamwise x-direction, vertical y-direction and spanwise z-direction, respectively. D_x and D_y are the number of grid points used to discretize the cylinder in the x- and y- directions, respectively. Δ_{min}/D represents the smallest grid spacing used near the cylinder. Finally, L_z is the spanwise domain size. We would like to point out that the grids

Table 1. Grid parameters

Case	N_xxN_yxN_z	D_xxD_y	Δ_{min}/D	L_z
LESH1	376x320x64	100x100	0.004	πD
LESH2	706x594x112	150x190	0.004	πD
LESH3	816x656x128	440x406	0.0013	$1D$

LESH1 and LESH2 are exactly the same as those employed in the LES of [6] and the DNS of [7] of the flow past a cylinder at a lower Reynolds number, namely 3900. The results obtained are generally in good agreement with those obtained by other authors and experiments. After an initial transient time of about $TU_\infty/D = 75$, statistics are gathered for about $\Delta TU_\infty/D = 200$. The spanwise direction is used to enhance our statistical sampling. All the simulations are computed on the Hitachi SR8000-F1 high performance computer of the Munich Leibniz Computer Centre. For case LESH3, as a typical example, we used 16 nodes of the computer. Domain decomposition is applied among the different nodes. A total mean performance of about 32 Gflops, or 2 Gflops per node, is achieved. It takes roughly 5 seconds to advance the whole flow field one time step. Around 20 GBytes of memory in total are necessary. 920000 time steps ($\Delta t = 0.0003D/U_\infty$) are needed in order to perform the entire simulation LESH3 ($T = 275D/U_\infty$). A total number of about 163000 CPU hours, all processors combined, or 53 days of computation in wall clock time are necessary.

3 Results

Table 2 presents an overview over drag and back pressure coefficients, mean recirculation lengths and separation angles obtained in our simulations. We compare our results to those of [1] and [5] and also to the experiment of Cantwell and Coles [2]. The results of [1], [5] contain lowest and highest parameter values, the latter corresponding to the best flow resolution chosen. The predominant feature of LESH1 and LESH2 is the overprediction of the mean recirculation length, a trend comparable to the DES of [5]. In fact, all the parameters of LESH1 and LESH2 lie in a range comparable to that computed by DES, with the exception of the mean separation angle which seems to be better predicted by the DES. In [2], the mean separation angle is obtained from the inflexion point of the pressure distribution along the surface of the cylinder which in fact does not exactly coincide with the actual separation point based on zero spanwise vorticity. The fact that the coarser grid used in LESH1 provides better agreement with the experiment as does LESH2 is most probably due to fortuitous error compensations. In the DES of [5], better agreement with the experiment is obtained with the finer grid at this Reynolds number, but at a lower Reynolds number, $Re = 5000$, a similar effect is observed as here, namely that the coarser grid showed better agreement with accepted experimental values than the finer one.

In all cases (DES, LESH1 and LESH2), the primary vortex formation process is obviously badly predicted. We recall here that the mean recirculation length which is intimately related to the formation length of the primary Kármán vortex, depends on the transition point to turbulence within the separated shear layers. After the separation from the surface of the cylinder, the shear layer has a thickness comparable to that of the attached boundary layer. This means that both in DES or in underresolved LES like LESH1 and LESH2, the dynamics of the separating shear layers is not properly predicted. This is especially true in the upper subcritical range where transition to turbulence occurs shortly downstream of separation. The underresolution of the boundary layer along the surface of the cylinder and of the detached shear layers leads to a vortex formation length that is too long. An erroneous prediction of L_r is sufficient to raise queries about the reliability of any other computed quantity.

In contrast, LESH3 predicts the right mean recirculation length and all the parameters lie within the range of those computed by [1], and in reasonable agreement with [2]. The back pressure coefficient shows a significant departure from the experiment of [2] by about 20%. This difference remains unexplained so far. When compared to the measurements, the separation angle also differs substantially. The angle at which the flow separates has a direct impact on the back pressure coefficient, as reflected in Tab. 2. The experiments available in the literature suggest a highly non-linear relationship between the separation angle and the Reynolds number in the high subcritical regime as pointed out in [1]. This could mean that the separation point is very sensitive to the boundary conditions like for example the level of turbulence in the inflow. In the numerical computations, we have a perfectly uniform and laminar inflow, which can never be achieved in an experiment. Another possible explanation is that the use of eddy-viscosity based SGS models has a stabilizing effect even in the laminar flow region, retarding separation. Figure 1 shows the distribution of the mean tangential velocity taken along a vertical plane at the apex of the cylinder ($\Theta = 90^0$). We compare LESH1 to LESH3.

Table 2. Mean flow parameters from DES, LES and experiments.

Data from	C_D	C_{Pb}	Θ	L_r/D
exp [2]	1.237	-1.21	77	0.44
DES [5]	0.87,1.08	-0.81,-1.04	77,78	1.1,1.5
LES [1]	1.22,1.45	-1.40,-1.76	92.6,96.4	0.34,0.57
LESH1	1.134	-1.22	90	0.98
LESH2	0.937	-0.980	90	1.47
LESH3	1.27	-1.45	96.7	0.44

The effect of boundary layer underresolution is clearly seen. The normal gradient of tangential velocity is lowered, the boundary layer is too thick near the wall and the flow accelerates less as it passes near the cylinder. This probably causes the separated shear layers to be more stable, thus retarding the primary vortex formation.

The mean streamwise velocity along the centreline of the cylinder is presented on Fig. 2. LESH1 and LESH2 strongly overpredict the recirculation length while LESH3, in Fig. 3, shows good agreement with the experiment of [2]. The streamlines of the mean flow of LESH3 are shown in Fig. 4. We note the symmetry of the recirculating region and conclude that sufficient statistical samples were taken. No secondary recirculation is noted.

Mean streamwise velocity profiles at $X/D = 1.0$ and $X/D = 3.0$ are shown, respectively, in Figs. 5 and 6. We note a good agreement with the experiment although a slight blockage effect is present in our simulation (overshooting streamwise velocity), due to the small computational domain ($10D$ in the vertical y-direction).

The mean crossflow velocity profile at $X/D = 1.0$ is shown in Fig. 7. Excellent agreement with the experiment is noted. The Reynolds shear stress $\overline{u'v'}$ profile at $X/D = 1.0$ is presented in Fig. 8. We note the higher peaks obtained in our simulation when compared to the experiment. The width of the wake, measured by the position at which the shear stress falls close to zero, is very well captured by our simulation.

RMS values of streamwise and crossflow velocity fluctuations along the centreline of the cylinder are shown in Figs. 9 and 10, respectively. We note a fair agreement of u' with the experiment while v' shows a significant departure from the measurements of [2]. Such a discrepancy is also noted by Breuer [1] in his fine grid simulations. The higher peak in v' reflects a behavior of the very near wake that is different from the experiment, causing also the lower back pressure coefficient that we obtained in our simulation. There is no clear explanation for this discrepancy, more numerical experiments are needed. We, nevertheless, point out that both in Breuer's and in our numerical experiments, the short spanwise extent of the computational domain along with an insufficient resolution in this direction inhibit the fluctuations in the spanwise direction and hence the energy transfer among the Reynolds stress tensor components.

In order to provide some insight into the complexity of the turbulent flow in the wake behind the cylinder we present snapshots of the instantaneous flow field. Figure 11 highlights instabilities in the separating shear layers in the form of Kelvin-Helmholtz vortices. They are easily identified as regions of low pressure. The footprints of these vortices can also be observed in the instantaneous crossflow velocity field. Figure 12 shows this field in the neighbourhood of the cylinder. The generation of Kármán vortices in the wake makes the instantaneous flow asymmetric as clearly seen on the upstream side of the cylinder. Kármán vortices have axes parallel to that of the cylinder

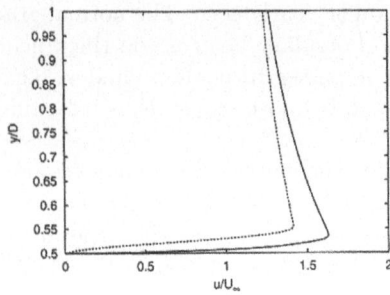

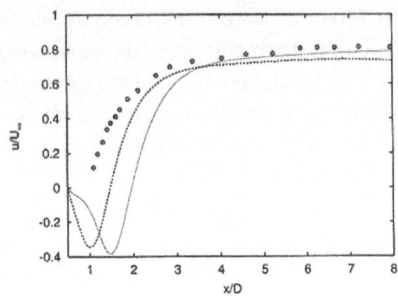

Fig. 1. Radial distribution of tangential velocity at $\Theta = 90^0$. —— LESH3, – – – – LESH1

Fig. 2. Mean streamwise velocity along the centreline of the cylinder. – – – – LESH1, ······· LESH2, Circles: Exp. [2]

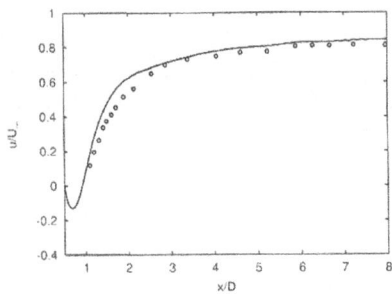

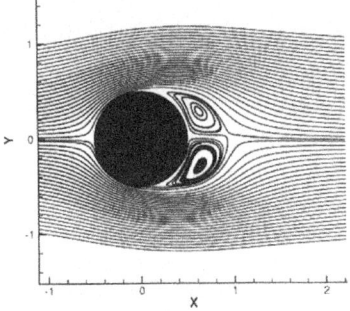

Fig. 3. Mean streamwise velocity along the centreline of the cylinder. —— LESH3, Circles: Exp. [2]

Fig. 4. Mean streamlines of LESH3 in the near wake.

and are shed alternatively from the upper and lower shear layers. They are interconnected by streamwise vortices, so-called braids. Figure 13 provides a perspective view of these coherent vortices.

4 Conclusions

The importance of an adequate resolution of the attached boundary layer along the surface of the cylinder and of the detached shear layers at high subcritical Reynolds numbers is demonstrated. Without a proper resolution, the near wake dynamics of the flow is highly unrealistic and strong overprediction of the recirculation length is observed. Without an accurate prediction of the recirculation length, serious concerns arise about the reliability of the results. The simulation LESH3, which properly resolved the attached boundary layer along the surface of the cylinder and the detached shear lay-

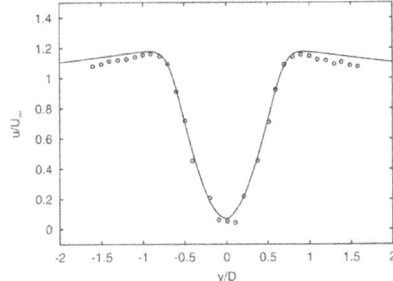

Fig. 5. Mean streamwise velocity profile at $x/D = 1.0$. ——— LESH3, Circles: Exp. [2]

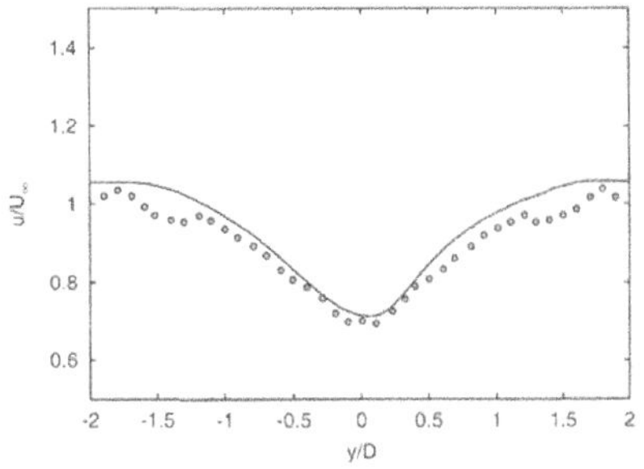

Fig. 6. Mean streamwise velocity profile at $x/D = 3.0$. ——— LESH3, Circles: Exp. [2]

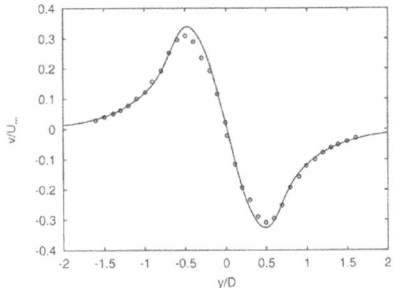

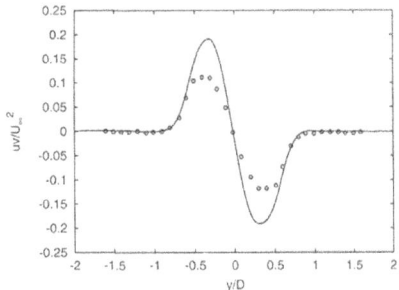

Fig. 7. Mean crossflow velocity profile at $x/D = 1.0$. ——— LESH3, Circles: Exp. [2]

Fig. 8. Reynolds shear stress profile at $x/D = 1.0$. ——— LESH3, Circles: Exp. [2]

ers, provided much more adequate predictions of the near wake behavior. The first order statistics are in very good agreement with the experiment of Cantwell and Coles [2]. The most notable deviations are observed in the back pressure coefficient and in the peak of crossflow velocity fluctuations. The same tendencies are also observed by Breuer [1]. Although no final explanation for these discrepancies can be given at the moment, we conjecture that the spanwise domain size and the resolution can be partly responsible for the deviation from the experiment. 2D simulations typically produce also a back pressure coefficient that is too low. In our case, we used a spanwise extent of only $1D$. More detailed studies of the effects of spanwise extent and resolution are thus needed.

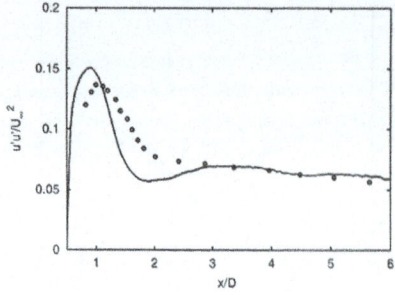

Fig. 9. Profile of the RMS streamwise velocity fluctuations along the centreline of the cylinder. —— LESH3, Circles: Exp. [2]

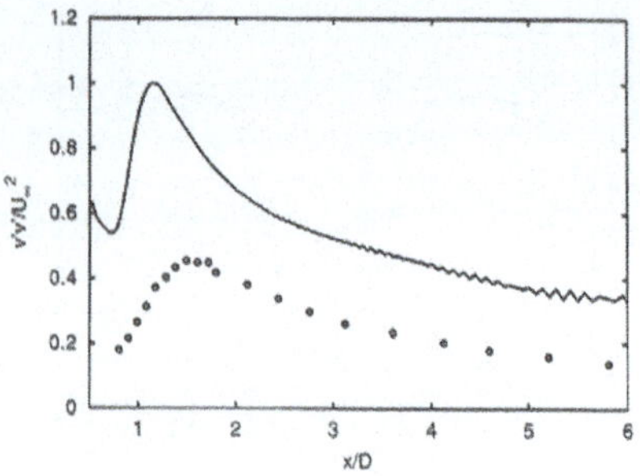

Fig. 10. Profile of RMS crossflow velocity fluctuations along the centreline of the cylinder. —— LESH3, Circles: Exp. [2]

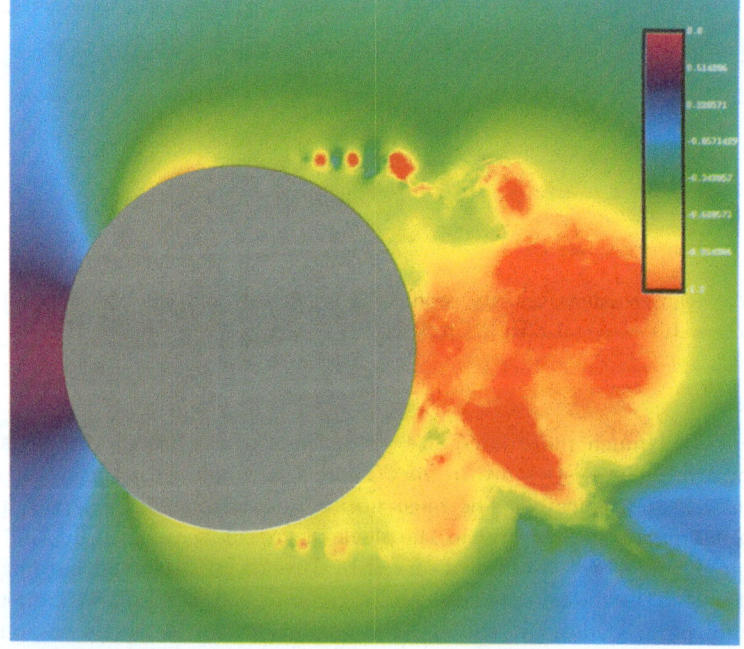

Fig. 11. Instantaneous pressure distribution in the near wake

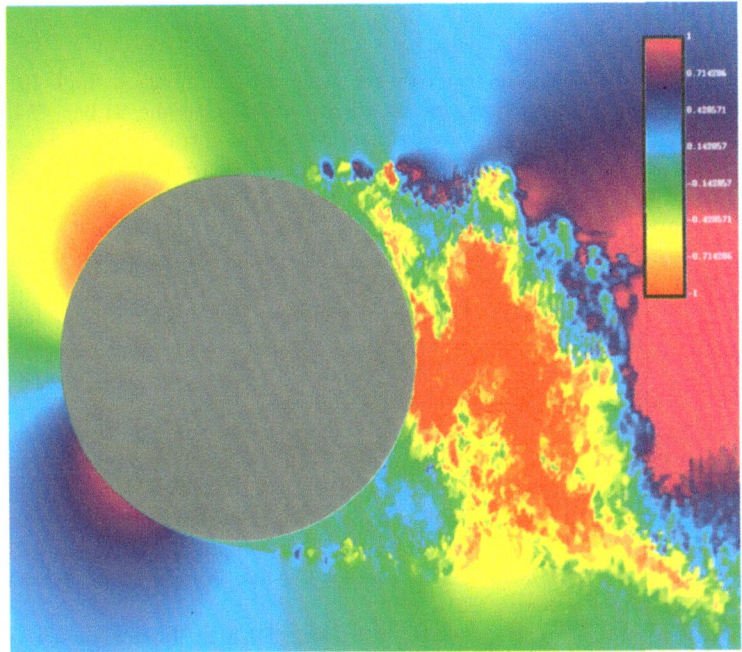

Fig. 12. Instantaneous crossflow velocity distribution in the near wake

Fig. 13. Isosurfaces of the pressure fluctuations $p'/(\rho U_\infty^2/2) = -0.5$; enlarged perspective view.

References

1. M. Breuer. A challenging test case for large eddy simulation: high Reynolds number circular cylinder flow. *Int. J. of Heat and Fluid Flow*, 21:648–654, 2000.
2. B. Cantwell and D. Coles. An experimental study of entrainment and transport in the turbulent near wake of a circular cylinder. *Journal of Fluid Mechanics*, 136:321–374, 1983.
3. M. Manhart, F. Tremblay, and R. Friedrich. MGLET: a parallel code for efficient DNS and LES of complex geometries. In *Proceedings of the Parallel CFD 2000*, Trondheim, Norway, May 22-25, 2000. NTNU.
4. M. Manhart and H. Wengle. Large-eddy simulation of turbulent boundary layer flow over a hemisphere. In Voke P.R., L. Kleiser, and J-P. Chollet, editors, *Direct and Large-Eddy Simulation I*, pages 299–310, Dordrecht, March 27-30, 1994. ERCOFTAC, Kluwer Academic Publishers.
5. A. Travin, M. Shur, M. Strelets, and P. Spalart. Detached-eddy simulations past a circular cylinder. *Flow, Turbulence and Combustion*, 63:293–313, 1999.
6. F. Tremblay, M. Manhart, and R. Friedrich. LES of flow around a circular cylinder at a subcritical Reynolds number with cartesian grids. In *Proceedings of the EUROMECH Colloquium 412 on LES of complex transitional and turbulent flows, Munich University of Technology, 4-6th October 2000 (full paper to appear)*. Kluwer Academic, 2001.
7. F. Tremblay, M. Manhart, and R. Friedrich. DNS of flow around a circular cylinder at a subcritical Reynolds number with cartesian grids. In *Proceedings of the 8th European Turbulence Conference, EUROMECH, Barcelona, Spain*, pages 659–662. CIMNE, 27-30th June 2000.
8. H. Werner and H. Wengle. Large-eddy simulation of turbulent flow over a square rib in a channel. In H.H. Fernholz and H.E. Fiedler, editors, *Advances in Turbulence*, volume 2, pages 418–423. Springer-Verlag, Berlin, 1989.

Numerical Simulation of Passively Controlled Turbulent Flows over Sharp–Edged and Smoothly Contoured Backward–Facing Steps

Jens Neumann and Hans Wengle

Institut für Strömungsmechanik und Aerodynamik
Universität der Bundeswehr München
85577 Neubiberg, Germany
hans.wengle@unibw-muenchen.de

1 Introduction

1.1 Background

In order to attain a certain desired behaviour of turbulent flow, such as drag reduction, minimization/maximization of mixing or the reduction of separation, the basic idea is to control the flow by interacting with the coherent (large-scale) structures. Efficient control requires deeper understanding of the underlying dynamics of such structures. Properly carried out numerical simulations can provide access to detailed three-dimensional and time-dependent information about the flow field, which then can be used to evaluate conclusions for an efficient control scheme.

For the present project, two different simulation concepts are applied. The so-called *Direct Numerical Simulation* (DNS) requires to resolve *all* relevant scales in the turbulent flow field and can therefore only be performed on large 'super'-computing facilities. However, the DNS concept is (still) restricted to relatively small Reynolds numbers. In the so-called *Large-Eddy Simulation* (LES) the effects of unresolved small-scale motions are modelled. Hence, about one order of magnitude less computing time is required compared with the DNS. However, as a drawback of the LES, the subgrid scale modelling issue arises. In addition, the flow close to rigid walls must be properly resolved to avoid application of wall functions or other approximations.

1.2 Goals

In our case, the purpose of the control of the turbulent flow is to reduce the size of the separation bubble which emerges downstream of either a sharp-edged or a smoothly contoured backward-facing step. In this paper, a *passive* control method (no external energy input) is applied using a small control fence upstream of the edge of the step to create large-scale coherent structures which influence the separation zone downstream of the step in the desired manner. Somewhere else, e.g. in [8], we use an *active* control method by forcing the flow with time-periodic blowing/suction through a narrow slot at a proper location. These studies are primarily motivated by similar experiments of Miau et al. [5] for the passive control and of Chun and Sung [1] for the active control of the separated flow region downstream of a sharp-edged backward-facing step. In order to reach a maximum reduction of the separation zone, we search for the optimum height and location of the passive fence and for the optimum forcing frequencies and optimum location for the blowing/suction forcing, respectively.

For a first overview, possible parameters were taken from preliminary LES runs on other computers. Our strategy was to solve well-defined configurations of turbulent flow over either a sharp-edged or a smoothly contoured step with DNS and highly resolved LES, respectively, on the Hitachi SR8000-F1 and use these reference solutions to validate LES runs on smaller computers (e.g. linux-cluster, origin). Subsequently, the LES concept can be applied with more confidence at higher Reynolds numbers.

2 Computational aspects

2.1 Algorithmic and programming issues

General The Navier-Stokes equations for an incompressible fluid are solved on a staggered and non-uniform cartesian grid. The code MGLET used for all the simulations of the project is vectorized and parallelized using MPI together with the domain decomposition concept. Apart from the experience of installations of this program on other computer platforms (origin, sp2, vpp700, t3d, linux-cluster), a pilot installation on the Hitachi SR8000-F1 was used as a starting point (carried out by Frederic Tremblay, a member of the cooperating research group of Prof. R. Friedrich at the Technical University of Munich).

Investigated flow cases In the first part of the project, reference DNS were performed for the uncontrolled flow (case S_DNS) and the passively controlled flow (case SF_DNS) over a sharp-edged step. The thickness, δ, of the oncoming turbulent boundary layer is smaller than the height, h, of the step ($\delta/h = 0.8$), and the Reynolds number is $Re_h = h\,U_\infty/\nu = 3000$ (based on the step height, h, the freestream velocity, U_∞ and the kinematic viscosity of the fluid, ν). The control fence with the height $0.33h$ is placed $4.0h$ upstream of the step edge. These parameters were selected according to the results of optimization LES using an adaptive feedback algorithm [6] (performed on other computing facilities).

The second part of the project involved highly resolved LES of uncontrolled (case R_LES) and passively controlled (case RF_LES) turbulent boundary layer flow over a smoothly contoured step at a Reynolds number of $Re_h = 9086$. For these flow cases, the oncoming boundary layer has a thickness of $\delta/h = 1.2$ and a Reynolds number, $Re_\theta = 1100$, based on the momentum thickness, θ. Experiments of Song and Eaton [9] were used to validate the results of the uncontrolled flow (case R_LES). Subsequently, LES on coarser grids were performed on other computing facilities. LES solutions on a relatively coarse grid are also available from Wasishto and Squires in [11]. In order to be able to compare the influence of the passive control on the sharp-edged step flow with the effects on the smoothly contoured step flow, the same fence setup (fence height, $h_f = 0.33h$, and fence position, $x_f = -4h$) was selected for case RF_LES. Note, the position of the onset of contour curvature (radius of curvature $6.05h$) is chosen as the origin in the streamwise direction ($X = 0$) in correspondence with the experiment [9], see Fig. 3.

Table 1. Grid parameters

case	S_DNS, SF_DNS	R_LES, RF_LES
grid points (N_x, N_y, N_z)	$(976, 160, 304)$	$(1008, 160, 306)$
domain size (L_x, L_y, L_z)	$(38.5, 5.0, 6.0)$	$(42.0, 5.0, 7.2)$
inflow position X_{in}	-18.2	-18.0

Grid generation In Table 1, the parameters for the two slightly different computational grids are summarized. All lengths are normalized with the step height, h, and the coordinate system is chosen such that (x, y, z) denote the streamwise, spanwise and vertical direction, respectively. Only in the homogeneous spanwise direction the grid spacing is equidistant for the sharp-edged step flow ($\Delta Y^+ = 5$)[1] and for the smoothly contoured step flow ($\Delta Y^+ = 14$). For the sharp-edged step flow, the size of the computational domain and its corresponding spatial resolution are the minimum required to perform a DNS at this Reynolds number ($Re_h = 3000$).

Discretization For the time-advancement, a (standard) second-order order leap-frog discretization is applied. In order to improve the spatial accuracy, a newly developed fourth-order compact-scheme, described in [4], is used for the sharp-edged step flow cases. Even though the higher-order method needs about 50% more computing time, the gain in accuracy corresponds to using about a factor of 2 more grid points in each coordinate direction. A (standard) second-order central differencing scheme is applied for the smoothly contoured step flow cases.

Rigid bodies Rigid bodies (step cuboid, vorticity generators, control fence) are created by simply blocking out the corresponding grid cells and applying no-slip boundary conditions at the cell faces where the solid domain is separated from the fluid. For the simulations of the rounded step flow, a special

---

[1] using the friction velocity, U_τ, of the uncontrolled flow at a reference location upstream of the step for the conversion into wall units.

algorithm developed by Tremblay et al. [10] was applied to handle smoothly curved surfaces within a cartesian grid.

Boundary conditions For the sharp-edged step flow cases, small vorticity generators are blocked out close to the inflow plane. Then, together with a constantly prescribed inflow velocity profile, a turbulent boundary layer is created which is fully developed in the region of interest starting some step heights upstream of the backward-facing step. The shape and lateral distance of the vorticity generators were taken from preliminary runs to match the desired boundary layer properties at a reference location. For the smoothly contoured step flow, Manhart's [3] method (rescaled fluctuations from a downstream position applied at the inflow plane) was used for the inflow boundary layer generation. For all investigated flow cases, slip conditions were applied at the top face of the computational domain (sufficiently far away from the region of interest). A zero-gradient outflow condition, or alternatively, a third order extrapolation was used at the outflow cross section (at $X \approx 20$).

Solution strategy After grid generation, the uncontrolled flow simulation was started from a flow field with zero velocities. Marching forward in time, the simulation soon resembles characteristic flow features of a turbulent boundary layer, or of a separation zone downstream of the step. However, from other simulations it is known that the flow needs to convect approximately 5 times through the entire domain in main flow direction to reach a balanced-out statistical steady-state. Then, after the fifth convection cycle, samples of the three-dimensional flow field (all velocity components and pressure) are taken at equidistant time intervals. By time-averaging these samples and in addition, by averaging in the homogeneous y-direction, the *mean* flow field is obtained, and as soon as this mean field has reached stable, i.e. time-independent values, the *fluctuating* velocity field can be evaluated. Finally, from these fluctuating quantities the desired higher-order statistics is calculated. For the passively controlled cases, a fully developed instantaneous flow field of the uncontrolled case was taken as initial stage.

2.2 Performance

A detailed analysis of the approximate runtime statistics for the sharp-edged step flow cases is provided in Table 2. From this data we can draw some conclusions:

- The plain backward-facing step DNS was conducted with 435 000 time steps, the passively controlled DNS with 760 000 time steps. The difference is caused – apart from the initial phase – by the fact that the plain case could be performed at a 1.5 times higher time step ($\Delta t = 0.0015$) compared with the controlled case ($\Delta t = 0.001$).

Table 2. Required resources for the sharp-edged step flow cases

	time steps	time interval[2]	CPU hours
case S_DNS			
initial phase	170.000	200	100.000
mean flow field	100.000	150	25.000
second-order statistics	165.000	250	35.000
case SF_DNS			
initial phase	180.000	180	25.000
mean flow field	180.000	180	25.000
second-order statistics	400.000	400	55.000

– A value of $1.7 * 10^{-6}$ node-sec per time step per grid point results from the total of 1 195 000 performed time steps within the 265 000 CPU hours (8 CPU per node) on grids with 47 million grid points.
– The relatively poor performance during the initial phase of the plain backward-facing step DNS can be explained by the flow physics itself, i.e. the evolution of a fully turbulent flow field from a zero flow field is much more expensive than the alteration of an already fully developed turbulent field (e.g. by adding the control fence). Furthermore, a number of algorithmic improvements were implemented during the initial runs.

Typically, we used 4 nodes of the Hitachi SR8000-F1 for one flow case and an average performance of about 2.2 GFlops per node was achieved. About 15 GBytes of memory in total were required.

3 Results

In the following, some results from the simulations performed on the Hitachi SR8000-F1 are discussed. A more comprehensive analysis of the DNS results of the passively controlled flow over a sharp-edged backward-facing step can be found in Neumann and Wengle [7,8].

Mean flow From the streamlines of the mean flow field (see Fig. 1), the influence of the passive control fence on the mean flow in the recirculating flow region downstream of the sharp-edged step is obvious. The mean flow streamlines in the vicinity of the step edge are already inclined due to the control. So, the location of the maximum height of the upper bound of the recirculation region moves towards the step corner, thus

---

[2] One time unit ($\Delta T = 1$) equals the time that is needed to travel one step height, h, at free stream velocity, U_∞.

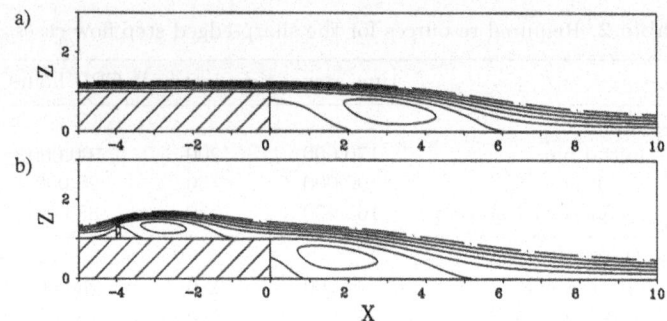

Fig. 1. Streamlines of the mean flow field: a) case S_DNS, b) case SF_DNS

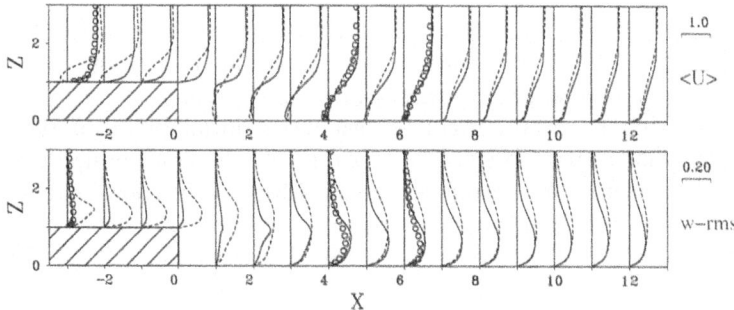

Fig. 2. Profiles of the mean streamwise velocity, $\langle U \rangle$, and of the vertical velocity fluctuations, w_{rms}: case S_DNS (full line), case SF_DNS (dashed line), Experiments of Jovic and Driver [2] (circles)

– reducing significantly the size of the secondary corner eddy,
– reducing the size of the entire recirculation region, i.e.
– reducing the mean reattachment length by 13% ($6.2h \rightarrow 5.4h$).

Figure 2 shows the spatial evolution of the mean streamwise velocity component, $\langle U \rangle$, and of the vertical velocity fluctuations, w_{rms}, for the sharp-edged backward-facing step flow. Due to the control, the intensity of the fluctuations is enhanced downstream of the step, in the region $0.0 < X < 2.0$. Further downstream, the redeveloping boundary layer exhibits similar properties compared with the uncontrolled case, except in the region above step height. Experimental data of Jovic and Driver [2] (at $Re_h = 5100$, $Re_\theta = 610$ and with an oncoming boundary layer thickness of $\delta/h = 1.2$) are included in this figure for comparison with our uncontrolled case S_DNS. Despite of the different Reynolds numbers and oncoming boundary layer properties the differences are not very large.

Instantaneous flow In Fig. 3, the spanwise vorticity, ω_y, is rendered (in arbitrarily chosen $x - z$ cross-sections) for the uncontrolled and for the passively

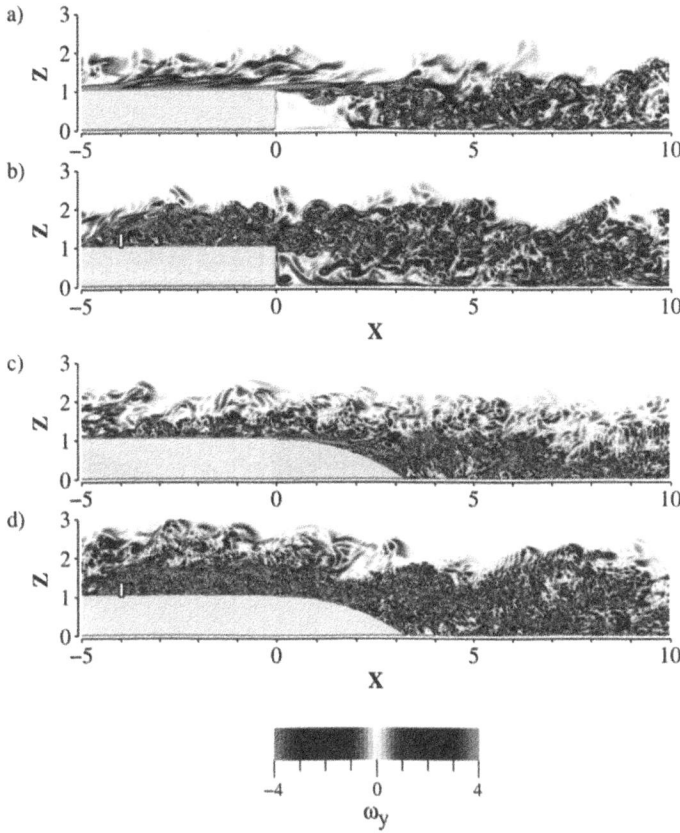

Fig. 3. Instantaneous snapshots of the spanwise vorticity, ω_y: a) S_DNS, b) SF_DNS, c) R_LES, d) RF_LES

controlled cases of the sharp-edged and smoothly contoured step flows. The creation of increased turbulent activity by the passive control fence in the boundary layer region upstream of the step becomes visible. Downstream of the sharp-edged step, it can be seen that due to the control the region close to the step foot ($0 < X < 2$) is now filled with turbulent flow, i.e. the secondary recirculation bubble with very low vorticity is nearly washed out. For the smoothly contoured step flow, the distinct tongue of high values of $|\omega_y|$ after separation of the boundary layer (case R_LES: $1 < X < 2$) has disappeared in case RF_LES. Figure 3 furthermore exhibits the impact of the different Reynolds numbers on the flow. Clearly, the size of the turbulent structures rendered by coherent areas of identical values $|\omega_y|$ is decreased for the higher Reynold number cases R_LES and RF_LES ($Re_h = 9086$).

Figure 4 gives an impression of the instantaneous spatial distribution of reverse flow regions (coloured red) which are bounded by the $U = 0$ isosur-

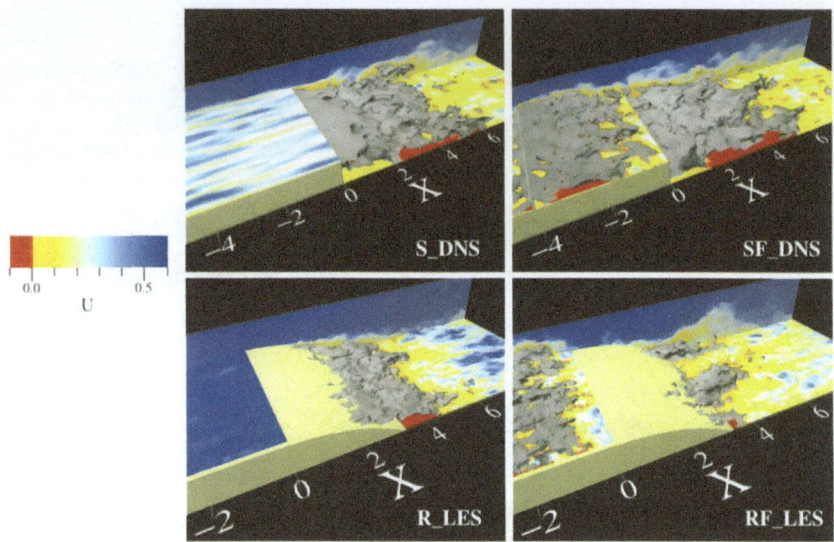

Fig. 4. Instantaneous snapshots of the streamwise velocity, U: isosurface $U = 0$ (gray)

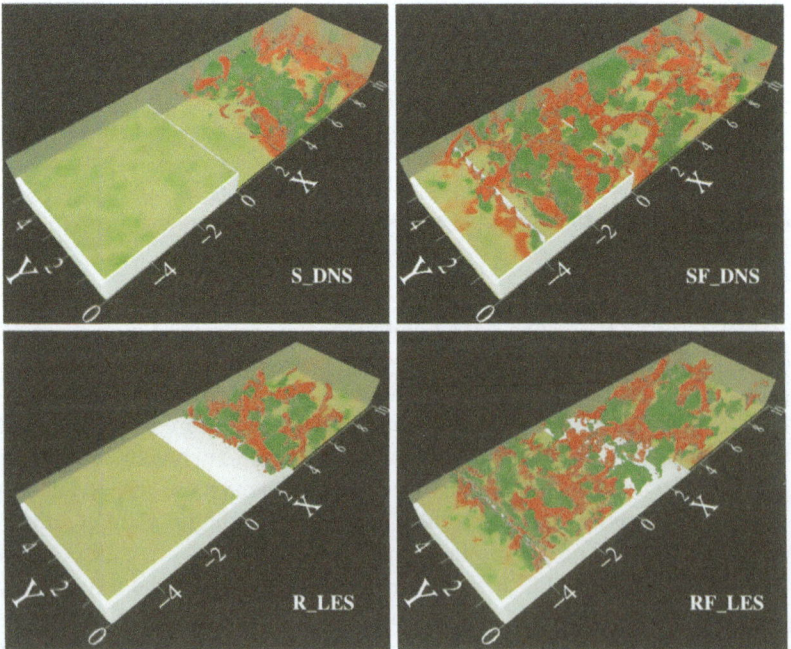

Fig. 5. Instantaneous snapshots of the pressure fluctuations, p': isosurfaces $p' = -0.25$ (red), $p' = +0.25$ (green)

face.[3] The recirculation region downstream of the control fence ends already upstream of the step edge (case SF_DNS). For the smoothly contoured step flow, the separation is not fixed anymore by a sharp edge, hence there is an instanteneous separation line varying in spanwise direction. However, the local instantaneous *reattachment* position shows a greater spanwise variation than the local instantaneous *separation* position. This can be explained with the formation of (strongly three-dimensional) large-scale structures in the separated flow region. For the case RF_LES, the spanwise structure of the backflow region is disrupted into separated backflow packets which can reach a considerable height above the bottom wall. The instantaneous streamwise velocity of the oncoming boundary layer of the uncontrolled flow cases (S_DNS, R_LES) at a horizontal plane ($Z = 1.02$) exhibits the enlongated shapes of typical structures ('streaks').

In Fig. 5, the influence of the passive control fence at $X = -4.0$ can be seen by looking at snapshots of the pressure fluctuations. For the uncontrolled flow case S_DNS, the region of high turbulent activity begins approximately 2 step heights downstream of the step edge. Due to the control fence (case SF_DNS), coherent structures are created already upstream of the step edge and they are convected into the separation region downstream of the step. Comparing the DNS cases (S_DNS and SF_DNS) we conclude in general that the pattern of the coherent structures downstream of the step edge appears to be more organized in the uncontrolled case (S_DNS). For the smoothly contoured step flow, already in the uncontrolled case R_LES the structures emerging downstream of the step exhibit a stronger three-dimensionality.

4 Conclusions

Reference data sets were provided by DNS and highly resolved LES of passively controlled turbulent boundary layer flow over a sharp-edged and a smoothly contoured backward-facing step. The Hitachi SR8000-F1 was found to be a suitable environment to handle a series of large-sized numerical simulations, in our flow case on grids with up to 50 million grid points. It was observed that a passive control fence placed 4 step heights upstream of the step significantly alters the characteristics of the instantaneous and of the mean flow field in the recirculation region downstream of the step. Due to the control, the mean reattachment length of the sharp-edged step flow is decreased by 13%. The flow over a smoothly contoured step exhibits an increased complexity of the separation process. The flow downstream of separation (in the recirculation bubble and after reattachment) seems to have a more pronounced three-dimensional structure in comparison to the sharp-edged step flow. Preliminary results show that depending on the height of

---

[3] Please note that the step-like shape of the rounded step part is used for graphical reasons only.

the control fence about a 50% reduction of the size of the separation bubble can be achieved.

Acknowledgement. This work was supported by the Deutsche Forschungsgemeinschaft (DFG) under grant number We 705/7, and by the Leibniz Computing Center (LRZ) of the Bavarian Academy of Sciences (Hitachi SR8000-F1).

References

1. K.B. Chun and H.J. Sung. Control of turbulent separated flow over a backward-facing step by local forcing. *Exp. Fluids*, 21:417–426, 1996.
2. S. Jovic and D. Driver. Reynolds number effect on the skin friction in separated flows behind a backward-facing step. *Exp. Fluids*, 18(6):464–467, 1995.
3. M. Manhart. Direct numerical simulation of turbulent boundary layers on high performance computers. In E. Krause and W. Jaeger, editors, *High performance Computing in Science and Engineering 1998*. Springer Verlag, 1999.
4. A. Meri, H. Wengle, M. Raddaoui, P. Chauve, and R. Schiestel. Large-eddy simulation of non-equilibrium inflow conditions and of the spatial development of a confined plane jet with co-flowing streams. In W. Rodi and D. Laurence, editors, *Engineering Turbulence Modelling and Experiments 4*, pages 197–206. Elsevier, Amsterdam, 1999.
5. J.J. Miau, K.C. Lee, M.H. Chen, and J.H. Chou. Control of separated flow by a two-dimensional oscillating fence. *AIAA Journal*, 29:1140–1148, 1991.
6. J. Neumann and H. Wengle. Active control of turbulent separated flows using large-eddy simulation. In B.J. Geurts, R. Friedrich, and O. Metais, editors, *Direct and Large-Eddy Simulation IV*, pages 427–434. Kluwer Academic Publishers, 2001.
7. J. Neumann and H. Wengle. DNS and LES of passively controlled turbulent backward-facing step flow. In *Proc. of the IUTAM Symposium on 'Unsteady Separated Flows'*, Toulouse, France, April 8-12 2002.
8. J. Neumann and H. Wengle. Passive versus active control of backward-facing step flow: DNS/LES and POD analysis. In I.P. Castro, P.E. Hancock, and T.G. Thomas, editors, *Advances in Turbulence IX*, pages 573–576. CIMNE, Barcelona, 2002.
9. S. Song and J.K. Eaton. Experimental study on non-equilibrium turbulent boundary layer with separation, reattachment, and redevelopment. In *Proc. of the 2nd Int. Symp. on Turbulence and Shear Flow Phenomena*, pages 27–31 (Vol. II), Stockholm, Sweden, June 27-29 2001.
10. F. Tremblay, M. Manhart, and R. Friedrich. LES of flow around a circular cylinder at a subcritical reynolds number with cartesian grids. In R. Friedrich and W. Rodi, editors, *Advances in LES of Complex Flows*, pages 133–150. Kluwer Academic Publishers, 2002.
11. B. Wasistho and K.D. Squires. Numerical investigation of the separated flow over a smoothly contoured ramp. In *Proc. of the 2nd Int. Symp. on Turbulence and Shear Flow Phenomena*, pages 405–410 (Vol. III), Stockholm, Sweden, June 27-29 2001.

Parallel Single- and Multiphase CFD-Applications Using Lattice Boltzmann Methods

Manuel Schulz[1], Jonas Tölke[2], Manfred Krafczyk[2], and Ernst Rank[1]

[1] Lehrstuhl für Bauinformatik, Technische Universität München, 80290 München, Germany. *mschulz/rank@bv.tum.de*
[2] Institut für Computeranwendungen im Bauingenieurwesen, Technische Universität Braunschweig, 38106 Braunschweig, Germany. *toelke/kraft@cab.bau.tu-bs.de*

1 Introduction

The lattice Boltzmann equation [1] is a particularly effective model for the simulation of complex problems like transient 3D single- or multi-phase flows. We give a short overview with respect to efficient data structures and put special emphasis on vectorization aspects and the parallelization for a LB kernel based on topologically unstructured grids. Three engineering applications are presented: The unsaturated multi-phase flow in soil probes, two-phase flow in a bioreactor and turbulent flow over a surface mounted cube.

2 Lattice–Boltzmann model for immiscible binary fluids

The first lattice gas model for immiscible binary fluids was proposed by Rothman & Keller [2] and the corresponding Lattice-Boltzmann model was proposed by Gunstensen & Rothman [3]. Grunau [4] modified the model for binary fluids with different density ratios and viscosities on a triangular lattice in two dimensions. We describe shortly a three-dimensional nineteen velocity lattice Boltzmann model (D3Q19) for immiscible binary fluid flow [5].

The Lattice-Boltzmann equation for each immiscible phase l (red or blue fluid in the subsequent discussion) is

$$f_a^l(t+\Delta t, \boldsymbol{x}+\boldsymbol{\xi}_a \Delta t) - f_a^l(t, \boldsymbol{x}) = \Delta t \, \Omega_a^l(t, \boldsymbol{x}), \qquad l = r, b \quad a = 0, \ldots, 18, \quad (1)$$

where the basic variables f_a^l are particle distributions propagating with discrete velocities

$$(\boldsymbol{\xi}_a)_{a=0,\ldots,18} = c \begin{pmatrix} 0 & 1 & -1 & 0 & 0 & 0 & 0 & 1 & -1 & 1 & -1 & 1 & -1 & 1 & -1 & 0 & 0 & 0 & 0 \\ 0 & 0 & 0 & 1 & -1 & 0 & 0 & 1 & -1 & -1 & 1 & 0 & 0 & 0 & 0 & 1 & -1 & 1 & -1 \\ 0 & 0 & 0 & 0 & 0 & 1 & -1 & 0 & 0 & 0 & 0 & 1 & -1 & -1 & 1 & 1 & -1 & -1 & 1 \end{pmatrix}$$

on a regular lattice { $e_a = \boldsymbol{\xi}_a \Delta t$, $a = 0, \ldots, 18$ }. The collision operator $\Omega_a^l(t, \boldsymbol{x})$ models the interaction between particles and has to be constructed in such a way that the moments ρ and $\rho\boldsymbol{u}$ (density and momentum)

$$\rho^r = \sum_a f_a^r, \quad \rho^b = \sum_a f_a^b, \quad \rho = \rho^r + \rho^b, \quad \rho\boldsymbol{u} = \sum_a \boldsymbol{\xi}_a(f_a^r + f_a^b) \qquad (2)$$

obey the Navier-Stokes equations [6] in the low Mach- and Knudsen-number limit. The collision operator is composed of three parts

$$\Omega_a^l = \Omega 3_a^l \{\Omega 1_a^l + \Omega 2_a^l\}, \qquad (3)$$

where $\Omega 1_a^l$ is the operator responsible for the correct hydrodynamics, $\Omega 2_a^l$ is the operator responsible for the generation of surface tension and $\Omega 3_a^l$ represents the 'recoloring step' which mimics the separation mechanism of the two phases. $\Omega 1_a^l$ and $\Omega 3_a^l$ operate on the local set of distributions, whereas $\Omega 2_a^l$ depends on the densities of the neighboring nodes.

The order parameter in the system of a binary mixture is

$$\Phi = \frac{\rho^r - \rho^b}{\rho^r + \rho^b}, \qquad (4)$$

and values $\Phi = 1$, -1 and 0 correspond to a purely red fluid, a purely blue fluid, and the interface, respectively. A more detailed description of this model can be found in [7].

3 A data structure based on indirect addressing

Equation (1) is an explicit Finite Difference scheme for the discrete Boltzmann equation [7] and can be decomposed in a collision step (RHS of (1)), where the local set of distributions is modified, and a propagation step (LHS of (1), where each distribution f_a moves along their direction e_a to a neighboring node. For porous media, where the volume of the flow domain is small compared to the volume of its bounding box, we discretize only the fluid domain using topologically unstructured grids (i.e. uniform cartesian grids with 'holes') [11]. Having different types of nodes like 'fluid' and 'boundary condition' nodes, which require a different treatment by the collision operator, we store the nodes with respect to the boundary condition type in a 1-D vector as depicted in Fig. 1 to optimize data flow. For each phase l a vector F and an additional copy array F_copy of the length num_propagations = 19 × num_nodes are allocated.

Propagation then implies copying distributions to specific locations in the one-dimensional array F_copy. The corresponding destinations are precomputed and stored in an array target of length num_propagations. The algorithm for the propagation is

```
for(i=0 ; i<num_propagations ; i++)  F_copy(target(i)) = F(i)
```

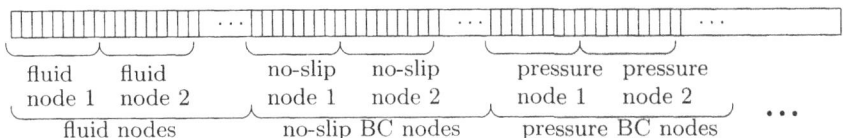

Fig. 1. Arrangement of nodes within the one-dimensional array **F**

This is a favorable statement for vector computers like the Hitachi, since appropriate parts of the arrays **F** and `target` are loaded into cache using a prefetch instruction, whereas for the random access in the array **F_copy** data is preloaded directly from main memory to special registers using a preload instruction.

4 Parallelization strategy

4.1 Domain Decomposition

In our application the load per domain (i. e. process) is roughly comparable to the number of nodes assigned to that domain and the amount of communication between two domains (processes) is comparable to the number of grid links intersected by the common boundary of the two domains. A domain decomposition, which minimizes the load-imbalance and the amount of communication between processes, is as important as an efficient data-structure. The relevance of interprocess communication is even more important in a multiphase- than in a single-phase application, because the algorithm does not allow message hiding during the data exchange of the density as explained later (Fig. 4). During the development of the code, the public domain libraries METIS and PARMETIS were employed, which use a graph partitioning algorithm. METIS only shows moderate performance on vector machines and for problems with more than 20 Mio. grid nodes the calculation of the partitions result in inacceptable memory consumption and computing times on a Linux cluster.

Therefor a geometric domain decomposer DIVIDE was integrated in the simulator's preprocessor, which is based on a recursive bipartitioning algorithm [8]. The resource requirements of this domain decomposer are rather small and depend on the outer dimensions of the volume to be decomposed. Even for large problems like an 300^3 grid, only 56000 bytes of memory are allocated for the domain decomposition.

The evaluation of the two domain decomposers for typical test cases shows, that the load-imbalance obtained with DIVIDE is comparable to the METIS result. In Fig. 2 decompositions of a porous medium using DIVIDE and METIS are shown.

Fig. 2. Domain-decomposition for 8 subdomains using DIVIDE and METIS

4.2 Communication

The collision operator in (1) only depends on local values for single-phase problems, whereas for multi-phase problems the densities of the neighboring nodes are needed. So the communication algorithm is different for single- and multiphase problems.

Single phase problems Communication can be optimized by computing the collision term of subdomain interface nodes (see Fig. 3) first followed by a subsequent initiation of their propagation to the corresponding subdomains by using asynchronous non-blocking message passing calls of the underlying MPI library [17]. After this initiation the collision and propagation of the subdomains inner nodal distributions (being by far the computationally dominant part) is done. Thus there is additional time left for the completion of the message transfer. Figure 4 shows a flow chart of the approach.

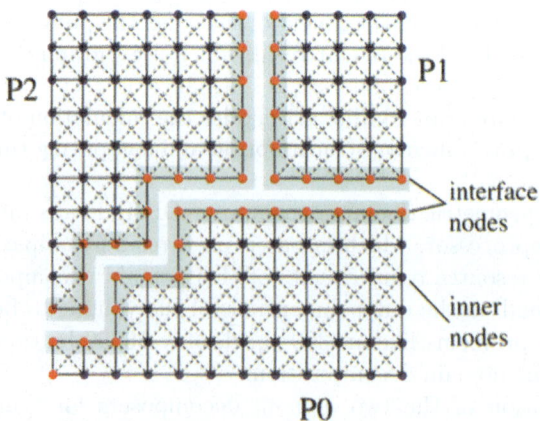

Fig. 3. Cut through the decomposed flow domain with inner and outer nodes

Multi phase problems As shown in Chapter 2, the collision operator depends on the densities of the neighboring nodes. These densities are not available until the propagation is completed. A detailed analysis of the algorithm reveals that communication hiding is only partially possible (Fig. 4).

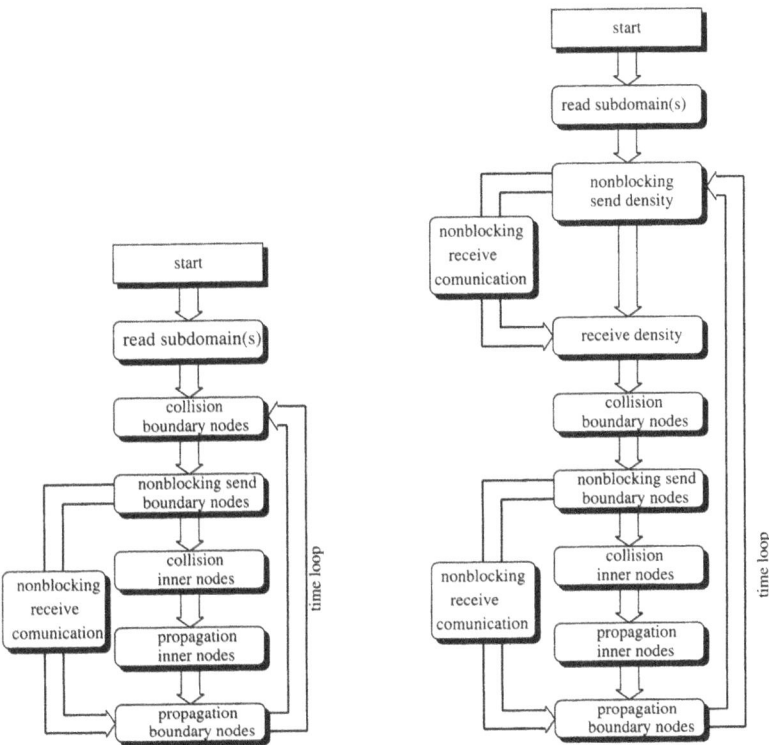

Fig. 4. Flowchart for a single (left) and multiphase (right) Lattice-Boltzmann application

4.3 Parallel Efficiency

For single phase simulations, a parallel efficiency of more than 0.9 even for small systems (memory use of 25%) was achieved (Fig. 5). The determination of speedup behavior for multiphase applications is far more difficult for various reasons. First, the number of nodes in the interface region is strongly related to the update rate (up to 30%), because only for these nodes the multiphase collision operator $\Omega 2$ and $\Omega 3$ are evaluated in addition to $\Omega 1$ (Sect. 2). For most of the simulations the interface surface and location varies in time, implying an additional load-imbalance in the course of the computation. Furthermore, the amount of communication due to the exchange of density values depends on the location of the interface at the outer nodes. Considering these problems, the efficiency values in Fig. 5 are very encouraging.

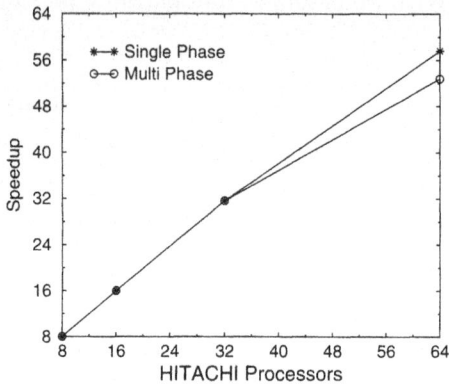

Fig. 5. Speedup for the single and multiphase application

5 Applications

5.1 Multiphase flow in porous media

The reliable prediction of macroscopic unsaturated flows in soils, which is an important part of a variety of environmental problems, still remains a challenging task. The transport of two or more immiscible fluids is described on a macroscale by the multiphase flow equations [12], which depend on effective parameters like the permeability, the permeability-saturation relationship or the pressure-saturation relationship. These parameters depend on the properties of the fluids and the detailed geometry of the porous medium and are usually obtained from expensive and difficult experiments using small soil probes.

In the framework of the MUSKAT project (multi-scale transport in porous media) [18] these effective soil parameters are obtained by direct simulations of single and multiphase flow on the pore scale. The pore geometry of small specimens of soil is reconstructed using computer tomography. An example of a digitized geometry is shown in Fig. 6. These data are transfered to the Lattice Boltzmann CFD simulator to compute integral effective parameters for the digitized soil probes. In Fig. 7 the resulting permeability-saturation and the pressure-saturation relationship for the geometry in Fig. 6 are shown. In a second step the flow on the macroscale (continuum) scale is simulated by solving the multiphase flow equations with multi-grid methods [12] using

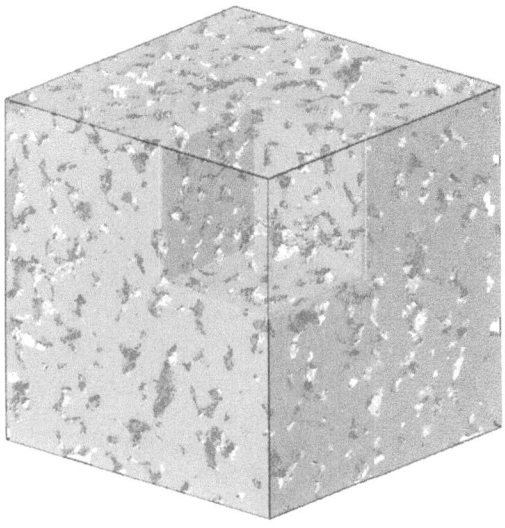

Fig. 6. Geometry of soil sample

the effective parameters determined by the microscale simulations. These results are then compared to solutions obtained by experiments [19]. In Fig. 9 the results of the simulation and the experiment for a multi-step outflow are shown. The notable differences between simulation and experiment are subject to ongoing research.

5.2 Air-water flow in a laboratory-scale waste water batch reactor

In the field of waste water treatment research a lot of laboratory-scale biofilm reactors are operated to better understand the relevant processes and, as a long-term goal, to optimize a real world plant by a suitable upscaling of the obtained results. One important phase of operation is the aeration of the bioreactor, where air bubbles are induced to activate certain biological processes. One important goal is to optimize the spatio-temporal homogeneity of nutrients and oxygen with respect to carrier body size distribution, reactor shape and/or inflow nozzle geometry. The dynamics of this two-phase system is completely determined by the geometric configuration of the flow, the inflow boundary conditions, the wetting properties of the media, and the following set of dimensionless parameters: Reynolds number, Weber-number, viscosity and density ratios of the two phases. An estimation for these parameters for a typical laboratory-scale reactor is given in Table 1. The geometric configuration is obtained from polydisperse sphere packings generated by molecular dynamics type simulations [9]. Figure 10 gives an impression of the transient dynamics in the system including coalescence and breaking of

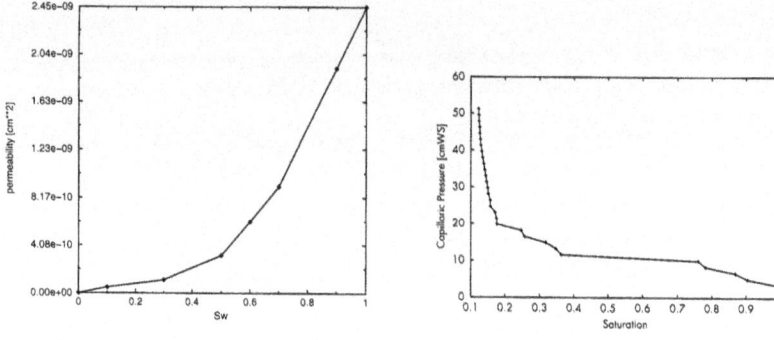

Fig. 7. Permeability-Saturation   **Fig. 8.** Presssure-Saturation

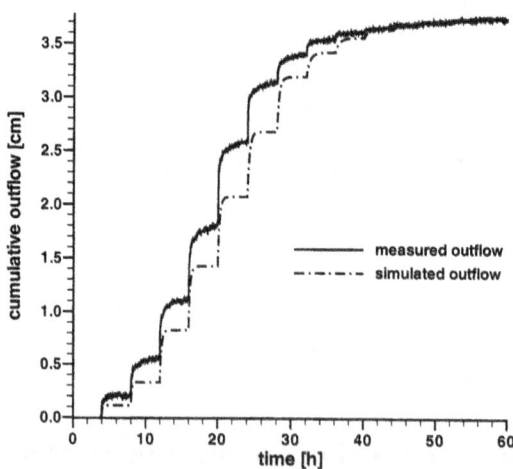

Fig. 9. Comparison of the simulated and measured multi-step outflow experiment

bubbles. The modeled reactor contains approximately 600 spheres and 36 air nozzles at the bottom. The system is discretized by a $101 \times 101 \times 245$ grid. The dimensionless parameters are given in Table 1 and differ by about one order of magnitude to those of the real world system. The simulation captures important features observed in the experiment, e.g. an increased permeability close to the cylinder walls and the existence of preferential flow paths for the bubbles. The simulation time of 210 000 time steps was sufficient to reach a dynamic equilibrium for quantities such as time-averaged spatial air concentration. A detailed discussion of the results is beyond the scope of this paper and can be found in [7].

Table 1. Parameters for the real laboratory scale reactor and the simulation

	ρ_r / ρ_b	μ_r / μ_b	Re	We
Simulation	4	4	3.5	0.1
Laboratory scale reactor	850	40	20	0.01

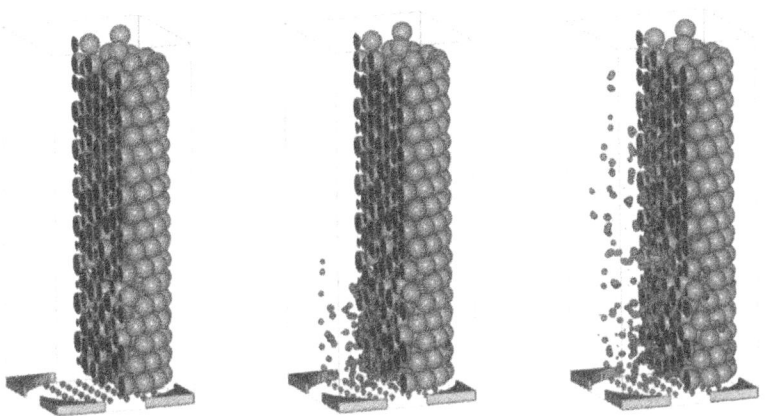

Fig. 10. Rising air bubbles in a model bioreactor

5.3 Large-Eddy simulation of turbulent flows

A Large-Eddy extension for LB-models A few years ago it was realized [13] that LB-models could easily be extended by algebraic turbulence models and thus could be utilized as a simulation tool for a variety of engineering flows. The basic idea of a Smagorinsky type LES approach is to compute a local eddy (turbulent) viscosity from locally filtered strain rates which adds to the molecular viscosity and models the influence of the unresolved spatiotemporal scales on the resolved scales [14].

It is important to note that for LB models the eddy viscosity can be computed from purely local (nodal) quantities in contrast to other schemes which typically have to compute numerical derivatives of the velocity field.

Description of the test case and numerical results The test problem under consideration is a flow around a surface mounted cube in a channel at $Re = 40000$. Details of the configuration as well as numerical and experimental results can be found in [15, 16].

From Fig. 11 we can identify at least seven structural properties of the flow all of which are reproduced in the simulation:

1. upstream stagnation point: distance to cube 1.040 (experiment) vs. 1.03 (simulation)
2. horseshoe vortex downstream: see Fig. 12

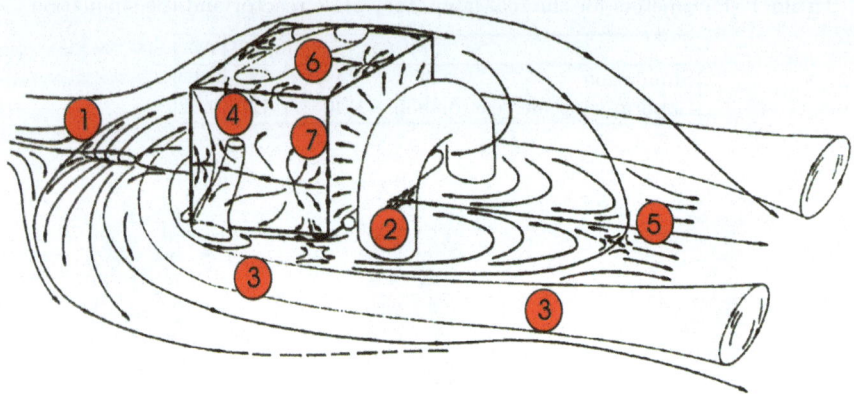

Fig. 11. Experimentally observed flow structures of the flow around a surface mounted cube at Re=40000 [15]). Markers indicate specific flow patterns

3. trumpet vortex pair starting in front of the cube (Fig. 13) and extends to the outflow boundary.
4. symmetric pair of vortices at the vertical middle of the cubes side walls (Fig. 12).
5. downstream stagnation point: distance to cube 1.612 (experiment) vs. 1.72 (simulation, Fig. 12).
6. in accordance to the experiment no attachment of the flow could be observed on top wall of the cube in the simulation. The vertical extension of the vortex coincides to within 10 %.
7. on the right and left side walls of the cube we find in accordance to the experiment a pair of vortices symmetric to a plane parallel to the ground plane and dividing the cube into two halfs (Fig. 13).

The numerical resolution was $301 \times 61 \times 211$ grid nodes (i.e. $dx = dy = dz = \frac{1}{30}$ of the cubes side length) and the total number of DOF was about 6×10^7 requiring about 1,3 GByte of memory. The computation time for 67 time periods in units of $\dfrac{\text{cube height}}{\text{mean inflow velocity}}$ was about 12 hours on *one* node of the Hitachi SR8000 located at the Leibniz-Rechenzentrums Munich. The computational performance was measured to be about 3.8 GFlops using the auto-paralellizing compiler and COMPAS directives.

Although the resolution of the simulation was very coarse (no-slip BC at walls, boundary layer thickness $\delta \simeq \frac{1}{\sqrt{Re=40.000}} = 0.005 \ll dx = dy = dz$), the results show that the present LB-LES model is capable of simulating complex turbulent flows. Further validation studies of grid dependence and convergence rates are under way.

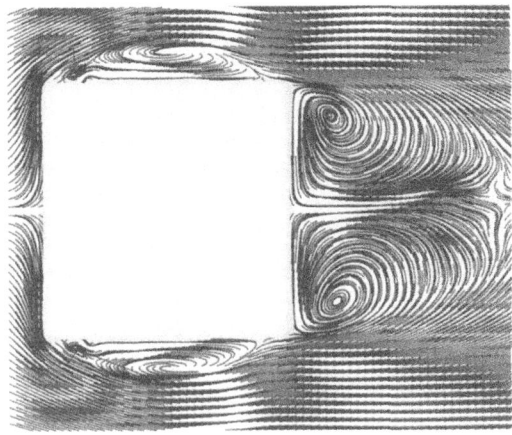

Fig. 12. Top view of a cut through the right horseshoe vortex and the front and back side vortex at half cube hight

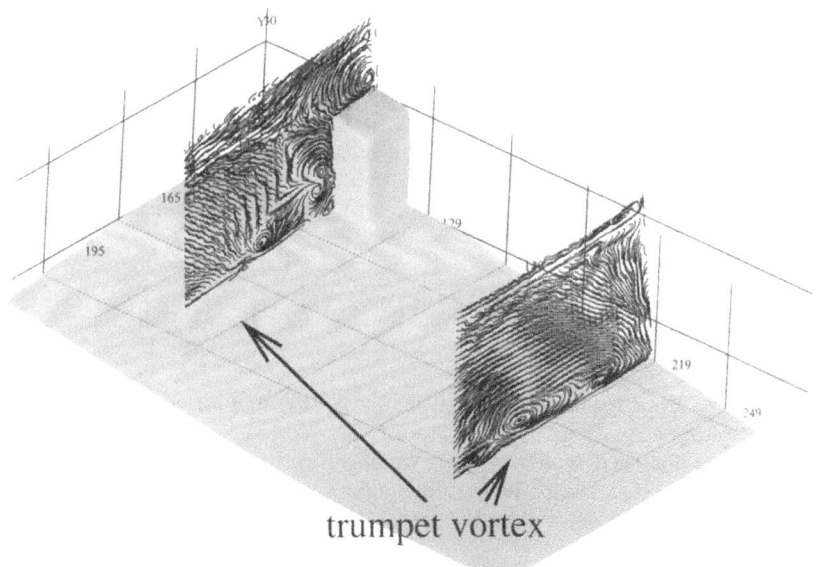

Fig. 13. symmetric trumpet vortex (see also Fig. 11)

6 Discussion

In the past years, the Lattice-Boltzmann method was developed as a valuable complementary approach to classical CFD techniques for solving the Navier-Stokes equations. The algorithmic structure and the local nature of the formulation leads to a powerful parallel application tool to solve various kinds of single and multiphase flow problems on pipeline or pseudo pipeline processors like the HITACHI SR-8000 F1 or the Fujitsu VPP700/52. The authors developed a data structure based on a topologically unstructured grid which allows the minimization of memory requirements, while retaining an excellent performance with respect to vectorization of standard LB-implementations on supercomputers. Due to extensive communication hiding an almost linear parallel speedup is achieved.

The simulation of flow through complex geometries like porous media, where the usage of high performance computing is essential, is the primary advantage of this method. Additionally it has been shown, that lattice Boltzmann methods are a valuable alternatives to classical methods and experiments for the computation of turbulent flow problems.

Acknowledgement. The authors would like to thank the HPC group at the Leibniz-Rechenzentrum for valuable technical support. The project is funded by the Deutsche Forschungsgemeinschaft (Ra 624/7-1).

References

1. Succi, S. (2001): The Lattice Boltzmann Equation for Fluid Dynamics and Beyond Oxford University Press, ISBN 0198503989
2. Rothmann, D. H. & Keller, J. M. (1988): Immiscible Cellular Automaton Fluids. *J. Stat. Phys.* **52**, pp. 1119–1127.
3. Gunstensen, A. K. & Rothman, D. (1991): Lattice Boltzmann model of immiscible fluids. *Phys. Rev.* A **43**, pp. 4320–4327.
4. Grunau, D., Chen, S. & Eggert, K. (1993): A Lattice Boltzmann Model for multiphase flow. *Phys. Fluids* A **5**, pp. 2557–2561
5. J. Tölke, M. Krafczyk, M. Schulz, E. Rank (2002): Lattice Boltzmann Simulations of binary fluid flow through porous media, Phil. Trans. R. Soc. Lond. A, Vol. 360, No. 1792.
6. Quian, Y. H., d'Humieres D., Lallemand P. (1992): Lattice BGK models for Navier Stokes equations. Europhysics Letters, Vol. 17, No. 6, pp. 479–484.
7. Tölke, J. (2001): Gitter-Boltzmann-Verfahren zur Simulation von Zweiphasenströmungen. Ph.D. thesis, Lehrstuhl Bauinformatik, TU München, München.
8. Schulz (2002): Ph.D. thesis in preparation, Lehrstuhl Bauinformatik, TU München, München.
9. Ristow, G. H. (1994): Granular Dynamics: a Review about recent Molecular Dynamics Simulations of Granular Materials. *Annual Reviews of Computational Physics I*, pp. 275–308. World Scientific, Singapore.
10. Karypis, G., Kumar, V. (1998) Multilevel Algorithms for Multi-Constraint Graph Partitioning.
    http://www-users.cs.umn.edu/~karypis/publications/partitioning.html
11. M. Schulz, M. Krafczyk, J. Tölke and E. Rank (2001): Parallelization strategies and efficiency of CFD computations in complex geometries using Lattice Boltzmann methods on high-performance computers. *High Performance Scientific and Engineering Computing*, pp. 115–122. Springer Verlag ISSN 1439-7358.

12. Helmig,R. (1997) Multiphase Flow and Transport Processes in Subsurface. Springer Verlag.
13. S. Hou et al. (1996): A Lattice-Boltzmann subgrid model for high Reynolds number flows. Fields Inst. Comm., No. 6, pp. 151–165.
14. J. Smagorinsky (1963): Mon. Weath. Rev.,91 (3), p. 99
15. R. Martinuzzi (1992), Experimentelle Untersuchung der Umströmung wandgebundener, rechteckiger, prismatischer Hindernisse, PhD thesis, LSTM, University of Erlangen, Germany
16. W. Rodi et al. (1997): Status of Large–Eddy Simulation: Results of a Workshop, J. of Fluids Eng., 119,2, pp. 248-262
17. http://www-unix.mcs.anl.gov/mpi
18. http://www.inf.bauwesen.tu-muenchen.de/forschung/muskat/muskat_eng.html
19. Graf, H., Schulz, V., Vogel, H.J. und Roth, K., (2001): Experimenteller Ansatz zur Untersuchung von Mehrphasenfluss in porösen Medien. Mitt. Dtsch. Bodenkundl.

Models of Type Ia Supernova Explosions

Jens Carsten Niemeyer, Martin Reinecke, and Wolfgang Hillebrandt

Max-Planck-Institut für Astrophysik
Karl-Schwarzschild-Str. 1
85748 Garching, Germany

Abstract. Type Ia supernovae have become an indispensable tool for studying the expansion history of the universe, yet our understanding of the explosion mechanism is still incomplete. We describe a generally accepted explosion scenario, sketch the most relevant physics, and report recent advances in multidimensional simulations of Chandrasekhar mass white dwarf explosions.

1 Introduction

Type Ia supernovae, i.e. stellar explosions which do not have hydrogen in their spectra, but intermediate-mass elements, such as silicon, calcium, cobalt, and iron, have recently received considerable attention because it appears that they can be used as "standard candles" to measure cosmic distances out to billions of light years away from us. Moreover, observations of type Ia supernovae seem to indicate that we are living in a universe that started to accelerate its expansion when it was about half its present age. These conclusions rest primarily on phenomenological models which, however, lack proper theoretical understanding, mainly because the explosion process, initiated by thermonuclear fusion of carbon and oxygen into heavier elements, is difficult to simulate even on supercomputers.

From the point of view of cosmology, in particular the planned use of SNe Ia as high-precision tools to map out the equation of state of the universe, some of the most urgent questions that supernova theorists and observers must answer are: a) What are the progenitors?, b) Is there only one class of explosions or many?, c) What is the physics that governs the diversity of SNe Ia?, and d) How robust are the correlations between peak brightness, light curve shape, and spectral features with respect to, say, multidimensional mixing? This is a challenging program that, apart from cosmology, involves a lot of fascinating physics.

Much progress has been made in recent years in the field of multidimensional explosion models. We give a very brief overview of our adopted explosion scenario in Sect. 2 (for more details and references, see [Hillebrandt & Niemeyer(2000)]) and describe our approaches to simulating Chandrasekhar mass deflagration models in two and three dimensions in Sect. 3. Sect. 4 gives a brief overview over the resources required for the performed simulations.

2 Adopted explosion scenario

The lack of hydrogen and presence of silicon in SN Ia spectra, the rate of the light curve decline powered by decaying nickel, and the inferred minimum age of some SN Ia progenitors are all consistent with thermonuclear explosions of C+O white dwarf stars [Hoyle & Fowler(1960)]. In order to trigger the explosion, the star is believed to accrete matter from a binary companion until critical conditions are reached. The various proposed explosion scenarios below differ mainly in the rate and composition of the accreted material, the mass of the white dwarf when it explodes, the location of the ignition, and the propagation mode of the burning front.

The strong temperature dependence of the nuclear reaction rates, $\dot{S} \sim T^{12}$ at $T \approx 10^{10}$ K, confines the nuclear burning to microscopically thin layers that propagate either conductively as subsonic deflagrations ("flames") or by shock compression as supersonic detonations. Both modes are linearly unstable to spatial perturbations. In the nonlinear regime, the burning fronts are either stabilized by forming a cellular structure or become fully turbulent – either way, the total burning rate increases as a result of flame surface growth. Neither flames nor detonations can be resolved in explosion simulations on stellar scales and therefore have to be represented by numerical models.

Given the overall homogeneity of SNe Ia, the good agreement of parameterized 1D models with observed spectra and light curves, and their reasonable nucleosynthetic yields, the bulk of normal SNe Ia is generally assumed to consist of exploding C+O white dwarfs that have reached the Chandrasekhar mass, M_{ch}, by accretion of hydrogen or helium that burns stably to carbon and oxygen. Flame ignition takes place near the center following roughly 10^3 years of core convection. There is no clear identification of natural progenitor systems, but supersoft X-ray sources (SSXS) look relatively promising [Livio(2000)].

In our simulations we concentrate on the subsonic propagation mode of the thermonuclear flame (deflagration), which in a SN Ia is governed by the following physical phenomena:

Once ignited, a subsonic thermonuclear flame becomes highly convoluted as a result of turbulence produced by the Rayleigh-Taylor instability (buoyancy) and the secondary Kelvin-Helmholtz instability (shear) along the flame front. It continues to burn through the star until it either transitions into a detonation or is quenched by expansion.

By far the most simulations to date are spherically symmetric, ignoring the multidimensionality of the flame and treating the turbulent flame speed S_t as a free parameter. These studies essentially agree that good agreement with the observations is obtained if S_t accelerates up to roughly 30 % of the sound speed.

In multiple dimensions, the problem of simulating turbulent deflagrations has two aspects: the representation of the thin, propagating surface separating hot and cold material with different densities, and the prescription of the local propagation velocity $S_t(\Delta)$ of this surface as a function of the hydro-dynamical state of the large-scale calculation with numerical resolution Δ. Our solution to this problem is sketched in Sect. 3; for a different approach see [Khokhlov(2000)].

Most authors agree that the turbulent flame speed decouples from mi-crophysics on large enough scales and becomes dominated by essentially uni-versal hydrodynamical effects, making the scenario intrinsically robust. A noteworthy exception is the location and number of ignition points that sig-nificantly influences the explosion outcome and may be a possible candidate for the mechanism giving rise to the explosion strength variability. Other possible sources of variations include the ignition density and the accretion rate of the progenitor system. All of these effects may potentially vary with composition and metallicity and can therefore account for the dependence on the progenitor stellar population.

3 Multidimensional M_{ch}-Models

In this section, we describe multidimensional simulations of what we consider the best model for the majority of SN Ia events, i.e. the pure turbulent deflagration model of Sect. 2 (see [Reinecke et al.(2002)Reinecke, Hillebrandt & Niemeyer] for details). Our basic assumptions are as follows: the initial model is a cold white dwarf with the Chandrasekhar mass, consisting of equal amounts of carbon and oxygen. Flame ignition starts in the inner ~ 150 km of the star and the initial flame geometry acts as our principal free parameter. No deflagration-detonation-transition is assumed to occur.

3.1 Modeling of the turbulent combustion front

The initial mixture consists of ^{12}C and ^{16}O at low temperatures. Because of the electron degeneracy the fuel temperature is nearly decoupled from the rest of the thermodynamic quantities, and since temperature is not used to determine the initial reaction rates, its exact value is unimportant.

When the flame passes through the fuel, carbon and oxygen are converted to ash, which has different compositions depending on the density of the unburned material. At high densities a mixture of ^{56}Ni and α-particles in nuclear statistic equilibrium (NSE) is synthesized. At lower densities burning only produces intermediate mass elements, which are represented by ^{24}Mg. Once the density drops below 10^7 g cm^{-3}, no burning takes place.

In the material burned to NSE, the proportion of ^{56}Ni and α-particles changes depending on density and temperature even after the flame has pro-cessed the material.

The transition densities from NSE to incomplete burning, as well as from incomplete burning to flame extinction were derived from data of a W7 run provided by K. Nomoto. This approach is rather phenomenological, and since these densities can have a potentially large impact on the simulation outcome it will have to be re-examined in a thorough manner.

The numerical representation of the thermonuclear reaction front (i.e. the location where the "fast" reactions take place) is described in detail in [Reinecke et al.(1999b)Reinecke, Hillebrandt, Niemeyer, Klein & Gröbl]. The flame front is associated with the zero level set of a function $G(\mathbf{r}, t)$, whose temporal evolution is given by

$$\frac{\partial G}{\partial t} = -(\mathbf{v}_u + s_u \mathbf{n})(-\mathbf{n}|\nabla G|), \tag{1}$$

where $\mathbf{v}_u$ and s_u denote the fluid and flame propagation velocity in the unburned material ahead of the front, and $\mathbf{n}$ is the front normal pointing towards the fuel. In our case, s_u identified with by the effective turbulent flame speed on the scale of the grid resolution, $S_t(\Delta)$ (see below). The advection of G caused by the fluid motions is treated by the piecewise parabolic method which is also used by our code to integrate the Euler equations. After each time step, the front is additionally advanced by $s_u \Delta t$ normal to itself.

This equation is only applied in the close vicinity of the front, whereas in the other regions G is adjusted such that

$$|\nabla G| = 1. \tag{2}$$

The source terms for energy and composition due to the fast thermonuclear reactions in every grid cell are determined as follows:

$$X'_{\text{Ashes}} = \max(1 - \alpha, X_{\text{Ashes}}) \tag{3}$$
$$X'_{\text{Fuel}} = 1 - X'_{\text{Ashes}} \tag{4}$$
$$e'_{\text{tot}} = e_{\text{tot}} + q(X'_{\text{Ashes}} - X_{\text{Ashes}}), \tag{5}$$

where α is the volume fraction of the cell occupied by unburned material; this quantity can be determined from the values of G in the cell and its neighbors. The quantity q represents the specific energy release of the total reaction.

All multidimensional simulations of exploding white dwarfs share the problem that it is impossible to resolve all hydrodynamically unstable scales. The consequence is that the simulated thermonuclear flame can only develop structures on the resolved macroscopic scales, while the real reaction front will be folded and wrinkled on much finer scales. Simply neglecting the surface increase on sub-grid scales would lead to an underestimation of the energy generation rate, which is not acceptable; therefore a model for a turbulent flame speed S_t is required to compensate this effect.

For the case of very strong turbulence it has been shown that S_t decouples from the laminar flame speed and is proportional to the turbulent velocity fluctuations v'. In our simulations, $v'(\Delta)$ is determined by using the sub-grid model introduced to SN Ia simulations by [Niemeyer & Hillebrandt(1995)]. For the presented calculations a few corrections were applied (see [Reinecke et al.(2002)Reinecke, Hillebrandt & Niemeyer]).

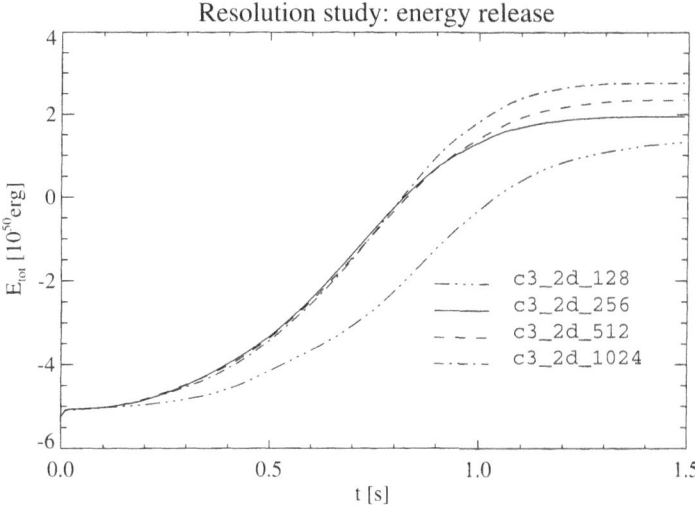

Fig. 1. Time evolution of the total energy for the initial flame geometry c3_2d and different resolutions. During the early and intermediate explosion stages there is excellent agreement between the better resolved simulations.

3.2 Two-dimensional resolution study

To study the robustness of our code with respect to a change of the numerical resolution, simulations were performed with grid sizes of 128^2, 256^2, 512^2 and 1024^2 cells, whose corresponding resolutions in the uniform inner part of the grid were $2 \cdot 10^6$ cm, 10^6 cm, $5 \cdot 10^5$ cm and $2.5 \cdot 10^5$ cm. The initial flame geometry (called c3_2d) used for all these calculations is identical to the setup C3 presented by [Reinecke et al.(1999a)Reinecke, Hillebrandt & Niemeyer]: the matter within a radius of $1.5 \cdot 10^7$cm from the stellar center was incinerated, and the surface of the burned region was perturbed to accelerate the development of Rayleigh-Taylor instabilities.

Figure 1 shows the energy release of the models; except for the run with the lowest resolution, the curves are nearly identical in the early and intermediate explosion stages. Simulation c3_2d_128 exhibits a very slow initial energy increase and does not reach the same final level as the other models. Most likely this is due to insufficient resolution, which leads to a very coarsely discretized initial front geometry and thereby to an underestimation of the flame surface. From this result it can be deduced that all supernova simulations performed with our code should have a central resolution of 10^6 cm or better.

Overall, our model for the turbulent flame speed appears to compensate the lack of small structures in the front very well.

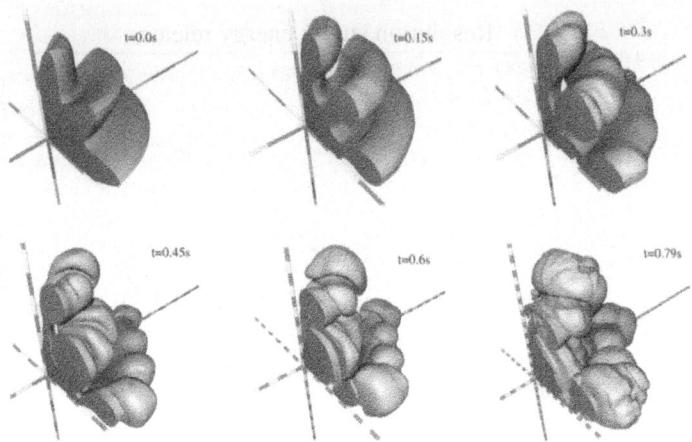

Fig. 2. Snapshots of the flame front for a centrally ignited three-dimensional scenario. One ring on the coordinate axes corresponds to 10^7cm.

3.3 Three-dimensional simulation

In order to compare two- and three-dimensional simulations directly, a 3D calculation was performed using the same initial conditions as given in Sect. 3.2. For this purpose the initial two-dimensional flame location was rotated by 90 degrees around the z-axis and mapped onto the three-dimensional Cartesian grid consisting of 256^3 cells with a central resolution of 10^6 cm. Only one octant of the white dwarf was simulated and mirror symmetry was assumed with respect to the coordinate planes.

The initial configuration, as well as snapshots at later times, are shown in Fig. 2. Obviously, the initial axisymmetry is lost after $0.2 - 0.3$ s, although no explicit perturbation in φ-direction was applied to the front. This happens because the initial flame geometry cannot be mapped perfectly onto a Cartesian grid and therefore the front is not transported with exactly the same speed for all φ. During the next few tenths of a second, these small deviations cause the formation of fully three-dimensional Rayleigh-Taylor-mushrooms, leading to a strong convolution of the flame. As expected, this phenomenon has a noticeable influence on the explosion energetics; this is illustrated in Fig. 3. Before the loss of axial symmetry in c3_3d_256, the total energy evolution is almost identical for both simulations, which strongly suggests that the two- and three-dimensional forms of the employed turbulence and level set models are consistent, i.e. that no errors were introduced during the extension of these models to three dimensions. In the later phases the 3D model releases more energy as a direct consequence of the surface increase shown in Fig. 2.

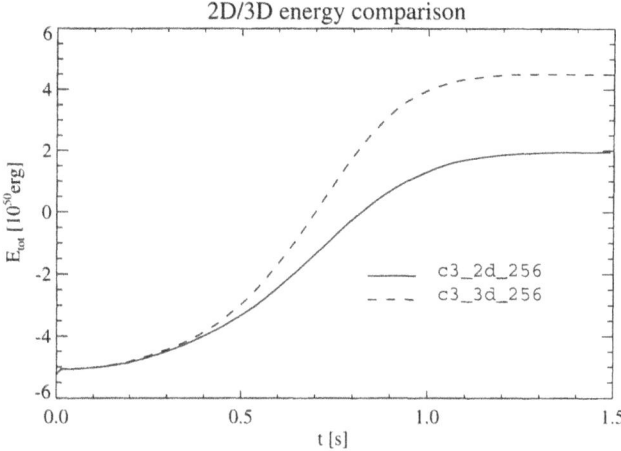

Fig. 3. Comparison of the explosion energy for identical initial conditions and resolution in two and three dimensions. After the loss of axial symmetry (at $t \approx 0.3\,\mathrm{s}$) the larger flame surface in the three-dimensional model leads to more vigorous burning.

3.4 Discussion and conclusions

The white dwarf becomes unbound in all of our models, which implies that no recontraction (and hence no pulsational detontation) occurs. Nevertheless only the three-dimensional model results in a powerful enough explosion to qualify as a typical SN Ia; the two-dimensional scenarios are too weak to accelerate the ejecta to the speeds observed in real events and produce too little nickel to power a standard SN Ia light curve.

The 3D calculation is a good candidate for typical SN Ia explosions, at least with respect to explosion strength and remnant composition. The produced nickel mass of 0.53 $M_\odot$ falls well into the range of $\approx 0.45 - 0.7\,M_\odot$ determined by [Contardo et al.(2000)Contardo, Leibundgut & Vacca] for several typical events, and it can be deduced from the amount of 0.18 $M_\odot$ of "magnesium" in the ejecta that enough intermediate mass elements were synthesized to explain the observed spectral features.

Qualitatively, our results for the explosion energetics are in relatively good agreement with recent simulations performed by [Khokhlov(2000)], employing different numerical models and initial conditions. We interpret this as an inherent robustness of the Chandrasekhar mass deflagration scenario.

This development is a major step towards constructing self-consistent models for type Ia supernovae. The remaining free parameters are chosen according to our best understanding of the unresolved physics without any reference to obtaining "good" explosions. As there will always be relevant unresolved scales in this problem, we need to keep improving our understanding by performing numerical experiments of turbulent thermonuclear combustion on microscopic and intermediate scales. Nevertheless, our results make us optimistic that multidimensional models will soon allow us to understand how type Ia supernovae really work.

4 Computational demands

While the two-dimensional calculations could be carried out on workstations or on single nodes of parallel computers, the resources required for the three-dimensional runs exceeded the limits of most supercomputers available to us, especially the high-resolution calculations that are presently carried out. The curently running simulation consists of 768^3 grid cells, requires a total memory of approximately 50GB, several 100 GB of storage space for checkpoint and output files, and will consume about 500–1000 hours on 64 nodes of the Hitachi SR8000-F1.

Such highly resolved simulations are necessary to further investigate the parameter space for the initial conditions of the explosion, since it is believed that the thermonuclear runaway will take place in many small bubbles in the central region of the white dwarf; these could not be represented properly on a relatively coarse grid.

Acknowledgement. We thank Bruno Leibundgut and Paolo Mazzali for helpful discussions.

References

Contardo et al.(2000)Contardo, Leibundgut & Vacca. Contardo, G., Leibundgut, B., Vacca, W. D.: 2000, *A&A* **359**, 876

Hillebrandt & Niemeyer(2000). Hillebrandt, W., Niemeyer, J. C.: 2000, *Ann. Rev. A&A* **38**, 191

Hoyle & Fowler(1960). Hoyle, F., Fowler, W. A.: 1960, *ApJ* **132**, 565

Khokhlov(2000). Khokhlov, A. M.: 2000, "Three-Dimensional Modeling of the Deflagration Stage of a Type Ia Supernova Explosion", Astro-ph/0008463, submitted to ApJ

Livio(2000). Livio, M.: 2000, Cambridge: Cabridge University Press, 33

Niemeyer & Hillebrandt(1995). Niemeyer, J. C., Hillebrandt, W.: 1995, *ApJ* **452**, 769

Reinecke et al.(1999a)Reinecke, Hillebrandt & Niemeyer. Reinecke, M. A., Hillebrandt, W., Niemeyer, J. C.: 1999a, *A&A* **347**, 739

Reinecke et al.(2002)Reinecke, Hillebrandt & Niemeyer. Reinecke, M. A., Hillebrandt, W., Niemeyer, J. C.: 2002, *A&A* **386**, 936

Reinecke et al.(1999b)Reinecke, Hillebrandt, Niemeyer, Klein & Gröbl. Reinecke, M. A., Hillebrandt, W., Niemeyer, J. C., Klein, R., Gröbl, A.: 1999b, *A&A* **347**, 724

Direct Numerical Simulation of Boundary Layer Separation along a Curved Wall with Oscillating Oncoming Flow

Jan Wissink and Wolfgang Rodi

Institute for Hydromechanics
University of Karlsruhe
Kaiserstr. 12
76128 Karlsruhe, Germany
wissink@ifh.uni-karlsruhe.de

Abstract. A direct numerical simulation has been performed in which the small scale fluctuations generated by a laminar separation bubble interact with a second separation bubble further downstream. Because of the oscillating oncoming flow, the size of the upstream separation bubble periodically varies, while the downstream bubble disappears when the inflow acceleration is close to its maximum. The fluctuations produced in the upstream bubble are damped by accelerating flow as they are convected downstream. The surviving disturbances are not strong enough to completely inhibit a second separation of the boundary layer downstream.

Key words: DNS, Flow Separation, Turbulence

1 Introduction

In the presence of a strong enough adverse pressure gradient, a laminar boundary layer over a solid surface will separate to form a laminar separation bubble (LSB). Since the separated shear layer is very unstable, it will usually undergo rapid transition to turbulence. Further downstream, the turbulence generated by the LSB may inhibit a second boundary layer separation. For instance, in order to prevent flow around an airfoil from separating near mid-chord, a well known strategy is to force the boundary layer to become turbulent immediately downstream of the leading edge. This so-called tripping of the boundary layer can be performed by provoking a small separation bubble just downstream of the leading edge.

Direct Numerical Simulation (DNS) of laminar separation bubble flow with constant free-stream has been previously performed by Alam and Sandham [1], Maucher *et al.* [4], Spalart and Strelets [5] and Wissink and Rodi [7].

To generate an adverse pressure gradient, required for the formation of a separation bubble, Alam and Sandham as well as Spalart and Strelets used suction through the upper boundary, while Maucher et al. applied a special boundary condition for the free-stream velocity. The computational geometry employed in Wissink and Rodi is similar to the one employed in this paper (see Fig. 1), except for the application of a free-slip boundary condition at the upper wall. The curved shape of the upper wall was used to generate an adverse pressure gradient along the flat plate. Because the free-slip boundary condition at the curved upper wall prevents the boundary layer from separating, a LSB is only found along the lower flat plate. Alam and Sandham report that the separated shear layer undergoes transition via oblique modes and Λ-vortex induced breakdown. It was shown that a LSB flow is absolutely unstable if the reverse flow is of the order of $15 - 20\%$ of the free-stream velocity. Like in Maucher et al. disturbances were explicitly introduced in the boundary layer before it separates to trigger transition. Maucher et al. (see [4] and the references therein) studied amplification rates of unstable modes to elucidate early stages of transition. In contrast to the other researchers, Spalart and Strelets [5] and Wissink and Rodi [7] relied on disturbances present in any LSB simulation to trigger transition. Wissink and Rodi [7] rely on the fact that their LSB flow is absolutely unstable. Transition takes place via a Kelvin-Helmholtz instability of the separated shear layer. The vortices that are shed quickly become three-dimensional. Compared to Spalart and Strelets, Wissink and Rodi employed a relatively low Reynolds number. As a result they were able to detect traces of vortex pairing of Kelvin-Helmholtz vortices before their structure was completely destroyed by the production of turbulent kinetic energy. In Spalart and Strelets no such pairing was observed. A further simulation of LSB flow with oscillating instead of steady main flow was performed by Wissink and Rodi [6]. Except for the free-slip boundary conditions at the upper wall and the Reynolds number, the geometry as well as the amplitude ($a = 0.20$) and the period ($T = 0.61$) of the oscillation are all identical to what is employed in the present paper. Because of the absence of separation at the upper wall, along the lower flat plate periodically a LSB was formed with a very strong reverse flow of up to 150% of the free-stream velocity. Owing to the oscillating oncoming flow, the location of separation was found to move back and forth along the plate. Each period a new separation bubble was formed, while the old one gradually moved downstream. In the present paper we will focus on the separation along the curved upper wall owing to the application of no-slip boundary conditions in this simulation.

1.1 Computational aspects

The computational domain (see Fig. 1) has been chosen in accordance with experiments performed by Prof. Hourmouziadis' group at the TU-Berlin (see for instance [3]). Along the upper curved wall as well as along the flat plate no-slip boundary conditions are employed. At the lower boundary for $x/L < 0$,

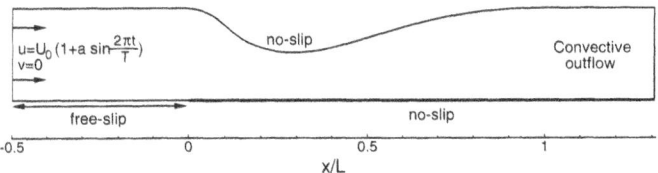

Fig. 1. The computational domain

free-slip boundary conditions are prescribed, while in the spanwise direction periodic boundary conditions are applied. The spanwise size of the computational domain is $l_z = 0.08$. At the inlet, an oscillating flow is prescribed with period $T = 0.61$ and amplitude $a = 0.20$, while at the outlet a convective boundary condition is applied. The Reynolds number, $Re = 120\,000$, of the flow is based on the mean inflow velocity U_0 and L (see Fig. 1). The 3D incompressible Navier-Stokes equations have been discretised on a $1606 \times 310 \times 128$ curvi-linear mesh using a cell-centred, second-order accurate finite-volume method in space combined with a three-stage Runge-Kutta method in time. Conservation of mass is enforced by solving a Poisson equation for the pressure using the SIMPLE approach. The grid is slightly stretched in the y-direction. Along the upper wall the distance between the wall-nearest grid point and the wall varies is less than $1.42 \times 10^{-4}L$, while along the lower wall this distance is less than $1.44 \times 10^{-4}L$. For more details on the numerical scheme see [2].

The computation has been performed on the Hitachi SR8000-F1 at LRZ in München using 256 processors (32 nodes). The speed that the simulation reached on the computer was approximately $0.246 GFlop/s$ per node which corresponds to 20.5% of the peak performance. The number of timesteps per period was 7680, resulting in a CFL number ranging from 0.5 to 0.9. To obtain a well developed flow field the simulation had to run for 10 periods. Subsequently phase-averaging was performed for another 10 periods. For the phase-averaging, each period T was divided in 256 equal phases.

2 Results

The existence of disjunct separation bubbles along the upper wall is illustrated in Fig. 2 (left and right). In Fig. 2 (left) the black line, corresponding to $u_\tau = 0$, identifies separation. The gray contours in the background correspond to the streamwise pressure-gradient. The separation bubble upstream of $x/L = 0.2$ remains present irrespective of the phase. Even during the inflow acceleration, for $0 \le \phi < \frac{1}{4}$ and $\frac{3}{4} < \phi < 1$, in a small strip inside the domain $0 < x/L < 0.1$ an adverse pressure gradient is found that is strong enough to maintain this separation. At separation, the Reynolds number based on the displacement thickness (Re_{δ^*}) varies quite significantly. At $\phi \approx 0.57$ it reaches a minumum of $Re_{\delta^*} \approx 328$, while for $0.04 < \phi < 0.29$ Re_{δ^*} reaches values in excess of 1500. At $\phi = 0.18$, just before the inflow is about to decelerate, a new region with an adverse pressure gradient appears upstream of the separation-point near $x/L \approx -0.096$. The adverse pressure gradient increases in strength as the inflow acceleration slowly turns into an inflow deceleration. As a result, at $\phi \approx 0.30$, the boundary layer separates such that a total of two separation bubbles upstream of $x/L = 0.2$ are found. The new separation bubble grows in the upstream as well as in the downstream direction until $\phi \approx \frac{5}{8}$, just before the end of the inflow acceleration at $\phi = \frac{3}{4}$. For larger ϕ its streamwise extent decreases. At $\phi \approx 0.49$ the two separation bubbles merge. The separation bubble downstream of $x/L = 0.2$ is only present for $0.11 < \phi < 0.92$. When the acceleration of the inflow is close to its maximum the bubble vanishes. The maximum streamwise extent of this bubble is approximately $0.125L$, compared to a maximum extent of $0.29L$ for the bubble upstream of $x/L = 0.2$. Again Re_{δ^*} varies with phase. It reaches a minimum of $Re_{\delta^*} \approx 210$ at $\phi \approx 0.43$, while at $\phi \approx 0.87$, just before the separation bubble vanishes, a maximum of $Re_{\delta^*} \approx 850$ is reached. Due to the persistent flow acceleration, the streamwise pressure gradient in the region around $x/L = 0.2$ is found to remain favourable during all phases.

Note that any non-smoothness in the phase-averaged results, such as may be observed in this and other figures, is most likely due to the limited amount of phase-averaging. Averaging over significantly more phases would have made the simulation too costly.

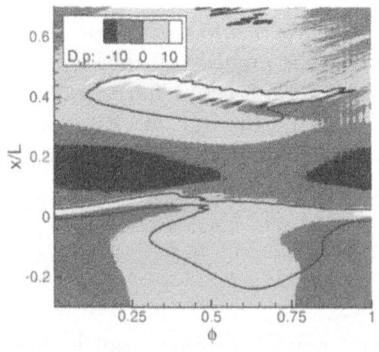
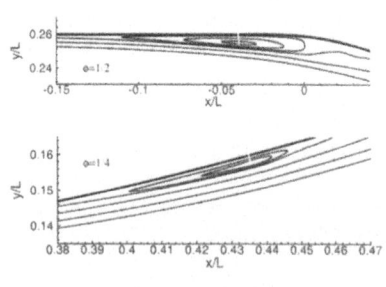

Fig. 2. Left: Contours of the phase-averaged stream-wise pressure gradient along the upper wall as a function of the phase ϕ. The black line corresponds to $u_\tau = 0$ and identifies separation. Right: The two separation bubbles obtained along the upper wall.

In Fig. 2 (left), it was illustrated that the upstream separation bubble remains present during all phases. As a consequence, disturbances may be

entrained inside the bubble for a considerable time. Since no explicit disturbances have been added to the flow, the separation bubble needs to be
absolutely unstable in order to generate turbulent flow. In Fig. 3, it is shown

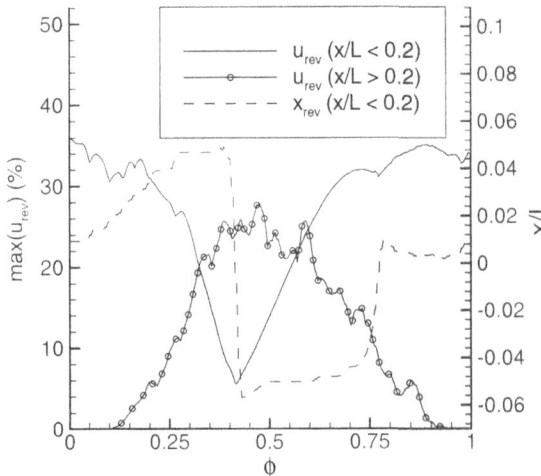

Fig. 3. Maximum reverse flow as a percentage of the free-stream velocity in the
streamwise direction and the location of the maximum of the upstream bubble.

that in the separation bubble upstream of $x/L = 0.2$ a maximum reverse
flow of more than 20% of the free-stream velocity is reached for $0 \leq \phi < 0.32$
and $0.56 < \phi < 1$. During these phases the re-circulating flow is absolutely
unstable [1] and self-sustained turbulence can exist. At $\phi \approx 0.42$ the maximum reverse flow reaches a minimum value of approximately 5.6% of the
free-stream velocity. The corresponding graph of x_{rev} shows that the jump
in the slope at this minimum coincides with a sudden change in the spanwise location of max u_{rev}. At this phase the reverse flow in the newly formed
separation bubble becomes stronger than the reverse flow in the old bubble.
As already illustrated in Fig. 2 (left), the separation bubble downstream of
$x/L = 0.2$ vanishes in an interval around $\phi = 0$. The maximum reverse flow
in the downstream bubble is found to exceed 20% of the free-stream velocity
for $0.32 < \phi < 0.61$.

In Fig. 4, contours of the phase-averaged fluctuating kinetic energy (k) are
plotted at four phases $\phi = 0, \frac{1}{4}, \frac{2}{4}, \frac{3}{4}$. The plots show that the LSB upstream
of $x/L = 0.2$ does produce fluctuating kinetic energy. At $\phi = \frac{1}{4}$ there is a
definite concentration of k near the bubble, indicating that the production of
k is quite large. This agrees with Fig. 3 which shows that at $\phi = \frac{1}{4}$ the LSB is
absolutely unstable. Convection of k becomes important at $\phi = \frac{2}{4}$ indicated

by the elongated area of concentrated fluctuations along the wall downstream of the bubble. According to Fig. 3, the flow in the bubble itself is no longer absolutely unstable, so we expect no or only minor production of turbulence. This is confirmed by comparing the plot at $\phi = \frac{1}{4}$, where the amount of k inside the bubble is quite high, and the plot at $\phi = \frac{2}{4}$ where the concentration of k inside the bubble is much smaller. At the same time the fluctuations produced at earlier phases reach the downstream separation bubble. While the fluctuations are convected along the wall through the narrowing passage, they are damped by the accelerating flow. As mentioned in the introduction, oncoming fluctuations might prove to be very effective in inhibiting boundary layer separation. However, as already observed in Fig. 2 (left), a separation bubble downstream of $x/L = 0.2$ *is* found to be present during the deceleration of the inflow velocity. Still, the oncoming disturbances generated by the LSB upstream of $x/L = 0.2$ are expected to affect the size of this bubble as well as the amount of fluctuations/turbulence produced inside it. The relatively small streamwise extent of the downstream bubble supports part of this expectation. The results presented Fig. 4 illustrate that for all phases, $\phi = \frac{1}{4}, \frac{2}{4}$, and $\frac{3}{4}$, where the downstream bubble exists, a local maximum of k is found inside this bubble. At $\phi = \frac{2}{4}$ the concentration of k inside the bubble is highest. This is to be expected since this is the only instance where the flow is absolutely unstable and thus a large production of k can occur. The reverse flow at the other two phases, $\phi = \frac{1}{4}$ and $\phi = \frac{3}{4}$, is too small for the flow to be absolutely unstable. Here the concentration of k inside the bubble most likely stems from entrained fluctuations with at most a minor contribution of local production. Owing to convection of k, further downstream the level of free-stream turbulence along the upper wall is found to periodically increase and decrease, though according to Fig. 4, it never falls below a minimum of $k = 0.01$.

In Fig. 5, snapshots of iso-surfaces of the spanwise vorticity are displayed to identify the boundary layer. In each plot the boundary layer upstream of the left separation bubble is found to be laminar. Like in [6], the flow is found to become three-dimensional after the separated boundary layer starts to roll up. At the same time three-dimensional structures appear. The fluctuations produced by the bubble upstream of $x/L = 0.2$ are shown to disturb the boundary layer downstream. So we may conclude that the disturbances are not only convected in the free stream (see Fig. 4) but also inside the boundary layer. It is clear to see that during one period the point of separation as well as the origin of the fluctuations moves back and forth. As also shown in Fig. 2 (left), the separation bubble remains present during all phases. Downstream of $x/L = 0.2$, disturbances convected from upstream are amplified inside the separation bubble (see the snapshots at $t/T = 10.500, 10.667, 11.334$) and the flow becomes turbulent. As the inflow accelerates, the downstream separation bubble vanishes while the small scale fluctuations are convected downstream.

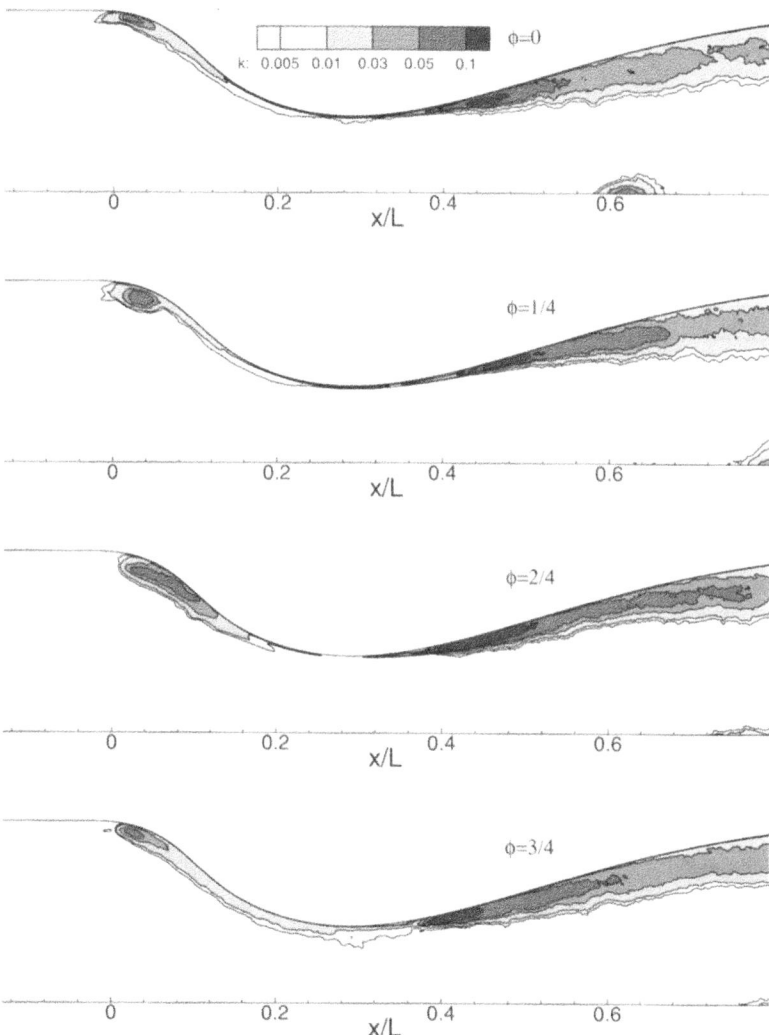

Fig. 4. Contour plots of the phase-averaged fluctuating kinetic energy near the upper wall

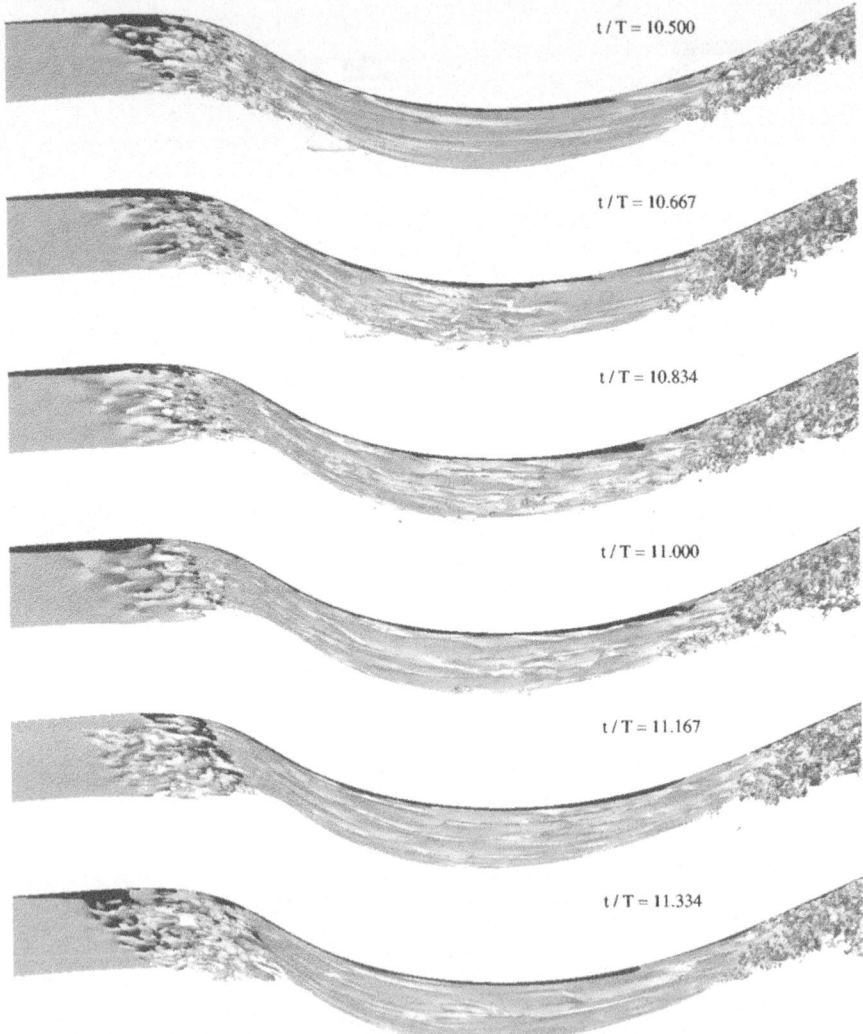

Fig. 5. Six snapshots showing the evolution of iso-surfaces of the spanwise vorticity at $\omega_z = 120$ during one period

In Fig. 6, a close-up is presented of the v-velocity contours in a plane inside the boundary layer, almost parallel to the upper wall. It is illustrated that the upstream LSB merely generates rather large scale disturbances inside the boundary layer, while downstream of $x/L \approx 0.4$, the flow is turbulent. It was already mentioned that just before $\phi \approx 1$, the downstream separation bubble vanishes. As this happens, at $t/T \approx 10.8$, a turbulent spot is found to appear inside the re-attached boundary layer. The four subsequent snapshots show that the spot grows as it moves downstream before it finally merges with

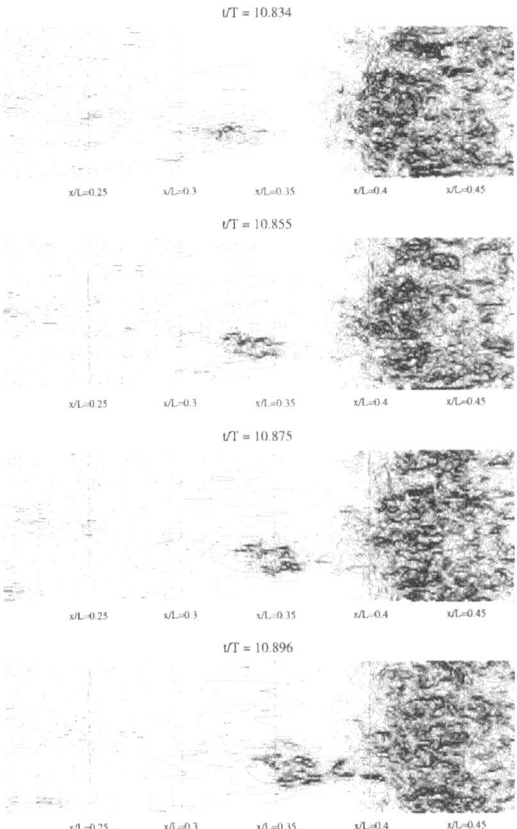

Fig. 6. Four snapshots showing v-velocity contours in a plane parallel to the upper wall, at a distance of approximately $0.001L$, 37 equally spaced contours are plotted min. value: -1.0, max. value: 1.0.

the turbulent region downstream of $x/L = 0.4$. Upstream of the turbulent spot, the boundary layer appears not to be susceptible to the fluctuations produced by the LSB upstream of $x/L = 0.2$. That is, though the fluctuations clearly disturb the boundary layer, the disturbances do not manage to trigger turbulence.

3 Discussion and Conclusions

The need for computational resources by a direct numerical simulation of turbulent flow is huge. Thanks to the availability of a high-performance computer, it was possible to perform a DNS of separation bubbles along an upper curved wall of a channel under the influence of oscillating flow. Owing to the shape of the upper wall, two disjunct separation bubbles could be identified. The upstream bubble behaves like a typical laminar separation bubble where fluctuations are produced virtually simultaneously with the roll-up of the shear layer (see also Spalart [5] Wissink and Rodi [6,7]). Unlike in [6], where the LSB is also influenced by oscillating flow, the separation bubble does not completely move downstream shortly after the inflow accelerates. Only the point of separation moves back and forth in the streamwise direction. Even though the bubble is not absolutely unstable during the whole period, enough of the disturbances entrained into the bubble survive until the re-circulating flow becomes absolutely unstable. The small scale fluctuations produced in the bubble are convected downstream along the upper wall. It is very likely that the downstream bubble is highly affected by these oncoming disturbances. For example the small streamwise extent of the bubble might be a direct consequence. The downstream separation bubble vanishes when the acceleration of the inflow surpasses a certain level. At the same time, the re-attached boundary layer is found to be unstable and oncoming disturbances can trigger the formation of a turbulent spot. The production of turbulence inside the bubble picks up immediately after the bubble reappears.

Downstream of the upstream LSB, disturbances are not only found to be convected in the free-stream flow but also inside the boundary layer. This makes it impossible to differentiate between the role the disturbed boundary layer plays in the break up of the shear layer on top of the downstream bubble and the role of the free-stream turbulence. It is very likely that it is a mixture of the two. The effect of boundary layer disturbances on a LSB was already investigated by several researchers (see for instance [4] and [1]). To complement this research there is a definite need to make a separate study of the effect of free-stream turbulence on a LSB.

Acknowledgement. The authors wish to thank the German Research Foundation (DFG) for funding this project and the steering committee of the supercomputing facilities in Bavaria (HLRB) for granting computing time on the Hitachi SR8000 F1.

References

1. Alam, M., Sandham, N.D. (2000): Direct Numerical Simulation of 'short' Laminar Separation Bubbles with Turbulent Reattachment. J. Fluid Mech., **410**, 1–28
2. Breuer, M., Rodi, W. (1996): Large Eddy Simulation for complex turbulent flows of practical interest. In: Hirschel, E.H. (ed) Flow Simulation with High-Performance Computers II, Notes in Numerical Fluid Mechanics, **52**, Vieweg Verlag, Braunschweig.
3. Lou, W., Hourmouziadis, J. (2000): Separation Bubbles under Steady and Periodic Unsteady Main Flow Conditions. In: The 45th ASME International Gas

Turbine & Aeroengine Technical Congress, Paper nr. 2000-GT-0270, Munich Germany.

4. Maucher, U., Rist U., Kloker, M., Wagner, S. (2000): DNS of laminar-turbulent transition in separation bubbles. In: Krause, E., Jäger, W. (eds) High-Performance Computing in Science and Engineering. Springer Berlin Heidelberg

5. Spalart, P.R., Strelets, M.Kh. (2000): Mechanisms of Transition and Heat Transfer in a Separation Bubble. J. Fluid Mech., **403**, 329–349

6. Wissink, J.G., Rodi, W. (2002): DNS of a Laminar Separation Bubble in the Presence of Oscillating Flow. In: Braza, M., Hirsch, C., Hussain, F. (eds) IUTAM Symposium on Unsteady Separated Flows, April 8-12 Toulouse, France.

7. Wissink, J.G., Rodi, W. (2002): DNS of Transition in a Laminar Separation Bubble. In: Castro, I.P., Hancock, P.E.. (eds) Advances in Turbulence IX, Proceedings of the Ninth European Turbulence Conference.

Part III

Biosciences

Kurt Binder

Johannes-Gutenberg-Universität Mainz
Institut für Physik
Staudingerweg 7
55099 Mainz, Germany

High Performance Computing in Biosciences and Chemistry presents particular challenges due to the multiscale character of the problems that one has to deal with in the corresponding computer simulations. Often the task is to explain the biological function of particular biological molecules. The study of retina protein rhodopsin, that has a central role for the vision process in the eye, that will be presented in this chapter, is a good example: length scales range, from the scale of chemical bonds to the mesoscale structure, where flexible and stiff-parts of the macromolecule form a particular arrangement, that fits into the biological membrane properly, and provides the right environment for the chromophore; i. e. the light absorbing part of the protein.

Of course, the protein structure would not be stable outside its natural environment (the biological membrane, which itself is surrounded by water containing a physiological concentration of salt [NaCl] as well), and hence there is a spatial inhomogeneity on large length scales that must not be ignored. On the other hand, the irradiation with light in the visible range causes the cis-trans isomerization at a particular bond of the chromophase and thus triggers the first step of a multi-step biochemical process, as a result of which finally a signal is transported via a neuron to the brain.

The multiscale structure of problems of this kind is evident: the electronic structure of the chromophore needs to be considered in detail for this process, but the environment also is important – depending on the distance from the reacting electrons, a different level of details is required. In the work on rhodopsin that will be described below, this challenge is met by a suitable combination of quantum-mechanical/molecular mechanics (QM/MM) methods with standard molecular dynamics simulation for the environment that mimicks the biological membrane: the complete model combines 24 000 atoms in total, but the quantum-mechanical part (that uses the restricted open-Shell Kohn-Sham density functional method, able to deal with the excited states of the photo-chemical reaction "on the fly") is applied to the chromophore only. It should be emphasized, that such extensions of the Car-Parrinello "ab initio" molecular dynamics" methods are by no means standard, but really constitute a new methological development as well! Thus, the availability of top-rank supercomputers such as provided by the HLRB are crucial for making progress in the formulation of hybrid new methods that allow to bridge the scales from the subatomar distances relevant in the electronic structure to the nanoscopic scale of large molecular groups and the mesoscopic scales of biopolymers embedded in membranes. Note that this range of length scales corresponds to a corresponding spread of time scales – from the sub-picosecond scale of the chemical raction to the nanosecond scale necessary to equilibrate the structure of large macromolecules and still much larger time-scales are necessary to deal with atomistically described membranes.

Such problems are quite typical when one deals with the chemistry and physics of soft matter, which not only includes biological problems as described above, but many more problems involving materials formed from synthetic polymers, colloidal systems that find also widespread applications (paints, cosmetics, food industry) but to a large extent are still incompletely understood on an atomistic level, etc. This general area of research is rapidly growing and very important, and the high performance computing problems posed by this field to a large extent are still unsolved (see e. g. the recent book "Multiscale Computational Methods in Chemistry and Physics", edited by A. Brandt, J. Bernholc, and K. Binder, IOS Press, Amsterdam 2001). Many of these multiscale problems involve even larger scales and slower processes as discussed above. This is true, e. g., for the neuronal dynamics that takes place in the brain, which happens on the millisecond time-scales and involves truly macroscopic length scales. In this case, it would not make any sense to approach the problem by a variant of molecular modelling, but rather a suitably coarse-grained effective abstract model of such a complex system needs to be set up. As will be described in the paper "Simulation of Neuronal Map Formation in the Primary Visual Cortex", one has again to deal with a system involving many degrees of freedom, dealing with several layers (which are modelled as two-dimensional 64 x 64 lattices) of neurons that dynamically interact with each other. The large number of time steps needed to simulate the complex pattern formation process make also for problems of this kind the availability of supercomputers such as the Hitachi SR8000 a necessity.

QM/MM Study of Rhodopsin

Ute F. Röhrig[1,2], Christel Nonnenberg[2], Irmgard Frank[2], Leonardo Guidoni[1], and Ursula Röthlisberger[1]

[1] Laboratory of Inorganic Chemistry, ETH Zürich
   8093 Zürich, Switzerland
[2] Department Chemie, Butenandtstr. 5-13, Haus E
   81377 München, Germany *frank@cup.uni-muenchen.de*

1 Introduction

The investigation of the first step of vision is a long-standing problem in theoretical chemistry [1]. It has been text-book knowledge for a long time [2, 3, 4], that irradiation of the retina protein rhodopsin with light in the visible range induces a *cis-trans* isomerization at the C_{11}-C_{12} double bond of the chromophore, i.e. of the light-absorbing part of the protein (Fig. 1). However, it is extremely difficult to study the reaction mechanism in full detail on experimental basis. The consecutive signal transduction cascade that finally leads to a neuron signal also still poses many questions.

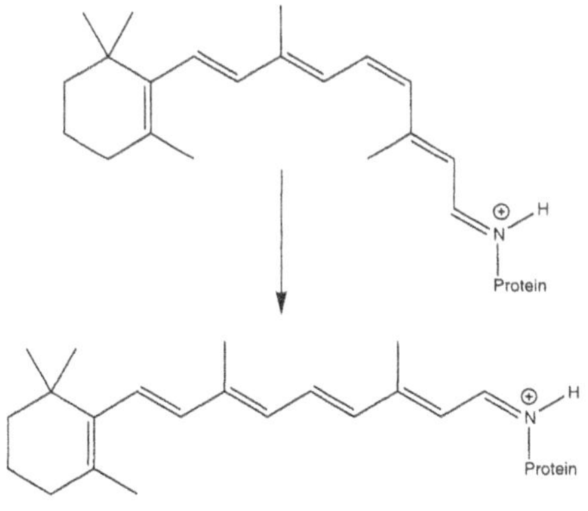

Fig. 1. The first step of vision: *cis-trans* isomerization of the rhodopsin chromophore

Experimental investigations are challenging for several reasons. First, sufficient amounts of bovine rhodopsin have to be isolated. Then, it is difficult to obtain good crystal structures of a membrane protein since the three-dimensional protein structure is determined by the specific membrane environment. Finally, a sample of rhodopsin undergoes the *cis-trans* isomerization just one single time, i.e. outside a living organism it never returns to the original state. For these reasons, most experimental and also many theoretical studies have been performed on bacteriorhodopsin, one of the light-absorbing proteins of the *halobacterium salinarium* [5,6,7]. In this case, accumulation and isolation of the protein are relatively simple, and the protein undergoes an optical cycle which renders it of high interest for technical applications [8].

Nevertheless, the uniquely fast and efficient reaction in the eye (quantum yield $\approx 67\%$, reaction time < 200 fs [9]) is of high interest, and many issues raised by experiments suggest a theoretical approach to the problem. However, theoretical investigations of such a complicated system must rely on a good experimental basis. The first high resolution crystal structure of rhodopsin published in August 2000 [10] provided the impetus to start a collaboration with the aim to model the first step of vision on a first-principles basis.

2 Methodology

The theoretical investigation of rhodopsin bears three major challenges:

- The electronic structure of the chromophore: The chromophore (the protonated Schiff base of retinal) is quite large for an *ab-initio* investigation. CASSCF calculations have been performed but yield up to now no consistent view [11,12,13,14,15].
- The environment: The *cis-trans* isomerization depends strongly on the environment. This is for example indicated by the fact that in bacteriorhodopsin a different isomerization takes place (C_{13}-C_{14} *trans-cis* instead of C_{11}-C_{12} *cis-trans*), although the chromophore is similar and only the protein environment differs.
- The dynamics: A condensed phase isomerization differs significantly from a gas phase isomerization. While in gas phase the isomerization can be described relatively easily on the basis of the potential curve that is obtained when varying the isomerization angle, in condensed phase the interactions with the environment during the isomerization play an important role. The dynamics of this environment and the response of the chromophore to it cannot be neglected. The sterically demanding *cis-trans* isomerization expected in gas phase cannot occur in the protein, since it would result in collisions with the surrounding.

Our approach consists in using first-principles molecular dynamics (MD) for the chromophore in combination with a classical MD scheme for the protein. The protein is modeled in a membrane-mimetic environment. We plan to describe the electronic structure of the excited state with the restricted open-shell Kohn-Sham (ROKS) method [16,17,18].

First-principles molecular dynamics according to Car and Parrinello (CPMD) [19,20] is nowadays a widely used tool to describe the dynamics of molecular systems in the ground state [21]. Using the Kohn-Sham Hamiltonian [22,23,24,25] with standard density functionals [26,27,28], the methodology is applicable to a great variety of systems. The use of a plane-wave basis set for the electronic wavefunction allows the description of gas phase and condensed phase on an equal footing. In contrast to classical MD, the quantum-chemical description of chemical bonds renders the simulation of chemical reactions feasible. CPMD uses the on-the-fly approach, i.e. only the points that are reached during the dynamics are computed instead of the complete potential surface. In this manner, it is possible to simulate chemical reactions in complex systems without any initial knowledge about the reaction mechanism [29,30].

CPMD is computationally expensive compared to classical simulations. This puts limitations on the size of the reactive system and the time scale of an unconstrained simulation, i.e., in a simulation in which no a priori knowledge of the reaction pathway is used. The rhodopsin photoreaction is fast enough to be observable on the accessible (picosecond) timescale. However, the environment is too large for a complete quantum chemical description. For this chemically unreactive part of the system, a classical description is sufficient. The hybrid quantum mechanics/molecular mechanics (QM/MM) code developed at the ETH Zurich is able to describe a quantum chemical system embedded in a classical environment [31, 32] and allows thus MD simulations of the complete system.

In order to describe the photochemical reaction in the protein, it is necessary to combine the QM/MM code with an approach that is able to cope with excited states 'on the fly'. The treatment of excited states is more difficult than that of ground states; electron correlation is very important. For the use with molecular dynamics, an excited-state method has to be simple to use, numerically stable and computationally affordable. In an MD simulation, some 10000 points on a potential energy surface are calculated, and it is not practicable to readjust the input parameters for every point. Furthermore it is desirable that a wavefunction (and not just the energy) for the excited state is computed, and that this wavefunction has a simple structure. Finally, the method should be self-consistent.

A method that essentially fulfills these conditions has been developed recently for the computation of first excited singlet states [16]. The restricted open-shell Kohn-Sham (ROKS) method uses a single spin-adapted function instead of a single determinant to represent the wavefunction [17,18]. A self-consistent scheme was derived using the Kohn-Sham approximation for the exchange-correlation part of the energy. Application to aldehydes, ketones and imines yield very good results [16]. First excited-state MD simulations have been performed for photoisomerization reactions [16,33,34].

3 Results

3.1 Protein modeling

The experimentally determined structure of rhodopsin (Fig. 2) is not stable outside a membrane, but modeling a lipid membrane is difficult due to the many slow degrees of freedom that have to be relaxed. Since the simulation of the photoreaction and of the consecutive motion within the protein it is not necessary to have a very accurate description of the membrane, we use a membrane-mimetic environment consisting of n-octane surrounded by water containing a physiological concentration of sodium chloride.

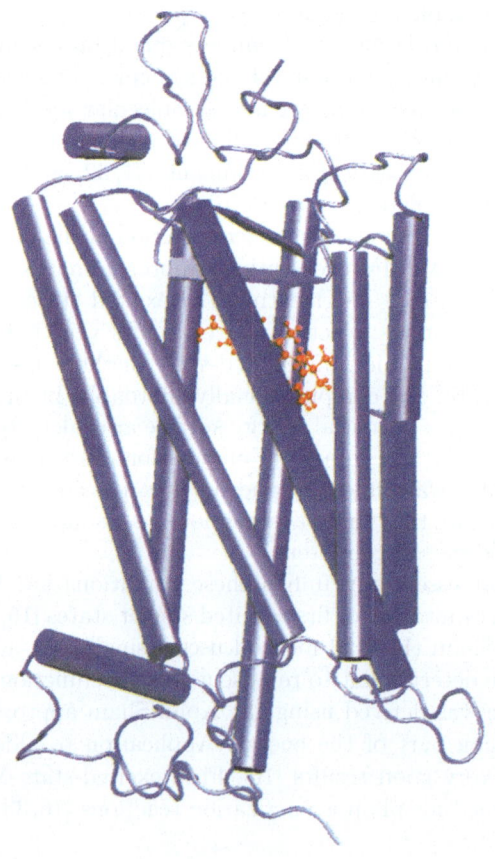

Fig. 2. The experimental protein structure

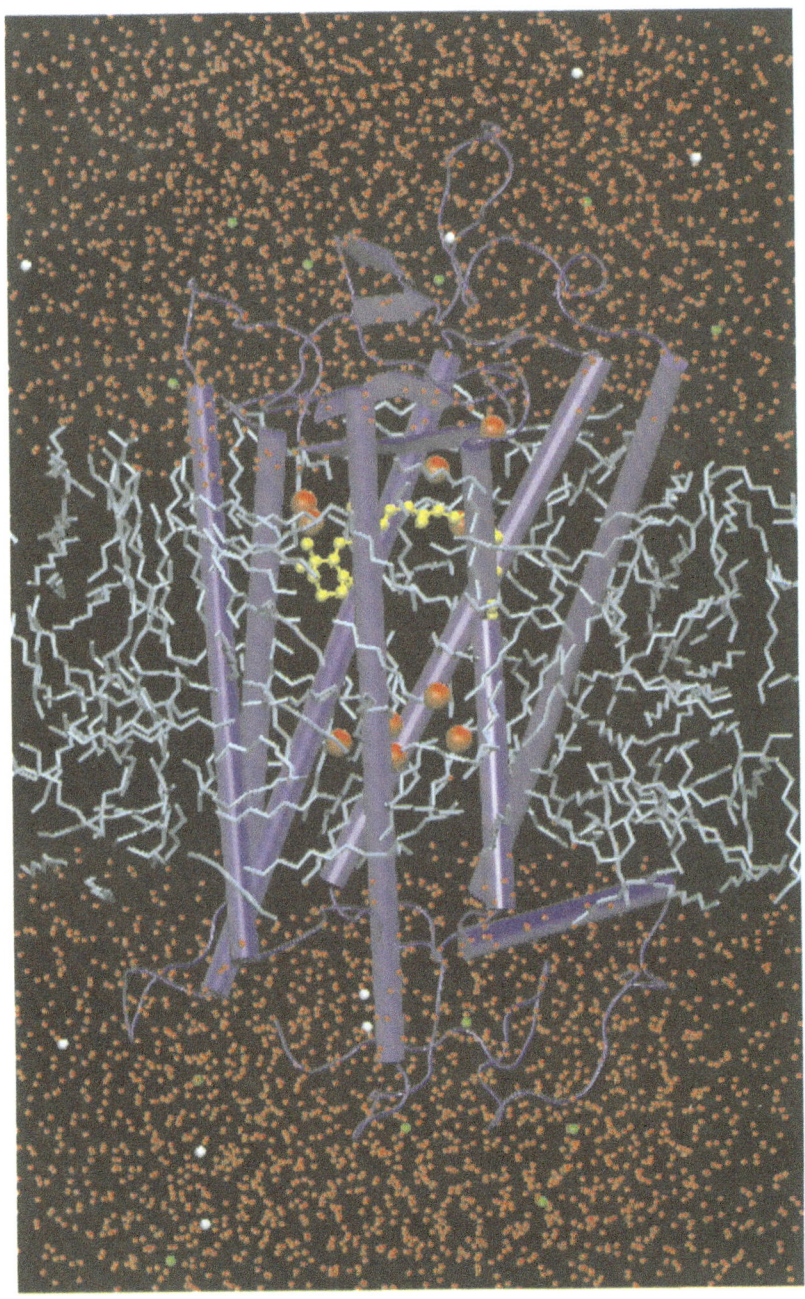

Fig. 3. Model of rhodopsin plus environment

While on this basis it is not possible to describe the full dynamics of the membrane, the model is able to maintain the structure of the protein on the time scale of several nanoseconds. This has been demonstrated for other membrane proteins [35,36], and we observed the same behaviour in both classical [37] and QM/MM simulations. With both approaches only vibrational motion around the equilibrium position was observed. The complete model contains ≈ 24000 atoms (see Fig. 3); we consider all 72000 atomic degrees of freedom in our dynamics. Breaking and formation of bonds is possible in the QM part only. That is, the chromophore is chemically reactive, while the apoprotein, the membrane and the water molecules are inactive.

3.2 Electronic structure

In previous gas phase calculations using ROKS [33,34], it was found that the photoisomerization of the isolated chromophore of rhodopsin exhibits a high barrier in the excited state. This barrier is significantly lowered by adding a single negatively charged amino acid from the protein binding pocket. A linear correlation between the C_{11}-C_{12} bond length and the isomerization barrier was found and indicates that in the protein the barrier is small (< 3 kcal/mol) [34].

In parallel to the calculations on the large chromophore, we are presently performing calculations on smaller conjugated molecules. These systems exhibit a $\pi \rightarrow \pi^*$ electronic transition, in contrast to the molecules investigated earlier [16]. Here, additional difficulties emerge from the fact that the two orbitals involved in the transition belong to the same spatial symmetry. This can eventually cause an unphysical rotation of the two orbitals with the minimization algorithm used up to now [38]. The effect is shown in Fig. 4.

The rotation leads to a collapse of the singlet energy to the triplet energy, i.e. the singlet-triplet gap is minimized by unphysical localization of the orbitals. Using a modification of the optimization algorithm we are meanwhile able to avoid this problem in a fully self-consistent calculation for small molecules like butadiene. In very recent calculations we were able to perform a Born-Oppenheimer MD simulation of the cis-trans isomerization of butadiene with the correct orbitals.

In future work we plan to achieve this also for the more complicated electronic structure of the rhodopsin chromophore. From first calculations it is evident that in this case the effect of the rotation is much smaller than in the case of butadiene (butadiene: ≈ 35 kcal/mol; rhodopsin: ≈ 1 kcal/mol). Nevertheless this difference might influence the reaction rate substantially. Before starting the excited-state MD simulation, it must be assured that during the simulation the proper orbital rotation is maintained. Furthermore, we want to combine the modification of the optimization algorithm with the more efficient Car-Parrinello MD scheme in order to reduce the computational cost.

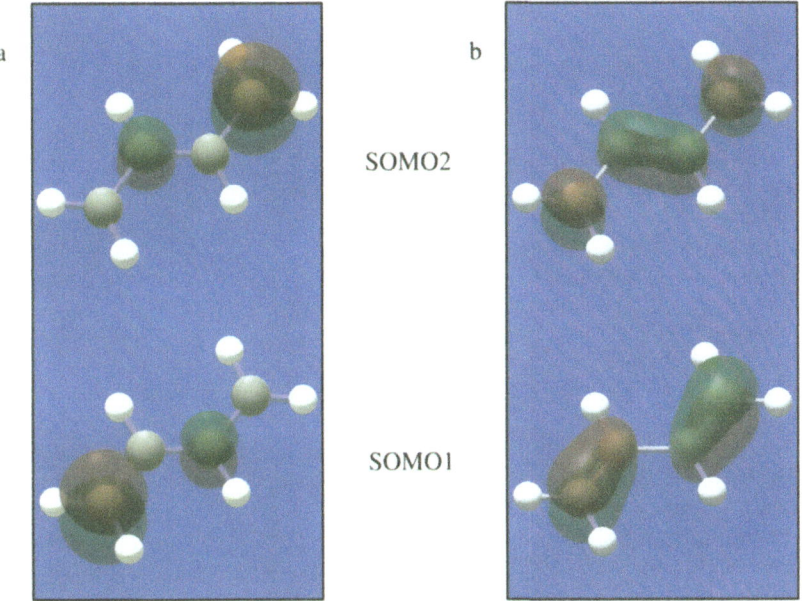

Fig. 4. Frontier orbitals of butadiene. a) unphysical localization. b) correct molecular orbitals.

4 Summary and Outlook

We have combined the restricted open-shell Kohn-Sham method with a QM/MM scheme, which brings very attractive applications into reach. One of the most interesting problems in this field is the photoisomerization of rhodopsin. We have developed a model system for rhodopsin in a membrane mimetic environment and have shown that it is stable without constraints in a classical MD simulation on a nanosecond timescale, and in a QM/MM simulation on a picosecond timescale. Our next aim is to simulate the photoreaction that initiates the process of vision.

References

1. Du, P., Davidson, E.R. (1990) J. Phys. Chem., **94**, 7013
2. Mathies, R.A., Lugtenburg, J. (2000) Handb. Biol. Phys. **3**, 55
3. Hellingwerf, K.J., Hoff, W.D., Crielaard, W. (1996) Molecular Microbiology **21**, 683
4. Birge, R.R. (1981) Ann. Rev. Biophys. Bioeng. **74**, 5669
5. Humphrey, W., Xu, D., Sheves, M., Schulten, K. (1995) J. Phys. Chem. **99**, 14549
6. Logunov, I., Schulten, K. (1996) J. Am. Chem. Soc. **118**, 9727

7. Gai, F., Hasson, K.C., Cooper Mc Donald, J., Anfinrud, P.A. (1998) Science **279**, 1886
8. Kolodner, P., Lukashev, E.P., Ching, Y.-C., Druzhko, A.B. (1997) Thin Solid Films **302**, 231
9. Schoenlein, R.W., Peteanu, L.A., Mathies, R.A., Shank, C.V. (1991) Science **254**, 412
10. Palczewski, K., Kumasaka, T., Hori, T., Behnke, C.A., Motoshima, H., Fox, B.A., Le Trong, I., Teller, D.C., Okada, T., Stenkamp, R.E., Yamamoto, M., Miyano, M. (2000) Science **289**, 739
11. Vreven, T., Bernardi, F., Garavelli, M., Olivucci, M., Robb, M.A., Schlegel, H.B. (1997) J. Am. Chem. Soc. **119**, 12687
12. Garavelli, M., Celani, P., Bernardi, F., Robb, M.A., Olivucci, M. (1997) J. Am. Chem. Soc, **119**, 6891
13. Yamamoto, S., Wasada, H., Kakitani, K. (1998) J. Mol. Struct. (THEOCHEM) **451**, 151
14. Ben-Nun, M., Molnar, F., Schulten, K., Martinez, T.J. (2002) PNAS **99**, 1769
15. De Vico, L., Page, C.S., Garavelli, M., Bernardi, F., Basosi, R., Olivucci, M. (2002) J. Am. Chem. Soc. **124**, 4124
16. Frank, I., Hutter, J., Marx, D., Parrinello, M. (1998) J. Chem. Phys. **108**, 4060
17. Filatov, M., Shaik, S. (1998) Chem. Phys. Lett. **288**, 689
18. Filatov, M., Shaik, S. (1999) J. Chem. Phys. **110**, 116
19. Car, R., Parrinello, M. (1985) Phys. Rev. Lett. **55**, 2471
20. Parrinello, M. (1997) Solid State Commun. **102**, 107
21. Marx, D., Hutter, J. (2000): Ab Initio Molecular Dynamics: Theory and Implementation. In: Grotendorst, J. (ed) Modern Methods and Algorithms of Quantum Chemistry. Forschungszentrum Jülich, NIC Series. **1**, 301
22. Hohenberg, P., Kohn, W. (1964) Phys. Rev. B **136**, 864
23. Kohn, W., Sham, L.J. (1965) Phys. Rev. A **140**, 1133
24. Parr, R.G., Yang, W. (1989): Density Functional Theory of Atoms and Molecules. Oxford University Press, Oxford
25. Dreizler, R.M., Gross, E.K.U. (1990): Density Functional Theory. Springer, Berlin
26. Becke, A.D. (1988) Phys. Rev. A **38**, 3098
27. Lee, C., Yang, W., Parr, R.G. (1988) Phys. Rev. B **37**, 785
28. Perdew, J.P., Zunger, A. (1981) Phys. Rev. B **23**, 5048
29. Curioni, A., Sprik, M., Andreoni, W., Schiffer, H., Hutter, J., Parrinello, M. (1997) J. Am. Chem. Soc. **119**, 7218
30. Frank, I., Parrinello, M., Klamt, A. (1998) J. Phys. Chem. **102**, 3614
31. Laio, A., VandeVondele, J., Rothlisberger, U. (2002) J. Chem. Phys. **116**, 6941
32. Laio, A., VandeVondele, J., Rothlisberger, U. (2002) J. Phys. Chem. B **106**, 7300
33. Molteni, C., Frank, I., Parrinello, M. (1999) J. Am. Chem. Soc. **121**, 12177
34. Molteni, C., Frank, I., Parrinello, M. (2001) Comput. Mater. Science **20**, 311
35. Zhong, Q., Jiang, Q., Moore, P.B., Newns, D.M., Klein, M.L. (1998) Biophys. J. **74**, 3
36. Guidoni, L., Torr, V., Carloni, P. (1999) Biochemistry **38**, 8599
37. Röhrig, U., Guidoni, L., Rothlisberger, U. (2002) Biochemistry, in press
38. Goedecker, S., Umrigar, C.J. (1997) Phys. Rev. A **55**, 1765

Simulation of Neuronal Map Formation in the Primary Visual Cortex

Joachim Noll, Oliver G. Wenisch, and J. Leo van Hemmen

Theoretical Biophysics
Physics Department, Technical University of Munich
85747 Garching, Germany
owenisch@ph.tum.de

1 Introduction

For nearly all animals the perception of moving objects is of vital importance to get alarmed by an approaching predator or for detecting suitable prey for the animal itself. With regard to neuronal coding of visual-motion signals, the primary visual cortex is the brain area which contains specialized neurons that selectively respond to certain features of a stimulus perceived by the photo sensors in the retina of the eye. In this project we have focused on the investigation of the emergence of direction-selective cell responses, i.e., the response properties of neurons that are activated only or at least most strongly whenever a luminance step within the visual input is moving in a certain characteristic direction over the cell's receptive field in the retina. Such a cell thus shows a response preference for this particular direction of stimulus movement. Another important response specificity of visual neurons is their orientation selectivity, because the activation of a cell is also dependent on the orientation of a luminance step within its receptive field.

Furthermore, the interaction of neurons with neighboring neurons within the cortical network gives rise to so-called "neuronal maps". In these maps particular stimulus features such as orientation and direction preference are "mapped" in a typical manner onto the cortical surface so that nearby neurons differ only slightly in the response preferences but locations further apart can have completely different response properties.

These visual mappings appear quite similar in higher animals — a fact that has led to the notion that these maps represent a key element in the neuronal coding of an animal's environment. However, the emergence of the structure of these maps and the mechanisms behind this pattern formation process are not yet fully understood. We have therefore tried to investigate the performance and consequences of a hypothesis for this process based on "spike time-dependent learning" [1]. Doing the simulations on the Hitachi SR8000 Supercomputer at the Leibniz-Rechenzentrum has allowed us to study a new neuronal algorithm and consider additional details of biological learning in neuronal networks as compared to previous simulation studies in this area.

2 Implementation of Neuronal Dynamics

To model the orientation and direction selective cells of the visual cortex in the context of an artificial cortical network, we have implemented a simplified but nevertheless biologically detailed numerical simulation of the neuronal circuits along the visual pathway and the visual cortex. Because even the network of the visual cortex itself consists of several millions of neurons and neurobiology is still far from completely identifying all cortical circuits, it is evident that even with today's supercomputers one has to make a crude approximation of this network. To this end, we have assumed an architecture of the cortex consisting of two 2-dimensional layers — one containing excitatory, the other inhibitory neurons — of typically 64 by 64 neurons each. In order to cut things short, we will not go into the details regarding pre-cortical modeling but assume for the moment that sensory signals can also be considered in our simulations. In the following sections we will describe a scenario of simulations where we focus completely on *intra*-cortical pattern formation processes without explicitly modeling external inputs to the cortex, cf. Fig. 1.

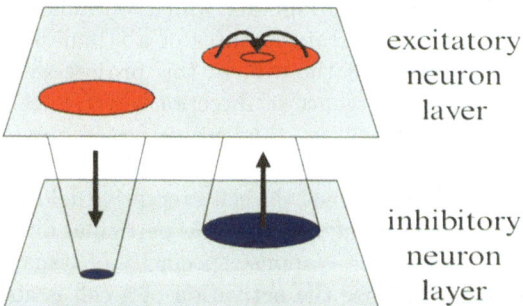

excitatory
neuron
layer

inhibitory
neuron
layer

Fig. 1. Two-layer architecture of the simulated neuronal network of a part of the visual cortex with recurrently interconnected excitatory and inhibitory neurons

Each neuron in the network layers is described by the *Spike-Response Model* [2]. This is a computationally simple but powerful neuron model (similar to the well-known *integrate-and-fire* type models) based on the fact that to excellent approximation a neuron's state is completely determined by its instantaneous membrane potential. The latter is the internal activation quantity where inputs from other neurons are summed. These inputs are modeled by first-order approximations [3] of cell responses to current injections into a non-linear conductance-based neuron description in the form of a so-called α-function; cf. Fig. 2. The inputs may add to or subtract from the momentary membrane potential according to whether the input neuron is excitatory or inhibitory. They are weighted by the corresponding synaptic strength of the synaptic connection between the input neuron and the receiving neuron. Based on its instantaneous membrane potential it is decided whether a neuron will emit a so-called action potential (*fire a spike*) or whether it stays "calm". When a spike is triggered in a real neuron a voltage pulse of about 1 ms duration and 0.1 V amplitude propagates along the "axonal tree" to synapses onto succeeding neurons. Thus inputs to more distant neurons are delayed by some milliseconds with respect to closer targets [4]. At a synapse a spike is transformed through the process of synaptic transmission into a post-synaptic membrane response $\epsilon(t)$, as shown in Fig. 2.

To model *spontaneous activity* within the network even in the absence of sensory inputs but, of course, with inputs from other brain areas that we do not want or even do not know how to model, we have assumed that spike emission is a stochastic process. It has been shown [5] that an inhomogeneous Poisson process is suitable for modeling cortical neurons taking the membrane potential as the time-varying spike probability density function. Another aspect implemented in our neuron model is refractoriness, i.e., the fact that a neuron shows a highly reduced tendency of spiking within a period of several milliseconds after the emission of an action potential. This is a means of limiting the firing rates of the neurons realized by subtracting a "refractory potential" $\eta(t)$ (see Fig. 2) from the neuron's membrane potential whenever it has fired during the last time step.

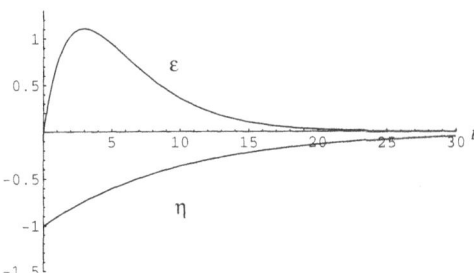

Fig. 2. Upper graph: post-synaptic response kernel in form of an α-function as it is produced by a pre-synaptic input spike. Lower graph: exponentially decaying negative refractory potential (scaled down by a factor of 20) following a spike event in the membrane potential to reduce probability of continued firing. Abscissa represents time after pre-/post-synaptic spike in milliseconds, whereas ordinate is given in arbitrary units of membrane voltage

Mathematically, this setup can be described by

$$V_i(t) = \sum_{j \in \left\{ \substack{\text{Inputs to} \\ \text{neuron } i} \right\}} \int_{-\infty}^{t} ds\ J_j(s)\epsilon(t-s)S_j(s) + \sum_{f \in \left\{ \substack{\text{Spikes of} \\ \text{neuron } i} \right\}} \eta(t - t_i^{(f)}).$$

Input spike trains $S_j(t) = \sum_{t_j^{(f)}} \delta(t - t_j^{(f)})$ to a particular neuron i, approximated in the simulation by sums of time-discretized δ-functions, come from a circular neighborhood of up to 7 grid positions distance in its own layer (i.e., excitatory or inhibitory) and from the under/overlaid layer of the opposite polarity (inhibitory/excitatory). In this way the network has a recurrent architecture, which takes into consideration the experimental finding that the visual cortex is also built up out of highly recurrent circuits. The spiking probability of neuron i in the interval $[t, t + \mathrm{d}t)$ is given by the sigmoidal activation function

$$P_i^f(t)\mathrm{d}t = \frac{\mathrm{d}t}{1 + \exp\left\{-\beta[V_i(t) - \theta]\right\}}$$

with β as a parameter controlling the steepness of activation and θ as threshold value adjusting the location of the steepest part of the activation function with respect to the membrane potential; see Fig. 3. A typical scenario of network activity that results from a setup as described above is shown in Fig. 4.

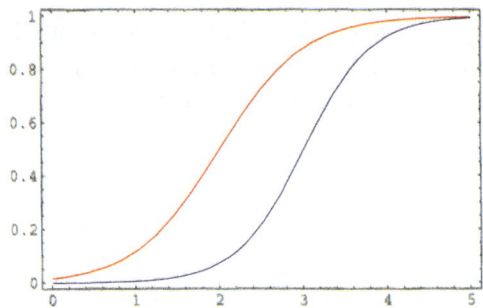

Fig. 3. Probability density function for spiking in dependence on the instantaneous membrane potential — the red curve stands for excitatory, the blue one is for inhibitory neurons

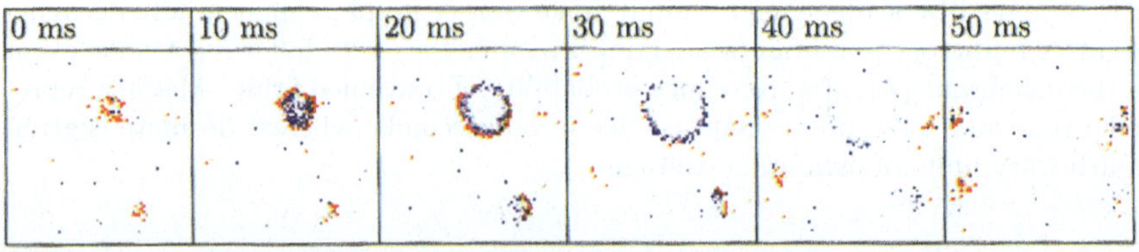

0 ms	10 ms	20 ms	30 ms	40 ms	50 ms

Fig. 4. Typical activity scenario as found for the described setup (see main text). Red dots represent excitatory spike events, blue dots code inhibitory spikes. Each frame shows the ongoing activity development of a 64 by 64 neuron cortex at 10 ms time intervals. The novelty of each spike event is visible by its corresponding color intensity. In the picture series a spontaneously emerging activity bubble can be seen which fades out as inhibition (blue) swaps excitation.

To make efficient use of the parallelization capabilities of the Hitachi SR8000 we split up the network simulation into a variable number of processes, each of which computed a vertical stripe of the 2-dimensional network. In this way, the computational and memory capacity of each node of the SR8000 is utilized in an optimal way for our purposes. Furthermore, because the lateral input range of a given neuron does not exceed one neighboring stripe width, we only have to communicate spike signals from the neurons within a given stripe to its directly neighboring stripes (and the corresponding Hitachi nodes) to the left and right and thus network traffic is reduced to a minimum. To minimize pollution of the network activities due to boundary effects we used periodic boundary conditions which do not conflict with the vertical striping. Coding itself was done in the C language and the MPI Message-Passing Library, which on the one hand offers the best portability possibilities to other architectures and on the other hand allows for specialized optimization according to the Hitachi processors.

3 Modeling Synaptic Dynamics

Up to now, we have only considered neuronal dynamics at a millisecond time scale but the synaptic coupling strengths by which inputs from other neurons influence a target neuron's membrane potential in varying magnitude have all been given a certain (constant) value. Synapses, however, are not static but, according to a result originating from Hebb's work [6], increase the coupling strength between two cells, if the activities of the corresponding pre- and post-synaptic cells are correlated, and decrease otherwise. Since the strengthening/weakening of synapses is a slow process on time scales of minutes to hours one speaks of long-term potentiation and depression (LTP/LTD).

Recently it has been predicted theoretically [1] and confirmed extensively by experiments (see for example [7, 8, 9]) that this is not the whole story but rather that the *exact* timing difference of the spiking activities of the pre- and post-synaptic neuron are important when trying to correctly describe the synaptic learning dynamics. It has turned out that synapses are potentiated when the pre-synaptic spike leads the post-synaptic spike by up to some ten milliseconds and that they are depressed when the pre-synaptic spike follows a post-synaptic one. In this way, synapses change most strongly only within a small time window centered around the post-synaptic spike so that only those synapses that contributed to the post-synaptic spike event by delivering an input *in time* get strengthened and those that sent their inputs too late get weakened. This time window (see Fig. 5) that governs spike time-dependent learning is therefore called the *learning window* and the resulting process spike timing-dependent plasticity (STDP). For computational reasons we approximated the learning window by an exponential function for the positive part and an α-function for the negative part.

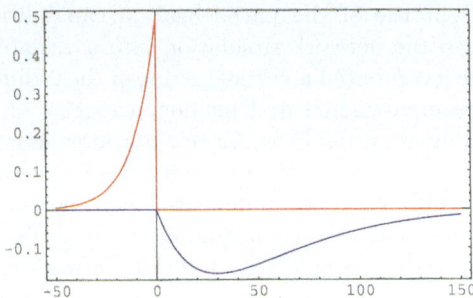

Fig. 5. Learning window according to which synapses between a pre- and a post-synaptic neuron are strengthened (red trace) or weakened (in blue) as determined by the time difference between pre- and post-synaptic spikes in milliseconds. This form of learning window corresponds to theoretical predictions [1] and recent experimental findings [9]

In this form the STDP learning rule can be implemented by two additional differential equations for each synapse: one representing a first-order leaky integrator of pre-synaptic input, the other representing a second-order leaky integrator of post-synaptic spikes. Consequently, input spikes that precede a post-synaptic one potentiate with exponentially decaying strength with regard to their time difference, whereas reverse spike order time differences yield α-function weighted depression. This implementation is highly efficient so that, in comparison to conventional simulations with spike-time unresolved learning rules, only slightly more computational power is required. In contrast to nearly all other simulations, we have considered a learning scenario where the network is left completely without any structured input by sensory signals and learning is driven by the cortex-intrinsic spontaneous activity modeled by the stochastic spiking of the neurons. In doing so, we have tried to explore whether (in a phase of neuronal development without structured input to the cortex, i.e., before eye opening or birth) maps of orientation and direction selectivity can emerge from a scenario of activity-dependent synaptic plasticity. We would like to stress, though, that in our opinion some kind of cooperation between intra-cortical interaction (as studied here) and feed-forward input originating from the retina is essential to map formation. To see how map formation comes about, we now started with the cortex.

In the present simulations only synapses between excitatory neurons were subject to this form of learning, all other synapses were kept constant up to corrections for overall excitation. Furthermore, synapses were always held within limits of minimal 0 and maximal 0.8 arbitrary units by correction of changes which reached across these limits. In the following figure we show exemplarily the map of the lateral synaptic coupling structures of a rectangle clipping of the neurons in the excitatory neuron layer.

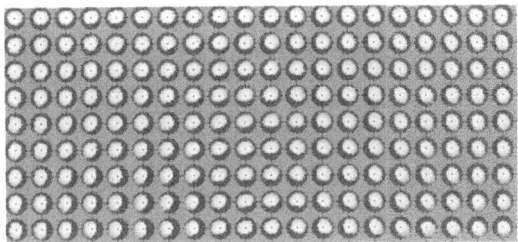

Fig. 6. Map of the lateral coupling structures between excitatory neurons after synaptic dynamics has converged. Each circular pattern represents the coupling strengths of the excitatory neuron at the corresponding cortex position to the excitatory neurons at neighboring positions. The strength of each synapse is coded in grey scales, where white refers to maximal and black to minimal strength. The directional asymmetry of the individual coupling patterns determines the direction preference of the respective neurons

Without going into the details of how to evaluate the selectivity and response preferences from a particular neuron's synaptic connection pattern, we show directly the color-coded distribution of response preferences for direction selectivity in Fig. 7. Starting from an isotropic lateral connectivity pattern for all neurons, the network changes its synaptic couplings driven by the network-inherent spontaneous activity as described above. It is obvious from the pictures that a certain structure, i.e., the *direction preference* map, gets more and more elaborated during the simulation of the learning dynamics, which typically lasts for 10 million time steps of a millisecond each. The evolving structure is similar to experimentally measured maps [10] and to the results of other simulations that do not take into account a learning rule as biologically detailed as we have used. With the simulations of this project we have therefore shown that the more detailed learning rule of spike timing-dependent synaptic plasticity is able to produce typical map structures in spiking neuronal networks, which is not at all evident ad hoc.

Furthermore, we could show that in this scenario the detailed form of the learning rule not only directly influences the development of selectivities of individual neurons but is also important with respect to the resulting map structure. With the simulations on the Hitachi SR8000 the dependence of the map structure on different parameters of the learning window could be studied. Because recurrent neural nets with this biological detail are practically intractable analytically the simulations allow a much deeper insight into the temporal evolution of this hotly discussed topic of neuronal coding.

Acknowledgement. The authors want to thank Armin Bartsch for providing the source code from his Spike-Response simulation program which significantly helped to speed up development time of the parallel simulation code.

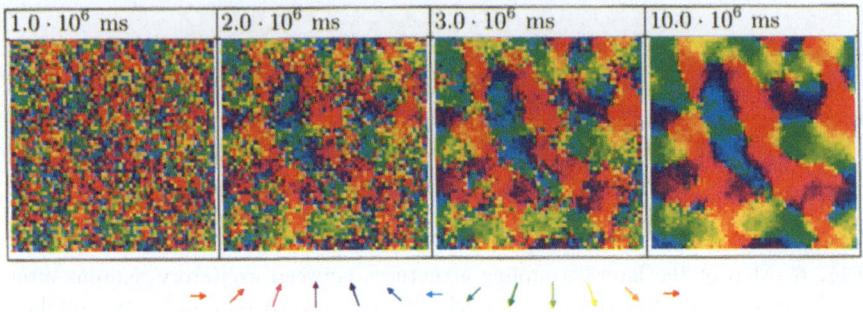

Fig. 7. Direction preference maps as computed after several time steps during the simulation of the map-formation process. From the colored arrows below the plots the direction preference of each neuron in the maps can be determined

References

1. W. Gerstner, R. Kempter, J. L. van Hemmen, and H. Wagner (1996): A Neuronal Learning Rule for Sub-Millisecond Temporal Coding. Nature, **383**, 76–81
2. W. Gerstner, and J. L. van Hemmen (1994): Coding and Information Processing in Neural Networks, in E. Domany, J. L. van Hemmen, and K. Schulten: *Models of Neural Networks II*, 1–93, New York. Springer
3. W. M. Kistler, W. Gerstner, and J. L. van Hemmen (1997): Reduction of the Hodgkin-Huxley Equations to a Single-Variable Threshold Model. Neural Computation, **9**, 1015–1045
4. V. Bringuier, F. Chavane, L. Glaeser, and Y. Fregnac (1999): Horizontal Propagation of Visual Activity in the Synaptic Integration Field of Area 17 Neurons. Science, **283**,695–699
5. T. W. Troyer, and K. D. Miller (1997): Physiological Gain Leads to High ISI Variability in a Simple Model of a Cortical Regular Spiking Cell. Neural Computation, **9**(5), 971–983
6. D. O. Hebb (1949): The Organization of Behavior. New York: Wiley.
7. H. Markram, and J. Luebke, and B. Sakmann (1997): Regulation of Synaptic Efficacy by Coincidence of Postsynaptic APs and EPSPs. Science, **275**:213–215
8. L. I. Zhang, H. W. Tao, C. E. Holt, W. A. Harris, and M.-M. Poo (1998): A Critical Window for Cooperation and Competition among Developing Retinotectal Synapses. Nature, **395**, 37–44
9. D. E. Feldman (2000): Timing-Based LTP and LTD at Vertical Inputs to Layer II/III Pyramidal Cells in Rat Barrel Cortex. Neuron, **27**, 45–56
10. M. Weliky, and W. H. Bosking, and D. Fitzpatrick (1996): A Systematic Map of Direction Preference in Primary Visual Cortex. Nature, **379**(6567), 725–728

Part IV

Chemistry

Walter Thiel

Max-Planck-Institut für Kohlenforschung
Kaiser-Wilhelm-Platz 1
45470 Mülheim an der Ruhr, Germany

Computational chemistry is a growing discipline. The combined advances in methodology, software, and hardware make ever larger parts of chemistry accessible to computation and simulation. Over the past decade, there has been much progress in the treatment of large complex systems, through methodological developments at different levels (ab initio approaches, density functional theory, and more approximate methods including hybrid quantum mechanical and molecular mechanical approaches), and most of the major codes have been parallelized to exploit modern hardware architectures. As a consequence of these advances, it has become quite common to combine experiment and computation in chemical research.

The computational demands in chemistry are essentially unlimited due to three factors. First, the diversity of chemistry: there are millions of different compounds that can be studied. Second, the quest for higher accuracy: there is a hierarchy of computational methods, ranging from classical force fields to highly correlated ab initio methods, that provide improved accuracy at often steeply rising costs. Third, the complexity of potential surfaces: with growing system size, there is a strong increase in the number of relevant structures that must be considered in optimization work, and in the simulation times that are needed in molecular dynamics studies to explore conformational space sufficiently.

While most of the everyday computational work in chemistry is done on local departmental hardware, the high-end projects require the use of high-performance computers. Traditionally, computational chemists have belonged to the major customers at most supercomputer centers. At HLRB the chemistry projects got off to a relatively slow start which was at least partly due to the limited availability of appropriate software on the Hitachi SR8000. This situation has improved in the meantime, and the chemistry projects are now taking a larger share of the available computation time (around 10 % in the first half of 2002). Currently, there are nine such projects, and several others in related fields that employ similar approaches. These projects involve a broad range of topics including homogeneous and heterogeneous catalysis, complexes and clusters of heavy elements, bioinorganic chemistry, and simulations of the first step of vision. As appropriate for a high-performance computing center, they are carried out at high and therefore computationally demanding levels (e.g., using Car-Parrinello molecular dynamics).

The current volume presents four contributions from chemistry. One of them comes from the Leibniz-Rechenzentrum and provides quantum-chemical benchmarks to assist potential users in the choice of the appropriate software, while the other three describe ongoing scientific projects. They address the structure and spectroscopy of models for enzyme cofactors (Neugebauer, Reiher, Hess), reaction pathways for oxidation on ruthenium dioxide surfaces (Seitsonen, Over), and the dynamics of vanadate complexes in aqueous solution (Bühl, Mauschick, Schurhammer). All of them bear a close relation to corresponding experimental studies and thus attest to the partnership between theory and experiment in solving complex chemical problems.

A User-Oriented Set of Quantum Chemical Benchmarks

Ludger Palm[1] and Frank Brechtefeld[2]

[1] Leibniz–Rechenzentrum der Bayerischen Akademie der Wissenschaften,
Barer Straße 21, D-80333 München
Ludger.Palm@lrz-muenchen.de
[2] Regionales Rechenzentrum Erlangen,
Martensstraße 1, D-91058 Erlangen
Frank.Brechtefeld@rrze.uni-erlangen.de

Users of our computing centres recurrently asked for comparisons of the machine performance for different quantum chemical programs. They are also interested in the scalability of parallel codes.

To answer these questions, we performed computations with Gaussian 98 [1], NWChem [2,3] and TURBOMOLE [4]. The Gaussian benchmarks were performed on different computers at the Leibniz Rechenzentrum, München, to compare the machines. The objective of benchmarking NWChem was mainly to find out how well it performs on a parallel relative to a single processor computer. The TURBOMOLE benchmark again compared several computers.

No effort is made to compare one program with an other. Every program had its own set of molecules. Although it would be nice to have a set of benchmarks comparing not only machines but also programs, we were guided by the chemical structures that our users are interested in.

1 Gaussian 98

Two sets of molecules have been used for the benchmarking of Gaussian 98: Firstly, a row of polyglycines with one to twelve glycine residues in the α–helical conformation was investigated. The structures have kindly been provided by Y.-D. Wu [5], Hongkong University of Science and Technology. Secondly, we used the set of alkane structures delivered with Gaussian as input for an exercise [6].

Due to their increasing size, the polyglycine structures were well suited for a systematic family of benchmarks. However, these structures are not minimized and they are by construction artificially constrained to the α–helix structure. As most users run geometry optimizations and then frequency computations, we wished to simulate these, too. Thus, we decided to use the alkane stuctures for optimizations and frequency calculations. These structures are near their minima and thus reflect everyday tasks better than the polyglycine structures.

Table 1 shows the range of chemical structures in the two benchmark sets.

Table 1. The chemical structures in the two Gaussian benchmark sets

benchmark set	stoichiometry	basis functions	electrons
polyglycines	C_2H_5NO–$C_{24}H_{38}N_{12}O_{12}$	70–796	32–362
alkanes	C_2H_6–$C_{10}H_{22}$	42–194	18–82

All calculations have been performed with the B3LYP [7, 8, 9] method as this is the method most often used by users asking for these benchmarks. The basis set was Pople's 6-31G* basis set. For the polyglycines, we also performed HF (Hartree–Fock) [10] and MP2 (Møller–Plesset perturbation theory to second order) [10] calculations with the same basis and again as single point computations, i.e., only for a single geometry, without geometry optimization.

The benchmarks have been run on the following computers at the Leibniz–Rechenzentrum: [1]

- SR8000: Hitachi SR8000-F1, one node with eight shared–memory processors,
- VPP: Fujitsu VPP700/52, one vector–processor used,
- Linux–1: Linux PC with one Pentium IV processor, 1500 MHz,
- Linux–4: Linux 4–fold shared–memory computer with four Pentium III processors of 700 MHz,
- IBM–1: IBM p690 with Power4 1.3 GHz CPUs, one processor used,
- IBM–4: IBM p690 with Power4 1.3 GHz CPUs, four shared–memory processors used.

It would have been interesting to use more than one node on the SR8000, but unfortunately, the mandatory parallelization tools are not available from Gaussian, Inc. Thus, the use of Gaussian 98 on the SR8000 is restricted to a single node.

1.1 Polyglycines

HF, B3LYP and MP2 computations were performed for the polyglycines as single point computations for the α–helical structures of Wu [5]. Table 2 shows the sum formulae, number of basis functions and number of electrons of the polyglycine structures.

---

[1] For a detailed description of the different computers please see http://www.lrz-muenchen.de/services/compute/hlr/.

Table 2. The stoichiometry, number of basis functions and number of electrons of the polyglycine structures

number of residues	stoichiometry	number of basis functions	number of electrons
1	C_2H_5NO	70	32
2	$C_4H_8N_2O_2$	136	62
3	$C_6H_{11}N_3O_3$	202	92
4	$C_8H_{14}N_4O_4$	268	122
5	$C_{10}H_{17}N_5O_5$	334	152
6	$C_{12}H_{20}N_6O_6$	400	182
7	$C_{14}H_{23}N_7O_7$	466	212
8	$C_{16}H_{26}N_8O_8$	532	242
9	$C_{18}H_{29}N_9O_9$	598	272
10	$C_{20}H_{32}N_{10}O_{10}$	664	302
11	$C_{22}H_{35}N_{11}O_{11}$	730	332
12	$C_{24}H_{38}N_{12}O_{12}$	796	362

Table 3. CPU times in seconds for the HF computations with Gaussian 98 on the polyglycine structures

residues	SR8000	VPP700	Linux–1	Linux–4	IBM–1	IBM–4
1	124	24	14	30	19	18
2	149	42	53	80	36	31
3	228	250	453	180	234	78
4	414	451	942	356	525	160
5	621	728	1677	611	962	285
6	930	1178	2700	1025	1660	487
7	1278	1595	3938	1539	2464	730
8	1717	2150	5332	2194	3371	1017
9	2195	2728	6964	3079	4500	1360
10	2782	3534	8889	4102	5814	1764
11	3436	4239	11057	5293	7225	2213
12	3976	5322	12797	6349	8349	2596

The starting point for quantum chemical computations for a long time was and still today often is a Hartree–Fock computation. Table 3 and Fig. 1 show the increase in cpu time for these computations as the structures get larger.

In Table 4 the cpu times for the B3LYP computations are collected. Figure 2 gives a graphical presentation of these numbers. Although the computation times are larger than that for the HF calculations, they are still in the same order of magnitude: the longest B3LYP computation lasts 17417 seconds on Linux–1 compared to 12797 seconds for the respective HF computation.

Table 4. CPU times in seconds for the B3LYP computations with Gaussian 98 on the polyglycine structures

residues	SR8000	VPP700	Linux–1	Linux–4	IBM–1	IBM–4
1	129	198	70	51	34	24
2	183	667	285	215	107	53
3	334	1560	913	364	380	113
4	573	2882	1973	768	855	241
5	837	4040	3084	1168	1449	404
6	1196	5788	4555	1754	2231	635
7	1613	8364	6215	2431	3223	919
8	2112	9162	8023	3317	4392	1274
9	2645	10835	10039	4304	5422	1665
10	3339	13100	12384	5456	7965	2074
11	4020	19354	14828	6760	9325	2514
12	4772	17135	17417	8228	10261	3147

Table 5. CPU times in seconds for the MP2 computations with Gaussian 98 on the polyglycine structures

residues	SR8000	VPP700	Linux–1	Linux–4	IBM–1	IBM–4
1	144	30	26	36	22	19
2	216	73	167	196	71	55
3	433	461	863	1641	501	249
4	1033	1648	2441	5952	1466	783
5	2456	2165	5806	12765	3224	1825
6	5262	4527	11857	20384	6433	3941
7	9353	8194	22028	49449	9991	6728

The cpu times for the MP2 computations are collected in Table 5 and displayed in Fig. 3. These computations have only been performed up to seven glycine residues as the computation times rise very quickly. While for a B3LYP calculation on seven glycine residues 6215 seconds are needed on Linux–1, the MP2 calculation needs 22028 on the same computer.

Figures 1, 2, and 3 show the well-known fact that the computation time rises dramatically as the system size increases. Single processor PCs soon come to a point where they must be run for weeks. Parallelization opens up new horizons for the treatment of much larger systems. The 4-fold shared memory Linux PC (Linux–4) performs very well for the HF and B3LYP computations. The SR8000 and the now new IBM Power4 computer (IBM–1 and IBM–4) perform even much better. In addition, these computers have bigger and faster caches, memories, and disks and can run the 64-bit version of Gaussian 98 to utilize them.

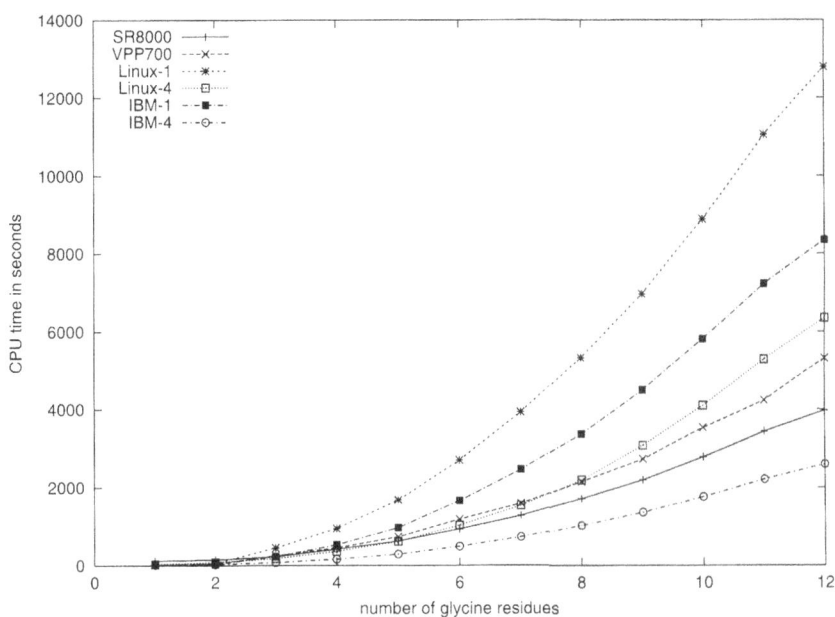

Fig. 1. CPU times in seconds for the HF computations with Gaussian 98 on the polyglycine structures

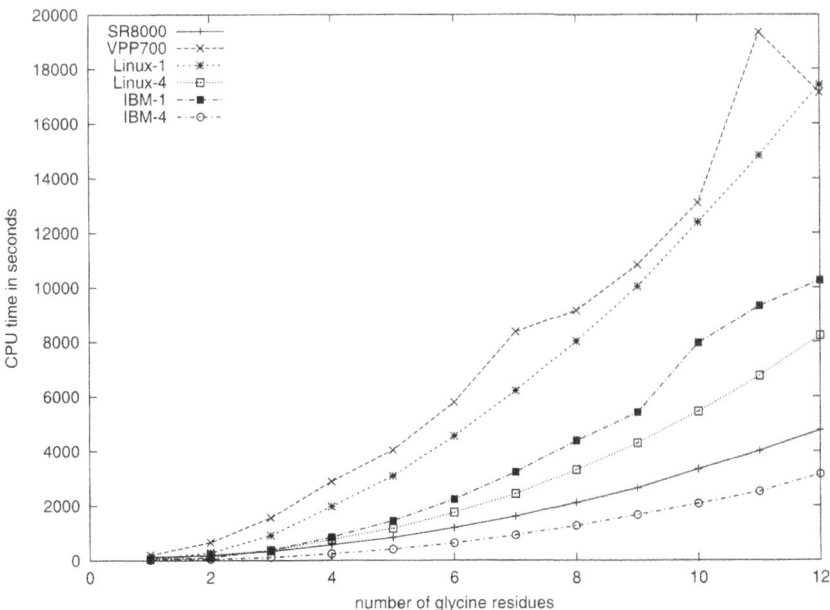

Fig. 2. CPU times in seconds for the B3LYP computations with Gaussian 98 on the polyglycine structures

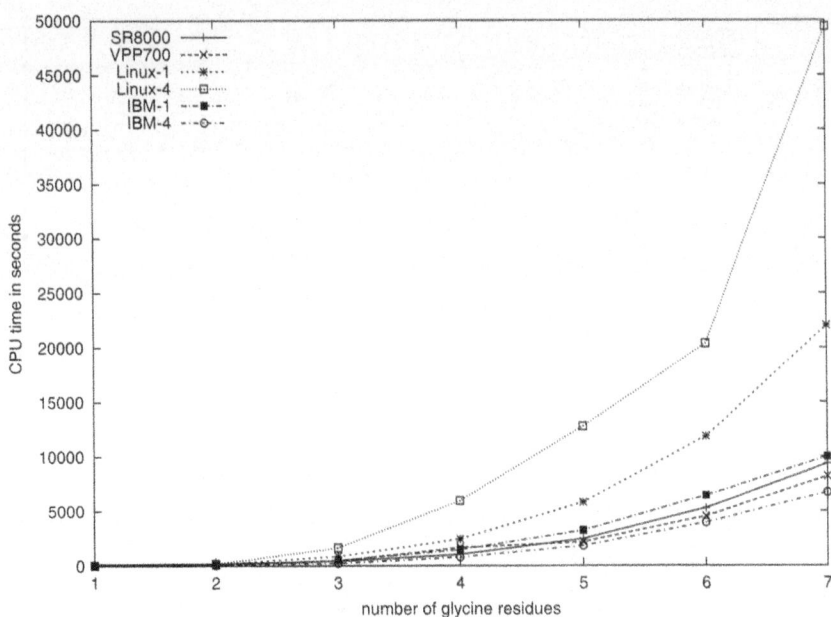

Fig. 3. CPU times in seconds for the MP2 computations with Gaussian 98 on the polyglycine structures

Figure 4 displays the speedup on the IBM p690 when four processors are used instead of one. The speedup was computed by the formula

$$speedup = \frac{\text{computation time for one processor}}{\text{computation time for four processors}}.$$

While the parallelization of MP2 is unsatisfactory, the speedup of the HF and B3LYP code is very good.

From Fig. 2 we conclude that the Fujitsu VPP700 vector processor is not well suited for B3LYP computations. It is performing even worse than a single Pentium PC and can not be recommended for B3LYP computations. Single processor Linux PCs are capable to perform HF and B3LYP computations even on molecular structures built up of about fifty non-hydrogen atoms containing more than 360 electrons and using about eight hundred basis functions. The scaling of Gaussian 98 in parallel HF and B3LYP computations on four-processor computers is excellent: even for a moderatly small system with two hundred basis functions and one hundred electrons, the speedup of 3.3 is near the ideal value of four.

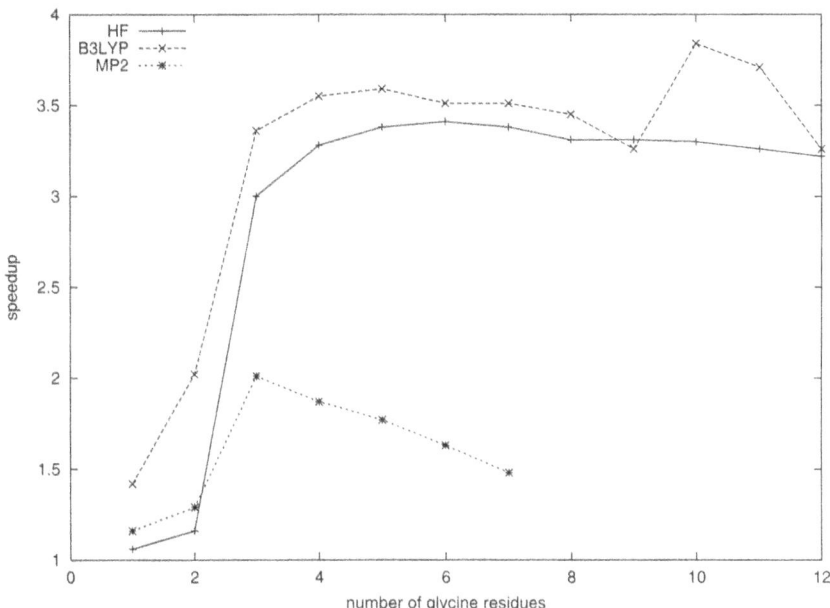

Fig. 4. Relative CPU times of four processors on the IBM p690 compared to the CPU times of a single processor with Gaussian 98 on the polyglycine structures

1.2 Alkanes

The alkane structures are near their minimum energy geometries. Optimizing these geometries and computing the frequencies resembles very much the usual task of todays computational chemist.

The alkane benchmarks have been performed on the same computers as the polyglycine benchmark except that the Fujitsu VPP700 was not used, as only the B3LYP method as implemented in Gaussian 98 was used for which the VPP shows very bad performance. Table 6 shows the sum formulae, number of basis functions and number of electrons of the alkane structures. Methan (n=1) was excluded from the benchmark set due to its special symmetry. Table 7 and Fig. 5 present the results of the benchmark computations.

Figure 6 displays the speedup on the IBM p690 when four processors are used instead of one. Again, as with the polyglycines (see Fig. 4), the speedup for the B3LYP computations with Gaussian 98 on the IBM p690 is near the ideal value of four.

Gaussian 98 allows the user to specifiy the amount of memory to be used by the program. We found that supplying too much memory may slow down the execution speed dramatically: The alkane benchmark set has been run with 100 MB (%Mem=100MB) and with 950 MB (%Mem=950MB) memory specified. Figure 7 shows the speedup if less memory is used:

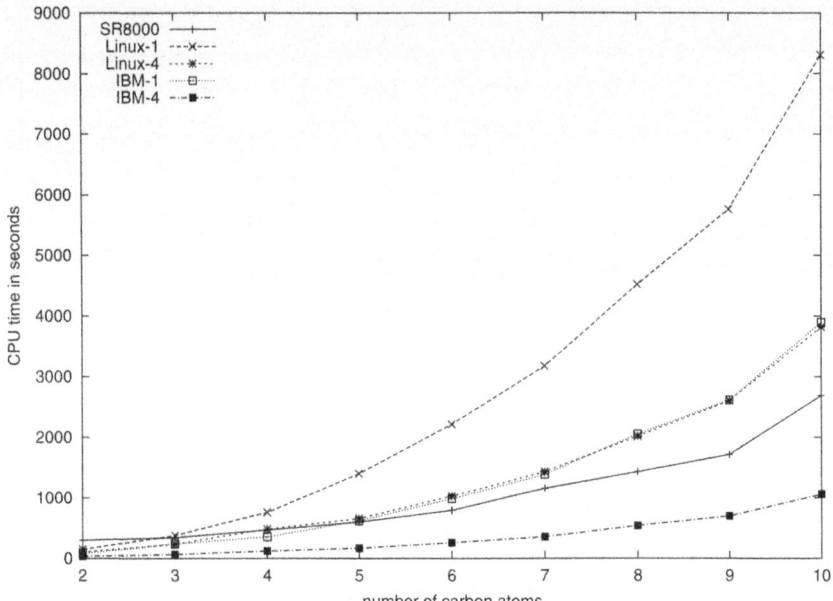

Fig. 5. CPU times in seconds for the B3LYP computations with Gaussian 98 on the alkane structures

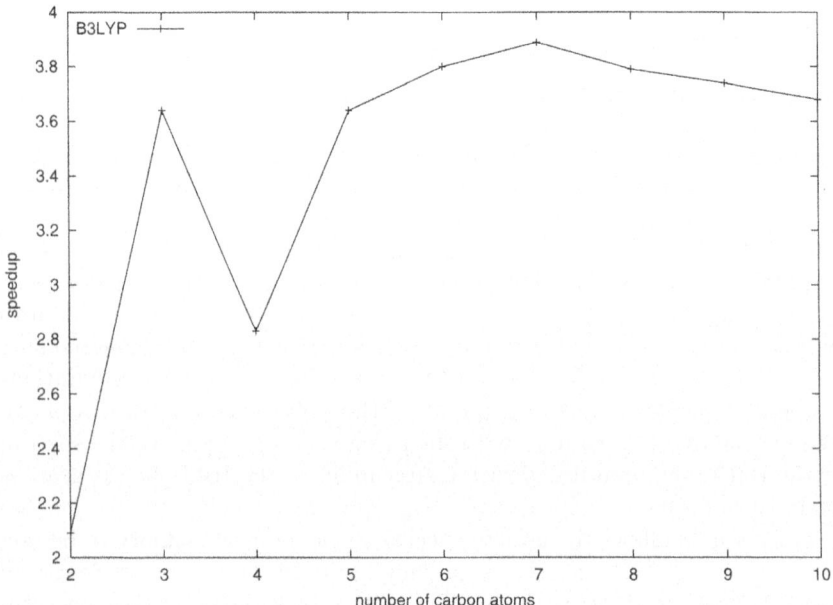

Fig. 6. Relative CPU times of four processors on the IBM p690 compared to the CPU times of a single proceesor with Gaussian 98 on the alkane structures

Table 6. The stoichiometry, number of basis functions and number of electrons of the alkane structures

number of carbons	stoichiometry	number of basis functions	number of electrons
2	C_2H_6	42	18
3	C_3H_8	61	26
4	C_4H_{10}	80	34
5	C_5H_{12}	99	42
6	C_6H_{14}	118	50
7	C_7H_{16}	137	58
8	C_8H_{18}	156	66
9	C_9H_{20}	175	74
10	$C_{10}H_{22}$	194	82

Table 7. CPU times in seconds for the B3LYP computations with Gaussian 98 on the alkane structures

carbons	SR8000	Linux–1	Linux–4	IBM–1	IBM–4
2	299	149	108	78	37
3	342	377	232	238	65
4	468	759	484	353	124
5	601	1402	659	621	170
6	793	2216	1029	987	259
7	1158	3176	1433	1388	356
8	1431	4518	2016	2052	540
9	1717	5761	2603	2616	699
10	2691	8309	3820	3898	1059

$$speedup = \frac{\text{computation time with 100 MB memory}}{\text{computation time with 950 MB memory}}.$$

In the case of the parallel computers (SR8000, Linux–4 and IBM–4) and small alkanes reducing the memory specified may double the execution speed! We suppose that it is better to rely on the operating system's memory management capabilities than to let the program or compiler decide how to store the data. Specifying 100 MB memory forces Gaussian to use disk storage. This disk storage, however, may be buffered in the additional computer memory, instead of writing to and reading from the disk physically. Anyway, we conclude that is worth adjusting ressource requests to find the optimal settings for your environment if you are going to compute many molecular structures of the same size and kind with the same method.

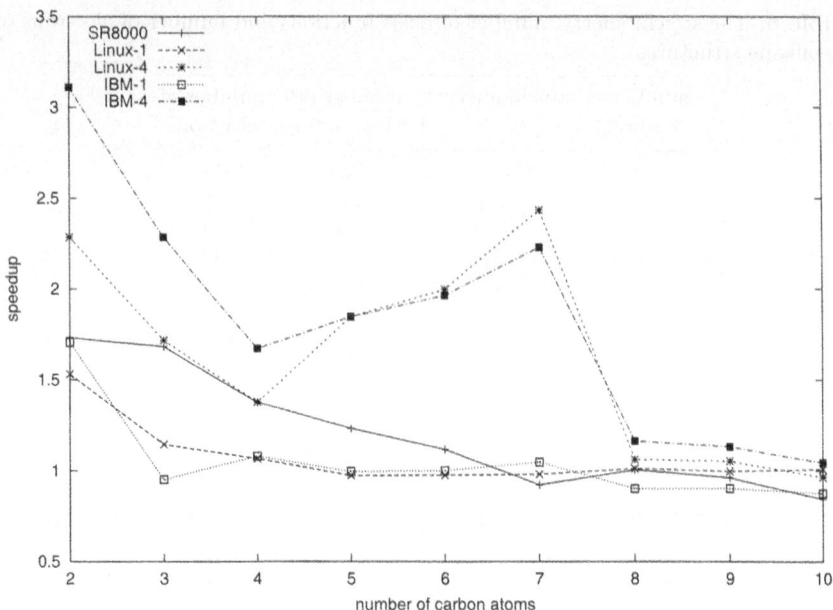

Fig. 7. Relative CPU times for different memory specifications with Gaussian 98 on the alkane structures (see text for details)

2 NWChem

NWChem [2, 3] has been developed especially for massively parallel computers. We were interested in finding out how well NWChem scales on the parallel Linux cluster and on the IBM p690. The benchmarks were performed with Version 4.0.1 of NWChem.

The following computers at the Leibniz–Rechenzentrum were used: [2]

Linux–1: Linux PC with one Pentium IV processor, 1500 MHz
Linux–8: A Myrinet cluster of four double Pentium III, 800 MHz, Linux computers, using Global Arrays over MPI
IBM–1: IBM p690 with Power4 1.3 GHz CPUs, one processor in use
IBM–4: IBM p690 with Power4 1.3 GHz CPUs, four shared–memory processors in use

The molecular structure , provided by Prof. B.A. Hess, Univ. Erlangen, has the stoichiometry $C_{12}H_9N_3FeF_{12}S_4$ and 300 electrons. The basis set contained 332 contracted basis functions. Single point density functional computations with the BP86 [8, 11] density functional were run. NWChem allows to fit the charge density in the two electron integrals by a linear combination of

---

[2] For a detailed description of the different computers please see http://www.lrz-muenchen.de/services/compute/hlr/.

basis functions. The computation time using this charge density fit (cdfit) was compared to the computation time without charge density fit (no-cdfit). The computations may be performed either by storing integrals on disk (nodirect) or using a direct algorithm, thus avoiding the storage of integrals.

Table 8 shows the cpu and wall clock times of the different runs, in addition the wall clock time per iteration is given as this allows better comparison among the different methods and computers.

Table 8. CPU and wall clock times in seconds for the BP86 computations with NWChem (for details, see text)

	Linux–1			
method	cpu time	wall clock time	iterations	wall clock time per iteration
no-cdfit-direct	125179.6	125353.7	18	6964.1
cdfit-direct	25894.4	25988.1	16	1624.3
cdfit-nodirect	8365.5	10401.9	14	743.0
	Linux–8			
method	cpu time	wall clock time	iterations	wall clock time per iteration
no-cdfit-direct	38390.8	38651.6	18	2147.3
cdfit-direct	4472.2	4589.1	16	286.8
cdfit-nodirect	4008.5	4135.3	16	258.5
	IBM–1			
method	cpu time	wall clock time	iterations	wall clock time per iteration
no-cdfit-direct	39787.2	40034.8	18	2224.1
cdfit-direct	10677.8	12571.8	16	785.7
cdfit-nodirect	10158.1	10320.3	16	645.0
	IBM–4			
method	cpu time	wall clock time	iterations	wall clock time per iteration
no-cdfit-direct	9757.0	9815.3	18	545.2
cdfit-direct	3026.8	3085.0	16	192.8
cdfit-nodirect	2865.8	2920.2	16	182.5

Several conclusions may be drawn from these results: First, we see a remarkable acceleration if the charge density fit is used: while without the charge density fit the computation takes 6964.1 seconds on the Linux–1 or

2147.3 on the Linux–8, it takes only 743.0 or 258.5 with charge density fit. Next, we see an excellent scaling for parallel execution: the speedup on four processors on the IBM–4 is near four relative to the IBM–1; also, the eight 800 MHz Pentium processors needing 258.5 seconds compare well with a single 1500 MHz Pentium processor, that needs 743.0 seconds for the non-direct computations with charge density fit. The combination of both methods leeds to a dramatic reduction in computer time: while the BP86 computation without charge density fit on the single processor 1500 MHz Linux–1 took 6964.1 seconds per iteration, the eight 800 MHz processors of Linux–8 take 258.5 seconds when charge density fit is exploited. Thus, a 27-fold acceleration is achieved at an affordable price.

In conclusion, NWChem seems to be well suited for the parallel computation of systems like $C_{12}H_9N_3FeF_{12}S_4$ with density funtional methods like BP86 on Linux clusters as well as the IBM p690.

3 Turbomole

The wall clock times of different TURBOMOLE benchmarks have been measured. The serial programs `ridft` (s. Table 9) and `dscf` (s. Table 10) and the parallel program `dscf_mpi` (s. Table 12) were used. The computers used are described in Table 11. For the MPI parallelized program `dscf_mpi` `$numprocs` denotes the number of slaves. The number of processors is given by `$numprocs`+1. Figure 10 displays the scaling of the saal-bench-hf benchmark.

For the fe5f benchmark series the molecule $Fe(N_2)'N_HS_{4CF3}'$ with $'N_HS_{4CF3}' - H_2 = 2,2'$-bis(2-mercapto-1,2-bis-trifluormethylethenyl-thio)-diethylamine is used (see Fig. 8). 688 SCF basis functions, composed from 722 cartesian basis functions are used to describe the molecular structure of the 300 electrons. The molecule $Fe_3O_7N_{33}C_{33}H_{27}$ is used for the saal benchmarks (see figure 9). The wavefunction is constructed from 1148 SCF basis functions (1227 cartesian basis functions) to describe 590 electrons. For further informations see [12].

Table 9. Program `ridft`

Benchmark	Computer	TM Version	run time
fe5f-ctt	ibmsmp	5.1	22m
	tcda	5.1	28m
	Pentium4 1.7GHz	5.1	28m
	mssgi1	5.1	34m
	Pentium4 1.5GHz	5.1	34m
	UIII 750MHz	5.1	43m
	SR8000	5.1	55m

Table 10. Program `dscf`

Benchmark	Computer	TM Version	run time
fe5f-86ct	mssgi1	5.1	329m
	tcdx01	5.1	237m44s
	tcdx01	5.4	482m58s
	tcda12	5.4	346m18s
fe5f-b3ct	mssgi1	5.1	419m
	tcda25	5.1	259m02s
	tcdx01	5.1	347m20s
	tcda12	5.4	414m19s

Table 11. Information on the computers

Computer	Operator	Description
`mssgi1`	RRZE	SGI Origin 3400, 28 CPUs, 56GB Memory, MIPS R14000 Processor with 500 MHz
`romulus`	TU Dresden	SGI Origin 3800, 128 CPUs, 64GB Memory, MIPS R12000 Processor with 400 MHz
`tcda`	Prof. Hess, Univ. Erlangen	Cluster with 43 double processor Athlon 1800+ machines, connected by fast ethernet and another double processor Athlon 1800+ file server with raid. Each double processor has 1GB memory. The configuration was purchased for less than 100000 EUR in December 2001.
`ibmsmp`	LRZ München	IBM p690, 8 CPUs, 32GB Memory, Power4 Processor with 1.3 GHz
`psi`	Max-Planck-Gesellschaft	The new system includes six 32-way Regatta compute nodes (eServer p690), equipped with 1.3 GHz Power 4 processors, with a peak performance of 166 GFlop/s and 96 GB of main memory per node, adding up to an aggregated performance of 1 TFlop/s and 1/2 TB of main memory. Another 16-way Regatta node will be dedicated to fileserving. See also `http://www.rzg.mpg.de/`.

A comparison of the measured run times shows that TURBOMOLE performs especially well on the `ibmsmp`. Also, the scaling with an increased number of processors on the `psi`, another IBM p690, is very good: the run time drops from 101 minutes with four processors to 44 minutes with eight processors, 26 minutes for 16 processors down to 17 minutes for 32 processors.

The TURBOMOLE benchmark studies were done in collaboration with Prof. B.A. Hess, Univ. Erlangen. The authors thank B.A. Hess for his help.

Table 12. Program `dscf_mpi`

Benchmark	Computer	TM Version	$numprocs	run time
fe5f-cttscfmpi	mssgi1	5.1	16	50m18s
	mssgi1	5.1	12	59m
	mssgi1	5.1	8	76m
	mssgi1	5.1	4	136m
	mssgi1	5.1	3	178m
	mssgi1	5.4	3	198m
	ibmsmp	5.4	3	89m
	tcda03	5.4	8	75m
	tcda01	5.4	16	43m53s
	psi	5.4	3	101m
	psi	5.4	7	44m
	psi	5.4	15	26m
	psi	5.4	31	17m
saal-bench-hf	mssgi1	5.1	3	45933s
	mssgi1	5.1	8	18154s
	romulus	5.1	16	11975s
	romulus	5.1	24	8557s
	romulus	5.1	32	6968s
	romulus	5.1	40	5832s
	tcda	5.1	2	61581s
	tcda	5.1	4	32212s
	tcda	5.1	8	17515s
	tcda	5.1	16	10379s
	tcda	5.1	24	8274s
	tcda	5.1	32	7022s
	tcda	5.1	40	6307s
saal-bench-bp86	mssgi1	5.1	4	19002s
	romulus	5.1	7	13612s
	mssgi1	5.1	8	10385s
	romulus	5.1	15	7125s
	mssgi1	5.1	15	6438s
	romulus	5.1	31	4156s
	romulus	5.1	43	3366s
	tcda	5.1	4	19997s
	tcda	5.1	8	10535s
	tcda	5.1	16	6685s

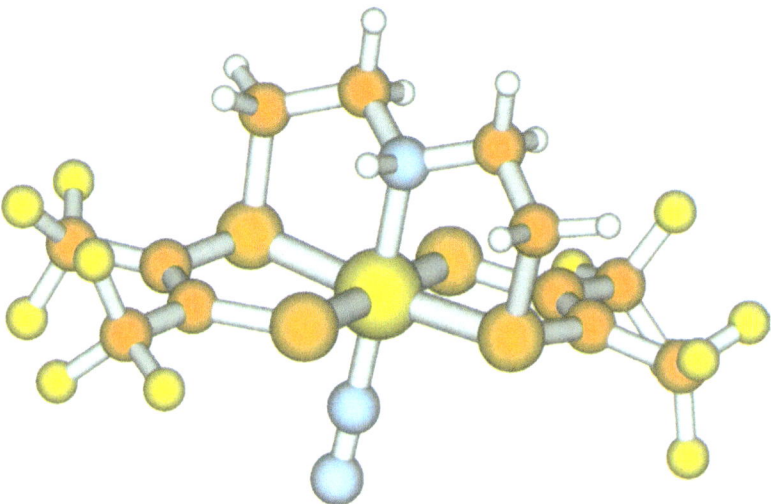

Fig. 8. Molecule used for the fe5f benchmarks

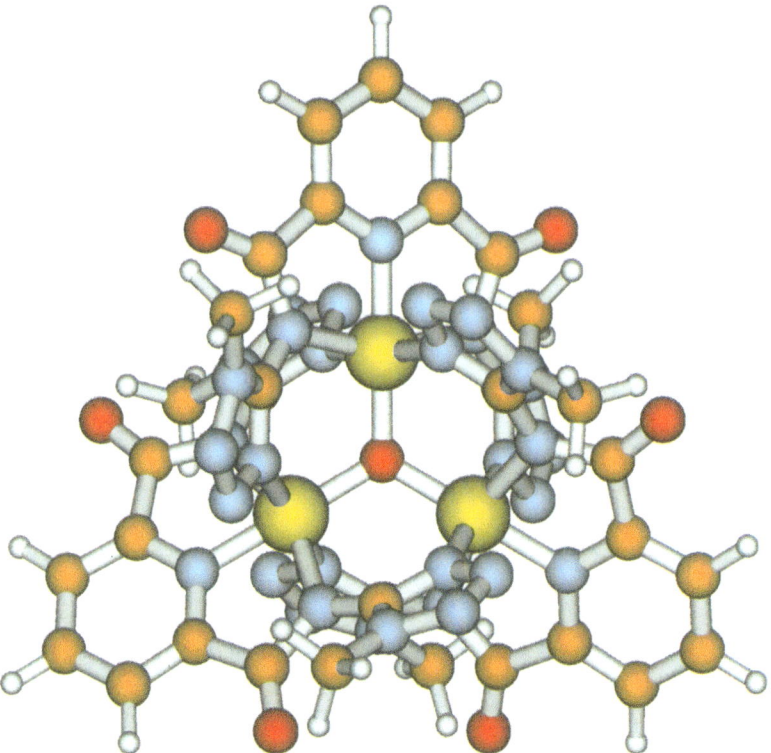

Fig. 9. Molecule used for the saal benchmarks

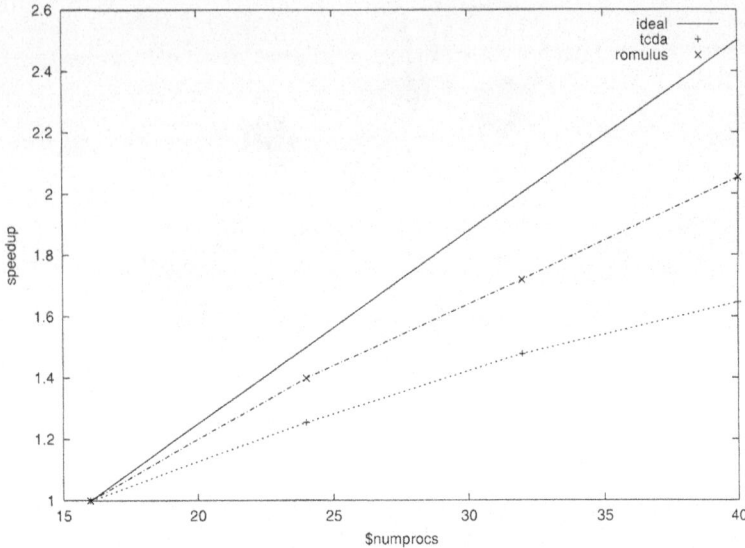

Fig. 10. Scaling of the saal-bench-hf benchmark on **tcda** and **romulus**

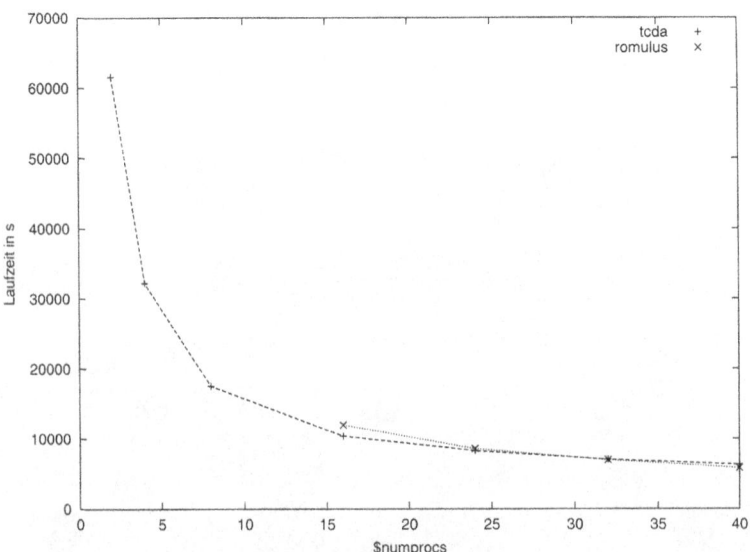

Fig. 11. Run time of the benchmark saal-bench-hf on **tcda** and **romulus**

References

1. Gaussian 98, Revision A.11, M. J. Frisch, G. W. Trucks, H. B. Schlegel, G. E. Scuseria, M. A. Robb, J. R. Cheeseman, V. G. Zakrzewski, J. A. Montgomery, Jr., R. E. Stratmann, J. C. Burant, S. Dapprich, J. M. Millam, A. D. Daniels, K. N. Kudin, M. C. Strain, O. Farkas, J. Tomasi, V. Barone, M. Cossi, R. Cammi, B. Mennucci, C. Pomelli, C. Adamo, S. Clifford, J. Ochterski, G. A. Petersson, P. Y. Ayala, Q. Cui, K. Morokuma, P. Salvador, J. J. Dannenberg, D. K. Malick, A. D. Rabuck, K. Raghavachari, J. B. Foresman, J. Cioslowski, J. V. Ortiz, A. G. Baboul, B. B. Stefanov, G. Liu, A. Liashenko, P. Piskorz, I. Komaromi, R. Gomperts, R. L. Martin, D. J. Fox, T. Keith, M. A. Al-Laham, C. Y. Peng, A. Nanayakkara, M. Challacombe, P. M. W. Gill, B. Johnson, W. Chen, M. W. Wong, J. L. Andres, C. Gonzalez, M. Head-Gordon, E. S. Replogle, and J. A. Pople, (2001): Gaussian, Inc., Pittsburgh PA
2. D. E. Bernholdt, E. Apra, H. A. Fruchtl, M.F. Guest, R. J. Harrison, R. A. Kendall, R. A. Kutteh, X. Long, J. B. Nicholas, J. A. Nichols, H. L. Taylor, A. T. Wong, G. I. Fann, R. J. Littlefield and J. Nieplocha (1995): Parallel Computational Chemistry Made Easier: The Development of NWChem. Int. J. Quantum Chem. Symposium **29**, 475-483
3. R. J. Harrison, J. A. Nichols, T. P. Straatsma, M. Dupuis, E. J. Bylaska, G. I. Fann, T. L. Windus, E. Apra, J. Anchell, D. Bernholdt, P. Borowski, T. Clark, D. Clerc, H. Dachsel, B. de Jong, M. Deegan, K. Dyall, D. Elwood, H. Fruchtl, E. Glendenning, M. Gutowski, A. Hess, J. Jaffe, B. Johnson, J. Ju, R. Kendall, R. Kobayashi, R, Kutteh, Z. lin, R. Littlefield, X. Long, B. Meng, J. Nieplocha, S. Niu, M. Rosing, G. Sandrone, M. Stave, H. Taylor, G. Thomas, J. van Lenthe, K. Wolinski, A. Wong, and Z. Zhang (2001): NWChem, A Computational Chemistry Package for Parallel Computers, Version 4.0.1. Pacific Northwest National Laboratory, Richland, Washington 99352-0999, USA
4. Ahlrichs, R. (2002): TURBOMOLE: Quantum Chemistry Group, University of Karlsruhe, Germany
5. Wu, Y.-D., Zhao, Y.L. (2001): A Theoretical Study on the Origin of Cooperativity in the Formation of $3_{10}-$ and $\alpha-$Helices. J. Am. Chem. Soc., **123**, 5313–5319
6. Foresman, J.B., Frisch, Æ. (1996): Exploring Chemistry with Electronic Structure Methods. Gaussian, Inc., Pittsburgh, 2nd ed;
   (Gaussian tree /explore/exercise/2.07.com)
7. Becke, A.D. (1993): Density-functional thermochemistry. III. The role of exact exchange. J. Chem. Phys. **98**, 5648–5652
8. Becke, A.D. (1988): Density-functional exchange-energy approximation with correct asymptotic behavior. Phys. Rev. A **38**, 3098–3100
9. Lee, C., Yang, W., Parr, R.G. (1988): Development of the Colle-Salvetti correlation-energy formula into a functional of the electron density. Phys. Rev. B **37**, 785–789
10. Hehre, W.J., Radom, L., Pople, J.A., Schleyer, P.v.R. (1986): Ab Initio Molecular Orbital Theory. Wiley, New York
11. Perdew, J.P. (1986): Density-funtional approximation for the correlation energy of the inhomogeneous electron gas. Phys. Rev. B **33**, 8822–8824
12. M. Reiher, O. Salomon, D. Sellmann, B.A. Hess (2001): Dinuclear Diazene Iron and Ruthenium Complexes as Models for Studying Nitrogenase Activity. Chem. Eur. J. **7**, 5195–5202

Structure, Energetics, and Spectroscopy of Models for Enzyme Cofactors

Johannes Neugebauer, Markus Reiher, and Bernd A. Hess

Lehrstuhl für Theoretische Chemie
Universität Erlangen-Nürnberg
Egerlandstraße 3
91058 Erlangen, Germany
{Johannes.Neugebauer, Markus.Reiher, Hess}@chemie.uni-erlangen.de

1 Introduction

In this work we describe the determination of structure, energetics, and spectroscopy of metal complexes which are designed to emulate the action of enzyme cofactors by methods of theoretical chemistry in combination with techniques of high performance computing. Our study focuses on biomimetic metal complexes related to nitrogenase activity, which are characterized by two metal centres (typically Fe and Ru) in a sulfur-rich first coordination sphere.

Nitrogenase acts as a catalyst in the conversion of molecular nitrogen to ammonia, which is one of the fundamental reactions of nature. To date, it has not yet been possible to determine the mechanism of this reaction or to emulate the reaction by means of a similarly effective catalytic system in the laboratory if ambient conditions without presence of strongly reducing reactants are the boundary conditions. The latter goal is without doubt one of the grand challenges of contemporary chemistry.

In this article, we present results for the structure and vibrational spectra of the potential intermediates of the biomimetic reduction process modelled by dinuclear Fe(II) complexes, which have been experimentally characterized in great detail by D. Sellmann and co-workers [1, 2]. In particular, we have developed a numerical approach for the calculation of infrared and Raman intensities, which requires computational resources well beyond the scope of traditional computational equipment. Massive parallelization is employed in these calculations, making efficient use of the capabilities of high performance computing.

2 Aspects of high performance computing

In order to obtain vibrational spectra, we use an ansatz in which analytically determined gradients of the electronic energy, dipole moments, and components of the polarizability tensor are numerically differentiated with respect to nuclear Cartesian coordinates. The derivatives of the gradients lead to vibrational frequencies and normal modes of vibration by diagonalization of the mass-weighted second derivatives matrix, while derivatives of dipole moments and polarizabilities yield the infrared and Raman intensities, respectively, when transformed into the basis of the normal coordinates of the molecule. Although these derivatives are available in analytical form for some quantum chemical methods, the numerical approach has some striking features, a part of which will be discussed in this section.

All techniques described here are implemented in our program package SNF [3, 4], which sets up a hierarchical structure of programs as depicted in Fig. 1: A superordinate program (SNF) determines the displacements of the equilibrium structure and carries out the data evaluation when all calculations have been finished by the data collector SNFDC. The data collector SNFDC executes and controls all necessary single-point calculations by taking advantage of coarse-grained parallelization. Standard quantum chemistry program packages — in the present version, these are TURBOMOLE [5] and DALTON [6] — are used for the calculation of the raw data, i.e., for the calculation of the analytic gradient of the electronic energy and the (static and/or dynamic) polarizability for a given displacement structure. This is advantageous since it allows to extend our program easily to new methods by simply creating an additional interface. SNF takes full advantage of the molecular point group (Abelian and non-Abelian point groups are supported).

2.1 Parallelization

One major advantage of the numerical differentiation process is that this computational task is a well suitable for coarse-grained, data-decomposition parallel computing: For each of the $(n-1) \cdot 3N$ distorted molecular geometries, which are generated for the data collection for the numerical differentiation according to n-point central difference formulae (N is the number of atoms), a single-point calculation of the electronic energy, its gradient, the dipole moment, and the polarizability tensor has to be performed.

In our implementation, we use the following master–slave concept: The master process determines the parameters for these single point calculations, sets up the slave processes and copies only the basic input files, which are necessary to execute the single point calculations, to the nodes on which the slave processes are spawned. Depending on the type of machine the program is running on, this copy is achieved either through the MPI/PVM interface, or via remote copy. When a slave has finished its calculation, only the basic results necessary for the calculation of the vibrational spectra are copied back to the master process. This means that there is only very little communication between the master and the slave processes. There is no communication between the slaves at all so that our implementation is very efficient even on low-speed networks.

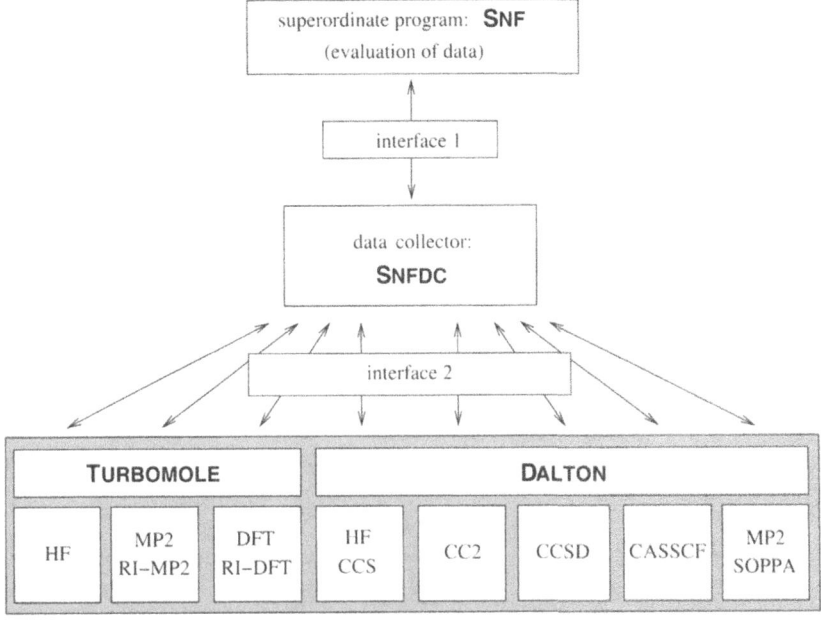

Fig. 1. Hierarchical structure of programs for the calculation of vibrational spectra.

The most satisfactory library to implement message passing between the master and the slaves is, of course, PVM since it permits dynamic spawning of the nodes taking part in the parallel virtual machine. Since, however, many parallel machines support only MPI, we have implemented our algorithm using PVM and different MPI implementations (LAM MPI, MPICH, MPI native to SMP machines), which can be selected using an automatic configuration script based on the autoconf facility before compilation of the source code. In case of the PVM version, the master process uses system calls to determine the load of all nodes available in the virtual machine. Slave processes are spawned on nodes whose load is below a predefined threshold. After completion of a task, the load is determined again and the node is released from the virtual machine if applicable. In this way an automatic load balancing is achieved. In addition, it is possible to spawn processes manually if desired. The master process also recognizes if some of the nodes are much faster than others, such that it might be advantageous not to spawn further processes on slow nodes, but to wait until a faster node has finished a slave job and can be used for another job. The slaves are monitored by the master using information which is communicated from the slave nodes in regular intervals, asynchronously with respect to the production calculation.

Although the *analytical* evaluation of derivatives is in general less computer-time demanding than our numerical approach, it is by far less suited for parallelization. The wall time needed in a numerical calculation can be much lower than for an analytical calculation, if a sufficiently large number of nodes is available. As a consequence, the calculation of vibrational spectra for large molecules becomes feasible in the framework we use, while it is out of reach of completely analytical approaches.

2.2 Restart facilities

The restart facilities built into our program are another noteworthy feature of considerable importance for practical calculations.

Numerical derivatives provide a straightforward way for restart by writing all data points necessary for calculating the derivatives to a restart file. Once the data are available for a given data point, they would not be affected by an unforeseen failure of the computer machinery. These data can also serve for secondary purposes (e.g., charge decomposition analyses based on atomic polar tensors [7]) in subsequent calculations.

Additionally, the program SNFDC provides a more fine-grained restart facility, which can be activated if the single point calculations themselves are very time-consuming. A crash of one of the slave processes would cause an unacceptable loss of data in these cases. The slave processes will then asynchronously copy the molecular orbital files to a safe location providing sufficient disk storage, which may also reside on a remote machine. If a system failure occurs, the start vector files are re-distributed among the slaves and the calculations can be restarted such that loss of CPU-time is minimized.

This procedure might cause problems if many slave processes try to copy their files to a remote location at the same time. We synchronize the remote copying by having the master distribute a predetermined (small) number of *tokens* to the slaves. Only those slaves are allowed to start the copy, which hold one of these tokens. When the files are copied, the tokens are given back to the master, which distributes them to the next slaves etc. This assures that the network load does not exceed a predetermined level.

3 Raman spectroscopy for large molecules: The characterization of biomimetic nitrogenase model complexes

3.1 Methodology

Raman intensities are usually calculated by taking the third derivative of the total electronic energy, once with respect to nuclear coordinates and twice with respect to the strength of a (static) electric field. Another possibility is the differentiation of the molecular polarizability with respect to nuclear co-ordinates, which is equivalent to the first method, since the components of the polarizability tensor can be understood as second derivatives of the electronic energy with respect to the electric field strength. If this method is employed using numerical differentiation of the polarizability, dynamic polarizabilities can also be used instead of static ones.

The latter method is implemented in our program package SNF and has been applied, e.g., to the determination of Raman intensities for the Buckminsterfullerene C_{60} using density functional methods [4] and for a set of smaller molecules in order to compare the accuracy of several different quantum chemical approaches (density functional theory, multi-configurational self-consistent field and coupled cluster theory) [8]. Within this approach, the molecular symmetry can be applied in order to reduce the number of single point calculations to be done, while each individual calculation has to be performed in C_1 symmetry.

It is often sufficient or desirable to calculate only Raman intensities for selected modes instead of for all $3N-6$ vibrational modes of a large molecule, which can be achieved if the normal coordinates of the molecule are already known. Therefore, the standard frequency analysis is performed as the first step. Then, we use displacements along selected mass-weighted normal coordinates $\mathbf{Q}_k$, for which (static and/or dynamic) polarizabilities are calculated [9].

The displacement may be written in terms of normalized displacement vectors for both the Cartesian and the mass-weighted normal modes,

$$s_{Q_k} \Delta \mathbf{Q}_k^{\mathrm{norm}} \hat{=} s_{Q_k} \Delta \mathbf{R}_k = s_{Q_k} |\Delta \mathbf{R}_k| \Delta \mathbf{R}_k^{\mathrm{norm}} = s_R \Delta \mathbf{R}_k^{\mathrm{norm}} \tag{1}$$

which leads to a step size s_{Q_k} for the numerical differentiation of

$$s_{Q_k} = s_R/|\Delta \mathbf{R}_k| = s_R \left(\sum_{i=1}^{3N} (Q_{k,i}^{\mathrm{norm}})^2/m_i \right)^{-1/2} \left(\frac{[\text{unit of length}]}{[\text{unit of mass}]^{1/2}} \right). \tag{2}$$

The numerical derivatives of the components of the polarizability tensor are obtained as

$$(\bar{\alpha}_{ij}')_k = \frac{\partial \alpha_{ij}}{\partial \mathbf{Q_k}} = \frac{\alpha_{ij}(\mathbf{R}_{\mathrm{eq}} + s_R \Delta \mathbf{R}_k^{\mathrm{norm}}) - \alpha_{ij}(\mathbf{R}_{\mathrm{eq}} - s_R \Delta \mathbf{R}_k^{\mathrm{norm}})}{2 \, s_{Q_k} |\Delta \mathbf{Q}_k^{\mathrm{norm}}|}, \tag{3}$$

if a three-point central differences formula is used; $\alpha_{ij}(\mathbf{R}_{\mathrm{eq}} \pm s_R \Delta \mathbf{R}_k^{\mathrm{norm}})$ are the components of the polarizability tensor of the displaced structures.

Raman intensities are usually given as Raman scattering factors,

$$S = 45a_k'^2 + 7\gamma_k'^2, \tag{4}$$

where a_k' and γ_k' contain the derivatives of the polarizability tensor components with respect to normal coordinates (cf. [10, 4]), or in terms of the Q-branch differential cross section for a scattering angle of $90°$ and incident light which is plane polarized perpendicular to the scattering plane,

$$\frac{d\sigma}{d\Omega} = \frac{\pi^2}{\epsilon_0^2}(\tilde{\nu}_{in} - \tilde{\nu}_k)^4 \frac{h}{8\pi^2 c\tilde{\nu}_k} \left(\frac{45a_k'^2 + 7\gamma_k'^2}{45}\right) \frac{1}{1 - \exp[-hc\tilde{\nu}_k/k_B T]}. \quad (5)$$

If Raman intensities are calculated for all vibrational modes by taking numerical derivatives of the polarizabilities with respect to Cartesian nuclear coordinates, they may also be easily re-calculated for isotope-substituted molecules by a transformation with the corresponding new matrix of eigenvectors of the Hessian. However, this cannot be done if only selected Raman intensities are calculated by taking numerical derivatives with respect to the normal coordinates, since the normal coordinates themselves depend on the nuclear masses.

Therefore, the combination of these two approaches allows us to obtain vibrational frequencies for the original molecule as well as for isotope-substituted molecules. Furthermore, we are able to determine the Raman intensities for all selected modes of interest also for very large molecules containing more than 100 atoms.

3.2 Raman spectra for large metal complexes

The dinuclear diazene-coordinating complex $[\{Fe`S_4`(PR_3)\}_2(N_2H_2)]$ **1** in Fig. 2 with 'S_4^{2-} = 1,2-bis(2-mercaptophenylthio)ethane(2−) is one of the model complexes important for nitrogen fixation. The vibrational properties of this complex have been studied by resonance Raman methods experimentally [11] and theoretically with the approach reviewed here [9]. A splitting of lines, which occured only in the resonance Raman spectrum but not in the infrared spectrum, has been observed which was attributed to a suggested photoisomerization process [11].

In [9] we emphasized that the potential photoisomerization process can be of significance to the biological fixation problem as a possible rearrangement mechanism for the conversion of **1(A)** to **1(B)** might involve abstraction of the hydrogen atoms, which can be accepted by the sulfur atoms in the ligand sphere acting as lewis bases, such that a linear dinitrogen species could be generated within the complex. This 'symmetric' transition state, which would represent an intermediate for the reduction of dinitrogen by a dinuclear complex like **1**, could switch back to the reactant as well as to the product **1(B)**. An alternative pathway by rotation of the metal fragments coordinating to the diazene moiety appears unlikely as large rotational barriers need to be overcome. These barriers are generated by strong hydrogen bonds (see below), which need to be broken in order to allow free rotation of the metal fragments. Furthermore, the photoisomerization occurs in the solid state at low temperatures, which does not favour extended spatial rearrangement of the dinuclear complex.

While we published results for complex **1** in [9], our focus is in this work on the influence of the outer ligand sphere on the vibrational spectra of type-**1** complexes. Modifications in the chelate ligand induce secondary effects on

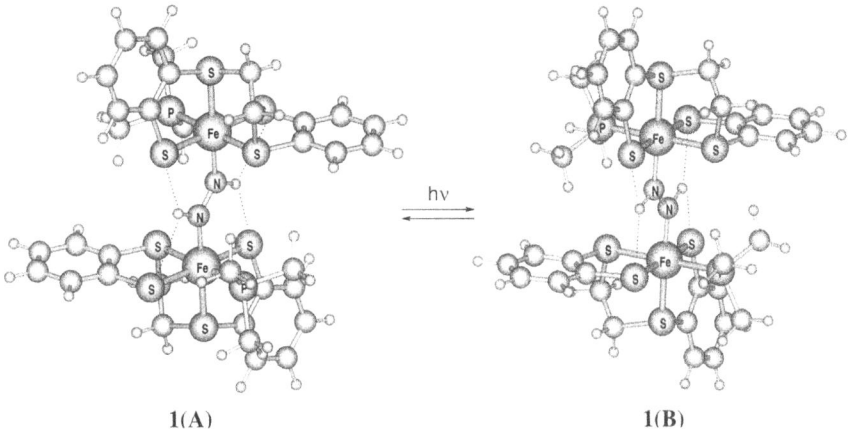

1(A)   **1(B)**

Fig. 2. BP86/RI/TZVP optimized structures of two isomers **1(A)** and **1(B)** with R=Me, which are involved in a postulated photoisomerization reaction.

reactivity, stability, structure, and spectra of the biomimetic complexes. The study of these effects is mandatory for the final goal of reactivity control through molecular architectures. Before we describe the structural changes introduced in the chelate ligand of **1**, we shortly mention the essential building blocks of the quantum chemical methodology:

For all calculations we used the density functional programs provided by the TURBOMOLE 5.1 suite [5]. The results are obtained from all-electron restricted Kohn–Sham calculations. We employ the Becke–Perdew functional dubbed BP86 [12, 13] as implemented in TURBOMOLE. In connection with the BP86 functional we always use the resolution-of-the-identity (RI) technique [14, 15]. For the efficient calculation of the polarizabilities for the displacement structures we used the ESCF module, which is capable of using the resolution-of-the-identity technique [16]. Ahlrichs' TZVP basis set [17] featuring a valence triple-zeta basis set with polarization functions on all atoms, which has proven to give reliable Raman intensities for buckminsterfullerene [18], was used throughout. The shared-electron number (SEN) method, which has been developed for the evaluation of hydrogen bonds in non-decomposable compounds, i.e., in compounds which cannot be decomposed into two parts such that the decomposition energy can be solely attributed to the broken hydrogen bond, is used for the evaluation of hydrogen bond energies. Details of the SEN method are described in [19].

In order to investigate the effect of substituents in the second ligand sphere, we optimized the structures of the two isomers of complex $[\{Fe`S_4'(PMe_3)\}_2(N_2H_2)]$ **2** in Fig. 3) with $`S_4^{2-}$ = 1,2-bis(2-mercaptophenylthio)phenylene(2−), which has also been synthesized (with PPr$_3$ as the phosphane ligands) [20, 21]. In **2** the two ethylene bridges of **1** are replaced by two phenylene bridges (indicated by two ovals in **2(A)**). In contrast with complex **1**, the two isomers **2(A)** and **2(B)** have been isolated and their structure could be characterized by x-ray analyses [20, 21]. A comparison of the experimental structures and our BP86/RI/TZVP optimized structures, which are, in general, in very good agreement, is given in Table 1. Since vibrational spectra are not yet available for complex **2**, a photoisomerization process as observed in the resonance Raman spectrum of **1** has thus not yet been detected. We have carried out a calculation of the Raman and infrared vibrational spectra of **2**.

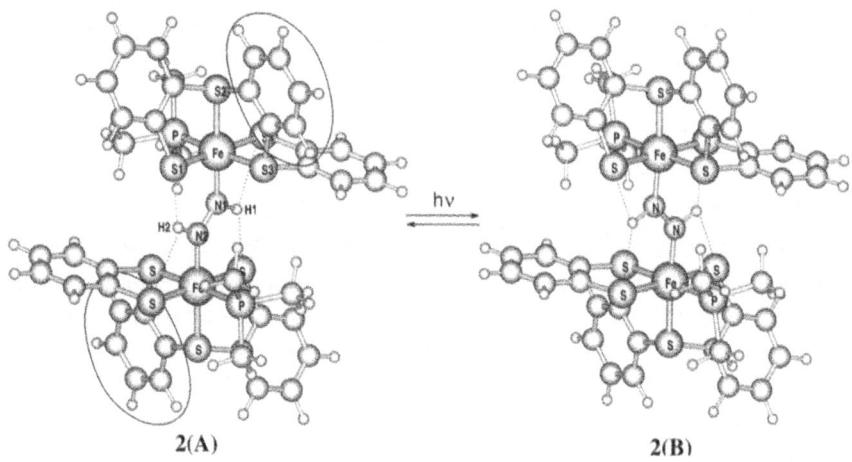

2(A)   **2(B)**

Fig. 3. BP86/RI/TZVP optimized structures of two isomers **2(A)** and **2(B)** with R=Me, which may be involved in a similar photoisomerization reaction like complex **1**.

While the ethylene brigde is in a staggered conformation, the rigid benzene ring forces the two sulfur atoms bound to this ring into the plane of the benzene carbon atoms.

As one would expect, the introduction of the two phenylene bridges has almost no (primary) effect on the bond lengths of the diazene ligand: the N−H bond lengths are 103.8 pm in **1(A)**, 104.3 in **1(B)**, 104.0 pm in **2(A)**, and 104.2 pm in **2(B)**; the Fe−N bond lengths are 189.1 pm in **1(A)**, 190.2 in **1(B)**, 189.5 pm in **2(A)**, and 190.5 pm in **2(B)**; the N−N bond lengths are 128.9 pm in [**1(A)** and in **1(B)** and 128.8 pm in **2(A)** and in **2(B)**.

However, there are pronounced secondary effects due to the induced structural changes of the chelate ligand: one of the original equatorial benzene rings in **1**, which is almost in plane with the {FeS$_3$P} fragment in **1** is subject to a deformation upon introduction of the rigid phenylene bridges towards the diazene moiety in **2**. This increases the differences in length and energy of the S$\cdots$H hydrogen bonds for isomer **(A)**: 267.5 pm and 6.0 kJ/mol for the longer hydrogen bond in **1(A)**, 249.9 pm and 8.3 kJ/mol for the shorter hydrogen bond in **1(A)**, while we have 269.9 pm and 5.7 kJ/mol for **2(A)** (longer bond) and 246.0 pm with 10.9 kJ/mol in **2(A)** (shorter bond). At the same time it decreases the differences in the hydrogen bonds for isomer **B**: 276.8 pm and 2.4 kJ/mol in **1(B)** (longer bond), 240.5 pm and 19.3 kJ/mol in **1(B)** (shorter bond); 272.2 pm and 17.4 kJ/mol in **2(B)** (longer bond), 244.9 pm and 3.2 kJ/mol in **2(B)** (shorter bond). (The SEN method [19] has been applied for the estimation of the hydrogen bond energies.)

Raman vibrational frequencies and intensities have been obtained for both isomers of the two complexes, **1(A)** and **1(B)** in [9] and **2(A)** and **2(B)** in this work, and are compared in Table 2. The accuracy of the harmonic wavenumbers is, in general, better than 6 cm^{-1} for the bands listed, when compared with the experimental data for **1**. The relative Raman scattering intensities were calculated from the differential cross sections. For their evaluation we used the wavelength of 613.33 nm, which is the wavelength of the laser employed in the experimental set up, for ν_{in} in (5). Polarizabilities have been calculated in the limit of a static perturbation, since intensities obtained from these data are known to be in good agreement with experimental intensities [22], while dynamic polarizabilities show poles near electronic excitation frequencies. We refrain from a discussion of the theoretically predicted Raman intensities in comparison to experimentally measured intensities, since these are not directly comparable because of the *resonance* Raman effect. A more detailed comparison of frequencies and intensities has been performed in our previous study [9].

It has been found in previous work [9] that the frequencies of the N−H bending mode at about 1500 cm^{-1} are influenced by the S$\cdots$H contacts for complex **1**: For the isomer with the strongest hydrogen bond (19.3 kJ/mol, **1(B)**) this peak appears at the larger wavenumber of 1512 cm^{-1} compared to 1449 cm^{-1} for isomer **1(A)**. Although the change of the hydrogen bond energies is only a few kJ/mol for complex **2** in comparison with **1**, we are able to verify this correlation between the hydrogen bond energies and the vibrational frequencies for this pair of isomers: The wavenumbers for the corresponding modes are 1501 cm^{-1} for **2(B)**, which exhibits the stronger S$\cdots$H hydrogen bond (17.4 kJ/mol), and 1455 cm^{-1} for **2(A)**. This bond energy for **2(B)** is slightly smaller than that of **1(B)**, and consequently, the wavenumber is smaller by 11 cm^{-1}. Considering the complexes **1(A)** and **2(A)** we find that the strongest hydrogen bond of 10.9 kJ/mol also corresponds to the larger wavenumber of 1455 cm^{-1} for **2(A)**, although the difference is only 6 cm^{-1} in this case.

This example shows that small modifications in the second ligand sphere can be applied for a fine-tuning of the bonding between the substrate (here modelled by the diazene moiety) and the complex via hydrogen bridges. This is of great importance, since neither too strongly nor to weakly binding model complexes are suited and applicable for the desired catalytic reduction process of dinitrogen under mild conditions.

Table 1. Comparison of experimental and calculated bond lengths in pm for complex **2**. Experimental values have been taken from [20].

	2(A)		2(B)	
	exp. (**2(A)**·2 toluene)	calc.	exp. (**2(B)**·4 CH$_2$Cl$_2$)	calc.
Fe−S1	231.9(2)	233.1	231.06(9)	233.9
Fe−S2	226.1(2)	226.5	227.49(8)	227.1
Fe−S3	226.2(2)	228.2	225.75(8)	228.3
Fe−S4	229.1(2)	232.1	230.84(9)	232.2
Fe−N	190.1(4)	189.5	191.3(2)	190.5
Fe−P	229.0(2)	227.6	228.71(8)	228.0
N−N	123.4(7)	128.8	128.4(4)	128.8
N−H	86.0	104.0	84.1(4)	104.2
S1−H2 and S1−H1	258.9	246.0	274.3(4)	272.2
S4−H1 and S4−H2	267.9	269.9	255.8(4)	244.9
S1−Fe−S4	174.60(6)	176.6	175.59(3)	177.0
S2−Fe−N	177.6(2)	179.3	173.12(7)	174.3
S3−Fe−P	176.99(7)	176.4	177.11(3)	177.2
S1−Fe−N	91.8(2)	91.8	87.53(7)	87.2
S1−Fe−P	96.10(6)	95.4	94.46(3)	94.3
P−Fe−N	90.0(2)	89.8	90.93(7)	89.6
Fe−N−N	133.5(5)	133.8	133.1(3)	132.2
S1−H2−N2 and N2−H2−S4	126.1	126.1	134.8(4)	129.7
N1−H1−S4 and S1−H1−N1	100.7	94.9	94.8(3)	91.7

Table 2. Comparison of calculated spectra for complexes **1** and **2**. Experimental wavenumbers $\tilde{\nu}_{exp}$ are given for comparison for **1** with R=Pr [11]. Note that the two bands at 1363 cm^{-1} and 1352 cm^{-1} are not resolved by experiment which yields only a single band at 1365 cm^{-1}. All wavenumbers are given in cm^{-1}.

sym.	$\tilde{\nu}_{exp}$	complex **1**; [9] isomer	$\tilde{\nu}_{calc}$	I_{rel}	complex **2**; this work isomer	$\tilde{\nu}_{calc}$	I_{rel}
a_g	637	**1(B)**	639	50.4	**2(B)**	626	165.5
a_g	663	**1(A)**	666	143.0	**2(A)**	664	16.3
a_g	1365	**1(B)**	1352	1434.4	**2(B)**	1358	1799.0
a_g	1365	**1(A)**	1363	1308.7	**2(A)**	1363	1530.5
a_g	1455	**1(A)**	1449	22.6	**2(A)**	1455	30.4
a_g	1509	**1(B)**	1512	100.0	**2(B)**	1501	142.0
a_g		**1(B)**	3111	54.0	**2(B)**	3126	84.3
a_g		**1(B)**	3112	22.5	**2(B)**	3127	37.8
a_g		**1(A)**	3191	<0.1	**2(A)**	3167	8.3

4 Hardware configurations

As has been mentioned in Sect. 2, the program package SNF is equally well suited for PC clusters and supercomputers, since it does not depend on high-speed communication between master and slave processes.

Clusters of Linux PCs are advantageous since they are inexpensive, and can be build using a heterogeneous set of computers, such that new machines can easily be integrated in an existing network. The Chair of Theoretical Chemistry at the Friedrich–Alexander-Universität Erlangen–Nürnberg runs a cluster of 82 Linux PCs and 6 Alpha workstations with 156 CPUs in total. At the moment, the core of this cluster consists of 44 Dual-Athlon 1800+ machines, of which one acts as a file server. The file server makes use of a 480 GB raid system and is connected to the switch of a Fast ethernet via a glass-fiber Gigabit ethernet connection. Most of the nodes are connected to the switch using Fast ethernet, whereas seven of them have a Gigabit ethernet connection as well. The Dual-Athlon machines are equipped with 1 GB of DDR work memory and 30 GB of disk storage each. This configuration was purchased for less than 100000 EUR.

Different aspects become important in case of supercomputers. It is in principle possible to use not only coarse-grained but also fine-grained paral-lelization techniques, since most of the quantum chemical programs used by SNF can be applied in a parallelized manner. Consider the Hitachi SR8000-F1 at the Leibniz-Rechenzentrum in Munich. The possibility of SMP within an SR8000 node suggests that every single point calculation be performed using 8-fold parallelization on one node. This further reduces the total wall time of the calculation and allows us to determine the vibrational spectra even for the largest molecules of interest in our studies of metal complexes modelling nitrogenase activity. First attempts to run SNF in this way on the SR8000 have been carried out successfully (after a bug in Hitachi's MPI im-plementation was fixed), but the parallelization of the TURBOMOLE package on 8 nodes has shown that the run time does not yet scale satisfactorily with the number of nodes — particularly when compared to timings on the Linux cluster.

References

1. Dieter Sellmann, Jörg Sutter. In Quest of Competitive Catalysts for Nitro-genase and Other Metal Sulfur Enzymes. *Acc. Chem. Res.*, **30**(11) (1997) 460–469.
2. Dieter Sellmann, Jürgen Utz, Nicole Blum, Frank W. Heinemann. On the function of nitrogenase FeMo cofactors and competitive catalysts: chemical principles, strcutural blue-prints, and the relevance of iron sulfur complexes for N_2 fixation. *Coord. Chem. Rev.*, **190-192** (1999) 607–627.

3. C. Kind, M. Reiher, J. Neugebauer, B. A. Hess. SNF — a program for quantum chemical calculations of vibrational spectra. University of Erlangen-Nürnberg, 1999 — 2002. www.chemie.uni-erlangen.de/hess/html/downloads.html.

4. J. Neugebauer, M. Reiher, C. Kind, B. A. Hess. Quantum Chemical Calculation of Vibrational Spectra of Large Molecules — Raman and IR Spectra for Buckminsterfullerene. *J. Comp. Chem.*, **23** (2002) 895–910.

5. Reinhart Ahlrichs, Michael Bär, Marco Häser, Hans Horn, Christoph Kölmel. Electronic Structure Calculations on Workstation Computers: The Program System Turbomole. *Chem. Phys. Lett.*, **162**(3) (1989) 165–169.

6. T. Helgaker, H. J. Aa. Jensen, P. Jørgensen, J. Olsen, K. Ruud, H. Ågren, A. A. Auer, K. L. Bak, V. Bakken, O. Christiansen, S. Coriani, P. Dahle, E. K. Dalskov, T. Enevoldsen, B. Fernandez, C. Hättig, K. Hald, A. Halkier, H. Heiberg, H. Hettema, D. Jonsson, S. Kirpekar, R. Kobayashi, H. Koch, K. V. Mikkelsen, P. Norman, M. J. Packer, T. B. Pedersen, T. A. Ruden, A. Sanchez, T. Saue, S. P. A. Sauer, B. Schimmelpfennig, K. O. Sylvester-Hvid, P. R. Taylor, O. Vahtras. DALTON, a molecular electronic structure program, Release 1.2.1, 2001.

7. J. Cioslowski. A New Population Analysis Based on Atomic Polar Tensors. *J. Am. Chem. Soc.*, **111**(22) (1989) 8333–8336.

8. J. Neugebauer, M. Reiher, B. A. Hess. Coupled Cluster Raman Intensities: Assessment and Comparison with Multi-Configuration and DFT Methods. *J. Chem. Phys.*, (2002) submitted.

9. M. Reiher, J. Neugebauer, B. A. Hess. Quantum chemical calculation of Raman intensities for large molecules: The photoisomerization of [{Fe'S$_4$'(PR$_3$)}$_2$(N$_2$H$_2$)] ('S$_4$'$^{2-}$ = 1,2-bis(2-mercaptophenylthio)ethane(2−)). *Z. Physik. Chem.*, (2002) accepted for publication.

10. Derek Albert Long. *Raman Spectroscopy.* McGraw-Hill, New York, 1977.

11. Nicolai Lehnert, Beatrix E. Wiesler, Felix Tuczek, Andreas Hennige, Dieter Sellmann. Activation of Diazene and the Nitrogenase Problem: An Investigation of Diazene-Bridged Fe(II) Centers with Sulfur Ligand Sphere. 2. Vibrational Properties. *J. Am. Chem. Soc.*, **119** (1997) 8879–8888.

12. A. D. Becke. Density-functional exchange-energy approximation with correct asymptotic behavior. *Phys. Rev. A*, **38**(6) (1988) 3098–3100.

13. John P. Perdew. Density-functional approximation for the correlation energy of the inhomogeneous electron gas. *Phys. Rev. B*, **33** (1986) 8822–8824.

14. Karin Eichkorn, Oliver Treutler, Holger Öhm, Marco Häser, Reinhart Ahlrichs. Auxiliary basis sets to approximate Coulomb potentials (Erratum: *ibid.* **242** (1995) 652). *Chem. Phys. Lett.*, **240** (1995) 283–290.

15. Karin Eichkorn, Florian Weigend, Oliver Treutler, Reinhart Ahlrichs. Auxiliary basis sets for main row atoms and transition metals and their use to approximate Coulomb potentials. *Theor. Chem. Acc.*, **97** (1997) 119–124.

16. F. Furche. ESCF, Universität Karlsruhe, 2001.

17. A. Schäfer, C. Huber, R. Ahlrichs. Fully Optimized Contracted Gaussian Basis Sets of Triple Zeta Valence Quality for Atoms Li to Kr. *J. Chem. Phys.*, **100** (1994) 5829.

18. Johannes Neugebauer, Markus Reiher, Carsten Kind, Bernd A. Hess. Quantum chemical calculation of vibrational spectra of large molecules — Raman and IR spectra for Buckminsterfullerene. *J. Comp. Chem.*, **23** (2002) 895–910.

19. Markus Reiher, Dieter Sellmann, Bernd Artur Hess. Stabilization of diazene in Fe(II)-sulfur model complexes relevant for nitrogenase activity. I. A new approach to the evaluation of intramolecular hydrogen bond energies. *Theor. Chem. Acc.*, **106** (2001) 379–392.

20. David C. F. Blum. *Synthese, Struktur und Reaktivität neuer Eisen-Komplexe mit S_4-, N_2S_2- und N_3S_2-Donoratomsätzen und biologisch relevanten Koliganden.* PhD thesis, Universität Erlangen-Nürnberg, 1998.

21. Dieter Sellmann, David C. F. Blum, Frank W. Heinemann. Transition metal complexes with sulfur ligands. Part CLV. Structural and spectroscopic characterization of hydrogen bridge diastereomers of $[\mu\text{-}N_2H_2\{Fe(PR_3)('tpS4')\}_2]$ diazene complexes ('tpS$_4$'$^{2-}$ = 1,2-bis(2-mercaptophenylthio)phenylene(2−)). *Inorg. Chim. Acta*, (2002) in press.

22. Carole Van Caillie, Roger D. Amos. Raman intensities using time dependent density functional theory. *Phys. Chem. Chem. Phys.*, **2** (2000) 2123–2129.

Ruthenium Dioxide, a Versatile Oxidation Catalyst: First Principles Analysis

Ari P. Seitsonen[1] and Herbert Over[2]

[1] Physikalisch-Chemisches Institut der Universität Zürich
Winterthurerstraße 190, 8057 Zürich, Switzerland `Ari.P.Seitsonen@iki.fi`
[2] Physikalisch-Chemisches Institut der Justus-Liebig-Universität Gießen
Heinrich-Buff-Ring 58, 35392 Gießen, Germany

1 Introduction

The Council for Competitiveness in Washington D.C. classifies catalysis as one of the technologies that is critical to international competitiveness of the U.S. economy. Due to the strong chemical industry in Germany this statement is equally valid for the German economy. So far, efficient catalysts are designed by chemical and engineer's intuition [1]. However, the development of future and more efficient catalysts is considered to rely on atomic-scale tailored materials. To accomplish this goal a close cooperation between experimentalist, theorists as well as engineers is mandatory. It requires first of all a detailed microscopic description of catalytic reactions on the atomic scale as it is accessible by large-scale density-functional [2, 3, 4] calculations. During the past few years, this concept has been pursued by the Topsoe company in collaboration with the universities in Lyngby and Aarhus [5].

Very often interesting questions are posed by the experiments, although the answers are too complex to be given solely by them. This situation calls for additional information by ab-initio calculations. The other way round is also frequently encountered: Ab-initio calculations raise new questions which trigger new experiments or a different interpretation of experiments. It is just this intimate collaboration of experiment and theory that has rendered our particular project on the catalytic activity of RuO_2 a success story in surface chemistry of oxide surfaces [6, 7]. In this paper we will concentrate on the discussion of the theoretical results gained by large-scale DFT calculations.

2 Calculational Details

In the DFT calculations we employed the generalized gradient approximation of Perdew et al. [8] for the exchange-correlation functional. We used a plane wave basis set with an energy cut-off of 60 Ry and ab-initio pseudopotentials in the fully separable form for O, C, and Ru [9]. The $RuO_2(110)$ surface was modelled by five double layers of RuO_2 with a (2×1) or a (1×1) unit cell. Consecutive slabs were separated by a vacuum region of about

16Å. O and CO adsorption on $RuO_2(110)$ were modelled by placing O and CO on both sides of the $RuO_2(110)$ slab. In the calculations we relaxed the positions of all the O, C, and the Ru atoms. The O-CO separation between the reacting particles defines the reaction coordinate. The transition state of each of these reaction pathways and the corresponding activation barriers were searched with a constrained minimization technique [10]. The transition state is identified when the O-CO distance reaches a value where the forces on the atoms vanish and the energy is a maximum along the reaction coordinate.

All these calculations are heavily computing time and storage demanding so that they require the use of massive parallel computer architectures such as provided by the Hitachi SR80000-F1.

3 What is known about the stoichiometric and reduced $RuO_2(110)$ surfaces?

The surface structure [11] of the pristine $RuO_2(110)$ surface is depicted in Fig. 1A. In bulk RuO_2 (rutile structure) the Ru atoms are attached to six O atoms, while the O atoms are (planarly) coordinated to three Ru atoms. At the surface of $RuO_2(110)$ coordinatively unsaturated sites (cus) are encountered. First, the so-called 1f-cus Ru atoms are attached to five instead of six O atoms. In order to emphasize the under-coordination of the surface Ru atom to oxygen atoms, we use the nomenclature 1f-cus Ru atoms. Second, the $RuO_2(110)$ surface exposes bridging O atoms, which are coordinated to two rather than to three Ru atoms. Both under coordinated surface species carry single dangling bonds, which determine the interaction of the $RuO_2(110)$ surface with the surrounding gas phase. Exposing a stoichiometric oxide surface to oxygen at room temperature leads to the stabilization of a weakly held oxygen species (O_{ot}) that is adsorbed O atom in terminal position above the 1f-cus Ru atoms (cf. Fig. 1B) [12].

The spectroscopic fingerprints of $RuO_2(110)$ are provided by high resolution electron energy loss spectroscopy (HREELS) [12] and high resolution core level spectroscopy (HRCLS) [13, 14] in combination with DFT calculations. The vibrational spectrum of as-grown $RuO_2(110)$ surface is dominated by a loss at 69 meV, that DFT calculations attribute to the stretching mode of the (under-coordinated) bridging O atom O_{br} against the Ru atoms underneath. The O_{ot} covered $RuO_2(110)$ surface leads to an additional loss at 103 meV in the HREEL spectrum.

To obtain a spectroscopic fingerprint of the catalytically most important 1f-cus Ru sites, the technique of HRCLS in combination with DFT calculations was applied [13]. The found emission at 280.76 eV is related to bulk-coordinated Ru atoms in $RuO_2(110)$, while the component at 280.47 eV is ascribed to 1f-cus Ru atoms on $RuO_2(110)$; the latter assignment was based on DFT calculations.

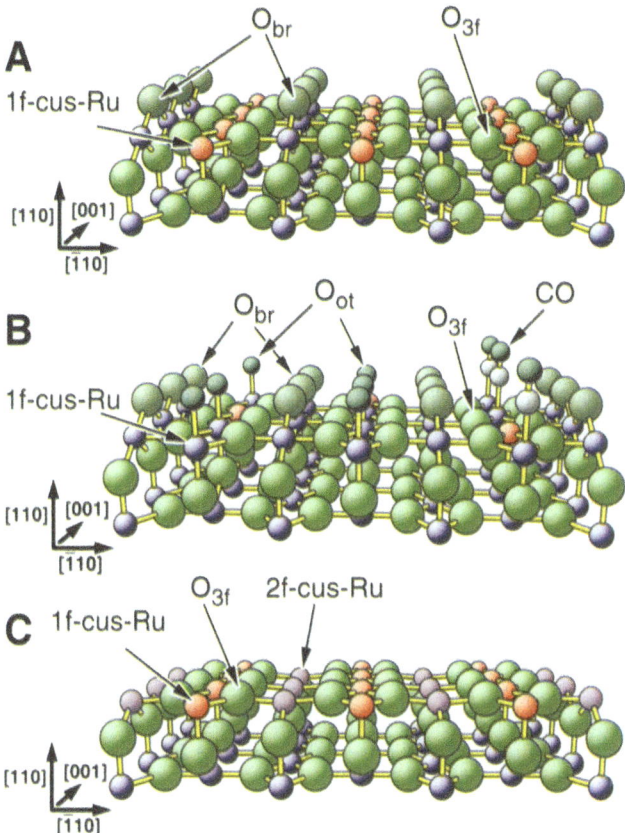

Fig. 1. Ball-and-stick model of the stoichiometric (A) and completely (mildly) reduced (C) RuO$_2$(110) surface. Large balls represent oxygen, and small balls represent ruthenium atoms of RuO$_2$(110). The pristine surface exposes under-coordinated oxygen (bridging O) and Ru atoms (1f-cus Ru). On the reduced RuO$_2$(110) surface, the under-coordinated bridging O atoms are removed, thus exposing two kinds of under-coordinated Ru atoms: 1f-cus Ru and 2f-cus Ru. The adsorption geometries of on-top O and on-top CO are indicated in panel (B).

If, on the other hand, the bridging oxygens are removed from the surface (for instance by reduction with CO or H$_2$), the mildly reduced RuO$_2$(110) surface exposes in addition to the 1f-cus Ru atoms also twofold under-coordinated Ru atoms (2f-cus Ru atoms; cf. Fig. 1C). The surface energy of the stoichiometric RuO$_2$(110) surface (γ = 70 meV/Å^2) is substantially lower than that of the mildly reduced RuO$_2$(110) surface (90.5 meV/Å^2 < γ < 145 meV/Å^2) [15]; we refer to mildly reduced RuO$_2$(110) when the only modification of the RuO$_2$(110) surface is due to the removal of bridging O atoms. Hence, the mildly reduced RuO$_2$(110) surface is even more active for the chemisorption of gas molecules than its stoichiometric counterpart. For instance, CO adsorption on the mildly reduced $RuO_2(110)$ surface proceeds via the symmetric bridge CO (binding energy: 1.85 eV), asymmetric bridge CO (both are bridging the 2f-cus Ru atoms) and only finally the on-top CO above the 1f-cus Ru atoms [16]. The binding energies of CO on the reduced $RuO_2(110)$ surface are by 0.5 eV higher than the pristine surface, where CO sits above the 1f-cus Ru atoms (cf. Fig. 1B: binding energy: 1.2 eV).

In general, experimental STM images do not show the positions of the atoms, since the STM probes the electronic structure near the Fermi energy rather than the total charge density that is related to the actual positions of the ion cores [17]. In constant current STM topographies the relative contribution of electronic and geometric effects can hardly be disentangled. Therefore, STM simulations on the basis of DFT calculations are needed. Using such STM simulations we identified the bridging O atoms on the pristine $RuO_2(110)$ surface in experimental STM images to be the rows of protrusions [18]. If the stoichiometric $RuO_2(110)$ surface is dosed with 0.5 L CO at 350 K, part of the bridging O atoms is removed via recombination with CO to form CO_2 [19]. This process was directly verified with STM [18].

4 Results and Discussion

4.1 DFT-calculated diffusion barriers of microscopic reaction steps

In general, the diffusion barriers of the reactants on the catalyst's surface governs the kinetics of the actual surface reaction. On metal surfaces the activation barriers for the diffusion of the reactants are quite low. In contrast, on oxide surface the diffusion barriers are quite high. For instance, the diffusion of on-top CO and on-top O on the stoichiometric $RuO_2(110)$ (cf. Fig. 1B) surface along the [001] direction exceeds 1.0 eV. This is the reason why one can image CO and O with STM even at room temperature.

The activation barriers for diffusion perpendicular to the close-packed row, i.e. along the $[\bar{1}10]$ direction, are substantially lower. For instance, the activation barrier for on-top O to hop into an adjacent vacancy in the bridging O row is only 0.7 eV. Equally facile is the diffusion of on-top CO into such vacancies; here the activation barrier is only 0.6 eV. Once a vacancy is created in the bridging O row, this vacancy can migrate along the rows. The actual activation barrier of this process is 0.9 eV.

4.2 DFT-calculated activation barriers for distinct reaction pathways

The first reaction pathway we examined was identical to that investigated first by Liu, Hu, and Alavi [20]. It corresponds to the initial reduction process of the stoichiometric RuO$_2$(110) surface by CO exposure at room temperature. In the initial state (cf. Fig. 2) the CO molecules reside in on-top positions above the 1f-cus-Ru atoms and the active oxygen on the surface is the bridging O atom. Our DFT calculations determines an activation barrier for this reaction pathway of 0.7 eV (cf. Fig. 3), which is smaller than 1.15 eV as found in [20].

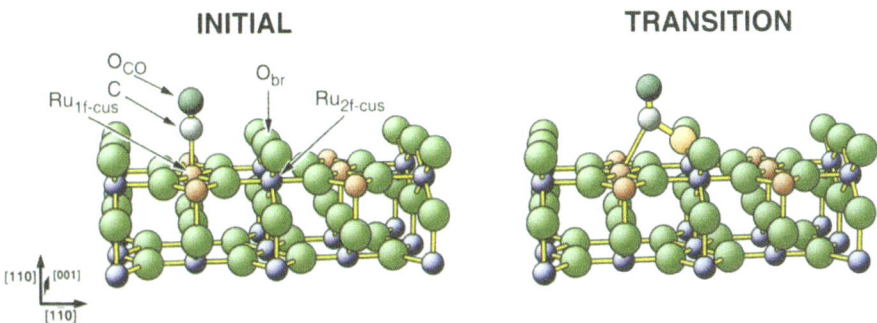

Fig. 2. A sketch of the initial and transition state for the CO oxidation over the stoichiometric RuO$_2$(110) surface.

However, we should note that the adsorption energies found by Liu, Hu and Alavi are also larger than ours. For instance, their CO, O$_{br}$ and O$_{3f}$ bonding energies are 1.66, 5.64 and 7.80 eV, while ours are 1.2, 4.6 and 5.8 eV, respectively. This may account for the observed differences in the activation barriers. The experimentally found activation energy of 0.9 eV is roughly reconciled by both theoretical studies. The atomic geometry of the transition state is depicted in Fig. 2. Both on-top CO and the adjacent O$_{br}$ atom are displaced from its initial adsorption site along the [$\bar{1}$10] direction. The CO molecule is, however, much farther inclined than the O$_{br}$ atom. This observation may be taken as a first indication that essentially the CO molecule has to be promoted in order to react with the bridging O atom.

A next set of reaction pathways is concerned with the partially reduced RuO$_2$(110) surface where part of the bridging O atoms have already been removed. For the initial state (cf. Fig. 4) we put the CO molecules in bridge position above the 2f-cus-Ru atoms consistent with a previous study [16]. Actually, the exact geometry is an quasi on-top position as predicted by DFT calculations [16]. The most obvious reaction pathway is that the CO molecule recombines with one of its adjacent bridging O atom. The activation barrier of 1.3 eV is, however, much higher than for the other pathways (cf. Fig. 3). Assuming an attempt frequency of 10^{13} Hz and a heating rate of 5 K/s the temperature-programmed recombination of bridging CO and bridging O over RuO$_2$(110) should reveal a maximum at 490 K, which agrees nicely

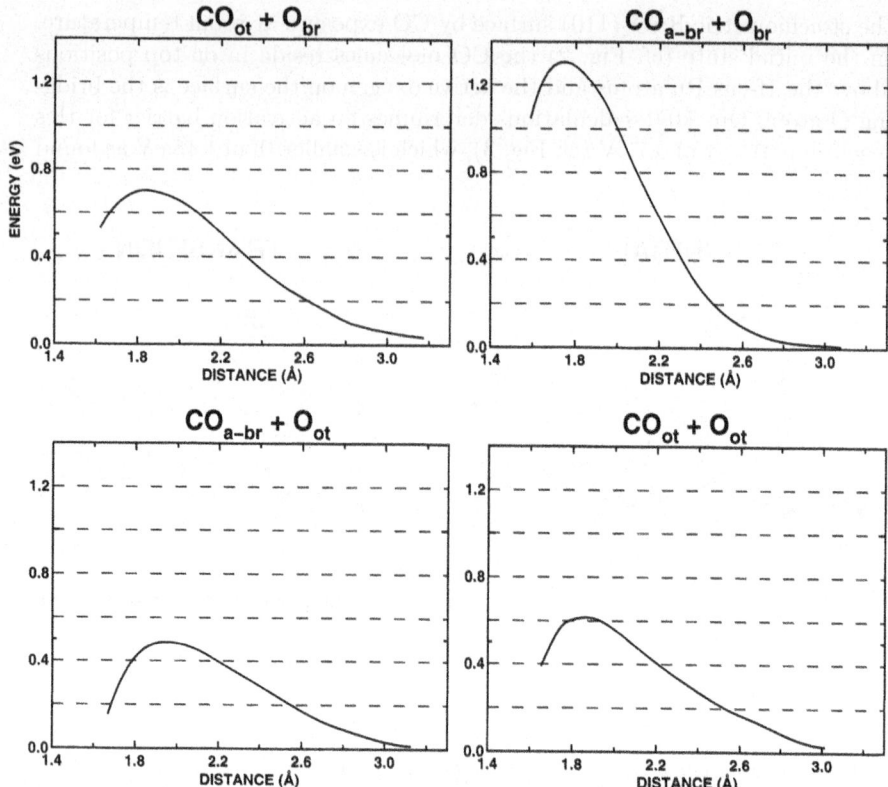

Fig. 3. Calculated energy diagrams for various reaction pathways of the reduction of RuO$_2$(110) by adsorbed CO molecules. The first reaction pathway (upper row, left) is that between a CO molecule (adsorbed above the 1f-cus-Ru atom in terminal position) and an adjacent bridging O atom. The activation barrier is 0.7 eV. The second reaction pathway (upper row, right) takes place on the partially reduced RuO$_2$(110), i.e. where part of the bridging O atoms have already been removed. The CO molecules sits above the 2f-cus-Ru in bridge position from where it recombines with an adjacent bridging O atom. The activation barrier is 1.3 eV. The third reaction pathway (lower row, left) considers the recombination between the bridging CO and the on-top O. Since the activation barrier is only 0.5 eV, self-poisoning of the reduced RuO$_2$(110) by strongly adsorbing CO molecules is prevented by oxygen post-exposure. the last reaction pathway (lower row, right) considers the recombination between on-top CO and on-top O, both sitting above neighboring 1f-cus-Ru atoms. The activation barrier is 0.6eV. The energy zero is defined by the adsorbed CO and the reactive oxygen in their initial positions.

with experimental data (maximum at 480 K) [21]. Hence, this particular reaction pathway is not important when the reaction temperature is kept at 300 K, although it may cause self-poisoning by strongly adsorbing bridging

CO molecules. The transition state for this reaction pathway is displayed in
Fig. 4. Both CO and O$_{br}$ are drawn to each other along the [$\bar{1}$10] direction
and the CO molecule is further displaced from its initial adsorption site than
the O$_{br}$ atom.

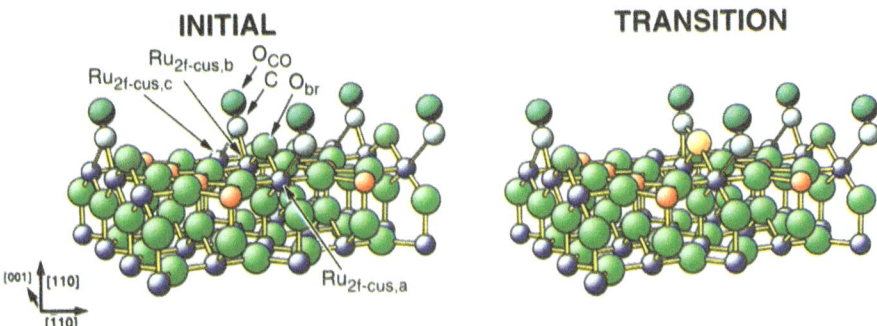

Fig. 4. A sketch of the initial and transition state for the CO oxidation over the
slightly reduced RuO$_2$(110) surface.

To comprehend on the above-discussed effect of self-poisoning of RuO$_2$(110)
by strongly adsorbed CO molecules over the 2f-cus-Ru atoms we explored the
reaction of bridging CO molecules with post-dosed oxygen that occupies the
on-top position above the 1f-cus-Ru atoms (cf. initial state in Fig. 5). The
activation barrier for the recombination reaction between the bridging CO
and on-top O turned out to be only 0.5 eV. Accordingly, bridging CO is easily removed by adsorbed on-top O atoms so that the RuO$_2$(110) surface is
prevented from self-poisoning by strongly adsorbing CO molecules.

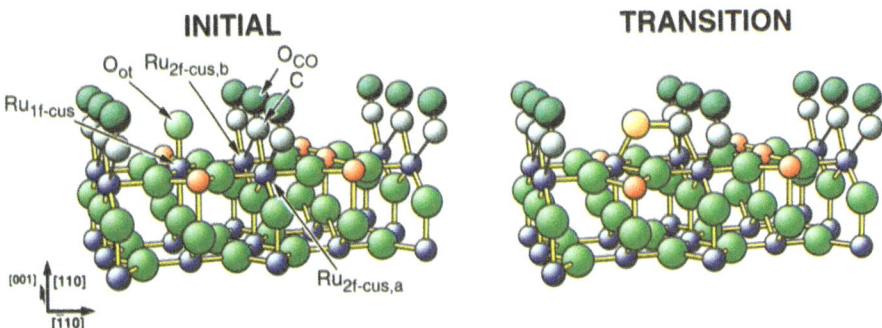

Fig. 5. A sketch of the initial and transition state for the CO oxidation over the
completely reduced RuO$_2$(110) surface.

Liu, Hu, and Alavi [20] pointed out that bridging O atoms are active in the oxidation of on-top CO (above the 1f-cus-Ru) atoms because the planar $2p_y$ orbital (y is referred to the [001] direction) of oxygen is pointing towards the CO molecules, thus establishing a kind of attractive interaction. If the CO molecules replace partly the bridging O rows, then the planar $2p_y$ orbital of the bridging O atoms cannot interact with the CO molecules and therefore are not promoting the CO oxidation reaction with bridging CO molecules. However, the on-top O has the ability to hybridize the planar 2p orbitals of oxygen in a way that they form a nascent bond with the bridging CO. This reduces the activation barrier for the oxidation reaction between the bridging CO and the on-top O. The actual transition state for the reaction pathway is shown in Fig. 5. Here the on-top O atom is displaced along the $[\bar{1}10]$ direction from the initial on-top adsorption site much farther than the bridging CO molecule.

The last reaction pathway addresses the recombination of on-top CO with on-top O. The low adsorption energy of on-top O and also recent HREELS measurements [19] suggested that the reaction pathway should be even more efficient than the others discussed so far. However, a detailed experimental study of this reaction pathway using isotope labelling experiments evidenced the opposite [22], i.e., on-top O is much less active than bridging O. Therefore we modelled the recombination of on-top CO and on-top O by DFT calculations (initial state cf. Fig. 6) Our DFT calculations estimate an activation barrier for this reaction pathway of 0.6 eV (cf. Fig. 3). However, this activation barrier is by 0.1 eV smaller than for the corresponding reaction between on-top CO and bridging O. Within the transition state theory this observation implies that the frequency factor of both reaction pathways must differ by 1-2 orders of magnitude. This may be explained by the specific configuration of the active O atoms. While the on-top O has the freedom to vibrate isotropically in plane, the frustrated translation modes of the bridging O atoms are stiff along the O-rows and soft perpendicular to the O_{br} rows. This effect directs the motion of the bridging O atom towards the CO molecule, as only this soft mode is sufficiently highly populated at 300-400 K. Accordingly, the effective attempt frequency to overcome the reaction barrier is higher for the bridging O atoms than for the on-top O case. The atomic geometry of the transition state is given in Fig. 6. Both CO and the adjacent O_{ot} atom are displaced from their initial adsorption sites along the [001] direction. The on-top CO molecule is, however, much farther inclined than the O_{ot} atom.

In order to determine whether the activation of the CO molecule or that of the active O species imposes the rate determining step, we decomposed the total activation barrier into three contributions. Two of them are due to the variation in the binding energies of the reactants, CO and O, when moving from the initial to the final state in the absence of the other reactant and the third one comes from the difference of the interaction energies between

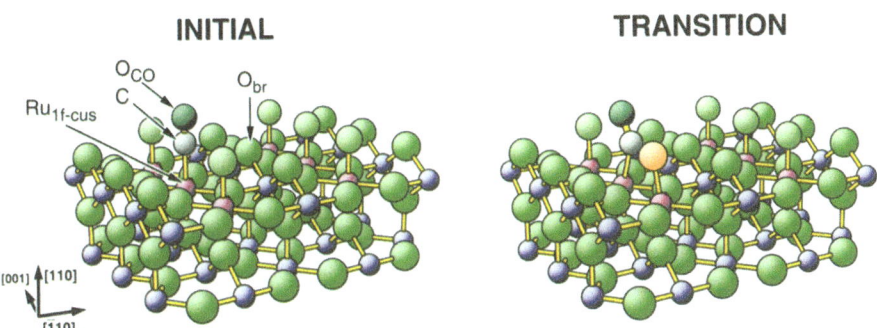

INITIAL   **TRANSITION**

Fig. 6. A sketch of the initial and transition state for the CO oxidation over the oxygen-rich RuO$_2$(110) surface.

the reactants in the initial and transition state. This approach was recently applied by Liu and Hu [23] for studying the general trends in the activation barriers of catalytic reactions on transition metal surfaces.

For the first reaction pathway considered here we determined a change in the CO chemisorption energy of 1.67 eV, while the active oxygen species suffered an energy penalty of 0.43 eV; for the other reaction pathways the values are compiled in Table 1. These values give evidence that indeed the activation of CO is the rate determining step for the recombination of on-top CO with bridging O atoms. Quite in contrast, the rate-determining step in the CO oxidation reaction over transition metal surfaces is due to the activation of the adsorbed oxygen atom [10, 20]. Since we already know the activation barrier of 0.7 eV, the change in the interaction energy between the CO molecule and the active O species is found to be attractive by 1.41 eV.

Table 1. Decomposition of the reaction energy (react. barr.) into the changes of the chemisorption energies of CO and O (ΔE_{CO}, ΔE_O) when moving from the initial to the final state in the absence of the other reactant and the corresponding variation of the interaction energies CO and O (ΔE_{int}).

	$CO_{ot} + O_{br}$	$CO_{a-br} + O_{br}$	$CO_{a-br} + O_{ot}$	$CO_{ot} + O_{ot}$
react. barr.	0.70 eV	1.30 eV	0.50 eV	0.62 eV
ΔE_{CO}	1.67 eV	1.02 eV	1.22 eV	1.53 eV
ΔE_O	0.43 eV	0.39 eV	0.84 eV	0.52 eV
ΔE_{int}	1.41 eV	0.11 eV	1.56 eV	1.61 eV

5 Summary: The complex mechanisms of reduction of RuO$_2$(110) by CO exposure

Assembling all the DFT-calculated details a self-consistent picture emerges of what happens during the CO oxidation reaction over RuO$_2$(110). CO molecules adsorb on the stoichiometric RuO$_2$(110) in terminal position above the 1f-cus-Ru atoms from where they recombine with bridging O atoms. This process reduces the RuO$_2$(110) surface and forms vacancies in the rows of bridging O atoms. The CO binding energy in such vacancies is by more than 0.5 eV stronger than over the 1f-cus-Ru atoms. Therefore, the ongoing reduction of RuO$_2$(110) surface is accompanied by the partial replacement of bridging O atoms by strongly adsorbed CO molecules. Around room temperature, the bridging CO molecules are not able to recombine with the adjacent bridging O atoms due to the high activation barrier of 1.3 eV. This process occurs only for sample temperatures above 450 K. However, on-top O, which forms by post-dosing oxygen to the reduced RuO$_2$(110) surface, can readily recombine with the bridging CO molecules. This process prevents the catalysts from self-poisoning by strongly adsorbed CO molecules.

On the other hand, the DFT calculations indicate that on-top CO and on-top O, both sitting above the 1f-cus-Ru atoms, can readily recombine. The determined activation barrier is by 0.1 eV smaller than that for the reaction between on-top CO and bridging O atoms. Since isotope labelling experiments give ample of evidence that on-top O is not very active in recombining with on-top CO molecules, this discrepancy may be resolved by the introduction of a 1-2 orders of magnitude lower attempt frequency for the on-top-CO/on-top-O reaction. The DFT-calculated reaction pathways reveal that the CO molecule has to be promoted in the catalytic recombination of CO and O on RuO$_2$(110). This is quite in contrast to the CO oxidation reaction over transition metal surfaces, where the adsorbed O atoms have to be promoted.

6 Outlook

We are currently optimising and adapting another program code which we plan to use on the Hitachi SR80000-F1. This will enable us to do real dynamical simulations, and these calculations are even more demanding for the computational capacity.

Acknowledgement. We thank the LRZ in Bavaria for providing massive-parallel computing time and support, which enabled us the present studies, in the Hitachi SR80000-F1.

References

1. G. Ertl; H. Knözinger; J. Weitkamp, (Eds.) Handbook of Heterogeneous Catalysis; Wiley: New York, 1997
2. P. Hohenberg; W. Kohn, Phys. Rev. **136** (1964) B864
3. W. Kohn; L. J. Sham, Phys. Rev. **140** (1965) A1133
4. R. M. Dreizler; E. K. U. Gross, Density functional theory; Springer-Verlag; Berlin, 1990
5. F. Besenbacher, I. Chorkendorff, B. S. Clausen, B. Hammer, A. M. Molenbroek, J. K. Norskov and I. Stensgaard, Science **279** (1998) 1913
6. H. Over, Y. D. Kim, A. P. Seitsonen, S. Wendt, E. Lundgren, M. Schmid, P. Varga, A. Morgante, G. Ertl, Science **287** (2000) 1474
7. H. Over, Appl. Phys. A **75** (2002) 37
8. J. P. Perdew, K. Burke, M. Enzerhof, Phys. Rev. Lett. **77** (1996) 3365
9. N. Troullier, J. L. Martins, Phys. Rev. B **43** (1991) 1993
10. A. Alavi, P. Hu, T. Deutsch, P. L. Silvestelli, J. Hutter, Phys. Rev. Lett. **80** (1998) 3650
11. Y. D. Kim, A. P. Seitsonen, H. Over, Surf. Sci. **465** (2000) 1
12. Y. D. Kim, A. P. Seitsonen, S. Wendt, J. Wang, C. Y. Fan, K. Jacobi, H. Over, G. Ertl, J. Phys. Chem. **105** (2001) 3752
13. H. Over, A. P. Seitsonen, E. Lundgren, M. Wiklund, J. N. Andersen, Chem. Phys. Lett. **342** (2001) 467
14. H. Over, A. P. Seitsonen, E. Lundgren, M. Smedh, J. N. Andersen, Surf. Sci. **504** (2002) 196
15. Y. D. Kim, H. Over, G. Krabbes, G. Ertl, Topics Catal. **14** (2001) 95
16. A. P. Seitsonen, Y. D. Kim, M. Knapp, S. Wendt, H. Over, Phys. Rev. B **65** (2002) 035413
17. J. Tersoff and D. R. Hamann, Phys. Rev. B **31** (1985) 805
18. H. Over, A. P. Seitsonen, E. Lundgren, M. Schmid, P. Varga, J. Am. Chem. Soc. **123** (2001) 11 807
19. C. Y. Fan, J. Wang, K. Jacobi, G. Ertl, J. Chem. Phys. **114** (2001) 10 058
20. Z.-P. Liu, P. Hu, A. Alavi, J. Chem. Phys. **114** (2001) 5957
21. S. Wendt, A. P. Seitsonen, Y. D. Kim, M. Knapp, H. Idriss, H. Over, Surf. Sci. **505** (2002) 137
22. S. Wendt, A. P. Seitsonen, H. Over, in preparation
23. Z.-P. Liu, P. Hu, J. Chem. Phys. **115** (2001) 4977

Theoretical Studies of Structures of Vanadate Complexes in Aqueous Solution

Michael Bühl, Frank T. Mauschick, and Rachel Schurhammer

Max-Planck-Institut für Kohlenforschung
Kaiser-Wilhelm-Platz 1
45470 Mülheim an der Ruhr, Germany
Fax: 0208/306 2996
buehl@mpi-muelheim.mpg.de

Abstract. Effects of thermal motion and interaction with a solvent on NMR chemical shifts of transition-metal complexes are studied with an approach based on molecular dynamics simulations and averaging of magnetic shieldings computed for snapshots along the trajectory. In some cases, for instance for iron cyanide complexes, large effects on the metal shielding are revealed, which can be traced back to changes in mean bond distances upon dynamical averaging in aqueous solution. For vanadate complexes in water, the effects are less pronounced, and amount to a few dozen ppm, irrespective of the density functional employed. The shielding of the ^{51}V resonance on going from the parent vanadate to peroxovanadate complexes is not reproduced in the simulations. The dynamics of a peroxovanadium-imidazole complex, a model compound for biomimetic oxidants, is studied in the gas phase and in water. In the pristine complex, rapid rotation of the imidazole ligand about the V-N bond occurs on the picosecond timescale. This complex does not coordinate hydrogen peroxide at vanadium. The implications for a possible mechanism of the oxidation reactions are discussed.

1 Introduction

Ever since the quantum-chemical computations of NMR chemical shifts became routinely feasible, they proved to be powerful probes for electronic and geometrical structure [1]. Ongoing progress in methodology and computer hardware allows the treatment of ever more complex and demanding systems, such as transition-metal complexes [2]. Important current developments are aimed at going beyond the treatment of substrates as vibrationless molecules at absolute zero and at allowing for specific interactions with the surrounding solvent. Both aspects can be addressed with molecular dynamics (MD) simulations by way of averaging magnetic shieldings computed for snapshots along the trajectory. This procedure has been pioneered by combining classical MD with quantum-chemical NMR computations in order to model the gas-to-liquid shift of ^{17}O in water [3]. Recently, we have extended this approach to assess thermal and solvent effects on transition-metal chemical shifts and have obtained interesting insights into the structure and dynamics of the solvation shells in aqueous vanadate systems [4]. The present paper describes recent progress that has been made in this area, and is organized as follows: Firstly, we give a brief general description of the methods that are being employed, followed by a discussion of the results for iron cyanide complexes and for peroxovanadium species in aqueous solution.

2 Methods

Central to the dynamical averaging of properties are density-functional based Car-Parrinello MD (CPMD) [5] simulations of the metal complexes, both in vacuo and in aqueous solution. In these simulations, periodic boundary conditions, suitable pseudopotentials and plane-wave basis sets are used. For a detailed description the reader should consult references [4], [6], and the literature cited therein. After short equilibration of typically 0.5 ps, snapshots are extracted from the trajectories and magnetic shieldings are computed using the established tools of density functional theory [2]. On the order of 40-50 snapshots are taken per picosecond and in the cases studied so far, a period of 1 to 2 ps has proven sufficient to converge the averaged shieldings. The NMR computations were carried out on local workstations at the MPI Mülheim. Almost all of the CPMD simulations were performed on the Hitachi SR-8000 of the LRZ, with typical resource requirements for a single job (usually up to 1 picosecond of simulation time) of 30 h on 16 - 32 nodes with 2 GB of memory. The experience with speed and turnaround times has been very good throughout. Since the setup of a simulation usually depends upon previous results and on a large number of test calculations, the use of a single workstation or smaller clusters thereof would lead to prohibitively long computation times. Thus, the present project could not have been pursued without access to a High Performance Computing facility.

3 Results and Discussion

3.1 Iron Cyanide Complexes

Iron complexes are not an integral part of the project pursued at the LRZ. However, simulations performed at LRZ for such complexes have contributed significantly to the field of computational transition-metal NMR in general. Hence, we will discuss these results first. The recent experimental observation that $\delta(^{57}\text{Fe})$ of aqueous $[\text{Fe(CN)}_5(\text{NO})]^{2-}$ (**1**) appears upfield from that of the parent $[\text{Fe(CN)}_6]^{4-}$ (**2**) by more than $\Delta\delta = 450$ [7] has prompted us to study thermal and solvent effects for these systems. For the tetraanion **2** in particular, strong interactions with the polar solvent water are to be expected. The results are summarized in 1. When $\delta(^{57}\text{Fe})$ is computed for isolated **2** in the

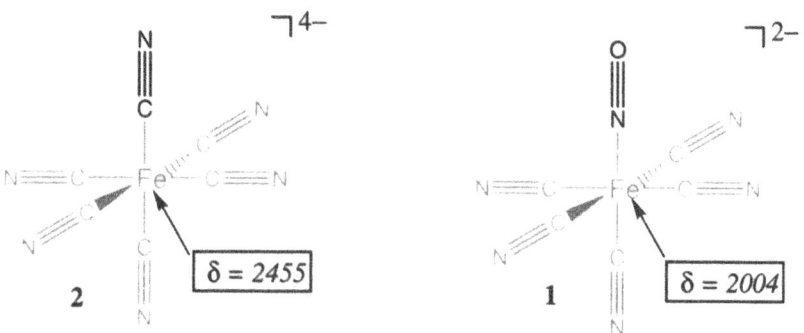

Fig. 1. Experimental ^{57}Fe chemical shifts of aqueous iron cyanide complexes

Table 1. ^{57}Fe chemical shifts, computed at the GIAO-B3LYP/DZ level and Fe-C distances (in Å, in parentheses), obtained from CPMD simulations with the BP86 functional.[a]

Level of approximation	$[Fe(CN)_6]^{4-}$ **(2)**		$[Fe(CN)_5(NO)]^{2-}$ **(1)**	
	δ	(r_{Fe-C})	δ	(r_{Fe-C})
δ_e // CP-opt	3240	(1.949)	2276	(1.958)
δ^{300K} // CPMD	3647	(1.964)	2458	(1.971)
δ^{300K} // CPMD/H$_2$O	1967	(1.907)	1997	(1.938)
δ^{300K} // BOMD/H$_2$O[b]	2593	(1.924)	2076	(1.943)
δ Experiment/H$_2$O[c]	2455		2004	

[a] From reference [6]. [b] Born-Oppenheimer MD (see text).
[c] From reference [7].

gas phase, either for the static equilibrium structure, δ_e, or for the dynamical average at 300K, δ^{300K}, large deviations from experiment are apparent, up to almost $\Delta\delta = +1200$ (first two entries in Table 1). Immersion in a periodic water box results in a dramatic shielding of the ^{57}Fe nucleus, which can be traced back to a significant decrease of the mean Fe-C bond distance (Table 1, values in parentheses). However, the final, CPMD-based $\delta(^{57}$Fe) value of **2** in water is much too strongly shielded compared to experiment, deviating by $\Delta\delta \approx -500$. It turned out that this discrepancy is an artifact of the limited box size, 11.5 Å, used in the periodic simulations. The repulsion between a tetraanion and its mirror images in the adjacent boxes leads to a spurious contraction of the iron-ligand bond and, thus, to an increased shielding of the theoretical $\delta(^{57}$Fe) value. When Born-Oppenheimer MD simulations are performed without periodic boundary conditions (employing the same density functional on the metal complex and a large water cluster described by a suitable force field), larger Fe-C distances and a higher $\delta(^{57}$Fe) value are obtained [8], the latter of which is in good accord with experiment (see last entry in Table 1). The same qualitative thermal and solvent effects as for **2** are obtained for nitroprusside **1**, albeit somewhat attenuated (Table 1). Both $\delta(^{57}\text{Fe})$ values derived from the CPMD and BOMD simulations in water are very similar and are both in good accord with experiment, suggesting that already for a dianion, periodic boundary conditions can be used with moderate box sizes.

3.2 Vanadate Complexes

^{51}V Chemical Shifts

In context with biomimetic oxidation reactions [9], we became interested in vanadate complexes with biogenic ligands [10]. The fact that these reactions are conducted in water prompted our first study on thermal and solvent effects in simple vanadates such as **3** and **4** [4]. Quite small such effects were obtained for $\delta(^{51}\text{V})$, on the order $\Delta\delta \approx -30$ (dynamical average in water vs. equilibrium value). What was not reproduced in the calculations is the observed shielding (by $\Delta\delta = -132$) of $\delta(^{51}\text{V})$ in **4** relative to that of **3**. It has been suspected that the BLYP functional, originally chosen because of its good performance for the description of liquid water, might not be the best choice for the transition metal complexes.

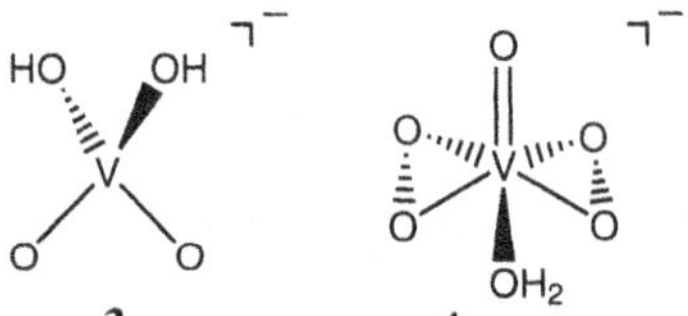

Scheme 1

Therefore, the CPMD simulations for **3** and **4** were repeated with the BP86 functional combination, which is known for its good performance in computational transition metal chemistry [11]. The results are summarized in Table 2. On going from the BLYP to the BP86 functional in the CPMD simulations, both δ_e and δ^{300K} of the isolated systems increase, whereas the δ value in aqueous solution decreases slightly. At no level is the observed $\delta(^{51}\text{V})$ sequence of **3** and **4** correctly reproduced. In order to identify density functionals that might be better suited for this purpose, the geometries of these two vanadates were optimized employing the PBE, HCTH, and B3LYP combinations [1]. In all cases, the ^{51}V nucleus in **4** was computed to be deshielded with respect to that of **3**, namely by $\Delta\delta = 44$, 107, and 4, respectively (δ_e values), at the GIAO-B3LYP level.[2] It is unlikely that inclusion of thermal and solvent effects would reverse this sequence. Further work is necessary to resolve this inconsistency between theory and experiment. These problems in

---

[1] For a description of the functionals and for the original citations see reference [11]

[2] The same trend is obtained when other GGA or hybrid functionals, such as BPW91, PBE, B1LYP, or mPW1PW91, are employed in the NMR calculations.

Table 2. Experimental and computed (GIAO-B3LYP) ^{51}V chemical shifts

Level of approximation	$H_2VO_4^-$ (3)		$VO(O_2)_2(OH_2)^-$ (4)	
	BLYP[a]	BP86	BLYP[a]	BP86
δ_e // CP-opt	-691	-659	-623	-588
δ^{300K} // CPMD	-717	-669	-564	-561
δ^{300K} // CPMD/H_2O[b]	-712	-715	-655	-682
Experiment/H_2O	-560		-692	

[a] Form reference [4]. [b] Average values from a 1 ps simulations
at ca. 300 K in a periodic water box.

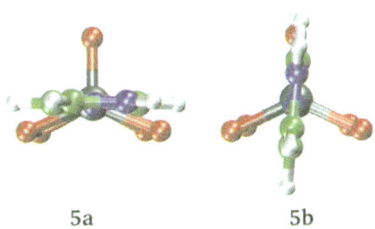

5a   5b

Fig. 2. Minima for **5**, viewed along the V-N axis (red: O, green: C, blue: N).

the description of the parent vanadate vs. peroxovanadate notwithstanding, trends within the same class of compounds should be reproducible. As a target system we chose $VO(O_2)_2(Im)^-$ (Im = imidazole) (**5** in Scheme 2) [12], which can be viewed as a model compound for vanadium-dependent haloperoxidases. Since related species have been found to be active in oxidation and epoxidation reactions [13], [14], we are currently investigating the possible activity of **5** and derivatives thereof in such reactions.

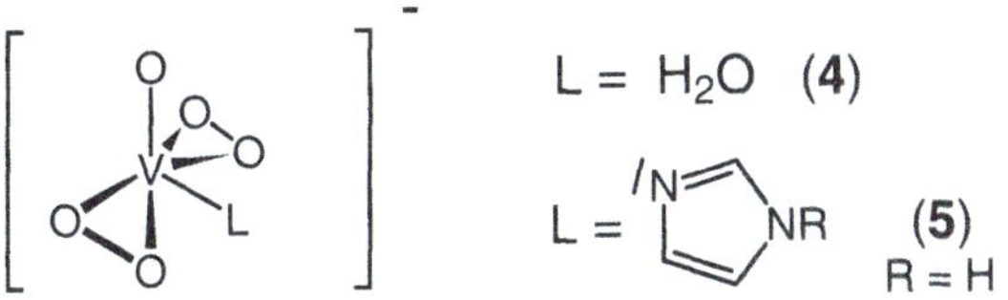

L = H_2O (**4**)

L = (imidazole structure) (**5**)
R = H

Scheme 2

Static computations with the BP86 functional afforded two minima with different orientation of the imidazole ligand (Fig. 2). The bisected isomer (**5a**) is slightly more stable than the eclipsed form (**5b**), by 0.5 kcal/mol. The transition state between both is only 0.01 kcal/mol higher than **5b**, indicative of a flat rotational potential of the imidazole group about the V-N bond. Despite the small energetic changes upon imidazole rotation, noticeable variations are revealed for the computed $\delta(^{51}V)$ value, which attains minimum and maximum values for bisected and eclipsed forms, respectively, and spans a range of more than 40 ppm (Fig. 3). In this context it is interesting to note that in vanadyl-imidazole complexes, a similar dependence of EPR properties on the ring orientation has recently been established [15]. From the very low barrier in **5**, essentially free rotation of the imidazole ligand is to be expected. When a CPMD simulation is started from isolated **5a** *in vacuo*, no such rotation is observed within 1 ps. When the simulation is started from the slightly higher-lying isomer **5b**, however, the imidazole moiety immediately (that is, already in the 0.5 ps equilibration period) commences rotating, and continues to do so for the total simulation time of 2 ps. The saw-tooth steps in Fig. 4, an illustration how an OVNC dihedral angle evolves with time, are not indicating discontinuities, but are a consequence of the definition of this angle ranging from $+180°$ to $-180°$. The rotational period of the imidazole unit is slightly longer than 1 ps. In the simulation starting from isolated **5a**, apparently, no initial momentum is present that would induce an immediate rotation. When **5a** is placed in aqueous solution, however, collisions with solvent molecules could provide such momenta. In fact, much larger amplitudes are found for the corresponding dihedral angle of **5a** in water, as compared to the gas phase (lower, bold curves in Fig. 5). No full rotation takes place, however, over a simulation time of 4 ps.[3] What is clearly visible is the decrease of the average V-N bond length on going from the gas phase into the solution (see upper curves in Fig. 5). A similar contraction had been found for the V-O bond to the coordinated water molecule in **4** [4]. It is most likely this change in bond length that is responsible for the computed shielding of the ^{51}V nucleus upon solvation in **4** and **5** (see Table 3). The difference between $\delta(^{51}V)$ computed

---

[3] Simulations starting from **5b** in water are in progress.

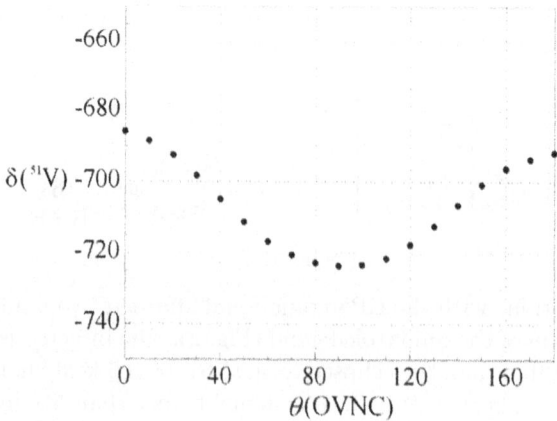

Fig. 3. Computed $\delta(^{51}V)$ of **5** as a function of the OVNC dihedral angle (rigid rotation starting from **5a**).

Table 3. ^{51}V chemical shifts (B3LYP level for BP86 optimized or simulated geometries) of $VO(O_2)_2(Im)^-$ (**5**).

Level of approximation	5a	5b
δ_e // CP-opt	-738	-669
δ^{300K} // CPMD	-722	-649
δ^{300K} // CPMD/H_2O	-833	n.a. [b]
Experiment/D_2O[a]		-744

[a] Form reference [12]. [b] Not yet available (in progress).

for the fully optimized isomers **5a** and **5b**, close to 70 ppm (δ_e values in Table 3), is slightly larger than that obtained from the rigid imidazole rotation in **5a**, ca. 40 ppm (cf. Fig. 3). Averaging over the MD trajectory starting from either minimum results in a deshielding of $\delta(^{51}V)$ with respect to the equilibrium values on the order of 20 ppm (compare first and second entry in Table 3), even though the imidazole ring rotates rapidly in **5b**, but not in **5a**. Eventually, averaging over long enough simulation times should afford the same δ^{300K} values for both starting structures. For gaseous **5**, this value will probably be closer to the entry for **5b** in Table 3 (because this shows the expected imidazole rotation). The MD simulation in water starting from **5b** is in progress. It is not unlikely that the aqueous environment will serve to slow down the rotation observed in the gas phase, due to water diffusing into the way of the rotating imidazole "wheel". In that case, the averaged δ^{300K} value in water would probably be close to the entry for **5a** in Table 3, that is, too strongly shielded by ca. 90 ppm with respect to experiment. Even

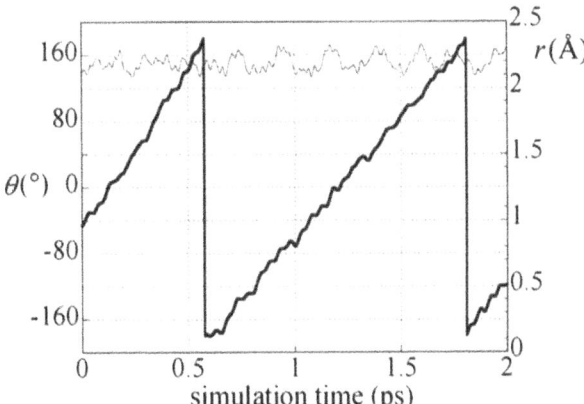

Fig. 4. Variations with time of the V-N distance r (top, right axis) and of a θ(OVNC) dihedral angle (bold, left axis) in a simulation starting from **5b** *in vacuo*.

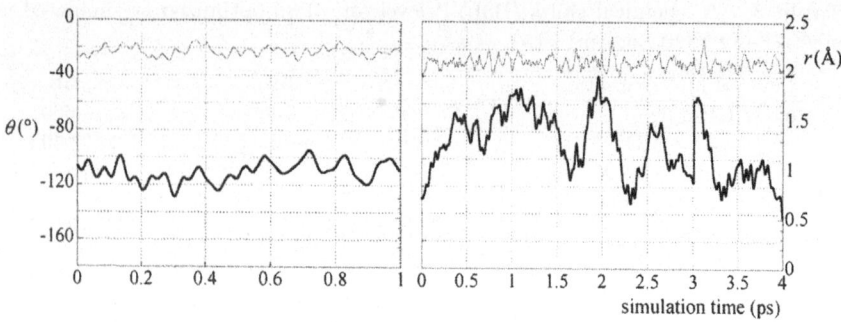

Fig. 5. Variations with time of the V-N distance r (upper curve) and of θ(OVNC) (lower curve) in simulations starting from **5a**; left: *in vacuo*, right: in water.

though, arguably, such a deviation is well within the accuracy of the methods employed, it is interesting to note that except for **4**, all $\delta(^{51}V)$ values of vanadates and peroxovanadates simulated so far are systematically too low (too negative) compared to experiment. Part of this systematic deviation could be due to the description of the standard, $VOCl_3$, which has not been simulated as the neat liquid used experimentally, but only in the gas phase.

Mechanism of Oxidation Reactions

As mentioned above, peroxovanadate complexes such as **5** are model compounds mimicking the oxidative chemistry of haloperoxidases and can be active oxidants, for instance for olefin epoxidation or sulfoxide formation with hydrogen peroxide [13], [14]. Borrowing from the proposed catalytic cycle for analoguous reactions involving related $MoO(O_2)_2L$ complexes [16], we studied hydrogen peroxide uptake and activation as the initial steps (highlighted box in Scheme 3). All attempts to locate a minimum with H_2O_2 coordinated via an oxygen atom to the vanadium center in **5** (cf. **6** in Scheme 3) failed. The additional ligand was always expelled from the coordination sphere of the metal and remained hydrogen-bonded to a peroxo moiety. The lowest minimum is displayed in Fig. 6. Consistent with this finding is the observation that in the solid state structure of **5** (with protonated imidazole as counterion and one crystal water), no donor atom is found close to the position trans to the oxo ligand [12].

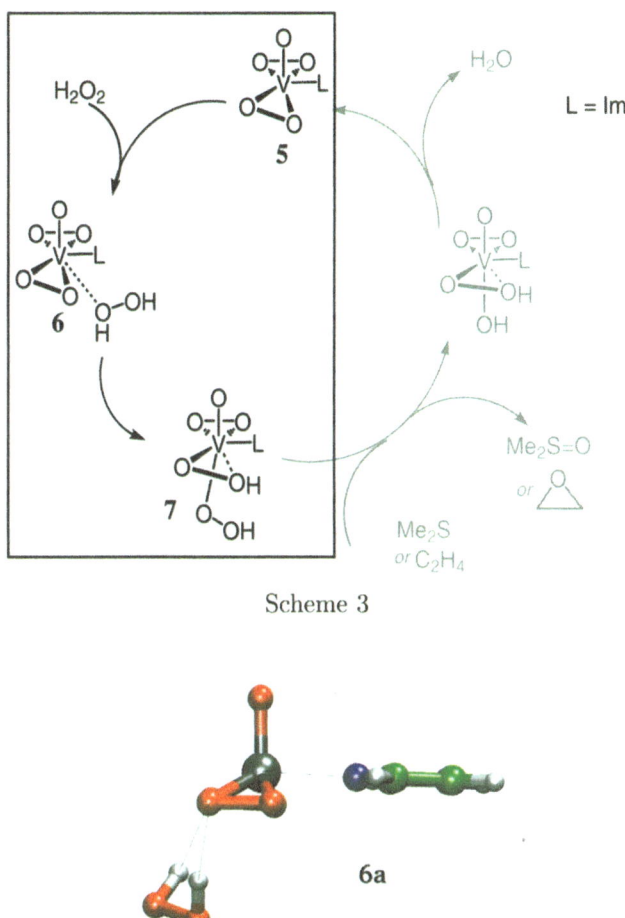

Scheme 3

6a

Fig. 6. BP86 optimized structure of $5 \cdot H_2O_2$ (side view)

Likewise, no minimum corresponding to **7** could be found. Optimizations starting from an arrangement as depicted in Scheme 3 resulted in dissociation of the imidazole ligand from the complex. It appears thus questionable at this point if the reaction sequence in Scheme 3 is correct. An alternative mechanism should be considered, where **5** itself can act as oxidizing agent by oxygen transfer from a peroxo ligand to the substrate, followed by re-oxidation of the resulting $VO_2(O_2)(Im)^-$ by hydrogen peroxide. Further theoretical work along these lines is necessary.

4 Conclusions

We have established a computational protocol for the estimation of thermal and solvent effects on transition-metal NMR chemical shifts. The procedure is based on MD simulations of the complex, typically in a periodic box filled with water as solvent, and averaging of the magnetic shieldings computed for snapshots along the trajectory. The simulations afford interesting insights into structural details of the solvated species, such as changes of geometrical parameters upon solvation. For anionic iron cyanide complexes, the mean Fe-ligand bond distances is found to be particularly sensitive to the medium. As this parameter is decisive for the ^{57}Fe chemical shift, remarkably large thermal and solvent effects on this property (exceeding 1000 ppm) are computed. For highly charged systems, specifically for a tetraanion, small box sizes in conjunction with periodic boundary conditions can lead to artifacts, which, apparently, do not come to the fore in case of singly or doubly charged species. For simple vanadate and peroxovanadate complexes, thermal and solvent effects on $\delta(^{51}V)$ are relatively small and do not depend critically on the particular density functional employed in the simulations. New results have been obtained for the dynamics of a peroxo vanadium imidazole complex, a model for biomimetic oxidants. In the gas phase, rotation of the imidazole ligand about the V-N bond is rapid, occuring on the picosecond timescale. In aqueous solution, this rotation is slowed down. The imidazole complex does not coordinate hydrogen peroxide at vanadium, a step hitherto believed to occur prior to catalytic substrate oxidation with this reagent. Hence, the mechanism for oxidations catalyzed by peroxovanadium systems may differ from that established for closely related molybdenum complexes. Further work is needed for a full mechanistic picture in the vanadium case, including the computational characterization of possible intermediates and transition states.

Acknowledgement. This work was supported by the Deutsche Forschungsgemeinschaft. M.B. thanks Prof. W. Thiel for continuous support. We gratefully acknowledge the CPU time allotment on the Hitachi SR-8000, without which the presented projects could not have been pursued.

References

1. See for instance NMR entries in: P. v. R. Schleyer,. N. L. Allinger, P. A. Kollman, T. Clark, H. F. Schaefer, J. Gasteiger, P. R. Schreiner (Eds.) (1998): Encyclopedia of Computational Chemistry. Wiley, Chichester
2. See for instance: a) M. Bühl (1997): Chem. Phys. Lett., **267**, 251–257; review: b) M. Bühl, M. Kaupp, V. G. Malkin, O. L. Malkina (1999): J. Comput.Chem., **20**, 91–105
3. V. G. Malkin, O. L. Malkina, G. Steinebrunner, H. Huber (1996): Chem. Eur. J., **2**, 452–457

4. M. Bühl, M. Parrinello (2001): Chem. Eur. J., **7**, 4487–4494
5. R. Car, M. Parrinello (1985): Phys. Rev. Lett., **55**, 2471–2474
6. M. Bühl, F. T. Mauschick, Phys. Chem. Chem. Phys., in press
7. C. Janak, T. Dorn, H. Paulsen, B. Wrackmeyer (2001): Z. Anorg. Allg. Chem., **627**, 1663–1668
8. M. Bühl, F. T. Mauschick, F. Terstegen, B. Wrackmeyer (2002): Angew. Chem. Int. Ed., **41**, 2312–2315; Angew. Chem., **114**, 2417–2420
9. See for instance: a) H. Mimoun, L. Saussine, E. Daire, M. Postel, J. Fischer, R. Weiss (1983): J. Am. Chem. Soc., **105**, 3101–3110; b) F. van de Velde, I. W. C. E. Arends, R. A. Sheldon (2000): Top. Catal., **13**, 259–265; c) W. Adam (Ed.) (2000): Peroxide Chemistry - Mechanistic and Preparative Aspects of Oxygen Transfer. Wiley-VCH, Weinheim
10. a) M. Bühl (1999): J. Comput. Chem., **20**, 1254–1261; b) M. Bühl (2000): J. Inorg. Biochem, **80**, 137–139
11. See for instance: W. Koch, M. C. Holthausen (2000): A Chemist's Guide to Density Functional Theory. Wiley-VCH, Weinheim
12. D. C. Crans, A. D. Keramidas, H. Hoover-Litty, O. P. Anderson, M. M. Miller, L. M. Lemoine, S. Pleasic-Williams, M. Vandenberg, A. J. Rossomando, L. J. Sweet (1997): J. Am. Chem. Soc., **23**, 5447–5448
13. J. Mukherjee, S. Ganguly, M. Bhattacharjee (1996): Ind. J. Chem. A, **35**, 471–474;
14. H. Glas, E. Herdtweck, G. R. J. Artus, W. R. Thiel (1998): Inorg. Chem., **37**, 3644–3646
15. T. S. Smith II, C. A. Root, J. W. Kampf, P. G. Rasmussen, V. L. Pecoraro (2000): J. Am. Chem. Soc., **122**, 767–775
16. A. Hroch, G. Gemmecker, W. R. Thiel (2000): Eur. J. Inorg Chem., 1107–1114

Part V

Solid-State Physics

Werner Hanke

Institut für Theoretische Physik und Astrophysik
Universität Würzburg
Am Hubland
97074 Würzburg, Germany

Solid-state physics has profited substantially from the high-performance supercomputing facilities of the HLRB in Munich. In the following, we give a
summary and appreciation of the corresponding projects:

The group around F. L. Gervasio and M. Parrinello considered the DNA
analysis on the basis of the so-called Car-Parrinello simulation technique. The
aim there was to simulate bio-polymers in experimentally relevant conditions.
Recently, there has been great interest in the nature of the electronic structure
of DNA because of its potential applications to nano-technology. The project
is based on the powerful Car-Parinello method to extract the electronic structure, the electronic polarization effects and inter-atomic forces. The latter
give us important clues on the conducting behavior of such molecules. In the
calculations, which have been performed in Munich at the HLRB, an important step was undertaken which allows to understand the above electronic
properties of DNA by making state-of-the-art density-functional calculations
on a system that has indeed been synthesized in the laboratory. Therefore,
the corresponding DNA system (a so-called dodecamer double strand crystallized in Z form) can now for the first time be used to make quantitative
comparison with experiment.

Another impressive example of the modern use of supercomputing comes
from the Augsburg group around D. Vollhardt and G. Keller. It is concerned
with solid-state many-body theory and, more specifically, with the metal-
insulator transitions and realistic modelling of strongly correlated electron
systems. Solid-state systems, such as the transition-metal oxides – a singularly important group of materials both from the point of view of fundamental
research and technological applications – may only be understood by explicit
consideration of the strong effective interaction between the conduction elec-

trons. These materials usually display a rich phase diagram as a function of temperature and an additional parameter, which can be pressure, doping of charge carriers, etc. The Augsburg project was concerned with the microscopic understanding of an important part of this phase diagram, namely the transition from the insulating to the metallic behavior, which many of the above materials display. In the last decade, a new approach for treating electronic lattice models, the dynamical mean-field theory (DMFT), has led to both analytical and numerical insights into the study of correlated electronic systems. This theory, which was developed by one of the authors, D. Vollhardt, in 1989, has recently been merged into another successful approach treating the electronic correlations, namely the local-density approximation (LDA) of density functional theory. This combined scheme makes possible the investigation of realistic systems close to a metal insulator transition (Mott-Hubbard transition) in strongly correlated materials. A variety of systems, such as V_2O_3 and other oxides have been investigated using the increased computer power of the Hitachi SR8000-F1, which allows to perform the DMFT + LDA calculations at experimentally relevant temperatures. The Augsburg project can be taken as an excellent example demonstrating that the calculation of physical properties of electronic many-particle systems made possible via high-performance computing (employing controlled approximations) is one of the most important achievements of modern theoretical solid-state physics. When successfully solved, it can replace the so far mostly used empirical search for improved material properties by a guided first-principle search.

The contributions of the group around W. Jahnke and P.-E. Berche (Leipzig) dealing with Monte-Carlo studies of three-dimensional bond-diluted ferromagnets also display the great impact of supercomputing on modern solid-state many-body problems. This problem deals, in particular, with the influence of quenched random disorder on phase transitions, here, of ferromagnetic character. Modelling the disorder by bond-dilution enables a test of the expected universality with respect to the type of disorder. Furthermore, this choice facilitates a quantitative comparison with recent high-temperature series expansions.

The contribution of A. Krug and A. Buchleitner (Dresden) deals with a theme from atomic physics, which, however, has also a strong overlap with solid-state questions. It is concerned with the microwave ionisation of Rydberg states of atomic hydrogen. These numerical high-performance studies present a new method, i. e. how highly excited alkali atoms can be elegantly simulated in the presence of an external field. This establishes a missing link between laboratory experiments on atomic hydrogen and various alkali atoms.

The project "Density functional calculation and inelastic neutron scattering of structural and dynamical properties in fluoride crystals" by the group around K. Schmalzl from Regensburg is a more standard electronic structure calculation using the already mentioned LDA of density functional theory. Here the aim has been to understand the electron-lattice coupling, i.e. phonons, of these complicated fluoride crystals. The project demonstrates convincingly, how the quantitative comparison with neutron scattering experiments allows to deduce important information on the inter-atomic effective forces in complex solid-state systems.

The project of A. Schmidt and F. Bechstedt (Jena) is concerned with a numerical calculation of the optical response of semiconductor surfaces and molecules. This calculation aims at extracting the so-called dynamical response function for semiconductors from electronic band structure information. This response function is important in that its poles determine the elementary excitations of the solids, such as electrons, phonons, excitons, plasmons etc. Here the concern is with the electron hole excitations measured in the optical response of solids. The method used is applicable to so-called weakly correlated electron systems, where the ratio of the Coulomb interaction energy over the kinetic energy establishes a small parameter, which can be used in standard perturbation theory. The corresponding Bethe-Salpeter equation can be solved by very efficient matrix inversion techniques and the corresponding elementary excitation extracted.

The studies of the Würzburg group around T. Eckl and W. Hanke deal with two topics from the microscopic theory of high-temperature superconductivity. One topic is concerned with the role of the so-called phase fluctuations: Although the binding of electrons into Cooper-pairs is essential and a prerequisite in forming the superconducting state, its remarkable properties – zero resistance and perfect diamagnetism – require phase coherence among the pairs as well. In conventional metals, Cooper pairs with short-range phase coherence survive no more than one degree above T_c. In underdoped high-T_c copper oxides, however, spectroscopic evidence for some form of pairing is found up to a temperature T^*, which is roughly 100 K above T_c. How this pairing and Cooper-pair formation are related, is a central unsolved problem in high-T_c superconductivity. In the Würzburg project first detailed numerical results of the dynamical properties of a model Hamiltonian are reported, which explicitly takes these phase fluctuations into account by using a new and very efficient Monte-Carlo technique. Comparision of the results with scanning-tunneling spectra of high-T_c superconductors, supports the idea that the celebrated pseudogap behavior observed in these and other experiments can be understood as arising from phase fluctuations of the pairing gap, whose amplitude forms on an energy scale set by a mean-field transition temperature T^* well above the actual superconducting transition. The second part of the project deals with two independent new Quantum-Monte-Carlo techniques, i. e. frozen phonon and response calculations, which are used to study the effects of electronic correlations on the so-called electron-phonon vertex function in the high-T_c superconductors. With the use of high-performance supercomputing this project aims at yet another unresolved issue in the high-T_c compounds. The pairing, which has just been mentioned above, forming the Cooper-pairs is undoubtedly mainly due to electronic correlations and magnetic forces in the high-T_c superconductors. However, as in any "low-T_c" BCS-type superconductor, there is also a contribution from the electron-lattice or electron-phonon interaction. The role of the latter contribution, whether it enhances or diminishes the electronically mediated pairing, is completely unclear up to now. This is due to the enormous difficulty to simultaneously deal with the strong correlations between the electrons and their interactions with dynamic lattice displacements, i. e. the phonons. The results obtained by the Würzburg group provide a first insight into the effects of strong electronic correlations on the electron-phonon interaction for certain relevant phonon branches.

Another project which is related to a certain extent with the above Würzburg project is from Th. Pruschke (Augsburg). It is concerned with the dynamical properties of correlated electrons in two and three dimensions using the Hubbard model and the so-called periodic Anderson model as prime examples. Both models play an important role, not only in the investigation e. g. of high-T_c superconductors, but also, more generally, of transition metal oxides and lanthanide compounds. The questions addressed in the project were to what extent so-called dynamical fluctuations influence the physics, for example, whether magnetic order renormalizes or destroys the Fermi liquid and whether superconductivity plays a role. Dynamical fluctuations stand here for electronic many-body effects beyond the conventional mean-field Hartree or Hartree-Fock pictures. The techniques used are Quantum-Monte-Carlo simulations in combinations with the already above mentioned dynamical mean-field theory (DMFT) for clusters. To efficiently use the structure of the Hitachi SR8000 for the DMFT-QMC computations, the Augsburg group exploited all of the parallelism that exists in their algorithm making the code inherently computationally efficient.

The project of G. Wellein (Erlangen), H. Fehske (Bayreuth) et al. deals with the interaction of spin, charge and lattice degrees of freedom in so-called Peierls-Hubbard systems. This project is concerned with the study of strongly correlated 1D electronic quantum systems on the basis of minimal microscopic models. Of particular concern are cooperative phenomena, which result out of the interaction of spin, charge and lattice degrees of freedom. As mentioned above for the Würzburg project, the physical challenge is to deal simultaneously with these competing degrees of freedom and the role of strong electronic correlations, which are not accessible to a small parameter and therefore to perturbation theory. Technically, this project uses exact diagonalization of large sparse matrices, which define a challenge for modern supercomputers.

In summary, in the field of supercomputing solid-state physics a variety of very interesting applications have been presented. They range from the more standard treatments of weakly correlated electron systems or band-electron systems, such as semiconductors, to strongly correlated materials, such as the high-T_c superconductors and technologically also very relevant other transition metal oxides. Beautiful other examples dealing with high-performance techniques, such as the Car-Parrinello simulation technique have been followed up to study clusters and surface structures of typically nano-systems. In these studies the aim is to understand system sizes of typically several thousand atoms. It is becoming more and more evident that these system sizes of intermediate length scales already contain salient fingerprints of both the structures and dynamics of the actual experimental system, which, of course, is realized in the thermodynamic limit or infinite-size limit.

Large Scale Car-Parrinello Simulation of Fully Hydrated DNA

Francesco Luigi Gervasio[1], Paolo Carloni[2], and Michele Parrinello[13]

[1] Centro Svizzero di Calcolo Scientifico
via Cantonale, 6928 Manno, Switzerland
[2] International School for Advanced Studies (SISSA/ISAS) and Istituto
Nazionale di Fisica della Materia (INFM)
via Beirut 4, 34014, Trieste, Italy
[3] Physical Chemistry, ETH Zurich
Hönggerberg, 8093 Zurich, Switzerland

Abstract. Density functional (BLYP/plane wave) calculations have been used to investigate the structure and the frontier orbitals of a fully hydrated crystalline DNA. Due to the size of the system (1194 atoms, 3,960 valence electrons requiring a total of 408,238 plane waves) this is one of the largest scale *ab-initio* simulations ever made. We find that the structure of water molecules around the DNA as well as their dipole moments are rather different from those of bulk water. The lowest conduction band is, surprisingly, found to be localized between the Na^+ counter-ions and the PO_4^- groups. This gives rise to a gap of only 1.28 eV. We have evaluated the anisotropic optical conductivity. At low frequency the conductivity along the DNA axis is dominated by the $\pi \rightarrow Na^+$ transitions, while the in-plane optical activity is mostly intrabase ($\pi \rightarrow \pi^*$ and $n \rightarrow \pi^*$). Our calculation demonstrates that hydration effects are fundamental for a proper understanding of DNA electronic properties.

Key words: DNA, electronic structure, plane waves, car-parrinello, ab initio

1 Introduction

DNA is doubtless one of the most important molecules from a bio-chemical point of view, as it is a fundamental molecule for life. Recently it has also attracted much interest as a possible nano-conductor [22, 30, 4, 24, 10, 34]. Indeed, DNA has several properties that make it attractive for a variety of applications. It is a stable polymer and can be handled easily and modified at will. Furthermore, its one-dimensional character and regular stacking of π-bases have suggested the possibility of using DNA as a nanoscale conductor. Unfortunately, experiments on this property have provided very contradictory results [10, 18].

Knowledge of the electronic structure could be of great help in understanding the conducting behaviour of DNA and also explain the mechanism of radiation-induced damage [23] and the importance of polarization and many-body effects on its structure and on that of the sourrounding water molecules.

However, due to size limitations, theoretical efforts to date have mostly been limited to the study of small fragments in the gas phase [19]. Only recently has an effort been made to study periodically infinite double strands [15]. However, in this instance a limited basis set was used and the solvation water and counterions, which are crucial for the stability of the DNA, were neglected [9]. It is our opinion that, in order not to miss important features of the electronic structure of this complicated molecule, it must be studied in conditions that can actually be synthesized and studied in the laboratory. In this study we make an important step towards understanding the electronic properties of DNA by making state-of-the-art density-functional calculations on a system that has been synthesized in the laboratory: a dodecamer double strand crystallized in Z form [6].

2 Methods

The system of choice was the Gua:Cyt dodecamer d(GpCpGpCpGpCpGpCpGpCpGpCp) (Fig. 1), a molecule that was synthesized and whose structure was solved by fibre diffraction X-ray crystallography [6]. This particular infinitely repeated biopolymer was chosen for three reasons, namely: among the crystal structures it has the smallest unit cell; it is very attractive for nanotechnology, since the absence of disorder in the base-pair sequence can in principle lead to a higher conductivity, and it is also interesting for radiation damage studies as it is rich in Gua bases, which are the most prone to UV-induced radical formation. This molecule has all the ingredients of an active DNA, namely the double strand, the counterions and the solvation waters.

The system crystallizes in the hexagonal space group P6(5)22, with cell dimensions a = b = 18.08 Å, c = 43.10 Å. The coordinates were taken from the Protein Data Bank [8]. The hydrogen atom positions, not resolved in the X-ray structure, were assigned by hand, respecting only the constraint of standard bond angles and bond lengths. Then the position of all the atoms was optimized with a quenched *ab-initio* molecular dynamics, which led to the spontaneous formation of a hydrogen bond network. Also the sodium counter-ions were not resolved in the crystal structure. After careful inspection their initial position was established in the following way. Twelve Na$^+$ were positioned in the middle of 12 symmetric solvent excluded cavities of 1.6 Å radius existing in the structure. The excluded volumes were calculated with the method of Connolly [13]. Since we could not find enough empty space in the structure to position the remaining 12 sodium ions, 12 symmetrical water molecules in the proximity of the phosphates were substituted with sodium ions. The position of the sodium ions was optimized with a quenched *ab-initio*

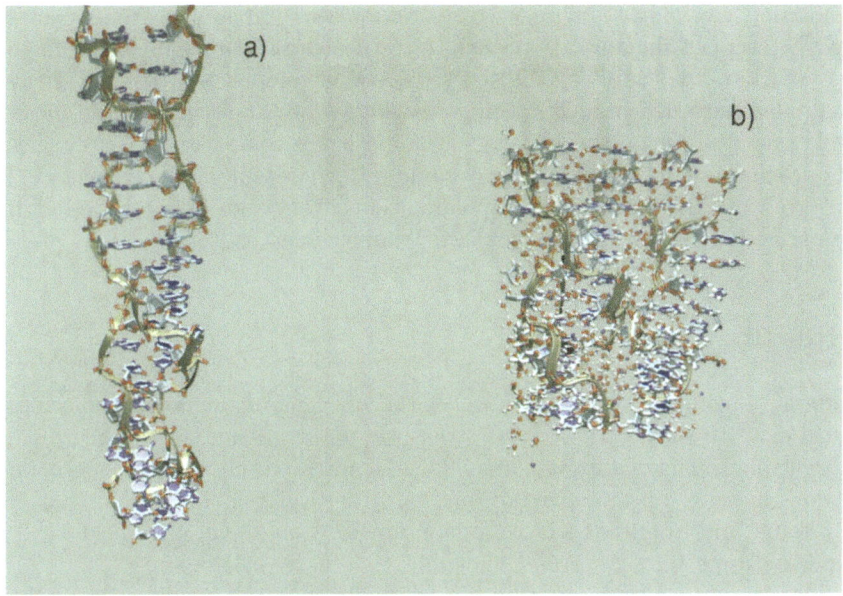

Fig. 1. View of the three-dimensional structure of the Gua:Cyt dodecamer [6] a) orthogonal to the **c** (z) axis. Water molecules, counterions and hydrogens have been removed for clarity. The sugar-phosphate backbone is represented as ribbons. b) View of the DNA, water and counter-ions in the periodic cell.

molecular dynamics in which all the symmetry restraints, with the exception of the periodic boundary conditions, were relaxed. Our model contained 654 heavy atoms and 540 hydrogen atoms (the molecular formula is: C_{228} N_{96} O_{144} P_{24} Na_{24} $H_{264}*138(H_2O)$).

The calculations were performed within the framework of density functional theory. Gradient corrected exchange and correlation functionals with the BLYP [7, 20] parametrizations were used. We treated 3,960 valence electrons explicitly, since in the case of sodium we treat explicitly the semicore states. The interaction between valence electrons and ionic cores was described by Martin-Troullier pseudopotentials [32]. The Kohn-Sham orbitals were expanded up to an energy cutoff of 70 Ry. The total number of PW treated was 408,238. The whole project ran for around 800,000 cpu-hours on the Hitachi SR8000 supercomputer of the Leibniz-Rechenzentrum München. We used a slightly modified version of the CPMD code [1]. The modifications were necessary to handle in a convenient way the very large restart files (more than 12 Gb each) on which the coordinates and plane wave coefficients were stored. The structure was optimized using a molecular dynamics/quenching scheme for ionic relaxation. The cell parameters were not optimized. We sample only the Γ point of the Brillouin zone, which given the cell size is an accurate approximation. Coulomb interactions are evaluated by the Ewald

summation method. No symmetry restriction was imposed on the calculation. We stopped the relaxation when the root-mean-square value of the force
was less than 10^{-3} au. The relaxed structure is compatible with the X-ray
data, which are determined within a resolution of 1.9 Å. In fact the mean
displacement of the optimized heavy atom positions compared the X-ray ones
is approximately 0.6 Å, and if we exclude water molecules it is less than 0.2
Å. The chemical bonding and the polarization effects were characterized with
the method of the maximally localized Wannier orbitals [21, 29].

3 Results and discussion

After the optimization the position of the oxygens of the water molecules
is still close to that of the crystal structure (as mentioned above the global
mean displacement is approximately 0.6 Å), but the resulting hydrogen-bond
network has an unusual structure and is rather different both from that of
bulk water and from that which were obtained by empirical force fields. (See
below and Fig. 2).

This leads to a wide distribution of dipole moments, ranging from the
value 1.7 D close to the gas phase of the molecules inside the helix to the
very high value of 3.8 D for a five-fold coordinated molecules that is found
in the water droplets between the major and the minor groove.

The spread of the dipole moments and the unusual water structure seems
to confirm the widely held view that water close to the nucleic acids (and
other biopolymers) has a different character from that in the bulk. This finding could have important consequences on the modelling of water-dissolved
biomolecules by molecular mechanics force fields.

We compared the hydrogen bond network obtained by full *ab-initio* optimization with that obtained with molecular mechanics calculations on the
same model system using the AMBER potential [14]. The optimized structure was obtained from the same initial configuration used for the *ab-initio*
optimization by a simulated annealing-quench procedure (6 ps of molecular
dynamics at 200 K followed by a quench to 0 K in 3 ps). The final structure,
while in general similar to the *ab-initio* one (in particular the position of the
oxygen atoms did not change much), has a different bonding pattern, particularly in the proximity of the water molecules that the *ab-initio* calculation
showed to be the most polarized compared to the bulk phase (Fig. 2).

We now turn to the description of the electronic properties. At the top
of the valence band we find twelve quasi-degenerate states (Fig. 3a, 3c). The
non-complete degeneracy is due to the residual disorder left in the structure.
These states have a π character and are mostly localized on the Gua nucle-
obases. The extent of localization is calculated by integrating the electron
density around the region surrounding the Gua bases and turns out to be
96 % of the total. The electron density of the individual states of this man-

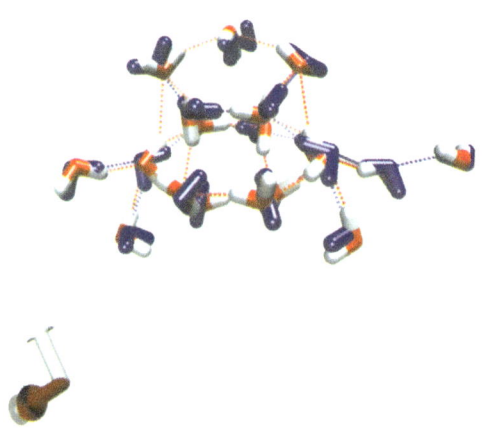

Fig. 2. Comparison of the H bond pattern obtained by *ab-initio* (white and red molecules) and molecular mechanics (blue molecules). The red axis points in the direction of the coil (Z).

ifold is non-uniformly spread over several Gua bases, being each time more concentrated on two (different) bases.

The summed density of the top 12 states is shown in Fig. 3 and has the full symmetry. The states immediately below the top of the valence band are relatively easy to assign to states localized also on the Gua. The first Cyt localized state is at 0.78 eV below the top. As we move further down the spectrum it becomes difficult to determine the nature of the states, as they become rather intermixed.

The DFT gap between empty and occupied states is particularly small, being only 1.28 eV. This small value is due to the presence of intercalating states between the top valence states and the first Cyt π^* states. The states at the bottom of the conduction band are *charge transfer* states where one electron has been moved outside the helix mostly on the Na^+ counterions and on the PO_4^- groups (Fig. 3b). This result is in agreement with the calculations of Enders et al. [16], who reaches similar conclusions about the reduced gap

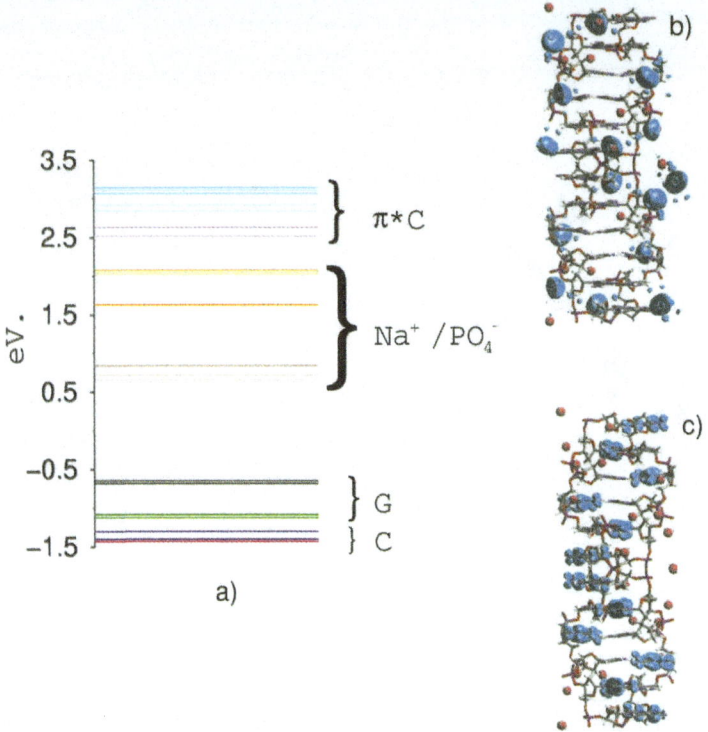

Fig. 3. a) Schematic level diagram around the Fermi level. The Fermi level positioned in the middle of the gap has been chosen as the zero of energy. b) Side and top view of the electronic density isosurface associated with the manifold of the 12 top states of the valence band. c) Top and side view of the electronic density isosurface associated with the manifold of the 12 low states of the conduction band. The isosurfaces represented have a value of 10^{-2} electrons Å^{-3}.

in the presence of counterions, and possibly with the experimental work of Tran et al. [27]

What happens if we remove the water molecules? To assess this issue we repeated the calculation by removing the water molecules but otherwise leaving the geometry of the DNA and counterions unchanged. The gap is much reduced. This reveals the electrostatic nature of the charge transfer states and the fundamental role of water in shielding the DNA from the electrostatic field of counterions.

To find the first excited states with a strong base character (Cyt π^* states) we must go up to 2.85 eV above the Fermi level, while the first π^* Gua state is at 3.18 eV. Since the gas phase values of the $\pi \rightarrow \pi*$ gap for Gua and Cyt are known, we can in the first approximation compare our calculated values to them. Our calculated gaps are 3.94 eV for Cyt and 3.82 eV for Gua, and those observed in experiments are 4.6 and 4.5 eV for Cyt and Gua, respectively [36]. This comparison shows that for these states the underestimation of the gap due to the use of DFT theory is less than 1 eV and that the relative values of the gaps are very well reproduced.

To obtain a quantity that in future can be measured experimentally we have calculated the diagonal elements of the optical conductivity tensor $\sigma_{\alpha\alpha}(\omega)$ using the Kubo-Greenwood expression [28, 2]:

$$\sigma_{\alpha\alpha}(\omega) = \frac{2\pi e^2}{3m_e^2 \Omega \omega} \sum_{v,c} | < \psi_v|p_\alpha|\psi_c > |^2 \delta(E_c - E_v - \hbar\omega)$$

where E_v, E_c and ψ_v, ψ_c are the energies and wavefunctions of the valence and conduction bands, respectively, and Ω is the volume of the supercell.

In Fig. 4 we plot the average optical conductivity $\sigma(\omega) = \frac{1}{3}Tr(\sigma_{\alpha\alpha}(\omega))$. This shows a sharp rise at about 4 eV due to $\pi \rightarrow \pi^*$ and $n \rightarrow \pi^*$ transitions described in the literature for the gas phase; [26,17,25,35] however before that

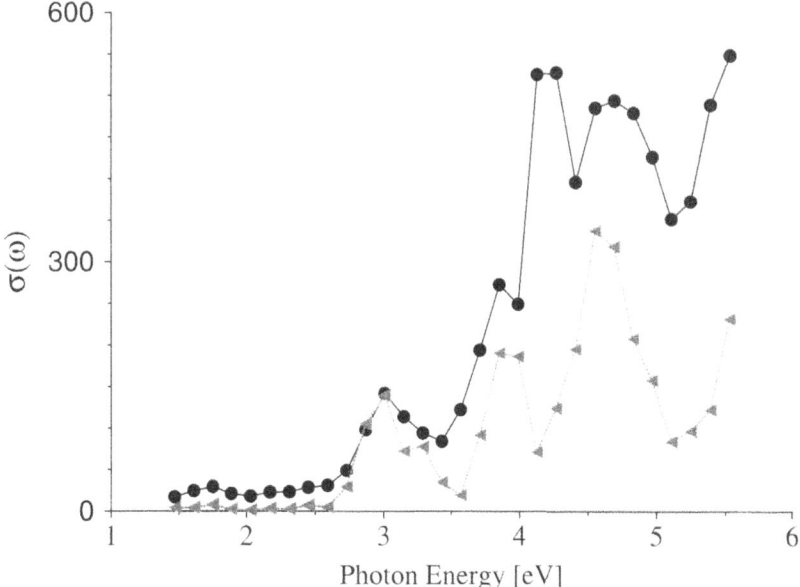

Fig. 4. Optical conductivity ($\Omega^{-1}cm^{-1}$) versus frequency (eV): a) $\sigma_{average}(\omega)$. The orange dotted line is the contribution to the total conductivity (black) due to excitations ending on the Na^+ electron transfer virtual states.

there is a tail of low-intensity states which extends to low energies. These are due to the $Gua \rightarrow Na^+$ kind of transitions. They have low intensity because the dipole moment matrix elements connect two states that are centered on different positions.

From the calculation of the optical activity in the strand direction we see that it is almost entirely due to the $Gua \rightarrow Na^+$ transition described here. In $\sigma_\perp(\omega)$, on the other hand, are also admixed the in plane $\pi \rightarrow \pi^*$ and $n \rightarrow \pi^*$ transitions. Note that the counterions Z coordinate lies between successive bases and therefore the transition dipole moments for the $Gua \rightarrow Na^+$ transitions have parallel as well as perpendicular components.

4 Conclusion and future perspectives

The present calculation is to our knowledge the largest Car-Parrinello simulation ever made and demonstrates what can nowadays be computed. Our finding on the structure of water around the DNA shows that polarization and possibly many-body effects are important and their weight should be carefully checked when simulating such systems with empirical (nonpolarizable) force fields.

The small value predicted for the electronic gap suggests the possibility that DNA can be doped, for instance by polyvalent counterions. One could also speculate that the contradictory experimental evidence on DNA conductions may be due to incidental doping.

Our calculations of the optical activity show great differences in the $\sigma_\perp$ and $\sigma_\parallel$ spectrum. These differences should be readily observed by experimental observation and are a mean to verify our findings.

From our calculations it emerges clearly that for a proper understanding of DNA electronic properties it is imperative to include solvation effects.

In the near future our group will study the dynamical behaviour of the electronic gap of this fundamental molecule. The structure of DNA dissolved in water (which, incidentally, is the condition in which most experiments have been performed) is highly disordered with fluctuations occurring on pico and nano-second time scales [5, 37]. These fluctuations are expected to have quite a large effect on the ability of DNA to transport charges, since the electronic overlap between neighbouring bases depends on the mutual orientation of the bases and on the disorder of the sourrounding water molecules [31, 3]. In the case of discotic materials such effects of dynamic fluctuations on the efficiency of the charge transport are considerable: the mobility of charge carriers is much lower in the liquid phase than in the crystalline phase [33]. The effects of dynamic structural fluctuations on charge transport in DNA have been so far only been discussed qualitatively or estimated theoretically using a tight binding Hamiltonian equation [12,11]. Furthermore the disorder of the sourrounding water and counter-ions has never been directly taken into account.

Since the energies of the frontier orbitals localized on the bases has been estimated to fluctuate by more than 0.2 eV from their mean value [12] under the effect of the disorder of the sourrounding water, their effect on the charge transport cannot be neglected.

We have made preliminary QM/MM calculations to study these effects.

The model molecule studied was the same as in the full quantum calculations, but only a Gua:Cyt couple was modelled as a quantum system, while all other atoms were simulated as classical particles with the Amber force field. We have observed very large fluctuations of the electronic gap (greater than 0.2 eV), coupled with the motion of counter-ions. Unfortunately the QM/MM simulations tend to overestimate such fluctuations. For this reason, i.e. to obtain a quantitative estimate of the electronic gap fluctuations, we plan to make Car-Parrinello molecular dynamics simulations at 300K of the whole DNA molecule. These simulations will be even larger than the one reported here, but due to the optimizations made on the CPMD code we expect approximately the same need of CPU hours per month as in the previous project phase.

Acknowledgement. We acknowledge a generous grant from the Leibniz-Rechenzentrum München on the Hitachi SR8000 supercomputer which has made this calculation possible.

References

1. *CPMD V3.5.* Copyright IBM Corp 1990-2001, Copyright MPI fuer Festkoerperforschung Stuttgart, 1997-2001.
2. Greenwood D. A. *Proc. Phys. Soc.*, 71:585, 1958.
3. Voityuk A. A., Siriwong K., and Roesch N. *Phys. Chem. Chem. Phys.*, 3:5421–5425, 2001.
4. A. P. Alivisatos, K. P. Johnson, T. E. Wilson, C. J. Loveth, M. P. Bruchez, and P. G. Schulz. *Nature*, 382:609–611, 1996.
5. Brauns E. B., Madaras M. L., Coleman R. S., Murphy C. J., and Berg M. A. *J. Am. Chem. Soc.*, 121:11644–11649, 1999.
6. C. Ban, B. Ramakrishnan, and M. Sundaralingam. *Biophys. J.*, 71:1215–1221, 1996.
7. A. D. Becke. *Phys. Rev. A*, 38:3098–3100, 1988.
8. F. C. L. Bernstein, T. F. Koetzle, G. J. B. Williams, E. E. Meyer Jr, M. D. Brice, J. R. Rodgers, O. Kennard, T. Shimanouchi, and M. Tasumi. *J. Mol. Biol.*, 112:535–542, 1977.
9. D. L. Beveridge and K. J. McConnell. *Current Opin. Struct. Biol.*, 10:182–196, 2000.
10. E. Braun, Y. Eichen, U. Silvan, and G. Ben-Yoseph. *Nature*, 391:775–778, 1998.
11. Dekker C. and Ratner M. A. *Phys. World*, 14:29–33, 2001.
12. Grozema F. C., Siebbeles L. D. A., Berlin Y. A., and Ratner M. A. *Chem. Phys. Chem.*, 3:536–539, 2002.

13. M. L. Connolly. *J. Appl. Cryst.*, 16:548–558, 1983.
14. Wendy D. Cornell, Piotr Cieplak, Christopher I. Bayly, Ian R. Gould, Kenneth M. Merz Jr., David M. Ferguson, David C. Spellmeyer, Thomas Fox, James W. Caldwell, and Peter A. Kollmann. *J. Am. Chem. Soc.*, 117:5179–5197, 1995.
15. P. J. de Pablo, F. Moreno-Herrero, J. Colchero, J. Gómez Herrero, P. Herrero, A. M. Baró, P. Ordejón, J. M. Soler, and E. Artacho. *Phys. REv. Lett.*, 85:4992–4995, 2000.
16. R. G. Enders, D. L. Cox, and R. R. P. Singh. *preprint at arXiv:cond-mat/0201404*, pages 1–23, 2002.
17. Zaloudek F., Novros J. S., and Clark L. B. *J. Am. Chem. Soc.*, 107:7344, 1985.
18. H. W. Fink and C. Schonenberger. *Nature*, 398:407–410, 1999.
19. P. Hobza and J. Sponer. *Chem. Rev.*, 99(11):3247–3276, 1999.
20. C. Lee, W. Yang, and R. G. Parr. *Phys. Rev. B*, 37:785–789, 1988.
21. N. Marzari and D. Vanderbilt. *Phys Rev. B*, 56:12847–12865, 1997.
22. C. A. Mirkin, R. L. Letsinger, R. C. Mucic, and J. J. Storhoff. *Nature*, 382:607–609, 1996.
23. D. T. Odom and J. K. Barton. *Biochemistry*, 40:8727–8737, 2001.
24. Y. Okata, T. Kobayashi, K. Tanaka, and M. Shimomura. *J. Am. Chem. Soc.*, 120:6165–6166, 1998.
25. Fulscher M. P., Serrano-Andres L., and Roos B. O. *J. Am. Chem. Soc.*, 119:6168–6176, 1997.
26. Fulscher M. P. and Roos B. O. *J. Am. Chem. Soc.*, 117:2089–2095, 1995.
27. Tran P., Alavi B., and Gruner G. *Phys. Rev. Lett.*, 85:1564–1567, 2000.
28. Kubo R. *J. Phys. Soc. Jpn.*, 12:570, 1957.
29. P. L. Silvestrelli, N. Marzari, D. Vanderbilt, and M. Parrinello. *Solid State Commun.*, 107:7, 1998.
30. J. J. Storhoff. *Chem. Rev.*, 99:1849–1862, 1999.
31. Fiebig T., Wan C., Kelley S. O., Barton J. K., and Zewail A. H. *Proc. Natl. Acad. Sci. USA*, 96:1187, 1999.
32. N. Troullier and J. L. Martins. *Phys. Rev. B*, 43:1993–2006, 1991.
33. van de Craats A. M., WArman J. M., de Haas M. P., Adam D., Simmerer J., Haarer D., and Schumacher P. *Adv. Mater.*, 8:823–826, 1996.
34. E. Winfree, F. Liu, L. A. Wenzler, and N. C. Seeman. *Nature*, 394:539, 1998.
35. Matsuoka Y. and Norden B. J. *J. Phys. Chem.*, 86:1378, 1982.
36. T. Yamada and H. Fukutome. *Bioplymers*, 6:43, 1968.
37. Liang Z., Freed J. H., Keyes R. S., and Bobst A. M. *J. Phys. Chem. B*, 104:5372–5381, 2000.

Metal-Insulator Transitions and Realistic Modelling of Correlated Electron Systems

Georg Keller[1], Dieter Vollhardt[1], Karsten Held[1 2], Volker Eyert[1], and Vladimir I. Anisimov[3]

[1] Center for Electronic Correlations and Magnetism, Theoretical Physics III
    Institute for Physics, University of Augsburg
    86135 Augsburg, Germany
[2] Physics Department, Princeton University
    Princeton, NJ 08544, USA
[3] Institute of Metal Physics
    Ekaterinburg, GSP-170, Russia

1 Introduction

The calculation of physical properties of electronic systems by controlled approximations is one of the most important challenges of modern theoretical solid state physics. In particular, the physics of transition metal oxides – a singularly important group of materials both from the point of view of fundamental research and technological applications – may only be understood by explicit consideration of the strong effective interaction between the conduction electrons in these systems. The investigation of electronic many-particle systems is made especially complicated by quantum statistics, and by the fact that the phenomena of interest (e.g., metal insulator transitions and ferromagnetism) usually require the application of non-perturbative theoretical techniques.

One of the most famous examples of a cooperative electronic phenomenon of this type is the transition between a paramagnetic metal and a paramagnetic insulator induced by the Coulomb interaction between the electrons, referred to as Mott-Hubbard metal-insulator transition. The question concerning the nature of this transition poses one of the fundamental theoretical problems in condensed matter physics [1, 2]. Correlation-induced metal-insulator transitions (MIT) are found, for example, in transition metal oxides with partially filled bands near the Fermi level. For such systems band theory typically predicts metallic behavior. The most famous example is V_2O_3 doped with Cr [3, 4, 5]; see Fig. 1.

While at low temperatures V_2O_3 is an antiferromagnetic insulator with monoclinic crystal symmetry, it has a corundum structure with a small trigonal distortion in the high-temperature paramagnetic phase. All transitions shown in the phase diagram (Fig. 1) are of first order. In the case of the transitions from the high-temperature paramagnetic phases into the low-temperature antiferromagnetic phase this is naturally explained by the fact that the transition is accompanied by a change in crystal symmetry. By contrast, the crystal symmetry across the MIT in the paramagnetic phase remains intact, since only the ratio of the c/a axes changes discontinuously. This is usually taken as an indication for the predominantly electronic origin of this MIT, caused by strong correlations.

In the last decade, a new approach for treating electronic lattice models, the dynamical mean-field theory (DMFT), has led to new analytical and numerical opportunities to study correlated electronic systems [6,7]. This theory, introduced by the work of Metzner and Vollhardt in 1989, is exact in the limit of infinite dimensions ($d = \infty$) [8]. In this limit, the problem is reduced to a single-impurity Anderson model with self consistency condition [9,10,11], allowing for a solution by quantum Monte-Carlo (QMC) simulations without a sign problem for one-band models (for multi-band models, see Ref. [12]),i.e. down to temperatures $T \sim 10^{-2}W$ (W: bandwidth).

Recently, the LDA+DMFT, a new computation scheme that merges electronic band structure calculations and the dynamical mean field theory, was developed [13,14,15,16]. Starting from conventional band structure calculations in the local density approximation (LDA) the correlations are taken into account by a Hubbard interaction term and a Hund's rule coupling term. The resulting DMFT equations are solved numerically with a parallelized auxiliary-field quantum Monte-Carlo algorithm. This scheme makes possible the investigation of real systems close to a Mott-Hubbard transition such as the MIT in V_2O_3 discussed above. In this paper, results on

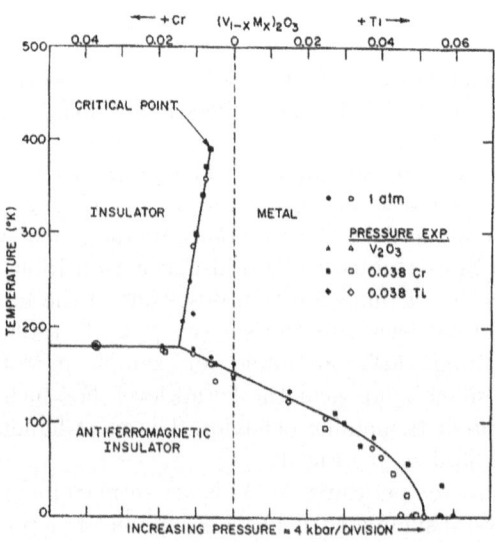

Fig. 1. Phase Diagram of V_2O_3 doped with Cr or Ti (temperature versus external pressure); from [4].

the correlation-induced metal-insulator transition and dynamic properties of multi-band Hubbard-type models at low temperatures obtained within the DMFT are presented. Calculations at experimentally relevant temperatures were made possible by the increased computer power of the Hitachi SR8000-F1.

2 Method and algorithm

2.1 General remarks

Although the one band Hubbard model is able to explain certain basic features of the Mott-Hubbard MIT and the phase diagram of V_2O_3, it cannot explain the physics of that material in any detail. Clearly, a realistic theory of V_2O_3 must take into account the complicated electronic structure of this system. In the high-temperature paramagnetic phase V_2O_3 has an electronic structure with a $3d^2$ V^{3+} state, where the two e_g-orbitals are empty and the three t_{2g}-orbitals are filled with two electrons. A small trigonal distortion lifts the triple degeneracy of the t_{2g}-orbitals, resulting in one non-degenerate a_{1g}-orbital and two degenerate e_g^π orbitals.

For the investigation of such realistic multi-band systems near a Mott-Hubbard MIT, the best available method is LDA+DMFT which has been developed recently [13, 14, 15, 16, 17, 18].

In the LDA+DMFT approach the LDA band structure, expressed by a one-particle Hamiltonian H_{LDA}^0, is supplemented with the local Coulomb repulsion U and Hund's rule exchange J:

$$\hat{H} = \hat{H}_{LDA}^0 + U \sum_m \sum_i \hat{n}_{im\uparrow}\hat{n}_{im\downarrow} + \sum_{i,m\neq\tilde{m},\sigma,\tilde{\sigma}} (V - \delta_{\sigma\tilde{\sigma}}J)\,\hat{n}_{im\sigma}\hat{n}_{i\tilde{m}\tilde{\sigma}}. \quad (1)$$

Here, i denotes the lattice site; m and $\tilde{m}$ enumerate the three interacting t_{2g} orbitals. The interaction parameters are related by $V = U - 2J$ which holds exactly for degenerate orbitals and is a good approximation for the t_{2g} orbitals. Furthermore, since the t_{2g} bands at the Fermi energy are rather well separated from all other bands we restrict the calculation to these bands (for details of the computational scheme see [16,17]). With this restriction only the LDA DOS of the three t_{2g} bands enter the LDA+DMFT calculation [17]. The LDA-calculated value of the Coulomb repulsion U has a typical uncertainty of at least 0.5 eV [16]. For this reason, we adjust U to yield a metal-insulator transition with Cr doping. *A posteriori*, we can compare whether the adjusted value is in the range of values obtained from a constrained LDA calculation.

2.2 Description of the algorithm

During the last ten years, DMFT has proved to be a successful approach to investigate strongly correlated systems with local Coulomb interactions

[7]. It becomes exact in the limit of high lattice coordination numbers [8, 19] and preserves the dynamics of local interactions. Hence, it represents a *dynamical* mean-field approximation. In this non-perturbative approach the lattice problem is mapped onto an effective single-site problem which has to be determined self-consistently together with the **k**-integrated Dyson equation connecting the self energy Σ and the Green function G at frequency ω. Within the LDA+DMFT scheme this implies:

$$G_{qlm,q'l'm'}(\omega) = \frac{1}{V_B} \int d^3k \left(\left[\omega 1 + \mu 1 - H_{\text{LDA}}^0(\mathbf{k}) - \Sigma(\omega) \right]^{-1} \right)_{qlm,q'l'm'}. \quad (2)$$

Here, 1 is the unit matrix, μ the chemical potential, $H_{\text{LDA}}^0(\mathbf{k})$ the LDA Hamiltonian derived in a LMTO-basis, $\Sigma(\omega)$ denotes the self-energy matrix which is non-zero only between the interacting orbitals, $[...]^{-1}$ implies the inversion of the matrix with elements n $(=qlm)$, $n'(=q'l'm')$, and the integration extends over the Brillouin zone with volume V_B.

The DMFT single-site problem depends on $\mathcal{G}(\omega)^{-1} = G(\omega)^{-1} + \Sigma(\omega)$ and is equivalent [11, 9] to an Anderson impurity model if its hybridization $\Delta(\omega)$ satisfies $\mathcal{G}^{-1}(\omega) = \omega - \int d\omega' \Delta(\omega')/(\omega - \omega')$. The local one-particle Green function at a Matsubara frequency $i\omega_\nu = i(2\nu + 1)\pi/\beta$ (β: inverse temperature), orbital index m ($l = l_d$, $q = q_d$), and spin σ is given by the following functional integral over Grassmann variables ψ and ψ^*:

$$G_{\nu m}^\sigma = -\frac{1}{\mathcal{Z}} \int \mathcal{D}[\psi]\mathcal{D}[\psi^*] \psi_{\nu m}^\sigma \psi_{\nu m}^{\sigma*} e^{A[\psi,\psi^*,\mathcal{G}^{-1}]}. \quad (3)$$

Here, $\mathcal{Z} = \int \mathcal{D}[\psi]\mathcal{D}[\psi^*] \exp(A[\psi,\psi^*,\mathcal{G}^{-1}])$ is the partition function and the single-site action A has the form (the interaction part of A is in terms of the "imaginary time" τ, i.e., the Fourier transform of ω_ν)

$$A[\psi,\psi^*,\mathcal{G}^{-1}] = \sum_{\nu,\sigma,m} \psi_{\nu m}^{\sigma*} (\mathcal{G}_{\nu m}^\sigma)^{-1} \psi_{\nu m}^\sigma$$

$$-\frac{1}{2} \sum_{m\sigma,m\sigma'}' U_{mm'}^{\sigma\sigma'} \int_0^\beta d\tau \, \psi_m^{\sigma*}(\tau)\psi_m^\sigma(\tau)\psi_{m'}^{\sigma'*}(\tau)\psi_{m'}^{\sigma'}(\tau)$$

$$+\frac{1}{2} \sum_{m\sigma,m}' J_{mm'} \int_0^\beta d\tau \, \psi_m^{\sigma*}(\tau)\psi_m^{\bar\sigma}(\tau)\psi_{m'}^{\bar\sigma*}(\tau)\psi_{m'}^\sigma(\tau) \ . \quad (4)$$

This single-site problem (3) has to be solved self-consistently together with the **k**-integrated Dyson equation (2) to obtain the DMFT solution of a given problem.

The QMC algorithm by Hirsch and Fye [20] is a well established method to find a numerically exact solution for the Anderson impurity model and allows one to calculate the impurity Green function G at a given $\mathcal{G}^{-1}$ as well as correlation functions. In essence, the QMC technique maps the interacting electron problem (3) onto a sum of non-interacting problems where the single particle moves in a fluctuating, time-dependent field and evaluates this sum by Monte Carlo sampling. To this end, the imaginary time interval $[0, \beta]$ of the functional integral equation (3) is discretized into Λ steps of size $\Delta\tau = \beta/\Lambda$, yielding support points $\tau_l = l\Delta\tau$ with $l = 1 \ldots \Lambda$. Using this Trotter discretization, the integral $\int_0^\beta d\tau$ is transformed to the sum $\sum_{l=1}^\Lambda \Delta\tau$ and the exponential terms in (3) can be separated via the Trotter-Suzuki formula [21], which is exact in the limit $\Delta\tau \to 0$. The single site action $\mathcal{A}$ of (4) can now be written in the discrete, imaginary time as

$$
\mathcal{A}[\psi, \psi^*, \mathcal{G}^{-1}] = \Delta\tau^2 \sum_{\sigma\, m\, l,l'=0}^{\Lambda-1} \psi_{ml}^{\sigma\,*} \mathcal{G}_m^{\sigma\,-1}(l\Delta\tau - l'\Delta\tau) \psi_{ml'}^\sigma
$$

$$
- \frac{1}{2}\Delta\tau \sum_{m\sigma,m'\sigma'}' U_{mm'}^{\sigma\sigma'} \sum_{l=0}^{\Lambda-1} \psi_{ml}^{\sigma\,*} \psi_{ml}^\sigma \psi_{m'l}^{\sigma'\,*} \psi_{m'l}^{\sigma'}, \qquad (5)
$$

where the first term was Fourier-transformed from Matsubara frequencies to imaginary time. In a second step, the $M(2M-1)$ interaction terms in the single site action $\mathcal{A}$ are decoupled by introducing a classical auxiliary field $s_{lmm'}^{\sigma\sigma'}$:

$$
\exp\left\{ \frac{\Delta\tau}{2} U_{mm'}^{\sigma\sigma'} (\psi_{ml}^{\sigma\,*}\psi_{ml}^\sigma - \psi_{m'l}^{\sigma'\,*}\psi_{m'l}^{\sigma'})^2 \right\} =
$$

$$
\frac{1}{2} \sum_{s_{lmm'}^{\sigma\sigma'} = \pm 1} \exp\left\{ \Delta\tau \lambda_{lmm'}^{\sigma\sigma'} s_{lmm'}^{\sigma\sigma'} (\psi_{ml}^{\sigma\,*}\psi_{ml}^\sigma - \psi_{m'l}^{\sigma'\,*}\psi_{m'l}^{\sigma'}) \right\}, \qquad (6)
$$

where $\cosh(\lambda_{lmm'}^{\sigma\sigma'}) = \exp(\Delta\tau U_{mm'}^{\sigma\sigma'}/2)$ and M is the number of interacting orbitals. This so-called discrete Hirsch-Fye-Hubbard-Stratonovich transformation can be applied to the Coulomb repulsion as well as the Z-component of Hund's rule coupling [12]. It replaces the interacting system by a sum of $\Lambda M(2M-1)$ auxiliary fields $s_{lmm'}^{\sigma\sigma'}$. The functional integral can now be solved by a simple Gauss integration because the Fermion operators only enter quadratically, i.e., for a given configuration $\mathbf{s} = \{s_{lmm'}^{\sigma\sigma'}\}$ of the auxiliary fields the system is non-interacting. The quantum mechanical problem is then reduced to a matrix problem

$$
G_{\tilde{m}l_1 l_2}^{\tilde{\sigma}} = \frac{1}{\mathcal{Z}} \frac{1}{2} \sum_{l} \sum_{m'\sigma',m''\sigma''}' \sum_{s_{lm''m'}^{\sigma''\sigma'} = \pm 1} [(M_{\tilde{m}}^{\tilde{\sigma}\mathbf{s}})^{-1}]_{l_1 l_2} \prod_{m\sigma} \det \mathbf{M}_m^{\sigma\mathbf{s}} \qquad (7)
$$

with the partition function $\mathcal{Z}$, the matrix

$$
\mathbf{M}_{\tilde{m}}^{\tilde{\sigma}\mathbf{s}} = \Delta\tau^2 [\mathbf{G}_m^{\sigma\,-1} + \Sigma_m^\sigma] e^{-\tilde{\lambda}_m^{\sigma\mathbf{s}}} + \mathbf{1} - e^{-\tilde{\lambda}_m^{\sigma\mathbf{s}}} \qquad (8)
$$

and the elements of the matrix $\tilde{\lambda}_m^{\sigma s}$

$$\tilde{\lambda}_{mll'}^{\sigma s} = -\delta_{ll'} \sum_{m'\sigma'} \lambda_{mm'}^{\sigma\sigma'} \tilde{\sigma}_{mm'}^{\sigma\sigma'} s_{lmm'}^{\sigma\sigma'}. \qquad (9)$$

Here $\tilde{\sigma}_{mm'}^{\sigma\sigma'} = 2\Theta(\sigma' - \sigma + \delta_{\sigma\sigma'}[m' - m] - 1)$ changes sign if $(m\sigma)$ and $(m'\sigma')$ are exchanged. For more details, e.g., for a derivation of (8) for the matrix $\mathbf{M}$, see [20, 9, 7].

Since the sum in (7) consists of $2^{\Lambda M(2M-1)}$ addends, a complete summation for large Λ is computationally impossible. Therefore the Monte Carlo method, which is often an efficient way to calculate high-dimensional sums and integrals, is employed for importance sampling of (7). Further details of the employed QMC algorithm can be found in [18].

Using a Markov process and single spin-flips in the auxiliary fields, the computational cost of the algorithm in leading order of Λ is

$$2aM(2M - 1)\Lambda^3 \times \text{number of MC-sweeps}, \qquad (10)$$

where a is the acceptance rate for a single spin-flip.

The advantage of the QMC method is that it is (numerically) exact. It allows one to calculate the one-particle Green function as well as two-particle (or higher) Green functions. On present workstations the QMC approach is able to deal with up to three *interacting* orbitals at temperatures of about 700 K. Very low temperatures are not accessible because the numerical effort grows like $\Lambda^3 \propto 1/T^3$. Since the QMC approach calculates $G(\tau)$ or $G(i\omega_n)$ with a statistical error, it also requires the maximum entropy method [22] to obtain the Green function $G(\omega)$ at real (physical) frequencies ω.

3 Results

In a first step, LDA calculations were performed for paramagnetic *metallic* V_2O_3 and paramagnetic *insulating* $(V_{0.962}Cr_{0.038})_2O_3$, respectively [23]. The LDA results for corundum V_2O_3 and $(V_{0.962}Cr_{0.038})_2O_3$ are very similar. In particular, the changes in crystal and electronic structure occuring at the transition are insufficiently reflected by the LDA calculations and the experimentally observed insulating gap is *missing* in the LDA DOS. It is generally believed that this insulating gap is due to strong Coulomb interactions which are not adequately accounted for by the LDA. This is where our LDA+DMFT(QMC) scheme sets in. Using this approach we can show explicitly that the insulating gap is indeed caused by electronic correlations.

The spectra obtained by LDA+DMFT(QMC) imply that the critical value of U for the MIT is about 5 eV [23]. Indeed, at $U = 4.5$ eV one observes pronounced quasiparticle peaks at the Fermi energy, i.e., characteristic metallic behavior, even for the crystal structure of $(V_{0.962}Cr_{0.038})_2O_3$, while at

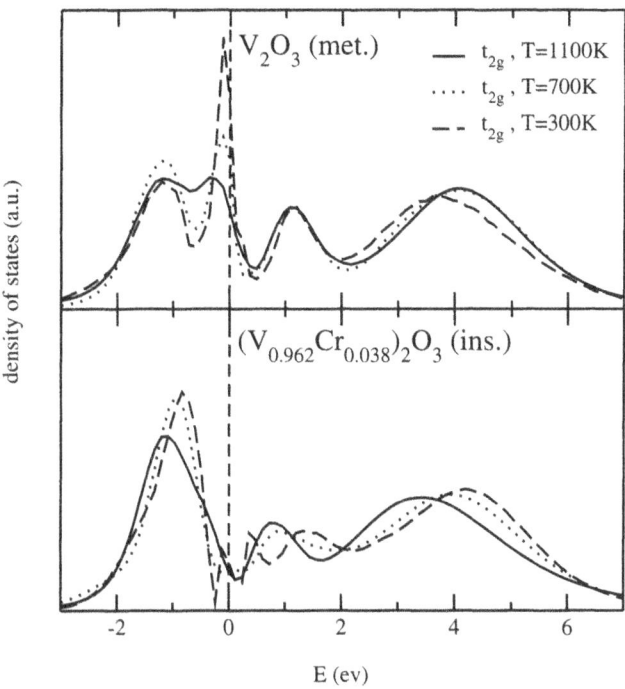

Fig. 2. LDA+DMFT(QMC) spectrum for metallic V_2O_3 and insulating $(V_{0.962}Cr_{0.038})_2O_3$ for $U = 5$ eV at $T = 1100$ K, $T = 700$ K and $T = 300$ K

$U = 5.5$ eV the form of the calculated spectral function is typical for an insulator for both sets of crystal structure parameters.

Whereas for the computations at $T = 1100$ K (which were done on a workstation), we only observe metallic-like and insulating-like behavior, with a rapid but smooth crossover between these two phases, the calculations done on the Hitachi SR8000-F1 at lower temperatures (Fig. 2) show more pronounced differences between the metallic and insulating phase; the smooth crossover is replaced by a sharp first order metal-insulator transition [24,25].

To compare with the photoemission spectrum of V_2O_3 by Schramme et al. [26] and Kim et al. [27] the LDA+DMFT(QMC) spectra are multiplied with the Fermi function at $T = 1100$ K and Gauss-broadened by 0.05 eV to account for the experimental resolution. In contrast to the LDA results, the theoretical results [23] for $U = 5$ eV are seen to be in good agreement with experiment (Fig. 3). We also note that the DOS is highly asymmetric with respect to the Fermi energy due to the orbital degrees of freedom. This is in striking contrast to the result obtained with a one-band model. The comparison between our results, the data of Müller et al. [28] obtained by X-ray absorption measurements, and LDA in Fig. 4 shows that, in contrast with LDA, our results not only describe the different bandwidths above *and* below the Fermi energy (≈ 6 eV and $\approx 2 - 3$ eV, respectively) correctly, but even resolve the two-peak structure above the Fermi energy.

Particularly interesting are the spin and the orbital degrees of freedom in V_2O_3. We find (not shown) that for $U \gtrsim 3$ eV the squared local magnetic moment $\langle m_z^2 \rangle$ saturates at a value of 4, i.e., there are *two* electrons with the same spin direction in the (a_{1g}, e_{g1}^π, e_{g2}^π) orbitals [23]. Thus, we conclude that the spin state of V_2O_3 is $S = 1$ throughout the Mott-Hubbard transition region. Our $S = 1$ result agrees with the measurements of Park et al. [29] and also with the data for the high-temperature susceptibility. Thus LDA+DMFT(QMC) provides a remarkably accurate microscopic theory of the strongly correlated electrons in the paramagnetic phase of V_2O_3 [23].

4 Summary

Applying the newly developed LDA+DMFT(QMC) scheme, which merges the conventional local density approximation (LDA) with DMFT in combination with QMC, to investigate the paramagnetic phase of V_2O_3 we find remarkable agreement with recent experiments. Indeed, LDA+DMFT(QMC) turns out to be a workable computational scheme which provides, at last, a powerful tool for future *ab initio* investigations of real materials with strong electronic correlations.

Future investigations will have to clarify the origin of the discontinuous lattice distortion at the first-order metal-insulator transition which leaves

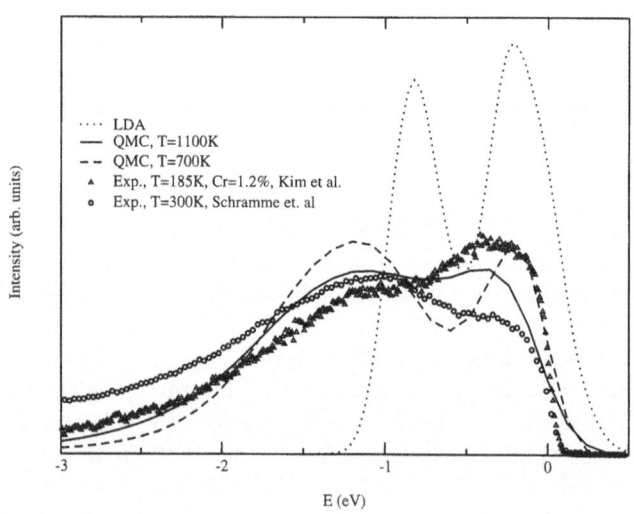

Fig. 3. Comparison of the LDA+DMFT(QMC) spectra [23] with the LDA spectrum and the photoemission experiments on metallic V_2O_3 by Schramme et al. [26](pure sample) and Kim et al. (Cr doped sample).

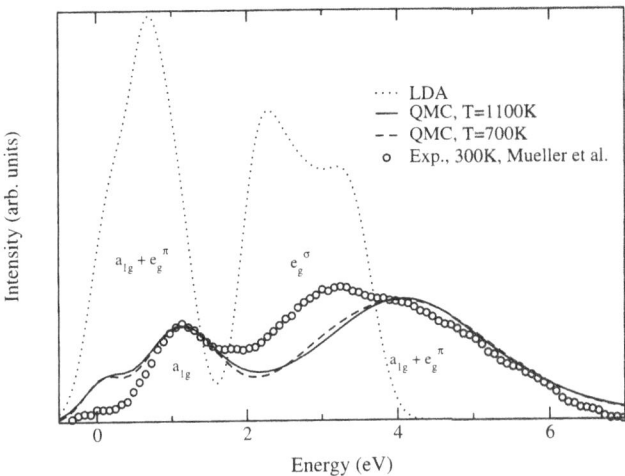

Fig. 4. Comparison of the LDA and LDA+DMFT(QMC) spectra at $T = 0.1$ eV (Gaussian broadened with 0.2 eV) [23] with the X -ray absorption data of Müller et al. [28].

the lattice symmetry unchanged. In particular, the metal-insulator transition might be the driving force behind the lattice distortion by causing a thermodynamic instability with respect to changes of the lattice volume and distortions. Such investigations at experimentally relevant temperatures are only possible on supercomputers like the Hitachi SR8000-F1. Eventually truly realistic investigations of transition metals and their oxides will require a genuine fusion of LDA with DMFT and the inclusion of all relevant electronic bands. As discussed above a larger number of bands and lower temperatures imply an enormous increase of numerical effort. This shows that substantial progress in the modelling of correlated electronic materials will only be possible with much faster computers.

We gratefully acknowledge support by the Leibniz-Rechenzentrum through HLRB-project h0531 and by the Deutsche Forschungsgemeinschaft through Sonderforschungsbereich 484.

References

1. N. F. Mott, Rev. Mod. Phys. **40**, 677 (1968); *Metal-Insulator Transitions* (Taylor & Francis, London, 1990).
2. F. Gebhard, *The Mott Metal-Insulator Transition* (Springer, Berlin, 1997).
3. D. B. McWhan and J. P. Remeika, Phys. Rev. **B2**, 3734 (1970).
4. D. B. McWhan et al., Phys. Rev. **B7**, 1920 (1973).
5. T. M. Rice and D. B. McWhan, IBM J. Res. Develop. 251 (May 1970).
6. D. Vollhardt, Investigation of correlated electron systems using the limit of high dimensions, in *Correlated Electron Systems*, Hrsg.: V. J. Emery, World Scientific, Singapore, 1993.

7. A. Georges, G. Kotliar, W. Krauth und M. Rozenberg, Rev. Mod. Phys. **68**, 13 (1996).
8. W. Metzner und D. Vollhardt, Phys. Rev. Lett. **62**, 324 (1989).
9. M. Jarrell, Phys. Rev. Lett. **69**, 168 (1992).
10. M. Jarrell und T. Pruschke, Z. Phys. B **90**, 187 (1993).
11. A. Georges und G. Kotliar, Phys. Rev. B **45**, 6479 (1992).
12. One limitation of QMC is that it is very difficult to deal with the spin-flip term of the Hund's rule coupling because of a "minus-sign problem" which arises in a Hubbard-Stratonovich decoupling of this spin-flip term, see K. Held, Ph.D. thesis Universität Augsburg 1999 (Shaker Verlag, Aachen, 1999).
13. V. I. Anisimov, A. I. Poteryaev, M. A. Korotin, A. O. Anokhin, and G. Kotliar, J. Phys.: Cond. Matt. **9**, 7359 (1997).
14. A. I. Lichtenstein and M. I. Katsnelson, Phys. Rev. B **57**, 6884 (1998).
15. M. B. Zölfl, Th. Pruschke, J. Keller, A. I. Poteryaev, I. A. Nekrasov, and V. I. Anisimov, Phys. Rev. B **61**, 12810 (2000).
16. I. A. Nekrasov, K. Held, N. Blümer, A. I. Poteryaev, V. I. Anisimov, and D. Vollhardt, Euro Phys. J. B **18**, 55 (2000).
17. K. Held, I. A. Nekrasov, N. Blümer, V. I. Anisimov, and D. Vollhardt, Int. J. Mod. Phys. B **15**, 2611 (2001).
18. For an introduction into LDA+DMFT, see K. Held, I.A. Nekrasov, G. Keller, V. Eyert, N. Blümer, A.K. McMahan, R.T. Scalettar, T. Pruschke, V.I. Anisimov, D. Vollhardt, Quantum Simulations of Complex Many-Body Systems: From Theory to Algorithms, eds. J. Grotendorst, D. Marx and A. Muramatsu, NIC Series Vol. 10 (NIC Directors, Forschungszentrum J?lich, 2002), p. 175.
19. E. Müller-Hartmann, Z. Phys. B **74**, 507 (1989); ibid. B **76**, 211 (1989).
20. J. E. Hirsch and R. M. Fye, Phys. Rev. Lett. **56**, 2521 (1986); M. Rozenberg, X. Y. Zhang, and G. Kotliar, Phys. Rev. Lett. **69**, 1236 (1992); A. Georges and W. Krauth, Phys. Rev. Lett. **69**, 1240 (1992); M. Jarrell, in *Numerical Methods for Lattice Quantum Many-Body Problems*, edited by D. Scalapino, Addison Wesley, 1997.
21. M. Suzuki, Prog. Theor. Phys. **56**, 1454 (1976)
22. M. Jarrell and J. E. Gubernatis, Physics Reports **269**, 133 (1996).
23. K. Held, G. Keller, V. Eyert, D. Vollhardt, and V. I. Anisimov, Phys. Rev. Lett. **86**, 5345 (2001).
24. M. J. Rozenberg, Phys. Rev. B **55**, R4855 (1997).
25. G. Moeller, Q. Si, G. Kotliar, M. J. Rozenberg, and D. S. Fisher, Phys. Rev. Lett. **74**, 2082 (1995); J. Schlipf, M. Jarrell, P. G. J. van Dongen, N. Blümer, S. Kehrein, Th. Pruschke, and D. Vollhardt, Phys. Rev. Lett. **82**, 4890 (1999); M. J. Rozenberg, R. Chitra and G. Kotliar, Phys. Rev. Lett. **83**, 3498 (1999); R. Bulla, Phys. Rev. Lett. **83**, 136 (1999).
26. M. Schramme, Ph.D. thesis, Universität Augsburg, 2000; M. Schramme et al. (unpublished).
27. H.-D. Kim, J.-H. Park, J. W. Allen, A. Sekiyama, A. Yamasaki, K. Kadono, S. Suga, Y. Saitoh, T. Muro, and P. Metcalf, cond-mat/0108044.
28. O. Müller, J. P. Urbach, E. Goering, T. Weber, R. Barth, H. Schuler, M. Klemm, S. Horn, and M. L. denBoer, Phys. Rev. B **56**, 15056 (1997).
29. J.-H. Park, L.H. Tjeng, A. Tanaka, J.W. Allen, C.T. Chen, P. Metcalf, J.M. Honig, F.M.F. de Groot, and S.A. Sawatzky, Phys. Rev. B **61**, 11 506 (2000).

Monte Carlo Studies of Three-Dimensional Bond-Diluted Ferromagnets

Pierre E. Berche[1], Christophe Chatelain[2,3], Bertrand Berche[2], and Wolfhard Janke[3]

[1] Groupe de Physique des Matériaux, Université de Rouen,
76821 Mont Saint-Aignan Cedex, France
pierre.berche@univ-rouen.fr
[2] Laboratoire de Physique des Matériaux
Université Henri Poincaré, Nancy I, BP 239
54506 Vandœuvre les Nancy Cedex, France
{chatelai, berche}@lpm.u-nancy.fr
[3] Institut für Theoretische Physik, Universität Leipzig
Augustusplatz 10/11, 04109 Leipzig, Germany
{christophe.chatelain, wolfhard.janke}@itp.uni-leipzig.de

1 Introduction

The influence of quenched, random disorder on phase transitions is of great importance in a large variety of fields [1], ranging from experiments with absorbed monolayers [2] in condensed matter physics to conceptual questions in non-perturbative quantum gravity [3].

For pure systems exhibiting a *continuous* phase transition, Harris [4] derived the criterion that random disorder is a relevant perturbation when the critical exponent of the specific heat of the pure system is positive, $\alpha_{\mathrm{pure}} > 0$. In this case one expects that the system falls into a new universality class with critical exponents governed by a "disordered" fixed point. For $\alpha_{\mathrm{pure}} < 0$ the behaviour of the pure system should persist, and $\alpha_{\mathrm{pure}} = 0$ is a special, marginal case. In two dimensions (2D) this scenario has been confirmed by various methods for many different systems [5]. In three dimensions (3D) extensive computer simulation studies have concentrated mainly on the *site*-diluted Ising model [5, 6].

If a pure system with a *first-order* phase transition is subject to quenched disorder, the transition is softened and may even turn into a continuous one [7]. This is always the case in 2D [8]; for numerical verifications see Refs. [9-13]. In higher dimensions, a tricritical point may appear at a finite concentration of impurities [14], separating "non-softened" first-order and "softened" second-order regimes [15]. Numerically such a scenario has recently been observed for the 3D *site*-diluted 3-state Potts model [16]. Since the first-order transition of the pure version of this model is very weak [17], however, the characterization of the tricritical point remained inconclusive.

In this report we give an overview on recent results obtained from extensive Monte Carlo (MC) computer simulations of the 3D 2-state (Ising) [18] and 4-state Potts [19-21] models with *bond*-dilution. The motivation to study the 4-state Potts model derives from the fact that, in the pure case, this model is known to exhibit a fairly strong first-order transition, such that a disorder-induced softening to a second-order transition would give clear support for the theoretical picture sketched above. Modeling the disorder by bond-dilution enables in the Ising case a test of the expected universality with respect to the type of disorder. Furthermore, for both models this choice facilitates a quantitative comparison with recent high-temperature series expansions [22, 23] for general random-bond q-state Potts models.

2 Model and Simulation Setup

The 3D bond-diluted q-state Potts model is defined by the Hamiltonian

$$-\beta H = \sum_{\langle ij \rangle} K_{ij} \delta_{\sigma_i, \sigma_j}; \quad \sigma_i = 1, \dots, q, \tag{1}$$

where the sum extends over all pairs of neighbouring sites on a cubic lattice of size L^3 with periodic boundary conditions, and the couplings K_{ij} are distributed according to the distribution

$$\wp(K_{ij}) = p\,\delta(K_{ij} - K) + (1 - p)\,\delta(K_{ij}), \tag{2}$$

where $K \equiv J/k_B T$ is the inverse temperature in natural units. The dilution parameter p is thus the concentration of magnetic bonds in the system, i.e., $p = 1$ corresponds to the pure case. Below the percolation threshold $p_c = 0.248\,812\,6(5)$ [24] one does not expect any finite-temperature phase transition since without any percolating cluster in the system long-range order is impossible.

The model (1), (2) was studied by means of large-scale MC simulations using the Swendsen-Wang (SW) cluster algorithm [25] in the regime of second-order transitions, and multicanonical simulations [26, 27] in the regime where the first-order transition of the pure 4-state Potts model persists, i.e., at weak dilution close to $p = 1$. Thermodynamic quantities were averaged over a large number of quenched disorder realisations, ranging between 2 000 and 5 000. The stability of the disorder averages has been checked by monitoring running averages as a function of the number of random samples. In fact, some care is necessary because too small a number of disorder realisations would lead to *typical* values rather than average ones [28], and these two values are different if the probability distribution over the disorder realisations exhibits a long tail.

3 Results

3.1 3D Bond-Diluted Ising Model

The phase diagram of the 3D bond-diluted Ising model has been obtained numerically from the locations of the maxima of a diverging quantity such as the magnetic susceptibility $\bar{\chi}_L$ depicted in Fig. 1. We focused on $\bar{\chi}_L$ because the stability of the disordered fixed point implies a negative specific-heat exponent in a random system [29]. The error in this quantity is hence typically larger than that in the susceptibility.

For an accurate determination of the maxima of the susceptibility we used the histogram reweighting technique with $N_{\mathrm{MCS}} = 2\,500$ MC sweeps (MCS) and between $N_s = 2\,500$ and $5\,000$ disorder realisations. The choice of N_{MCS} is justified by the increasing behaviour of the energy autocorrelation time τ_e as a function of p and L. At the critical point of a second-order phase transition one expects a finite-size scaling (FSS) behaviour $\tau_e \propto L^z$, where z is the dynamical critical exponent. From fits to the data shown in Fig. 2 we obtained values of $z \approx 0.27, 0.38, 0.41$, and 0.59 for $p = 0.4, 0.55, 0.7$, and 1.0 (the pure case), respectively. The critical slowing down thus weakens for the disordered model and becomes less and less pronounced when the concentration of magnetic bonds p decreases. The largest autocorrelation time observed for the disordered model was around $\tau_e \approx 9$ for $p = 0.7$ and $L = 96$. For each dilution and each size we thus collected at least 250 effectively uncorrelated measurements of the physical quantities, i.e., $N_{\mathrm{MCS}} > 250\,\tau_e$. On the other hand, for small $p \approx p_c$ it is necessary to increase the number of disorder realisations because of the vicinity of the percolation threshold.

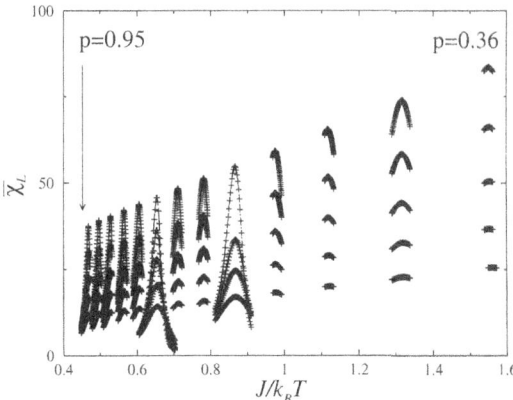

Fig. 1. The average magnetic susceptibility $\bar{\chi}_L$ of the 3D bond-diluted Ising model versus $K = J/k_BT$ for several concentrations p and $L = 8, 10, 12, 14, 16, 18$, and 20. For each value of p and each lattice size L, the curves are obtained by standard histogram reweighting of the simulation data at one value of K.

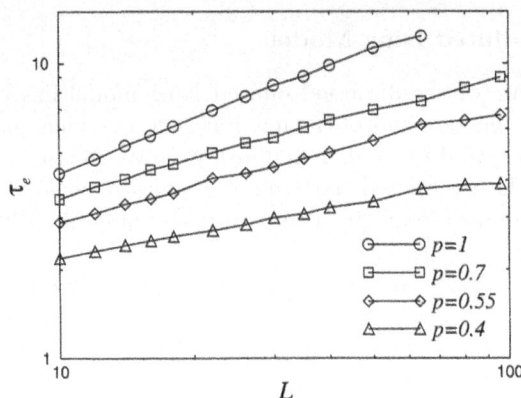

Fig. 2. The energy autocorrelation time τ_e versus the size L of the 3D bond-diluted Ising model with SW cluster-update dynamics for various concentrations of magnetic bonds p on a log-log scale. The pure case corresponds to $p = 1$.

Table 1. Evolution of the 3D Ising model susceptibility for $p = 0.7$ and $L = 96$ with the number of MC sweeps, N_{MCS}, for different disorder realisations, χ_j, and the average value over 2 500 realisations, $\bar{\chi}$.

N_{MCS}	χ_1	χ_2	χ_3	χ_4	χ_5	$\bar{\chi}$
100	1 268	720	1 141	939	833	1 058
500	1 272	1 520	1 223	1 029	953	1 210
1 000	1 262	1 544	1 205	1 068	911	1 219
1 500	1 282	1 433	1 277	1 047	915	1 227
2 000	1 332	1 441	1 221	1 073	917	1 235
2 500	1 358	1 484	1 234	1 012	1 014	1 234

In order to check the quality of the averaging techniques, we first studied the stability of the susceptibility versus the number of MC sweeps involved in the thermal average. For the largest size considered our results are given in Table 1 for different samples as well as for the disorder average. With 2 500 MCS, the accuracy of the results for a given sample is not perfect, of course, but the precision of the average over disorder is quite good on the other hand. The disorder average procedure has been investigated by computing the susceptibility χ_j for different samples, $1 \leq j \leq N_s$. As can be seen in Fig. 3, the dispersion of the values of χ is not very large because the fluctuations in the (running) average value disappear already after a few hundreds of realisations.

The phase diagram as obtained from the susceptibility maxima for the largest lattice size is shown in Fig. 4. We do not include the results from high-temperature series expansions [22] in this figure since they would just

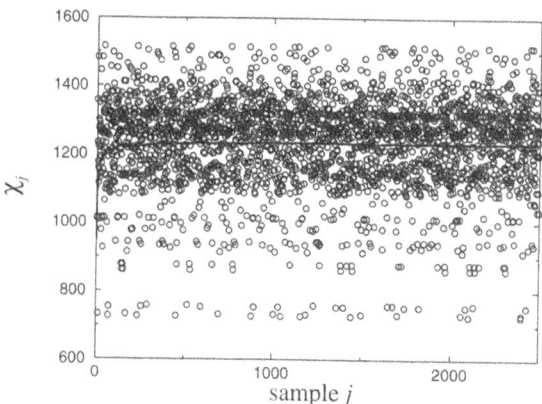

Fig. 3. Distribution of the susceptibility for the different disorder realisations of the 3D Ising model with a concentration of magnetic bonds $p = 0.7$ and $L = 96$. The (running) average value over the samples $\bar{\chi}$ is shown by the solid line.

fall on top of the MC data. For comparison we have drawn, however, a simple mean-field (MF) estimate of the transition point,

$$K_c^{\mathrm{MF}}(p) = K_c(1)/p, \tag{3}$$

with $K_c(1) = 0.443\,308\,8(6)$ [30]. This rather crude approximation only holds in the low-dilution regime, $p > 0.8$. On the other hand, the single-bond effective-medium (EM) approximation of Turban [31],

$$K_c^{\mathrm{EM}}(p) = \ln\left[\frac{(1 - p_c)e^{K_c(1)} - (1 - p)}{p - p_c}\right], \tag{4}$$

gives very good agreement with the simulated transition line over the full dilution range. Since $K_c(1)$ and p_c are input parameters this relation is trivially exact in the vicinity of both the pure system and the percolation threshold.

Due to the competition between different fixed points, one of the main problems encountered in previous studies of the disordered Ising model was the question whether one measures effective or asymptotic exponents. Although the change of universality class should happen theoretically for an arbitrarily low disorder, it can be very difficult to measure the new critical exponents because the asymptotic behaviour cannot always be reached practically. Another difficulty comes from the vicinity of the ratios γ/ν and β/ν in the pure and disordered universality classes. Indeed, for the 3D Ising model these values are:

$$\gamma/\nu = 1.966(6), \ \beta/\nu = 0.517(3), \ \nu = 0.6304(13) \quad \text{pure case [32]}, \tag{5}$$
$$\gamma/\nu = 1.963(5), \ \beta/\nu = 0.519(3), \ \nu = 0.6837(53) \quad \text{disordered case [33]. (6)}$$

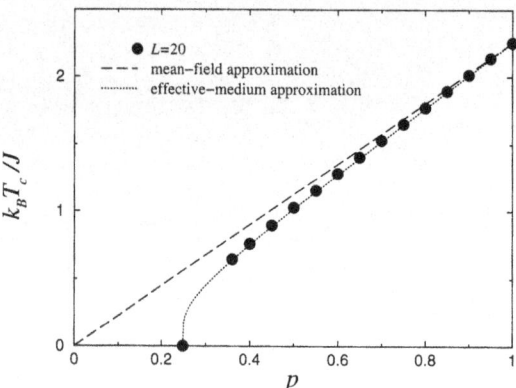

Fig. 4. Phase diagram of the 3D bond-diluted Ising model compared with the mean-field and effective-medium approximations (3) and (4), respectively.

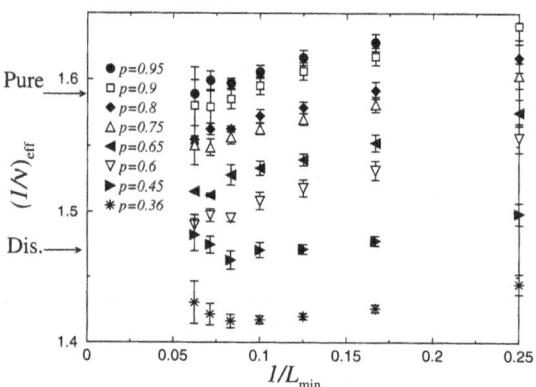

Fig. 5. Effective exponents $(1/\nu)_{\text{eff}}$ as a function of $1/L_{\min}$ for $p = 0.95, 0.9, 0.8, 0.75, 0.65, 0.6, 0.45$, and 0.36. The error bars show the standard deviations of the power-law fits. The arrows indicate the values of $1/\nu$ for the pure [32] and site-diluted [33] 3D Ising models, respectively.

Recent field theoretical determinations of critical exponents in the disordered case are presented in Refs. [34, 35], and for an excellent review of various experimental, theoretical and numerical estimates in the last two decades, see [6]. Thus, from standard FSS techniques, the critical exponent ν only will allow us to discriminate between the two fixed points. This exponent can be evaluated from the FSS behaviour of the derivative of the magnetisation w.r.t. temperature which is expected to behave as $d \ln \bar{m}/dK \propto L^{1/\nu}$. From this power-law behaviour, we have extracted the effective size-dependent exponent $(1/\nu)_{\text{eff}}$ which is plotted in Fig. 5 against $1/L_{\text{min}}$ for different bond concentrations p, with L_{min} denoting the smallest lattice size used in the fits.

We clearly see that in the regime of low dilution (p close to 1), the system is influenced by the pure fixed point. On the other hand, when the bond concentration is small, the vicinity of the percolation fixed point induces a decrease of $1/\nu$ below its expected disordered value. This is plausible since the percolation fixed point is characterized by $1/\nu \approx 1.12$ [24].

3.2 3D Bond-Diluted 4-State Potts Model

Let us now turn to the 4-state Potts model which exhibits in the pure case a rather strong first-order phase transition. In order to map out the phase diagram of the diluted model we considered all concentrations p in the interval $[0.28, 1]$ in steps of 0.04 and determined again the locations of the maxima of the susceptibility for a given lattice size L. The resulting phase diagram is depicted in Fig. 6, where we show for comparison again the simple mean-field prediction (3) and the effective-medium approximation (4), using $K_c(1) = 0.62863(2)$ [19]. On the scale of Fig. 6, estimates from high-temperature series expansions up to order 18 are hardly distinguishable, for a comparison see [23].

In a second step, the order of the phase transitions was investigated. Here, in order to satisfy our criterion $N_{\text{MCS}} > 250\,\tau_e$, the number of MC sweeps had to be increased up to $15\,000 - 30\,000$, which is rather large compared to the values used in the Ising case. A first indication is given by the FSS behaviour of the autocorrelation time τ_e at the transition point. A glance on the log-log plot of Fig. 7 shows a crossover around $p = 0.80$ from a power-law behaviour for strong disorder (small p) to a clear exponential behaviour for

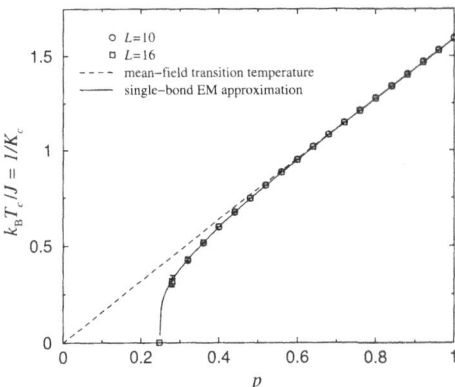

Fig. 6. Phase diagram of the 3D bond-diluted 4-state Potts model as obtained from MC simulations as well as the mean-field (dashed line) and single-bond effective-medium (solid line) approximations (3) and (4), respectively.

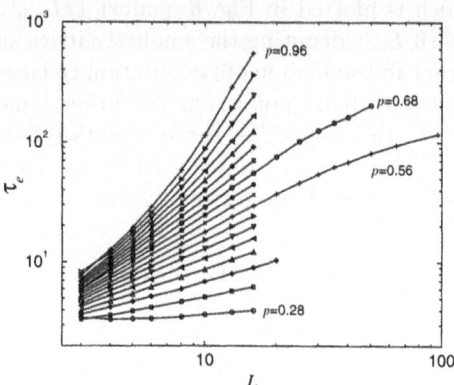

Fig. 7. Autocorrelation time τ_e of the energy at $K_c(p)$ versus lattice size L (p in steps of 0.04) for the 3D bond-diluted 4-state Potts model.

weak disorder ($p \approx 1$), as is typical for a first-order phase transition. In fact, in the latter case one expects $\tau_e \propto \exp(2\sigma_{od}L^2)$, where the (reduced) interface tension σ_{od} parameterizes the free-energy barrier separating the coexisting ordered and disordered phases.

In the first-order regime we performed multicanonical simulations and estimated the interface tension from

$$\sigma_{od} = \frac{1}{2L^2} \log \frac{P_{\max}}{P_{\min}}, \tag{7}$$

where $P_{\max}$ is the maximum of the probability density reweighted to the temperature where the two peaks are of equal height, and $P_{\min}$ is the minimum in between, see Fig. 8. The linear extrapolations of σ_{od} in $1/L$ in the lower part of Fig. 8 imply non-vanishing interface tensions only for $p = 0.84$ and above. For $p \le 0.76$, σ_{od} seems to vanish in the infinite-volume limit, being indicative of the expected softening to a second-order phase transition. The tricritical point would thus be located around $p = 0.76 - 0.84$, in good agreement with the estimate of $p = 0.80$ derived from the analysis of autocorrelation times.

To confirm the softening for $p \le 0.76$ we have performed a detailed FSS study at $p = 0.56$ with lattice sizes ranging up to $L = 96$ and the number of realisations varying between 2 000 and 5 000 [19]. The choice of $p = 0.56$ is motivated by our observation that in this range of dilutions the corrections to asymptotic FSS of the effective transition points are minimal, cf. Fig. 9.

The log-log plot for $\bar{\chi}_{\max}$ in Fig. 10 shows that for this quantity the corrections to asymptotic FSS seem to become quite small above $L = 30$, and fits of the form $a_\chi L^{\gamma/\nu}$ starting at $L_{\min} > 30$ yield $\gamma/\nu = 1.50(2)$. Using the data for $L < 30$ only, on the other hand, we obtained perfect fits assuming percolation exponents, $\gamma/\nu \approx 2.05$ [24], cf. Fig. 10. Similarly, the FSS of the quantity $(d\ln \bar{m}/dK)_{K_{\max}} \propto L^{1/\nu}$ gives for $L_{\min} > 30$ an estimate

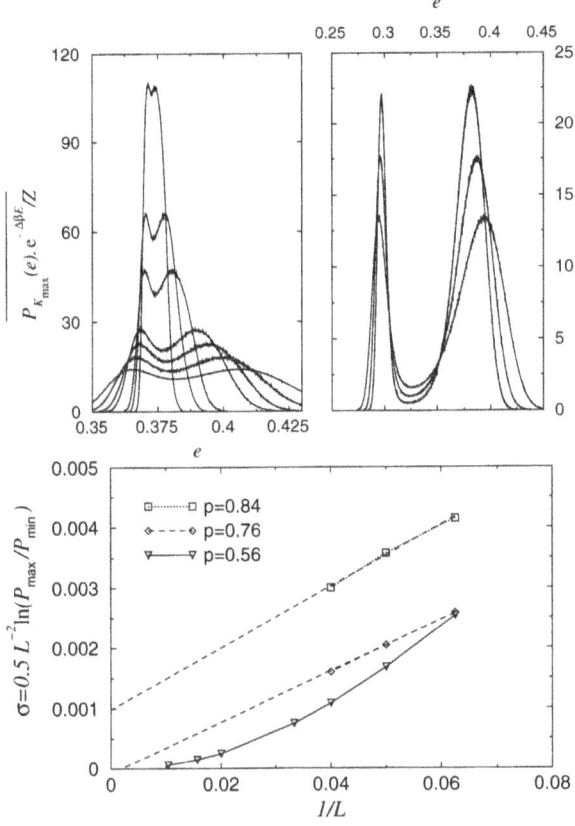

Fig. 8. Probability density of the energy of the 3D bond-diluted 4-state Potts model reweighted to equal peak height for $p = 0.56$ (top left) and $p = 0.84$ (top right). Interface tension versus inverse lattice size (bottom).

of the exponent $1/\nu = 1.33(3)$, consistent with the stability condition of the random fixed point ($1/\nu \leq D/2 = 1.5$) [29]. The same procedure was applied to the magnetization $\bar{m} \propto L^{-\beta/\nu}$, but here the associated critical exponent turned out to be not yet stable. We therefore also considered the FSS behaviour of higher (thermal) moments of the magnetization, $\overline{\langle \mu^n \rangle}$, which should scale with an exponent $n\beta/\nu$. The results for the first moments exhibit, however, again much stronger corrections to scaling than we observed for $\bar{\chi}$ or $d \ln \bar{m}/dK$, leading to quite a conservative final estimate of $\beta/\nu = 0.65(5)$. We nevertheless note that our results do not fit satisfactorily the scaling law $2\beta/\nu = d - \gamma/\nu$. The reason could be the strong corrections to scaling at the random fixed point which are hard to cope with for medium-sized systems [19].

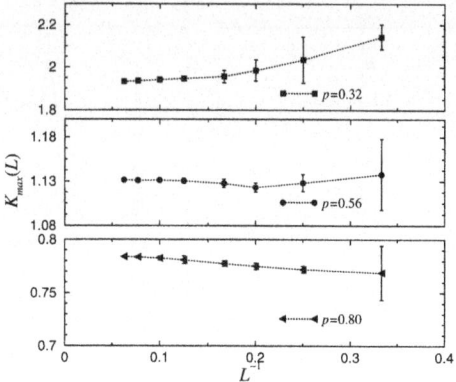

Fig. 9. FSS behaviour of the effective transition points for three different dilutions of the 3D 4-state Potts model as derived from the susceptibility maxima.

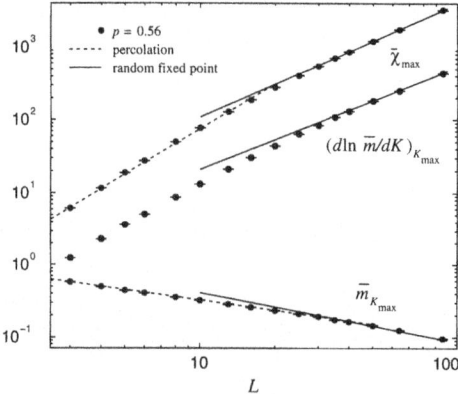

Fig. 10. FSS behaviour of the susceptibility, of $d \ln \bar{m}/dK$, and of the magnetization at $K_{\max}$ for the 3D bond-diluted 4-state Potts model at $p = 0.56$ (the quantities have been shifted in the vertical direction for the sake of clarity). The scaling behaviour for small lattice sizes below a crossover length scale is presumably governed by the percolation fixed point.

4 Conclusions

By performing large-scale Monte Carlo simulations we have investigated the influence of bond dilution on the critical properties of the 3D Ising and 4-state Potts models. In the 3D Ising case the universality class of the disordered model is modified by disorder but its precise characterization turned out be difficult because of the competition between the different fixed points which induce crossover effects, even for relatively large lattice sizes.

Applying similar techniques to the 3D 4-state Potts model we obtained clear evidence for softening to a continuous transition at strong disorder, with estimates for the critical exponents of $\nu = 0.752(14)$, $\gamma = 1.13(4)$, and $\beta = 0.49(5)$ at $p = 0.56$. The analysis of both the autocorrelation time and the interface tension leads to the conclusion of a tricritical point around $p = 0.80$.

Acknowledgement. We would like to thank Meik Hellmund and Loic Turban for helpful discussions. The mutual visits within this collaboration were financially supported by the joint PROCOPE exchange programme of the DAAD and EGIDE. C.C. thanks the EU network "EUROGRID: *Discrete Random Geometries: From solid state physics to quantum gravity*" for a post-doctoral position in Leipzig, and W.J. thanks the German-Israel-Foundation (GIF) for support. The numerical work would have been impossible without the computer-time grants 2000007 of the Centre de Ressources Informatiques de Haute-Normandie (CRIHAN), hlz061 of NIC, Jülich, and h0611 of LRZ, München. We are grateful to all institutions for their generous support.

References

1. Cardy, J. (1996): Scaling and Renormalization in Statistical Physics. Cambridge University Press Cambridge; chap. 8
2. Schwenger, L., Budde, K., Voges, C., Pfnür, H. (1994): Phys. Rev. Lett., **73**, 296
3. Janke, W., Johnston, D.A. (2000): Nucl. Phys. B, **578**, 681; J. Phys. A, **33**, 2653
4. Harris, A.B. (1974): J. Phys. C, **7**, 1671
5. Selke, W., Shchur, L.N., Talapov, A.L. (1994): In: Stauffer, D. (ed) Annual Reviews of Computational Physics I. World Scientific Singapore; pp. 17–54
6. For a recent overview, see Folk, R., Holovatch, Y., Yavors'kii, T. (2001): e-print cond-mat/0106468
7. Imry, Y., Wortis, M. (1979): Phys. Rev. B, **19**, 3580
8. Aizenman, M., Wehr, J. (1989): Phys. Rev. Lett., **62**, 2503
9. Chen, S., Ferrenberg, A.M., Landau, D.P. (1995): Phys. Rev. E, **52**, 1377
10. Picco, M. (1997): Phys. Rev. Lett., **79**, 2998
11. Jacobsen, J.L., Cardy, J.L. (1998): Nucl. Phys. B, **515**, 701
12. Chatelain, C., Berche, B. (1998): Phys. Rev. Lett., **80**, 1670
13. Olson, T., Young, A.P. (1999): Phys. Rev. B, **60**, 3428
14. Cardy, J., Jacobsen, J.L. (1997): Phys. Rev. Lett., **79**, 4063
15. For a review, see Cardy, J. (1999): Physica A, **263**, 215
16. Ballesteros, H.G., Fernández, L.A., Martín-Mayor, V., Muñoz Sudupe, A., Parisi, G., Ruiz-Lorenzo, J.J. (2000): Phys. Rev. B, **61**, 3215
17. Janke, W., Villanova, R. (1997): Nucl. Phys. B, **489**, 679
18. Berche, P.-E., Chatelain, C., Berche, B., Janke, W. (2002): Comp. Phys. Comm., **147**, 427
19. Chatelain, C., Berche, B., Janke, W., Berche, P.-E. (2001): Phys. Rev. E, **64**, 036120

20. Chatelain, C., Berche, P.-E., Berche, B., Janke, W. (2002): Nucl. Phys. B (Proc. Suppl.), **106&107**, 899
21. Chatelain, C., Berche, P.-E., Berche, B., Janke, W. (2002): Comp. Phys. Comm., **147**, 431
22. Hellmund, M., Janke, W. (2002): Comp. Phys. Comm., **147**, 435
23. Hellmund, M., Janke, W. (2002): Nucl. Phys. B (Proc. Suppl.), **106&107**, 923
24. Lorenz, C.D., Ziff, R.M. (1998): Phys. Rev. E, **57**, 230
25. Swendsen, R.H., Wang, J.S. (1987): Phys. Rev. Lett., **58**, 86
26. Berg, B.A. (2000): Fields Inst. Commun., **26**, 1
27. Janke, W. (1998): Physica A, **254**, 164
28. Derrida, B. (1984): Phys. Rep., **103**, 29; Aharony, A., Harris, A.B. (1996): Phys. Rev. Lett., **77**, 3700; Wiseman, S., Domany, E. (1998): Phys. Rev. Lett., **81**, 22
29. Chayes, J.T., Chayes, L., Fisher, D.S., Spencer, T. (1986): Phys. Rev. Lett., **57**, 2999; (1989): Comm. Math. Phys., **120**, 501
30. Talapov, A.L., Blöte, H.W.J. (1996): J. Phys. A, **29**, 5727
31. Turban, L. (1980): Phys. Lett. A, **75**, 307; J. Phys. C, **13**, L13
32. Guida, R., Zinn-Justin, J. (1998): J. Phys. A, **31**, 8103
33. Ballesteros, H.G., Fernández, L.A., Martín-Mayor, V., Muñoz Sudupe, A., Parisi, G., Ruiz-Lorenzo, J.J. (1998): Phys. Rev. B, **58**, 2740
34. Varnashev, K.B. (2000): Phys. Rev. B, **61**, 14660
35. Pelissetto, A., Vicari, E. (2000): Phys. Rev. B, **62**, 6393

Microwave Ionisation of Non-Hydrogenic Alkali Rydberg States

Andreas Krug and Andreas Buchleitner

Max-Planck-Institute for the Physics of Complex Systems
Nöthnitzer Str. 38
D-01187 Dresden
apk@mpipks-dresden.mpg.de, abu@mpipks-dresden.mpg.de

Abstract. We present a novel quantum treatment of highly excited alkali Rydberg states strongly driven by a coherent microwave field. Our approach allows realistic numerical experiments on various species of driven alkali atoms under laboratory conditions, and we apply it to lithium Rydberg atoms. We observe an unambiguous transition of the ionisation threshold from the regime of near-adiabatic driving into the domain of dynamical localisation. Thus, we establish the missing link between (laboratory) experiments on atomic hydrogen and various alkali atoms.

1 Introduction

Microwave driven Rydberg states of atomic hydrogen are one of the most prominent examples to study the influence of classically nonlinear motion on the dynamics of real, experimentally accessible quantum systems [1, 2]. For driving frequencies comparable to the highly excited electron's classical Kepler frequency, the classical dynamics turns chaotic leading to a diffusive energy gain of the electron, and finally to the ionisation of the atom on a finite time scale. However, due to the quantum nature of real atoms, the aforementioned classical, diffusive ionisation process is suppressed by quantum interference effects in a certain parameter range. This effect – known under the label dynamical localisation – is essentially equivalent to Anderson localisation [3], which explains the metal-insulator transition in disordered solids. The described scenario – i.e. the interplay between classically diffusive transport and ionisation in the presence of dynamical localisation – was predicted by theoretical investigations based on simplified (one-dimensional) models of the real atom, in order to handle the extremely high density of states in experimental situations, and agrees qualitatively well with experimental results on atomic hydrogen.

Despite their apparent similarity, however, little understanding has – until recently – been gained on microwave driven multi-electron atoms, since most of the theoretical work tackled the simpler atomic system of atomic hydrogen. In fact, laboratory experiments on alkali atoms showed strongly enhanced ionisation yields as compared to atomic hydrogen, when non-hydrogenic atomic states where initially populated [4, 5]. One reason for the theoretical/numerical complications experienced with non-hydrogenic, multi-electron atomic cores is the fact that such systems are indubitable three-dimensional objects. Hence, any theoretical approach for a realistic description of the experimental situation has to deal with the full dimensionality of the problem, and therefore with an extremely high density of bound states strongly coupled to the atomic continuum.

Sustained development in high-performance massive parallel computing techniques, together with the availability of large-scale supercomputing facilities, such as the HITACHI SR8000-F1, allows us nowadays to treat this problem in an exact manner, in its full complexity. It is the purpose of the present contribution to outline our algorithmic approach [6,7,8], and to present some of the recent results of our numerical experiment on microwave driven lithium atoms.

2 Theory and Numerical Implementation

Let us start with the non-relativistic Hamiltonian of an electron subject to the atomic potential V_{atom} and an intense, linearly polarised microwave field of frequency ω and amplitude $\mathbf{F}$,

$$H(t) = \frac{\mathbf{p}^2}{2} + V_{\text{atom}} + \mathbf{r} \cdot \mathbf{F} \cos(\omega t), \tag{1}$$

which we wrote in atomic units, employing the length gauge and the dipole approximation. Since the external field is periodic in time (with $T = 2\pi/\omega$, the microwave period), we can employ the Floquet theorem (a generalised version of the Bloch-theorem known from solid state physics). Hence, we search for the Floquet eigenstates $|\Psi_{\varepsilon_j}(\mathbf{r}, t)\rangle$ and associated quasi-energies ε_j which solve the Floquet eigenvalue problem:

$$\mathcal{H}|\Psi_{\varepsilon_j}(\mathbf{r}, t)\rangle = \varepsilon_j|\Psi_{\varepsilon_j}(\mathbf{r}, t)\rangle, j \in \mathbb{Z}, \tag{2}$$

with $\mathcal{H} = H - i\partial_t$ the Floquet-Hamiltonian. $\mathcal{H}$ acts on the extended Hilbert space of square integrable, time-periodic functions $\mathcal{L}^2(\mathbb{R}^3) \otimes \mathcal{L}^2(T)$, and its spectrum is invariant under a translation by ω. Thus, it suffices to consider only eigenvalues of $\mathcal{H}$ inside a given Floquet zone (the equivalent of the well-known Brillouin zone in solid state physics) of width ω.

As suggested by the time-periodicity of the problem, we introduce the Fourier decomposition of the Floquet eigenstates $|\Psi_{\varepsilon_j}(\mathbf{r}, t)\rangle = \sum_k \exp(-ik\omega t)|\Psi_{\varepsilon_j}^k(\mathbf{r})\rangle$, and obtain the following, time-independent set of coupled equations:

$$\left(-\frac{d^2}{dr^2} + \frac{\ell(\ell+1)}{r^2} - 2V_{\text{atom}}(r) - 2k\omega - 2\varepsilon_j \right) |\Psi_{\varepsilon_j,\ell}^k\rangle$$

$$+Fr\left(A_{\ell+1}\left(|\Psi_{\varepsilon_j,\ell+1}^{k-1}\rangle + |\Psi_{\varepsilon_j,\ell+1}^{k+1}\rangle \right) + A_\ell\left(|\Psi_{\varepsilon_j,\ell-1}^{k-1}\rangle + |\Psi_{\varepsilon_j,\ell-1}^{k+1}\rangle \right) \right) = 0,$$

$$A_\ell = \sqrt{\frac{\ell^2 - m^2}{4\ell^2 - 1}}, \quad k = -\infty, \ldots, +\infty. \tag{3}$$

The projection m of the angular momentum ℓ on the field polarisation axis (which we chose as the z axis) is conserved, and we will choose $m = 0$ in the following. In (3), k counts the number of photons that are exchanged between atom and field due to the coupling term $\mathbf{F} \cdot \mathbf{r} \cos \omega t$ in (1) (i.e., the non-diagonal terms in the second line of (3)) [9]. As a consequence of this field-induced absorption and emission of a photon, all bound states of the field free Hamiltonian $\mathbf{p}^2/2 + V_{\text{atom}}(r)$ are coupled to the atomic continuum. Therefore, the spectrum of (3) consists no longer of a discrete and a continuum part, but rather of resonances with finite life times $1/\Gamma_j$, embedded in the continuum. To separate these resonances from the continuum states, we use the method of complex dilation [10], i.e. we complexify the position and the momentum operator according to

$$r \to r \exp(\mathrm{i}\Theta), \quad p \to p \exp(-\mathrm{i}\Theta). \tag{4}$$

After this non-unitary transformation, the Floquet Hamiltonian – which will be represented in a real basis set – is represented by a complex symmetric rather than a hermitian matrix, with complex eigenvalues. Its spectrum has the following properties:

- The continuum states are rotated by an angle -2Θ from the positive energy axis into the complex plane. The resulting half lines are branching at the multi-photon ionisation threshold energies $K\omega$ (K an integer).
- The energies of the resonance states are given by $\varepsilon_j = E_j - \mathrm{i}\Gamma_j/2$ (with $E_j \in \mathbb{R}$, and $\Gamma_j \in \mathbb{R}^+$), where both, the positions E_j and the widths Γ_j, are independent of Θ, provided that the dilation angle is large enough to separate the resonances from the continuum. The corresponding eigenfunctions are square integrable, in contrast to the eigenfunctions of the undilated Hamiltonian, which are outgoing waves.
- Apart from exceptional values of F and ω, there are no real eigenvalues, since under periodic driving all atomic bound states turn into decaying states.

Here we will not extend on the details of complex dilation which are elaborated, e.g., in [11, 12].

So far, we did not specify the atomic potential V_{atom} in (1) and (3), and, hence, our discussion is valid for any atomic (or ionic) system with one active electron. To describe alkali atoms – for which a unique one-particle potential V_{atom}, valid for all $r \in \mathbb{R}$, does not exist – we generalise a variant of R-matrix theory, employed earlier [13] to describe alkali atoms in static fields, to the case of periodically driven systems. Accordingly, we split configuration space in three regions: In the internal region, $0 < r < r_{\text{core}}$, the dynamics of the highly excited electron is dominated by complicated multi-particle interactions. Outside the atomic core, but not far away ($r_{\text{core}} < r < r_0$) the valence electron's wavefunction can be written with the help of quantum defect theory, as a linear combination of regular and irregular Coulomb functions $s_{\ell,E}$ and $c_{\ell,E}$:

$$F_{\ell,E}(r) = \cos(\pi\delta_\ell)s_{\ell,E}(r) + \sin(\pi\delta_\ell)c_{\ell,E}(r), \tag{5}$$

where the δ_ℓ denote the angular momentum dependent quantum defects that specify the different alkali atoms, and which are known from highly accurate spectroscopic experimental data [14]. Finally, in the outermost region (i.e. for $r \geq r_0$), the potential V_{atom} can be described as a Coulomb potential $-1/r$. However, the eigenfunction in the outer region has to be matched smoothly to the wave function (5). This determines the phase shift of the alkali Rydberg electron's wave function as compared to the hydrogenic wave function, which is caused by a scattering of the valence electron off the multi-particle core.

To solve (3) for alkali Rydberg states, we thus have to substitute $V_{\text{atom}}(r) = -1/r$, and solve (3) in the range $r > r_0 > 0$. However, the operator d^2/dr^2 is no more hermitian on the interval (r_0, ∞). To enforce hermiticity in the reduced range, we add a surface term $-\delta(r - r_0)\left(\frac{\partial}{\partial r} + C_{\ell,\varepsilon_j}\right)$ to the Hamiltonian. In the constant term C_{ℓ,ε_j}, the matching between the outer and intermediate region, and thus the non-hydrogenic phase shift of the alkali wave function for large r is incorporated. Hence, we define C_{ℓ,ε_j} as the logarithmic derivative of the wave function $F_{\ell,E}(r) = F_{\ell,\varepsilon_j,k}(r)$, evaluated at energy $\varepsilon_j + k\omega$ and position r_0:

$$C_{\ell,\varepsilon_j,k} = \frac{1}{F_{\ell,\varepsilon_j,k}(r)}\frac{\partial}{\partial r}F_{\ell,\varepsilon_j,k}(r)\bigg|_{r=r_0}. \tag{6}$$

The complex dilated Floquet Hamiltonian amended by the core induced surface term is now represented in a real Sturmian basis set. These functions form a discrete basis set that represents both, the discrete and the continuum states, and they are ideally suited for Coulomb (-like) problems since they perfectly match the dynamical symmetry of the hydrogen atom [15]. This has to be paid by the minor disadvantage of the Sturmians being orthonormal with respect to a scalar product involving a factor $1/r$, instead of the usual scalar product on $\mathcal{L}_2(\mathbb{R}^3)$. Therefore, our eigenvalue problem (3) turns into a generalised eigenvalue problem

$$(\mathcal{A} - \varepsilon_i \mathcal{B}) \cdot \mathbf{x}_i = 0, \tag{7}$$

where both, $\mathcal{A}$ and $\mathcal{B}$, are non-diagonal, complex symmetric matrices.

Due to the high order of the multi-photon process that leads to the microwave ionisation of Rydberg states with principal quantum number $n_0 = 30, \ldots, 80$, these matrices become rather huge. More precisely, to obtain numerically converged results, we have to deal with matrices of dimension $280000, \ldots, 1010000$, and of bandwidth $4800, \ldots, 6300$, depending on the energy range we are interested in. However, to determine the ionisation probability

$$P_{\text{ion}}(t) = 1 - \sum_j w_j e^{-\Gamma_j t}, \tag{8}$$

(where the sum is understood to run over all eigenstates $|\Psi_{\varepsilon_j}\rangle$ in the chosen Floquet zone) we do not need all eigenvalues, but only those with a non-vanishing weight w_j of the associated atom-field eigenstates $|\Psi_{\varepsilon_j}\rangle$ projected on the atomic initial state. This means that typically less than 10000 eigenvalues and eigenvectors contain all the necessary spectral information. Since this is a small number as compared to the dimensions of the matrices, the diagonalisation of (7) is performed with a stable version of the Lanczos algorithm [16]. The latter is ideally suited to extract only a few close-by lying eigenvalues and associated eigenvectors of huge matrices.

In brief, the diagonalisation routine consists of the following parts:

- a Cholesky decomposition of $\mathcal{A}$ which is performed once per diagonalisation;
- the iterative generation of the Lanczos vectors that are orthogonal with respect to the matrix $\mathcal{B}$ (due to the use of Sturmian functions);
- at each Lanczos step a matrix inversion is needed, which is performed via a LDL^T decomposition;
- finally, the tridiagonal matrix which is constructed with the Lanczos algorithm is diagonalised, using a standard diagonalisation routine to obtain the eigenvectors $\mathbf{x}_j$ and eigenvalues ε_j.

Originally, the parallel version of this algorithm [17] was implemented on the CRAY T3E of the RZG Garching, i.e. on a distributed memory architecture. Later on, when our problem exhausted the CRAY T3E's capacity, we ported the code successfully to the HITACHI SR8000-F1 of the LRZ. At present, we do not take advantage of the hybrid architecture of the HITACHI, but use a massively parallel code which does not (yet) distinguish between intra- and internode communication.

3 Results

In the following, we shall present some recent results obtained in our numerical experiment on highly excited lithium atoms (specified by the quantum defects $\delta_{\ell=0} = 0.399468$, $\delta_{\ell=1} = 0.047263$, $\delta_{\ell=2} = 0.002129$, $\delta_{\ell=3} = -7.7 \cdot 10^{-5}$ [14]). We choose exactly the laboratory parameters of the experiment reported in [18], where the microwave ionisation of atomic hydrogen was studied. More precisely, we employ a linearly polarised microwave field with fixed frequency $\omega/2\pi = 36$ GHz, consider initial atomic states $n_0 = 28, \ldots, 80$, and fix the atom-field interaction time at $t = 327 \times 2\pi/\omega$. Since we are interested in core-induced, non-hydrogenic effects [6, 7, 8], we prepare the lithium atoms in states with the largest quantum defects, i.e. we choose $\ell_0 = m_0 = 0$ initial states. This is in contrast to the (laboratory) experiment on atomic hydrogen [18], where only the principal quantum number n_0 of the initial state was well-defined. Furthermore, we assume a constant microwave amplitude F experienced by the atoms during the atom-field interaction time t, thereby neglecting pulse-induced switching effects. These effects, however, are of minor importance in this kind of experiments [2, 19]. In a typical run of the numerical experiment we obtain the ionisation probability of several atomic initial states (by virtue of (8)) at fixed frequency, field amplitude, and atom-field interaction time t. Such result, i.e., the ionisation probability $P_{ion}(t)$ as a function of the principal quantum number n_0 is shown in Fig. 1. Since the external field, measured in units of the Coulomb field between electron and atomic core (which scales as n_0^{-4}), increases with increasing quantum number n_0, the higher principal quantum numbers tend to exhibit a larger ionisation yield, and P_{ion} globally increases with n_0. Besides that, all four curves show a typical threshold behaviour: below some quantum number n_{thr} (e.g. $n_{thr} \simeq 60$ at $F = 2.7 \cdot 10^{-9}$ a.u.) the ionisation probability remains rather flat. For $n_0 > n_{thr}$, the ionisation probability starts to increase rapidly, the external field is strong enough to induce strong coupling between a large number of atomic bound and continuum states. Put differently, beyond the perturbative regime the external field (with the cylindrical symmetry of a plane wave) destroys the spherical symmetry of the unperturbed initial atomic state, ultimately leading to a large ionisation probability. Note that the field-induced ionisation process is rather efficient, given the high order (15 ... 38 photons have to be absorbed by the electron before ionisation, in the parameter regime of Fig. 1) of the multi-photon transition amplitudes which connect the atomic initial state to the atomic continuum: in a perturbative picture no appreciable signal would be expected. Indeed, the field-induced continuum coupling is highly non-perturbative, and can be understood as a sequence of near-resonant one-photon transitions along a ladder of Rydberg states between the initial state and the continuum threshold [1].

Besides the threshold behaviour we also observe local structures on top of the global increase of P_{ion} with n_0, e.g. the local maxima at $n_0 = 63$ (at $F = 2 \cdot 10^{-9}$ a.u., ..., $2.7 \cdot 10^{-9}$ a.u.). Such local enhancement of the ionisation probability of a given initial state is caused by the complicated structure of the Floquet spectrum: Near-degeneracies ("avoided crossings") between two (or more) atom-field eigenstates at a given field amplitude reflect multi-photon resonances between atomic bound states. These enhance the coupling to the atomic continuum and thus also the observed total ionisation rate [20]. The overlap w_j of the initial state $|n_0 = 63, \ell_0 = m_0 = 0\rangle$ with such near-degenerate states is larger than the one of those $\ell_0 = 0$ states with principal quantum number $n_0 = 62$ and $n_0 = 64$, and thus we observe a local maximum at $n_0 = 63$ in Fig. 1 [8]. At $F = 3.1 \cdot 10^{-9}$ a.u., the avoided crossing is passed, and the states with $n_0 > 63$ experience a larger ionisation probability than the $n_0 = 63$ state. Similar effects, i.e. the appearance of local extrema on top of a global increase of P_{ion} were already reported in the discussion of laboratory experiments on the microwave ionisation of atomic hydrogen [2].

Given a sufficiently dense mesh of driving field amplitudes, we can now extract the "ionisation threshold" $F_{10\%}$ of a particular initial state $|n_0, \ell_0 = m_0 = 0\rangle$, from plots like Fig. 1. $F_{10\%}$ is defined – for historical reasons [2] – as that value of F which ionises 10% of the atoms for given ω and t. Fig. 2 shows the n_0-dependence of $F_{10\%}$ resulting from our numerical experiment on lithium, together with laboratory results on atomic hydrogen [18]. As mentioned above, the hydrogen experiment and our numerical experiment were performed with precisely the same parameters characterising the field and the atomic initial state. A closer inspection of Fig. 2 reveals the following:

- For low principal quantum numbers $n_0 < 54$, the ionisation threshold of lithium is significantly lower than the one of hydrogen, down to a factor of approx. four.
- For high principal quantum numbers $n_0 \geq 54$, atomic hydrogen and lithium exhibit essentially equal ionisation thresholds.
- Hydrogen and lithium results exhibit an algebraic decay of $F_{10\%}$ with $n_0^{-\gamma}$, with a clear change of the decay exponent at an element specific value of n_0.

The former two of these observations reconcile the apparently contradictory laboratory results on non-hydrogenic alkali Rydberg states and our intuitive picture of a Rydberg electron: As apparent from our exact numerical results, alkali Rydberg states do exhibit strongly enhanced ionisation for low principal quantum numbers, but mimic hydrogen dynamics at higher excitations. Comparison of the unperturbed Kepler frequency $\simeq n_0^{-3}$ of a n_0 hydrogen Rydberg state to the driving frequency immediately shows [6] that, indeed, alkali and hydrogen thresholds coincide as long as the driving frequency is larger than or comparable to n_0^{-3}, i.e., as long as $n_0 > 54$ in the example depicted in Fig. 2. Then, also in the alkali atom essentially hydrogenic transitions are driven by the external field, and the non-hydrogenic core potential remains largely immaterial for the ionisation yield. For lower n_0, hence larger Kepler frequency, the alkali dynamics is less stable with respect to the external perturbation, what can be understood in terms of the alkali level structure [6] which also provides the clue to item three in the above list: the change in the algebraic decay law expresses a transition from efficient field induced one-photon transitions (high n_0 values) between quasi resonant atomic field free states, to the low-n_0 regime (also characterised as the regime of near-adiabatic driving [5]), where the external field cannot drive resonantly any more any one-photon transitions between field free atomic states. Since the angular momentum degeneracy of the hydrogen spectrum is lifted by the low angular momentum states in the unperturbed alkali spectrum, the local average level spacing is smaller in lithium, and the transition from the high-n_0 to the low-n_0 regime occurs at lower n_0. As a matter of fact, all currently available experimental data on the microwave ionisation of non-hydrogenic alkali Rydberg states have been obtained in regime (II) and (III), what explains the experimentally observed discrepancy between hydrogen and alkali ionisation thresholds.

4 Conclusions and Outlook

With the help of the Floquet theorem, quantum defect theory and complex scaling of the Hamilton operator, we performed an exact numerical experiment on microwave driven alkali Rydberg states, without any adjustable parameters. An efficient parallel implementation of our approach on large supercomputers (CRAY T3E, and HITACHI SR8000-F1) allows to perform such calculations with precisely the parameters used in laboratory experiments. With these tools, we can now solve the puzzle of distinct hydrogenic and non-hydrogenic ionisation thresholds observed in experiments on both atomic species, a problem that remained unresolved for more than one decade. We could identify three regimes of the ionisation dynamics: For low principal quantum numbers (regime (III) and (II)), alkali atoms show enhanced ionisation rates as compared to atomic hydrogen, in agreement with previous experiments, while we predict that the ionisation thresholds of alkali and hydrogen atoms agree for high principal quantum numbers (regime (I)). State-of-the-art laboratory experiments [4, 21] can immediately verify our prediction on this transition between the different regimes.

Finally, our work proves that nowadays the exact treatment of complex atomic dynamics – without the application of simplified models nor the use of adjustable parameters – is indeed possible. In addition, we did not only show that such kind of project can be accomplished, but, moreover, our results clearly demonstrate that such a theoretical and numerical 'tour-de-force' is crucial for the comprehensive understanding of highly non-perturbative situations at high spectral densities.

Acknowledgement. CPU time was provided by the RZG of the Max-Planck-Gesellschaft, on a CRAY T3E, and by the LRZ of the Bayerische Akademie der Wissenschaften, on a HITACHI SR8000-F1. We acknowledge helpful discussions with D. Delande on the implementation of the R-matrix method in [13].

References

1. Casati, G., Chirikov, B. V., Shepelyansky, D. L., Guarneri, I. (1987): Relevance of classical chaos in quantum mechanics: the hydrogen atom in a monochromatic field. Phys. Rep. **154**, 77-123
2. Koch, P. M., van Leeuwen, K. A. H. (1995): The importance of resonances in microwave "ionization" of excited hydrogen atoms. Phys. Rep. **255**, 289-403
3. Fishman, S., Grempel, D. R. Prange, R. (1982): Chaos, quantum recurrences, and Anderson localization. Phys. Rev. Lett. **49**, 509-512.

4. Benson, O., Buchleitner, A., Raithel, G., Arndt. M., Mantegna, R. N., Walther, H. (1995): From coherent to noise-induced microwave ionization of Rydberg atoms. Phys. Rev. A **51**, 4862-4876

5. Mahon, C. R., Dexter, J. L., Pillet, P., Gallagher, T. F. (1991): Ionization of sodium and lithium Rydberg atoms by 10-MHz to 15-GHz electric fields. Phys. Rev. A **44**, 1859-2873

6. Krug, A., Buchleitner, A., (2001): Chaotic ionization of non-hydrogenic Rydberg states. Phys. Rev. Lett. **86**, 3538-3541

7. Krug, A., Buchleitner, A. (2002): Chaotic ionization of non-classical Rydberg states – Computational Physics beats experiment. Comp. Phys. Comm., in print

8. Krug, A. (2001): Alkali Rydberg States in Electromagnetic Fields: Computational Physics Meets Experiment. PhD Thesis, Ludwig-Maximilians-Universität, München, available at www.ub.uni-muenchen.de/elektronische_dissertationen/physik/Krug_Andreas.pdf

9. Shirley, J. H. (1965): Solution of the Schrödinger Equation with a Hamiltonian Periodic in Time. Phys. Rev. **138**, B979 - B987

10. Yajima, K. (1982): Resonances for the AC-Stark Effect. Comm. Math. Phys. **87**, 331-352

11. Ho, Y. K. (1983): The Method of Complex Coordinate Rotation and its Applications to Atomic Collision Processes. Phys. Rep. **99**, 1-68

12. Graffi, S., Grecchi, V., Silverstone, H. J. (1985): Resonances and convergence of perturbation theory for N-body atomic systems in external AC-electric field. Ann. Inst. Henri Poincaré **42**, 215-234

13. Halley, M. H., Delande, D., Taylor, K. T. (1993): The combination of R–matrix and complex coordinate methods: application to the diamagnetic Rydberg spectra of Ba and Sr. J. Phys. B **26**, 1775-1790

14. Lorenzen, C. J., Niemax, K. (1983):, Quantum Defects of the $n^2 P_{1/2,3/2}$ Levels in ^{39} KI and ^{85} RbI. Physica Scripta **27**, 300-305

15. Delande, D., Gay, J. C. (1986): The hydrogen atom in a magnetic field. Spectrum from the Coulomb dynamical group. J. Phys. B **19** L173-L178

16. Ericsson, T. Ruhe, A. (1980): The Spectral Transformation Lanczos Method for the Numerical Solution of Sparse Generalised Symmetric Eigenvalue Problems. Math. Comput. **35**, 1251-1268

17. Buchleitner, A., Taylor, K. T., Delande, D.: to be published

18. Galvez, E. J., Sauer, B. E., Moormann, L., Koch, P. M., Richards, D. (1988): Microwave Ionization of H Atoms: Breakdown of Classical Dynamics for High Frequencies. Phys. Rev. Lett. **61**, 2011-2014

19. Buchleitner, A., Delande, D., Gay, J. C. (1995): Microwave ionization of 3-dimensional hydrogen atoms in a realistic numerical experiment. J. Opt. Soc. Am. B **12**, 505-519

20. Buchleitner, A., Delande, D. (1995): Spectral Aspects of the Microwave Ionization of Atomic Rydberg States. Chaos, Solitons & Fractals **5**, 1125-1141

21. Noel, M. W., Griffith, M. W., Gallagher, T. F. (2000): Classical subharmonic resonances in microwave ionization of lithium Rydberg atoms. Phys. Rev. A **62**, 063401

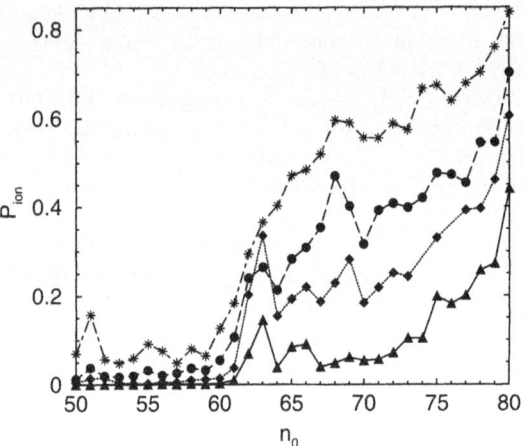

Fig. 1. Ionisation probability P_{ion} of lithium Rydberg states $|n_0, \ell_0 = m_0 = 0\rangle$ exposed to a $\omega/2\pi = 36$ GHz field, for an interaction time $t = 327 \times 2\pi/\omega$, and different values of the driving field amplitude $F = 2 \times 10^{-9}$ a.u. (triangles), 2.3×10^{-9} a.u. (diamonds), 2.7×10^{-9} a.u. (circles), and 3.1×10^{-9} a.u. (stars), as a function of n_0. Since the binding energy of the initial atomic state decreases with n_0, we observe a global increase of P_{ion} with n_0, for given F.

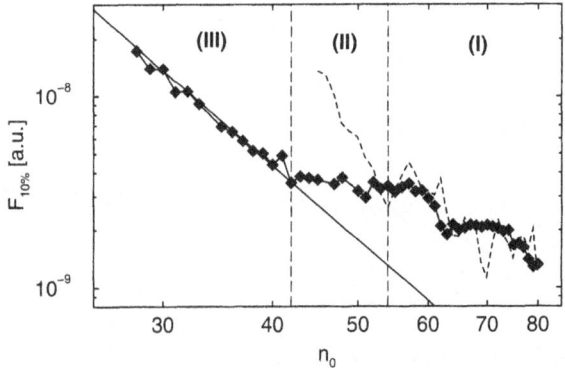

Fig. 2. Double-logarithmic plot of the numerical ionisation threshold $F_{10\%}$ of lithium Rydberg states $|n_0, \ell_0 = m_0 = 0\rangle$ (diamonds) as a function of n_0, compared to laboratory thresholds [2] of atomic hydrogen (dashed line). Both species were exposed to the same driving frequency $\omega/2\pi = 36$ GHz for the same interaction time $t = 327 \times 2\pi/\omega$, in the numerical and in the laboratory experiment, respectively. Note that the thresholds are manifestly different in the low n_0 regime (II,III), but essentially identical in the high n_0 regime (I). This explains the apparently enhanced ionisation of alkali Rydberg states conjectured on the basis of available laboratory data [4,5], which were all obtained in the low n_0 regime [8].

Density-Functional Calculation and Inelastic Neutron Scattering of Structural and Dynamical Properties in Fluoride Crystals

Karin Schmalzl[1,2], Dieter Strauch[1], H. Schober[2], Bruno Dorner[2], and A. Ivanov[2]

[1] Institut für Theoretische Physik
  Universität Regensburg
  93040 Regensburg, Germany
  dieter.strauch@physik.uni-regensburg.de
[2] Institut Laue-Langevin
  38042 Grenoble Cedex 9, France
  schmalzl@ill.fr

Abstract. We investigate electronic, structural, and dynamical properties of the superionic conductors LaF_3 and CaF_2 bulk fluoride systems using density-functional theory and pseudopotential methods with plane waves. The temperature dependence of the phonon frequencies (dispersion) and widths is investigated by inelastic neutron scattering.

1 Introduction

The fluoride crystals CaF_2 and LaF_3 are ionic conductors with a strong increase of the ionic conductivity at 1430 K and 1150 K, respectively, where the conductivity becomes comparable to the conductivity of a molten salt.

The origin of the ion transport mechanism is not completely clarified, but motional disorder in the fluorine sublattice and hopping over potential barriers is made responsible for it. The details of the conduction mechanism are believed to be different for the two materials. In order to add to the understanding of this mechanism we are investigating the lattice statics and dynamics.

We have performed *ab-initio* calculations to study the static and dynamical properties of the bulk fluoride systems CaF_2 and LaF_3. The results are used in conjunction with the interpretation of our recent preliminary experimental neutron-scattering data.

2 Static properties: Structure of CaF_2 and LaF_3

CaF_2 crystallizes in the fcc structure with one formula unit (three atoms) in the unit cell. It can be viewed as made up of tetrahedra and octahedra with Ca atoms at the corners and with the F atoms at the centers of the tetahedra. Figure 1 shows a cube containing four formula units; a complete octahedron is spanned by the Ca atoms on the face centers of the cube, while a tretrahedron is spanned by a Ca atom on a cube corner and the three adjacent Ca atoms on the face centers.

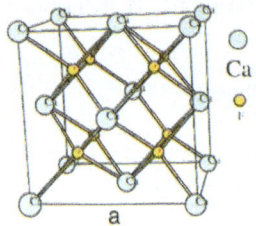

Fig. 1. Crystal structure of CaF$_2$.

The crystal structure of LaF$_3$ is still under debate [1]. Most authors agree on a trigonal structure with six formula units (24 atoms) in the unit cell and 7 structure parameters [2,3,4], see Figure 2. But also a hexagonal structure with only two formula units per elementary cell and three structure parameters has been proposed [5].

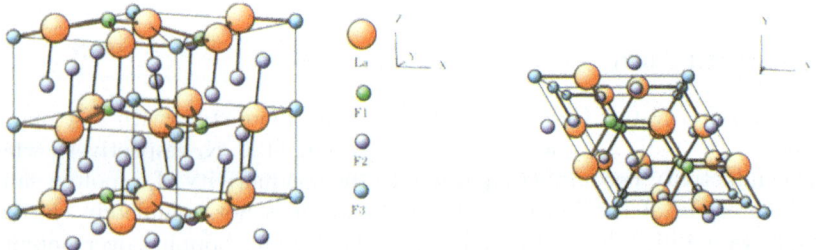

Fig. 2. The structure of LaF$_3$ (courtesy of A. Privalov) in side and top view.

3 Dynamical properties: Phonons

We combine experimental and theoretical methods in our investigations.

Since the ionic motion pertaining to the ionic conductivity is strongly anharmonic and thus strongly influenced by temperature, so is the ionic conduction a strongly temperature dependent process. Our investigations are thus focused on the thermal motion of the atoms in the ionic conductors. This has been done by investigating the temperature dependence of the dispersion curves of CaF_2 and LaF_3 by means of inelastic neutron scattering at the Institute Laue-Langevin in Grenoble, thereby extending the room-temperature data of [6] and [7], respectively.

The anharmonicity shows up in a change of phonon line shifts and widths at higher temperatures.

The experiments are accompanied by *ab-initio* calculations. Due to their reliablity these ab-initio results assist in assigning the symmetry to the various phonon dispersion branches and in analyzing the extent to which the fluorine ions take part in the various vibrations, at least at low temperatures where the weakly-anharmonic experimental data are closest to the purely-harmonic theoretical data.

4 Theory

Density-functional methods are used to compute the electronic, static, and dynamical properties of CaF_2 and LaF_3, the latter in the harmonic approximation. We used the method of full-potential linearized augmented plane waves (WIEN97 code [8]) and the pseudopotential method (ABINIT [9] and VASP [10] codes). The pseudopotential calculations have been done with Hartwigsen-Goedecker-Hutter pseudopotentials (ABINIT) [11] and ultrasoft pseudopotentials (VASP) [12].

The lattice parameters as the important part of the static properties have been calculated from the minimum of the total energy in different approximations (LDA, GGA), and the results are found to be in good agreement with literature and experimental data, see Table 1 below.

The dynamical properties have been calculated using the theoretical lattice constants, for some results see Table 2 below. In application to the experimental neutron scattering spectra, the phonon frequencies give the position of the bands, and the eigenvectors together with the frequencies determine the intensities. Since phonons with the same wavenumber can be observed for momentum transfer from different Brillouin zones the a-priori calculation can be (and is) used to preselect the zone with the highest scattering intensity for a given phonon. Also, the theoretical scattering intensities can be (and are) used to unravel unresolved bands. These considerations will be particularly helpful (if not necessary) in the case of LaF_3 in view of its complicated structure and the poorly resolved spectra of the huge number of close-lying branches (up to 72 phonon branches).

5 Results

5.1 Results for CaF_2

We have determined the lattice structure from the (numerical) minimum of the energy using the above-mentioned first-principles techniques. A precise

determination of the structure is needed as the staring point for the succeeding investigation of the dynamical properties. Like in other cases, the LDA
underestimates the value of the lattice constant, and the GGA overestimates
it, see Table 1. Hence frequencies are generally overestimated in LDA and
underestimated in GGA, see Table 2.

Table 1. CaF_2. Comparison of the calculated structure parameter a (in Å) with
the experimental value.

Exp. (300 K)	5.463		
WIEN97 (LDA, GGA(PBE), GGA(PW91))	5.333	5.493	5.482
ABINITv3.1.3 (LDA)	5.333		
Deviation [%]	-2.39	0.55	0.35
VASP (LDA, GGA(PW91))	5.172		5.44
Deviation [%]	-5.33		-0.42

Table 2. CaF_2. Comparison of different calculated Γ-point frequencies with experimental values.

	Exp.	WIEN97			ABINIT
	(300 K)	LDA	GGA (PBE)	GGA (PW)	LDA
frequency [THz]	7.71	8.358	7.156	7.247	8.4
deviation [%]		8.4	-6.0	-7.2	8.9
frequency [THz]	9.66	10.129	9.205	9.289	10.331
deviation [%]		4.9	-4.7	-3.8	6.9
frequency [THz]	13.9				14.593
deviation [%]					4.99

Figure 3 shows the results of the *ab-initio* calculation of the phonon dispersion curves for CaF_2 in the three main-symmetry directions in comparison
with our results of inelastic neutron scattering measurements at room temperature.

With three atoms per unit cell one obtains up to nine phonon branches
(fewer in the case of degeneracies). The results are in satisfactory agreement

with other experimental data [6] and literature. In particular, the calculated (harmonic) dispersion curves fit into the experimental (anharmonic) dispersion curves as measured for various temperatures and then extrapolated to low temperatures. This is shown in Figure 4 for the example of the Γ-point mode.

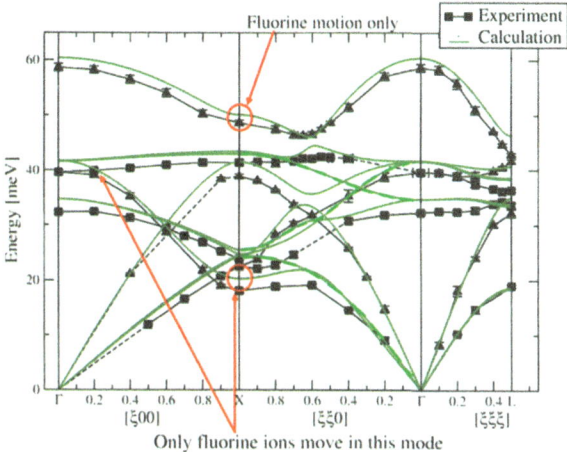

Fig. 3. CaF_2. Measured (symbols with connecting lines) and calculated phonon dispersion curves (at RT). Vertical error bars indicate the measured width of the phonon peak divided by 10. The three lightly-coloured branches indicate modes invisible in the scattering geometry of our experiment.

We have turned our particular attention to two phonon modes (at the X-point on the Brillouin-zone boundary) for which group theory shows that only the F^- ions move (indicated with a circle in figure 3). Figure 5 shows the corresponding raw experimental data. We have found an anomaly in the temperature dependence of the lower-frequency mode in the way that this mode shows an enormous decrease in frequency with increasing temperature. Other frequencies do not change as much, but all modes become very broad with temperature.

The width of the phonon peaks becomes comparable to the frequency at high temperatures. Even though the low-frequency mode has a much smaller number of decay channels, the FWHM of the low-frequency mode is comparable to that of the other modes, see Figure 6, indicationg a particularly strong anharmonicity.

5.2 Results for LaF$_3$

Since the reported difference between the (simple) hexagonal and (more complicated) trigonal structures is very small we had originally hoped that we

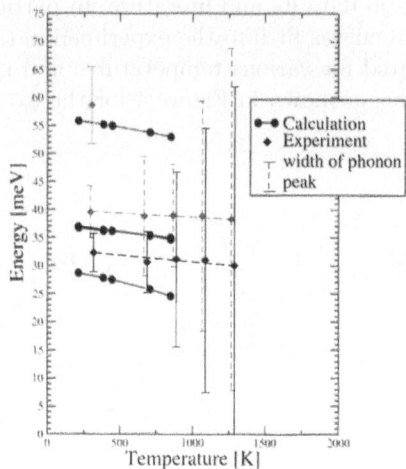

Fig. 4. CaF$_2$. Comparison of experimental Γ-point frequencies for different temperatures with the results of ab-initio calculations, the latter including (only) the effect of thermal expansion

could save computer time assuming the hexagonal structure and backfolding techniques. However, a calculation of the Γ-point phonon frequencies for LaF$_3$ in the hexagonal structure employing the frozen-phonon method have been found to be in disagreement with the measured frequencies [13,14]. Hence the much simpler hexagonal structure cannot be used for further calculations.

First sketchy phonon-frequency calculations for the trigonal structure have resulted in unstable phonon modes indicating that the published structure may not be stable and that a less symmetric structure with internal parameters in excess of the 7 parameters of the published structure is energetically favoured. A similar finding has been reported for the metal-hydrogen system YH$_3$ (a 'switchable mirror'), which originally had been assigned the same structure as proposed for LaF$_3$ and where small symmetry-breaking hydrogen displacements can lower the total energy of the system leading to a less symmetric structure [15]. In the case of LaF$_3$ the computations present, therefore, a challenge to computer speed and data storage in view of the complicated structure of this system. The structure optimization is still in progress.

The phonon frequencies at the Brillouin-zone center are known from light-scattering experiments [13, 14]. Phonon dispersion curves in the high-symmetry directions exist only in the lowest-frequency part of the spectrum [7]. We were able to follow the phonon branches to some higher frequencies, see Figure 7, and we have measured this part of the spectrum in one symmetry direction for several temperatures. One rather dispersionless

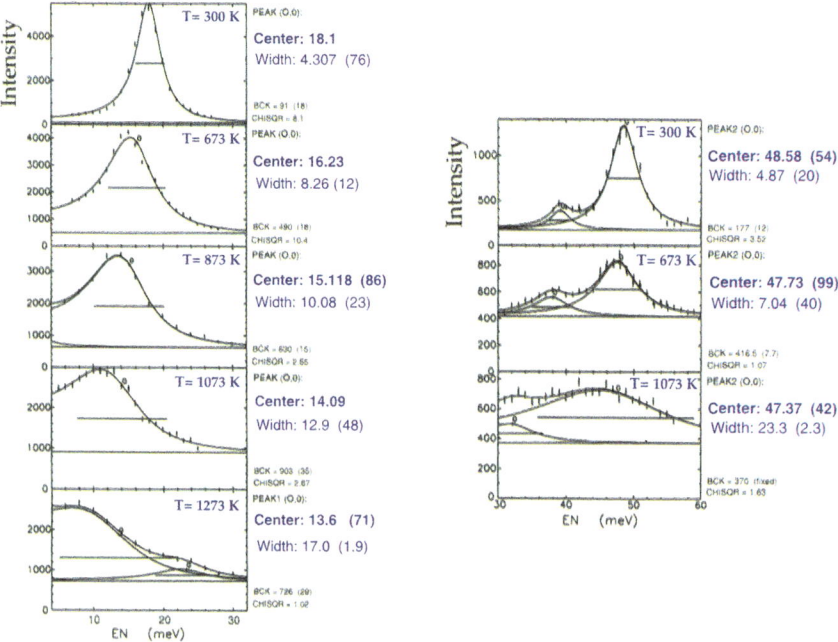

Fig. 5. CaF$_2$. Raw experimental inelastic neutron scattering data (points) and damped-harmonic-oscillator fits uncorrected for instrumental resolution. The left panel shows the spectra from to the lower-frequency mode, the right panel those from to the higher-frequency mode at the X-point on the zone boundary.

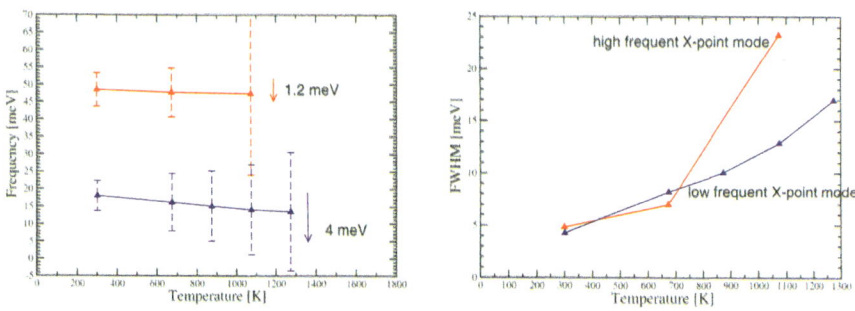

Fig. 6. CaF$_2$. Left panel: Frequency versus temperature; the error bars indicate the width of the measured phonon peak at the zone boundary. Right panel: FWHM versus temperature

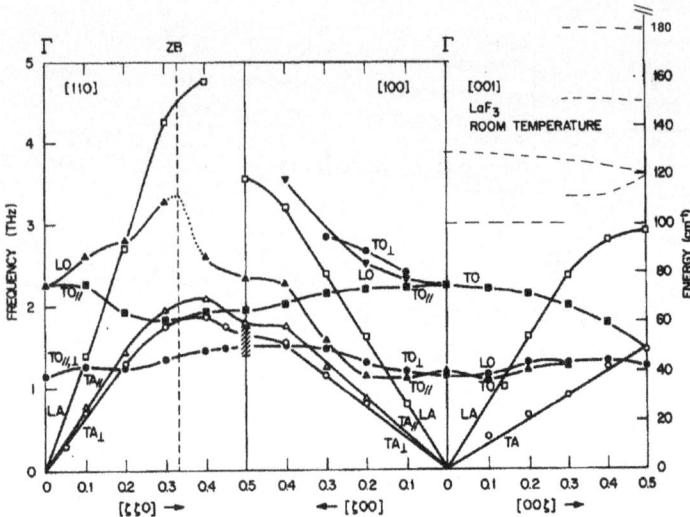

Fig. 7. LaF$_3$. Phonon dispersion curves measured with inelastic neutron scattering at RT; solid lines from [7]; broken lines indicate part of our own results (preliminary).

excitation branch stiffens with increasing temperature, in contrast to CaF$_2$ where the strongly temperature dependent phonon mode softens, and broadening occurs only for a part of the modes, again in contrast to CaF$_2$ where essentially all modes broaden. A calculation of the phonon frequencies will show whether or not the dispersionless excitation has a phononic origin.

Figure 8 shows the temperature dependence of the four branches in [00ξ]-direction from figure 7. The different branches are labelled corresponding to the notation there.

References

1. Jordan, W. M., Catlow, C. R. A. (1987): An investigation into the structural and transport properties of LaF$_3$. Cryst. Latt. Def. and Amorph. Mater. **15**, 81.
2. Mansmann, M. (1965): Die Kristallstruktur von Lanthantrifluorid. Z. Kristallogr. **122**, 375.
3. Zalkin, A., Templeton, D. H. (1985): Refinement of the trigonal crystal structure of lanthanum trifluoride with neutron diffraction data. Acta Cryst. **B 41**, 91.
4. Gregson, D., Catlow, C. R. A. (1983): The structure of LaF$_3$ - a single-crystal neutron diffraction study at room temperature. Acta Cryst. **B 39**, 687.
5. Schlyter K. (1952): On the crystal structure of fluorides of the tysonite or LaF$_3$ type. Arkiv Kemi **5**, 73.
6. Elcombe, M. M., Pryor, A. W. (1970): The lattice dynamics of calcium fluoride. Solid State Phys. **3**, 492.

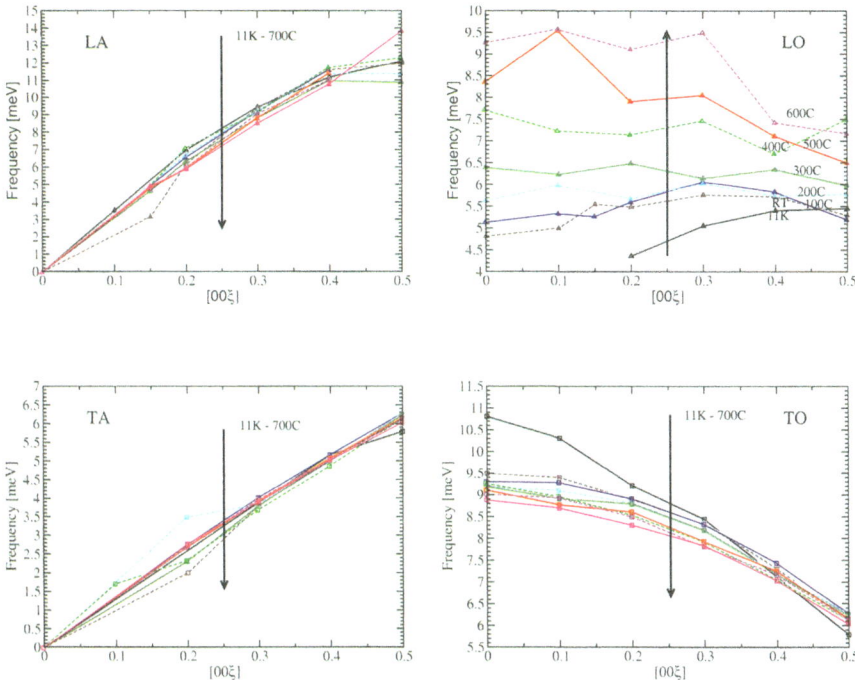

Fig. 8. The measured temperature dependence of the longitudinal acoustic (LA) and optic (LO) and the transverse acoustic (TA) and optic branch (TO) in [00ξ]-direction.

7. Dixon, G. S., Nicklow, R. M. (1983): Low frequency lattice vibrations of LaF$_3$ by neutron scattering. Solid State Commun. **47**, 877.
8. Blaha, P., Schwarz, K., Sorantin, P., Trickey, S. B. (1990): Full-potential, linearized augmented plane wave programs for crystalline systems. Comp. Phys. Comm. **59**, 399.
9. Gonze, X. (1997): First-principles responses of solids to atomic displacements and homogeneous electric fields: Implementation of a conjugate-gradient algorithm. Phys. Rev. **B 55**, 10337.
10. Kresse, G., Furthmüller, J. (1996): Efficient iterative schemes for ab initio total-energy calculations using a plane-wave basis set. Phys. Rev. **B 54**, 11169.
11. Hartwigsen, C., Goedecker, S. Hutter, J. (1998): Relativistic separable dual-space Gaussian pseudopotentials from H to Rn. Phys. Rev. **B 58**, 3641.
12. Kresse, G., Hafner, J. (1994): Norm-conserving and ultrasoft pseudopotentials for first-row and transition elements. J. Phys. Condens. Matter **6**, 8245.
13. Bauman, R. P., Porto, S.P.S. (1967): Lattice vibrations and structure of rare-earth fluorides. Phys. Rev. **161**, 842.

14. Cerdeira, F., Lemos, V., Katiyar, R.S. (1979): Stress dependence of the zone-center optical phonons of LaF$_3$. Phys. Rev. **B 19**, 5413.
15. van Gelderen, P., Kelly, P. J., Brocks, G. (2001): Phonon spectrum of YH$_3$: Evidence for a broken symmetry structure. Phys. Rev. **B 63**, 100301(R).

Optical Response of Semiconductor Surfaces and Molecules Calculated from *First Principles*

W. Gero Schmidt, Martin Preuß, Patrick H. Hahn, Kaori Seino, and Friedhelm Bechstedt

Computational Materials Science Group
Friedrich-Schiller-Universität
Max-Wien-Platz 1, 07743 Jena, Germany
W.G.Schmidt@ifto.physik.uni-jena.de

Abstract. *First-principles* pseudopotential calculations on the optical anisotropy of GaAs(001) surfaces and on the optical absorption of of DNA base molecules are presented. It is found that both electronic surface states as well as surface-perturbed bulk wave functions contribute to the optical anisotropy of GaAs(001). The latter contributions are modified by surface electric fields, giving rise to signals which are both reconstruction and electric field dependent. Pronounced differences in the absorption behavior of the DNA base molecules suggest the possibility of a base discrimination by means of single molecule spectroscopy.

1 Introduction

Optical spectroscopies, such as reflectance anisotropy spectroscopy (RAS) have become very important for the real-time monitoring of surface growth. In addition, they are very helpful to investigate the interaction between surfaces and simple adsorbates or complex molecules. However, the full potential of optical spectroscopy can only be realized, if it becomes possible to calculate optical spectra with true predictive power. This is complicated by the computational expense required to calculate numerically converged optical spectra. Much methodological progress has recently been made in the accurate calculation of optical spectra of small clusters [1], bulk crystals [2,3] and surfaces [4,5,6]. However, many questions still remain open.

One of these is the influence of electric fields on the surface optical response. Experimentally it has been known for a long time that electric fields, induced by, e.g., the pinning of the Fermi level at the sample surface (band bending) [7] or by a δ-doping layer [8], modify the RAS signal. The investigation of the mechanism behind the electric field induced modification of the optical signal may not only help to better understand the origin of specific RAS features, but should pave the way for applications, such as contact-less determination of the carrier concentration in a bulk material. Previous calculations [9,10] modeled the electric field induced line shape changes around the E_1 and $E_1 + \Delta_1$ critical point (CP) energies of GaAs based on the piezoelectric effect. However, the influence of the surface reconstruction on the field induced RAS changes or wider spectral ranges have not been considered. Here we present – to our knowledge for the first time – *first-principles* calculations on the interplay of surface geometry and surface electric fields in the reflectance anisotropy. The GaAs (001) surface with its large number of stoichiometry-dependent surface reconstructions (cf. Fig. 1) and its importance for III-V based optoelectronics has been chosen as model system.

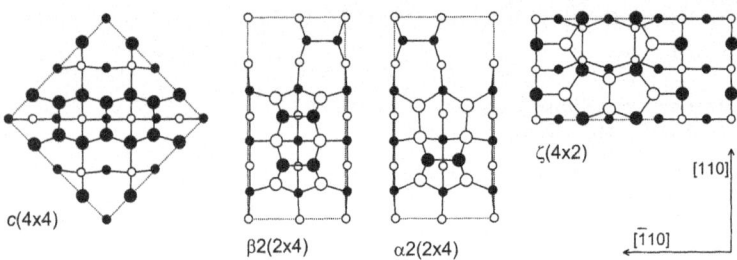

Fig. 1. Top view of relaxed GaAs(001) surface structures. Empty (filled) circles represent Ga (As) atoms. Positions in the uppermost two atomic layers are indicated by larger symbols.

Optical spectroscopy has also become a technique of common use for analyzing the adsorption of molecules on surfaces. The deoxyribose nucleic acid (DNA) base molecules adenine, guanine, cytosine and thymine form a particularly interesting class of molecules: due to their Watson-Crick complementarity they are well suited for molecular self-assembly processes, which are increasingly becoming a route to the fabrication of nanosize objects. Structures formed by these molecules may be used as templates to engineer macromolecules, molecular sieves or photonic materials with novel properties [11]. Other exciting applications which are presently being explored are nanomechanical devices [12] or DNA computing [13, 14]. The understanding of the optical spectra of isolated, gas-phase DNA base molecules is a necessary prerequisite to monitor self-assembly processes of these molecules by optical means. As a first step in that direction, we perform *ab initio* calculations of the absorption spectra of adenine, guanine, cytosine and thymine.

2 Computational Method

Density-functional theory in the local-density approximation (DFT-LDA) together with nonlocal norm-conserving pseudopotentials is used to determine the structurally relaxed ground state of the GaAs surface. A massively parallel, real-space finite-difference method is used to deal efficiently with the large unit cells needed to describe the surface. Thereby a real-space mesh is used to represent the wave functions, the charge density, and the ionic pseudopotentials. The density functional equations are discretized using a generalized eigenvalue form:

$$H_{mehr}[\psi_n] = \frac{1}{2}A_{mehr}[\psi_n] + B_{mehr}[V_{eff}\psi_n] = \epsilon_n B_{mehr}[\psi_n] \,,$$

where A_{mehr} and B_{mehr} are the components of the *Mehrstellen* discretization, which is based on Hermite's generalization of Taylor's theorem. It uses a weighted sum of the wave function and potential values to improve the accuracy of the discretization of the entire differential equation. Only nearest and next-nearest neighbor points are used in the discretization. This short-ranged representation of the equation leads to an efficient domain-decomposition-based implementation on massively parallel computers (for details see [15,16]). Typical jobs utilize 32 to 128 PEs. The solution of the discretized equations is obtained on a grid fine enough to accurately represent the pseudopotentials and the electronic wave functions. We use a multigrid iteration technique that accelerates convergence by employing a sequence of grids of varying resolutions. Both the Poisson and the density functional equations are solved on coarser grids and the resulting corrections are transferred to the fine grid. This provides excellent preconditioning for all length scales present in our system and leads to very fast convergence rates. The operation count to converge one wave function with a fixed potential scales linear with the number of grid points. Further details of the DFT-LDA calculations are those in [17].

The electronic structure obtained within DFT-LDA is used to calculate the optical response: within independent-particle approximation the microscopic dielectric function is obtained from the Bloch band eigenfunctions $|n\mathbf{k}\rangle$ characterized by band index n, wave vector $\mathbf{k}$, and energy eigenvalue $\varepsilon_n(\mathbf{k})$ using the Ehrenreich-Cohen Formula. For semiconducting systems the microscopic dielectric tensor is given by

$$\epsilon_{ij}(\omega) = \delta_{ij} + \frac{16\pi e^2 \hbar^2}{\Omega} \sum_{\mathbf{k},c,v} \frac{\langle c\mathbf{k}|v_i|v\mathbf{k}\rangle\langle v\mathbf{k}|v_j|c\mathbf{k}\rangle}{[\varepsilon_c(\mathbf{k}) - \varepsilon_v(\mathbf{k})]\left([\varepsilon_c(\mathbf{k}) - \varepsilon_v(\mathbf{k})]^2 - \hbar^2[\omega + i\eta]^2\right)}.$$

The matrix elements $\langle v\mathbf{k}|v_j|c\mathbf{k}\rangle$ of the velocity operator v_j with the valence and conduction states are calculated in real-space. This does not only allow for an efficient parallelization, but also for a classification of the optical excitation processes according to their spatial localization. Of particular importance for the predictive power of the calculated spectra is their full numerical convergence. For the surface calculations we use $\mathbf{k}$-point sets equivalent to 1024 points in the full (1×1) surface Brillouin zone and include all conduction states within an energy of 1 Ryd from the valence band maximum. Due to the memory requirements, calculations of this size can presently only be performed using massively parallel supercomputers such as the Hitachi SR8000.

From the dielectric tensor calculated for the supercells used to model the GaAs bulk and surface, we calculate the reflectance anisotropy according to the scheme devised by Del Sole and Manghi *et al.* [18,19]. In general, optical spectra are strongly modified by many-body effects such as self-energy corrections and electron-hole attraction [2,3,6,20]. However, RAS spectra are difference spectra, which are furthermore normalized to the bulk dielectric function. Due to the error cancellation, single-particle calculations within DFT-LDA are actually quite reliable in predicting surface optical anisotropies [21]. Therefore, we simply use the scissors-operator approach [22] to take self-energy effects into account. Excitonic and local-field effects are neglected.

A saw-tooth function added to the electrostatic potential is used to mimic the effect of an electric field. From the self-consistent solution of the Kohn-Sham equations the influence of the electric field on both the wave functions and the eigenvalues is obtained. The sign convention used here is such that the field points in the direction of the surface normal.

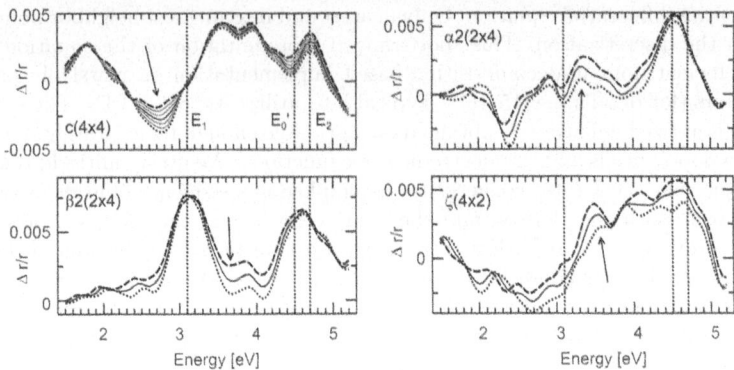

Fig. 2. Calculated RAS spectra for GaAs(001) surface reconstructions For GaAs(001)$c(4\times4)$ the field varies in steps of 0.005 V/Å. Thick solid/dotted/dashed lines correspond the field magnitudes of 0.0/-0.015/0.015 V/Å.

To calculate the structurally relaxed ground state of the DNA base molecules we use the VASP (Vienna Ab-initio Simulation Package) implementation of the DFT in generalized gradient approximation (GGA) [23]. This implementation allows for the use of non-norm-conserving ultrasoft pseudopotentials of the Vanderbilt type. These pseudopotentials are extremely efficient for the calculation of systems containing first-row elements. The optical spectra of these molecules were calculated from all-electron wave functions obtained by the projector-augmented wave (PAW) method [24].

3 Results and Discussion

3.1 Optical Anisotropy of GaAs(001) surfaces

The stoichiometry-dependent surface structures of GaAs(001) have been studied intensively over the last couple of years. There now seems to be consensus on the geometries of the four main reconstructions shown in Fig. 1: $c(4\times4)$ and $\beta2(2\times4)$ reconstructions occur for As-rich surfaces, a $\zeta(4\times2)$ geometry is characteristic for Ga-rich surfaces, and stoichiometric surfaces form the $\alpha2(2\times4)$ structure [25].

The thick solid lines in Fig. 2 show the RAS spectra calculated for these surface models. Obviously, the surface optical anisotropy is strongly related to the surface geometry. The spectra calculated for $c(4\times4)$ and $\beta2(2\times4)$ agree well with experiment [26, 27, 28]. The appearance of negative anisotropies for the $\alpha2(2\times4)$ structure at energies somewhat above 2 eV also agrees with the experimentally observed trend: annealing temperatures higher than those needed to prepare the

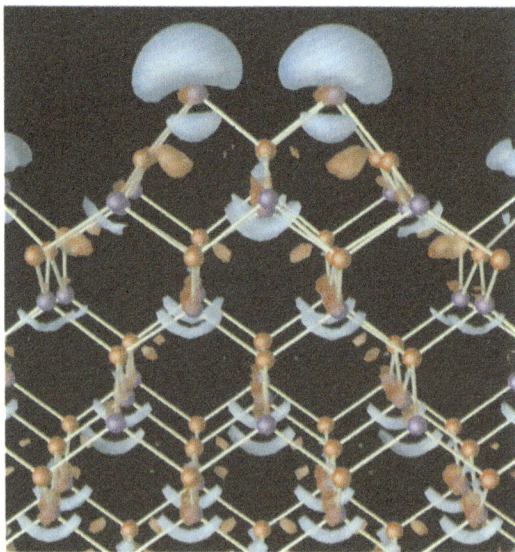

Fig. 3. Electron density transfer (red/blue isosurfaces indicate $\Delta\rho = \pm10^{-4}e/Bohr^3$) induced by an electric field of 0.015 V/Å in a slab modeling the GaAs(001)$\beta2(2\times4)$ surface. Red (blue) balls represent Ga (As) atoms.

GaAs(001)$\beta2(2\times4)$ surface lead to negative anisotropies for photon energies of about 2 eV [27]. The main experimentally observed RAS features for Ga-rich (4×2) reconstructed surfaces are negative anisotropies for photon energies below the E_1 CP and positive anisotropies for higher energies [26, 28]. They are reproduced by our calculation for the $\zeta(4\times2)$ surface.

The influence of external electric fields ranging from -0.015 to 0.015 V/Å on the calculated RAS spectra is also shown in Fig. 2. In all cases the magnitude and to some extent the energy positions of specific RAS features are modified. A remarkable outcome of our study is the strong reconstruction dependence of the field induced RAS changes. For the $c(4\times4)$ reconstructed GaAs surface, strong modifications of the RAS signal are observed for photon energies below 3 eV and at the E_0'/E_2 CP energies. The optical anisotropy of the (2×4) reconstructed $\beta2$ and $\alpha2$ surfaces, on the other hand, is mainly altered in the energy region between the E_1 and E_0' CPs. Finally, nearly the entire RAS spectrum of the Ga-rich $\zeta(4\times2)$ surface is modified by the application of an electric field. Our finding of a strong influence of the surface reconstruction on the field induced RAS is supported by recent experiments [29]. Although the influence of the electric field on the surface optical anisotropy is reconstruction-dependent, the spatial analysis of the origin of the RAS features shows that it is the optical signal from the layers underneath the surface, i.e. the so-called intrinsic anisotropy, that is affected by the field.

Our calculations indicate that the surface optical anisotropy for energies near the bulk CPs is a consequence of the anisotropically modified decay of the bulk Bloch states into the vacuum region. The surface induced deformations of the bulk-like wave functions are weighted differently by the x and y components of the transition operator, leading to an anisotropic optical response. Superimposed on the surface induced anisotropies of the electron wave functions are the electric field induced deformations. These are anisotropic too, as visualized in Fig. 3 for the electron density of the GaAs(001)$\beta 2(2\times 4)$ surface slab. We find that the changes of the combined density of states due to the electric field are negligible compared to the deformations of the wave functions resulting in modified transition matrix elements.

Our results on the influence of electric fields on the optical anisotropy of GaAs(001) surfaces cannot directly be compared with experiment. On one hand, this is due to computational shortcomings. We neglect the spin-orbit coupling and therefore cannot describe the $E_1/E_1 + \Delta_1$ splitting. The RAS features close to the $E_1/E_1 + \Delta_1$ energy are considerably affected by the spin-orbit interaction. Most measurements of the linear electro-optic effect focus specifically on these features. On the other hand, the comparison of experimental and simulated spectra is complicated by the facts that: (i) the magnitude of the surface electric field, induced by, e.g. δ-doping or space-charge layers is not known exactly, and (ii) most experiments are performed on surfaces, the geometries of which are not well characterized, e.g. oxidized surfaces.

In measurements done in air, Yang, Chen and Wong [30] found a linear relationship between the change in the RAS at the $E_1/E_1 + \Delta_1$ energy and the surface electric field. They determined a linear electro-optic coefficient of 0.46 Å/V. In Fig. 4 we show the change of the RAS signal for the four surface reconstructions at energy positions indicated in Fig. 2, i.e., slightly below or above the E_1 transition

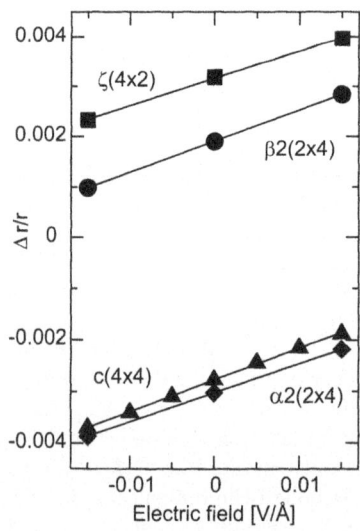

Fig. 4. Variation of the RAS calculated for GaAs(001) surface reconstructions at the photon energies marked by arrows in Fig. 2, for applied electric fields. The solid lines are a guide to the eye.

as a function of the electric field. Clearly, the calculated change of the RAS is linear, as observed experimentally. The slopes of the curves in Fig. 4 are about 0.07 Å/V, much smaller than the measured value of 0.46 Å/V. The fact that the field needed in the calculation to reproduce changes in the RAS comparable in magnitude to experiment is nearly one order of magnitude stronger than measured, is related to the large penetration depth of light. It amounts to about 170 Å for a photon energy of 3 eV [31], whereas our slab only models the uppermost 15 Å of the GaAs surface region.

By varying the distance between the δ-doping layer and the surface, Sobiesierski, Westwood and Elliot [8] have shown that the surface electric field reaches far into the bulk and modifies the optical response along the full light penetration depth. The incoming light decreases in intensity as it penetrates the sample, therefore the ratio between the light penetration depth and the slab thickness gives an upper bound for the scaling factor needed to relate the calculated and the measured field sensitivity of the RAS. The value of the electro-optic coefficient calculated here is thus in the expected range. This agreement may be fortuitous, however, given the neglect of spin-orbit coupling in our study. Nevertheless, because both experiment and calculations find a linear relation between the changes of the optical anisotropy and the surface electric field, we expect our results, although obtained for comparatively strong fields, to correctly describe the basic mechanisms responsible for the field induced RAS changes.

3.2 Optical Absorption of the DNA bases

Figure 5 shows the absorption spectra of adenine, guanine, cytosine and thymine calculated within DFT-GGA. In all four cases the calculated onset of absorption occurs for photon energies slightly below 4 eV. The oscillator strength of the lowest optical transition, as well as the absorption behavior for higher energies, however, are strongly molecule specific. Guanine and thymine are characterized by relatively few and energetically well separated (by about one eV) optical transitions, whereas the energetical separation of the absorption peaks is much reduced in the cases of adenine and cytosine. The spectra presented in Fig. 5 refer to spin-averaged, zero temperature calculations. The influence of the quantized vibrations and rotations at finite temperatures will considerably broaden the spectra, as will the interaction with solvent molecules.

Further modifications of the spectra are due to many-body effects. In particular, electronic self-energy and electron-hole attraction effects change the spectra considerably. For the delocalized electron gas, these effects are usually calculated using a Green's function technique [1,2,3,6]. In the present case, however, the localization of the electronic states allows for a numerically far less demanding treatment of these many-body effects: we investigated their influence by means of ΔSCF – also called constrained-DFT – calculations. Thereby the total-energy differences between the ground state and the optically excited states of the molecule are calculated. The electrons are allowed to relax, while the occupation numbers are constrained to the chosen configuration. We observe that the widening of the energy gaps due to the electronic self-energy is partially canceled by the gap narrowing due to the Coulomb attraction between electrons and holes. The net effect depends on the specific transition investigated and consists in an increase of the excitation energy

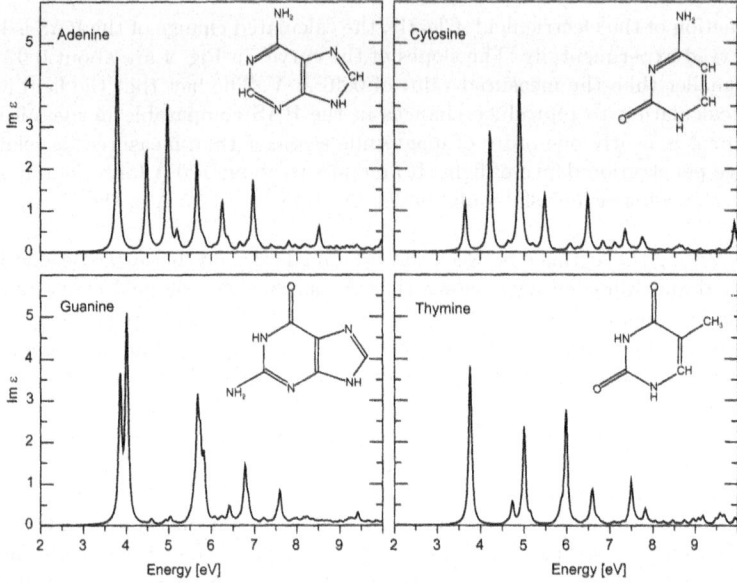

Fig. 5. Absorption spectra (in arb. units) of DNA base molecules calculated in the independent-particle approximation.

by one to several tenths of an eV. If in addition to the electronic relaxation also the geometry of the molecule is allowed to relax, the energy shifts between light absorption and emission (Stokes shifts) can be calculated. These shifts are considerable. For guanine, for instance, we calculate a Stokes shift of about 0.7 eV for the transition between the highest occupied and lowest unoccupied molecular orbitals (HOMO and LUMO, respectively). The molecular characters of the guanine HOMO and LUMO are depicted in Fig. 6. The HOMO consists of π orbitals formed by p_z states of carbon and nitrogen atoms. However, there are also some lone pair-like contributions from the oxygen and the outermost nitrogen. The LUMO is formed by non-bonding or antibonding combinations of N, C, and O p_z states.

4 Summary

We have investigated the effects of reconstructions and surface electric fields on the optical anisotropy of GaAs(001). We find geometry-dependent surface-structure signatures which explain very well the experimental observations. Surface-perturbed bulk wave functions modified by electric fields via the linear electro-optic effect are mainly responsible for field induced modifications of the surface optical anisotropy. These modifications are strongly reconstruction-dependent. They are small, however, compared to the RAS signals due to the anisotropy of the surface itself.

We also investigated the optical absorption behavior of the DNA bases adenine, guanine, cytosine and thymine. Strong differences between the optical spectra of

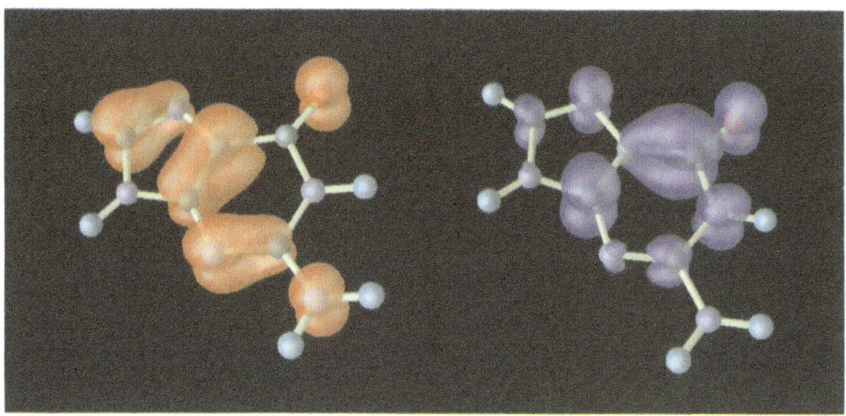

Fig. 6. Calculated charge density (isosurfaces $\rho = 0.05e/Bohr^3$) of the guanine HOMO (left) and LUMO (right).

the four molecules are observed. At finite temperatures these differences will be somewhat masked by the influence of vibrational and rotational excitations. Nevertheless, a base discrimination by means of single molecule spectroscopy appears possible.

We thank Dave Aspnes, Markus Pristovsek and Norbert Esser for many useful discussions. Grants of computer time from the Leibniz-Rechenzentrum München are gratefully acknowledged.

References

1. M. Rohlfing and S. G. Louie, Phys. Rev. Lett. **80**, 3320 (1998).
2. L. X. Benedict, E. L. Shirley, and R. B. Bohn, Phys. Rev. Lett. **80**, 4514 (1998).
3. S. Albrecht, L. Reining, R. Del Sole, and G. Onida, Phys. Rev. Lett. **80**, 4510 (1998).
4. W. G. Schmidt, N. Esser, A. M. Frisch, P. Vogt, J. Bernholc, F. Bechstedt, M. Zorn, T. Hannappel, S. Visbeck, F. Willig, and W. Richter, Phys. Rev. B **61**, R16335 (2000).
5. W. Lu, W. G. Schmidt, E. L. Briggs, and J. Bernholc, Phys. Rev. Lett. **85**, 4381 (2000).
6. P. H. Hahn, W. G. Schmidt, and F. Bechstedt, Phys. Rev. Lett. **88**, 016402 (2002).
7. V. L. Berkovits, I. V. Makarenko, T. A. Minashvili, and V. I. Safaro, Sov. Phys. Semicond. **20**, 654 (1986).
8. Z. Sobiesierski, D. I. Westwood, and M. Elliott, Phys. Rev. B **56**, 15277 (1997).
9. Y. H. Cheng and Z. Yang, Appl. Phys. Lett. **73**, 1667 (1998).
10. A. Lastras-Martinez, R. E. Balderas-Navarro, L. F. Lastras-Martinez, and M. A. Vidal, Phys. Rev. B **59**, 10234 (1999).

11. E. Winfree, F. Liu, L. A. Wenzler, and N. C. Seemann, Nature **394**, 539 (1998).
12. C. Mao, W. Sun, Z. Shen, and N. C. Seemann, Nature **397**, 144 (1999).
13. M. Ogihara and A. Ray, Nature **403**, 143 (2000).
14. Q. Liu, L. Wang, A. G. Frutos, A. E. Condon, R. M. Corn, and L. M. Smith, Nature **403**, 175 (2000).
15. E. L. Briggs, D. J. Sullivan, and J. Bernholc, Phys. Rev. B **54**, 14362 (1996).
16. J. Bernholc, E. L. Briggs, C. Bungaro, M. B. Nardelli, J. L. Fattebert, K. Rapcewicz, C. Roland, W. G. Schmidt, and Q. Zhao, phys. stat. sol. (b) **217**, 685 (2000).
17. W. G. Schmidt, S. Mirbt, and F. Bechstedt, Phys. Rev. B **62**, 8087 (2000).
18. R. Del Sole, Solid State Commun. **37**, 537 (1981).
19. F. Manghi, R. Del Sole, A. Selloni, and E. Molinari, Phys. Rev. B **41**, 9935 (1990).
20. M. Rohlfing and S. G. Louie, Phys. Rev. Lett. **83**, 856 (1999).
21. W. G. Schmidt, F. Bechstedt, and J. Bernholc, J. Vac. Sci. Technol. B **18**, 2215 (2000).
22. R. Del Sole and R. Girlanda, Phys. Rev. B **48**, 11789 (1993).
23. G. Kresse and J. Furthmüller, Comp. Mat. Sci. **6**, 15 (1996).
24. B. Adolph, J. Furthmüller, and F. Bechstedt, Phys. Rev. B **63**, 125108 (2001).
25. W. G. Schmidt, Appl. Phys. A **75**, 89 (2002).
26. I. Kamiya, D. E. Aspnes, L. T. Florez, and J. P. Harbison, Phys. Rev. B **46**, 15894 (1992).
27. A. I. Shkrebtii, N. Esser, W. Richter, W. G. Schmidt, F. Bechstedt, B. O. Fimland, A. Kley, and R. Del Sole, Phys. Rev. Lett. **81**, 721 (1998).
28. W. G. Schmidt, F. Bechstedt, K. Fleischer, C. Cobet, N. Esser, W. Richter, J. Bernholc, and G. Onida, phys. stat. sol. **188**, 1401 (2001).
29. M. Pristovsek and S. Tsukamoto, Bulletin of the APS **47**, 85 (2002).
30. Z. Yang, Y. H. Chen, and Y. Wong, Appl. Phys. Lett. **73**, 1520 (1998).
31. D. E. Aspnes and A. A. Studna, Phys. Rev. B **27**, 985 (1983).

Phase Fluctuations and the Role of Electron Phonon Coupling in High-T_c Superconductors

Thomas Eckl, Zhong-Bing Huang, Werner Hanke, and Enrico Arrigoni

Institut für Theoretische Physik und Astrophysik
Universität Würzburg
Am Hubland
97074 Würzburg, Germany

Abstract. Two issues, which play a key role in the present discussion of the microscopic mechanism of the pairing in high-T_c superconductors are investigated, employing large-scale computing. In the first part of this paper, the single-particle density of states and the tunneling conductance are studied for a two-dimensional BCS-like Hamiltonian with a $d_{x^2-y^2}$-gap and phase fluctuations. The latter are treated by a classical Monte Carlo simulation of an XY model. Comparison of our results with recent scanning tunneling spectra of Bi-based high-T_c cuprates supports the idea that the pseudogap behavior observed in these experiments can be understood as arising from phase fluctuations of a $d_{x^2-y^2}$ pairing gap whose amplitude forms on an energy scale set by T_c^{MF}, well above the actual superconducting transition.

We then apply this phase fluctuation model to recent reflectivity measurements, which have shown a violation of the in-plane optical integral in underdoped Bi2212 up to frequencies much higher than those expected by standard BCS theory [28,30]. The sum rule violation can be related to a loss of in-plane kinetic energy. We show that the above BCS-like Hamiltonian with a d-wave gap and phase fluctuations can explain this change of in-plane kinetic energy at T_c. Our model is also applicable for other superconductors where phase fluctuations should play a dominant role, i. e. with small charge carrier density like the organic superconductors.

In the second part of this paper, we investigate numerically the effects of electronic correlations on the electron-phonon vertex function in the one-band Hubbard model. Our simulations are based on a new numerically exact technique to extract the vertex, which is especially important for the case of interest, i. e. strong correlations, which cannot be controlled perturbatively. The simulations are performed both on one-dimensional and two-dimensional lattices. We find that the on-site Coulomb interaction suppresses the electron-phonon coupling effectively. In particular, the backward scattering with large phonon momentum is suppressed much more than the forward scattering with small phonon momentum. With decreasing the doping density, the electron-phonon coupling is reduced at all phonon momenta. In the weak-coupling regime, our numerical simulations are in good agreement with the Feynman diagram expansions.

1 Introduction

Intensive research has focused on the pseudogap regime, which is observed in the high-T_c cuprates below a characteristic temperature that is higher than the transition temperature T_c. It occurs in a number of different experiments as a suppression of low-frequency spectral weight [1,2,3,4,5,6,7]. This striking pseudogap behavior initiated a variety of proposals as to its origin [8,9,10, 11,12,13,14,15], since the answer to this question may be a key ingredient for the understanding of high-T_c superconductivity. At present, there is no agreement as to which of these proposals is correct. In part, this could, in principle, reflect the possibility that there are different pseudogap phenomena as well as a range of materials requiring different theoretical descriptions. In part, this is because of the difficulty in determining the consequences of the various theoretical proposals. A certain convergence has recently emerged in that measurements of the high-field phase diagram of the cuprates have provided compelling evidence that the transition from T_c into the higher-temperature pseudogap regime corresponds to a loss in phase rigidity, rather than a vanishing of the pairing amplitude [16].

In view of these developments, the first part of this paper aims at three key aspects: (i) To provide a quantitative understanding of the role of phase fluctuations in the standard model, i. e. the BCS model with a d-wave gap and phase fluctuations. This is the simplest minimal model, which is essentially exactly solved numerically on a large lattice, Monte Carlo (MC) averaging over many ($\sim 10^4$) configurations. This exact solution constitutes a useful guide for future more detailed studies in the high-T_c cuprates. Already in its minimal form, it can be used to pin down the role of phase-fluctuations in a variety of experiments. (ii) As an example, we focus in Sect. 2.1 on the pseudogap phenomena observed in scanning tunneling spectroscopy measurements (STM) [4,5,17] on $Bi_2Sr_2CaCu_2O_{8+\delta}$ (Bi2212) and $Bi_2Sr_2CuO_{6+\delta}$ (Bi2201). We explore the notion that the pseudogap observed in these experiments arises from phase fluctuations of the gap [11,12,13,14,15]. In this scenario, below a mean-field temperature scale T_c^{MF}, a significant $d_{x^2-y^2}$-wave gap amplitude is assumed to develop. However, the superconducting transition is suppressed to a considerably lower temperature T_c by phase fluctuations. In the intermediate temperature regime between T_c^{MF} and T_c, the phase fluctuations of the gap give rise to the pseudogap phenomena. The good overall agreement with the STM data serves to further strengthen the phase-fluctuation scenario as a *unified* description of the pseudogap.

(iii) We want to discuss recent experiments [28,30] about optical sum rule violation within the framework of our phase fluctuation scenario: In ordinary BCS superconductors, the optical conductivity is suppressed at frequencies within a range of twice the superconducting gap. The corresponding spectral weight W_{low} is transfered to the zero-frequency delta peak W_D (giving the superfluid weight D) [32], associated with the dissipationless transport in the superconducting state. This is the Glover-Ferrell-Tinkham (GFT) sum rule.

On the other hand, the *total* frequency integral of the optical conductivity is conserved, when decreasing the temperature across the superconducting transition, due to the f-sum rule [32], i. e. $W_{low+D}^{sc} = W_{low+D}^{n}$.

Recent measurements of the in-plane optical conductivity [28, 30], however, have shown a *violation* of the GFT optical sum rule for frequencies up to $2eV$ in underdoped Bi2212. That is, by entering the superconducting state, not only spectral weight from the microwave and far-infrared W_{low}, but also from the visible optical spectrum W_{high} contributes to the superfluid condensate W_D (and, thus to D), i. e. in contrast to ordinary BCS-superconductors, "a *color* change" is introduced. This unusual result has stimulated new discussion about an unconventional pairing mechanism in the high-T_c copper oxide superconductors (HTSC). The *color* change may be attributed to a reduction of *kinetic* energy [29] which drives the system to superconductivity, instead of the reduction of potential energy upon pairing as in in conventional BCS superconductors [32].

The full optical integral, when integrated over all frequencies and *energy bands*, is proportional to the carrier density over the bare mass

$$W_{low+D} + W_{high} = \int_0^\infty Re\ \sigma_{xx}(\omega)\ d\omega = \frac{ne^2}{2m}, \tag{1}$$

and, thus, is conserved. When the optical integral is restricted over a finite (low) range of frequencies Ω, in the HTSC typically of the order of eV, one can consider the weight W_{low+D} as stemming essentially from a single band around the Fermi energy:

$$W_{low+D} = \int_0^\Omega Re\ \sigma_{xx}(\omega)\ d\omega = (\pi e^2 a^2/2\hbar^2 V)E_K, \tag{2}$$

where σ is the single-band conductivity, a the lattice constant and V the unit cell volume. With this single-band assumption, the frequency integral of the optical conductivity is proportional to the the inverse mass tensor ($\frac{\partial^2 \epsilon_k}{\partial k_x^2}$, x being the direction in which the conductivity is measured), weighted with the momentum distribution n_k [31]. This quantity depends upon the band structure, being proportional to minus the kinetic energy $E_K = -E_{kin}$ for a tight-binding model, while for free electrons it is a constant given by the electron density divided by the effective mass.

In Sect. 2.2 of this paper, we show that – indeed – phase fluctuations contribute to a significant reduction of the in-plane kinetic energy at the Kosterlitz-Thouless transition temperature $T_{KT} \equiv T_c$ of the two-dimensional (2D) system, where the phase correlation length ξ diverges, with a magnitude comparable to recent experimental results. The physical reason for this finding is that, when long-ranged phase coherence develops, the Cooper pairs can delocalize *better* (i. e. in a phase-coherent manner) and the kinetic energy decreases. The kinetic energy loss can, therefore, be traced back to the development of *coherence peaks* in the density of states, when approaching the superconducting transition.

In the second part of this paper, i. e. Sect. 3, we want to discuss one important aspect of the pairing mechanism of high-temperature superconductivity, which is a controversial issue since its discovery 15 years ago. The central problem is how to describe correctly the interplay of electron-phonon interaction and strong electronic correlations [34]. On the one hand, the anomalous magnetic and transport properties of the cuprate oxides have shown the important role of strong electronic correlations and stimulated a large effort towards an unconventional, i. e. purely electronic superconductivity mechanism [35,34]. On the other hand, experiments also show clear phonon and lattice effects in these materials. Superconductivity-induced phonon self-energy effects [36], large isotope coefficients away from optimal doping [37,38], tunneling phonon structure [39,40] etc. give evidence of a strong electron-phonon coupling. Recently, photoemission data indicate a sudden change in the electronic single-particle dispersion near a characteristic energy scale [41, 42], which is possibly caused by coupling of quasiparticles to phonon modes. Furthermore, neutron scattering data suggest that the oxygen so-called half-breathing $(\pi, 0)$ phonon couples strongly to doped charge carriers and correspondingly, softens significantly with doping [43].

To understand the effects of strong electronic correlations on the electron-phonon interaction perturbatively, based on $1/N$ expansion (with $N = 2$ the actual physical situation) within slave-boson [44,45] and X operator [46,34] formalisms, several authors have calculated the electron-phonon vertex function in the one-band and three-band Hubbard models for certain electron-phonon coupling (ionic, on-site coupling). Their finding is that, the backward scattering with large phonon momentum is suppressed much more than the forward scattering with small phonon momentum. They argued that the renormalization of electron-phonon interaction accounts for not only the dominant d-wave pairing at small doping, but also a reduction of the transport electron-phonon coupling constant λ_{tr} [34,46]. In order to gain further insight into the electron-phonon coupling in the correlated systems without resorting to $1/N$ or slave-boson assumptions, in the second part of this paper, we calculate exactly the electron-phonon vertex function by using Grand Canonical Monte Carlo (GCMC) [47] in the one-band Hubbard model.

2 Phase Fluctuations

Our starting Hamiltonian is of a simple BCS form given by

$$H = K - \frac{1}{4} \sum_{i\,\delta} (\Delta_{i\,\delta} \langle \Delta_{i\,\delta}^{\dagger} \rangle + \Delta_{i\,\delta}^{\dagger} \langle \Delta_{i\,\delta} \rangle), \tag{3}$$

with the next neighbor hopping term

$$K = -t \sum_{\langle ij \rangle, \sigma} (c^\dagger_{i\sigma} c_{j\sigma} + c^\dagger_{j\sigma} c_{i\sigma}), \tag{4}$$

where $c^\dagger_{i\sigma}$ creates an electron of spin σ on the ith site and t denotes an effective nearest-neighbor hopping. The $\langle ij \rangle$ sum is over nearest-neighbor sites of a 2D square lattice, and in the pairing term δ connects i to its nearest-neighbor sites. The local d-wave gap,

$$\langle \Delta^\dagger_{i\delta} \rangle = \frac{1}{\sqrt{2}} \langle c^\dagger_{i\uparrow} c^\dagger_{i+\delta\downarrow} - c^\dagger_{i\downarrow} c^\dagger_{i+\delta\uparrow} \rangle = \Delta\, e^{i\Phi_{i\delta}}, \tag{5}$$

is characterized by the *fluctuating* phases

$$\Phi_{i\delta} = \begin{cases} (\varphi_i + \varphi_{i+\delta})/2 & \text{for} \quad \delta \text{ in x-direction} \\ (\varphi_i + \varphi_{i+\delta})/2 + \pi & \text{for} \quad \delta \text{ in y-direction,} \end{cases} \tag{6}$$

and by a spatially constant amplitude Δ. We neglect the relative bond phase fluctuations between $\delta = \hat{x}$ and $\hat{y}$ as well as amplitude fluctuations and consider only the pair *center of mass* phase fluctuations, which are the relevant low-energy degrees of freedom, in a situation in which the superfluid density is small, like in the underdoped cuprates.

One could, of course, add a next-nearest-neighbor hopping t' and a chemical potential term μ which, however, for simplicity, we have set equal to zero. We will assume that, below a mean field temperature T_c^{MF}, there is a finite value of $\Delta \simeq 2T_c^{MF}$. The detailed temperature dependence of Δ is not central. The important point for our calculations is simply that a $d_{x^2-y^2}$-gap amplitude of order $2T_c^{MF}$ in magnitude forms as T drops below T_c^{MF}.

Since we are only interested in the temperature region $T \gtrsim T_c$, one can, in accordance with [21], convincingly argue that quantum phase fluctuations imply large relative number, i. e. charge fluctuations and (because of bad screening) large Coulomb energies. Therefore the fluctuations of the phase φ_i are predominantly determined by a classical XY free energy,

$$F[\varphi_i] = -J \sum_{\langle ij \rangle} \cos(\varphi_i - \varphi_j) \ . \tag{7}$$

In principle, the coupling energy J can be considered as arising from integrating out the high-energy degrees of freedom of the underlying microscopic system. Here, we will proceed phenomenologically, neglecting the temperature dependence of J and simply use it to set the Kosterlitz-Thouless transition temperature T_{KT} equal to some fraction of T_c^{MF}. Specifically, for the present calculations we will set $T_{KT} \simeq \frac{1}{4} T_c^{MF}$.

Our physical picture is that the XY-action arises from integrating out the shorter wavelength fermion degrees of freedom, including those responsible for the formation of the local pair amplitude and the internal $d_{x^2-y^2}$ structure of the pair. Thus, the *scale* of the XY-lattice spacing is actually set by the pair coherence length ξ_0. In our work, we have chosen Δ so that $\xi_0 \sim \frac{v_F}{\pi \Delta} \sim 1$. In this case, the phase configurations φ_i calculation can be carried out on the same $L \times L$ ($L = 32$) lattice that is used for the diagonalization of the Hamiltonian. This allows the Kosterlitz-Thouless phase correlation length ξ to grow over a sufficient range as T approaches T_{KT} and minimizes finite size effects. Thus, we are always in the limit where the phase correlation length ξ is larger than the Cooper pair size ξ_0, when the temperature T is below the mean-field critical temperature $T_c^{MF} \equiv T^*$.

2.1 Density of States and Tunneling Conductance

The calculation of the density of states for an $L \times L$ periodic lattice now proceeds as follows [23,24]. A set of phases $\{\varphi_i\}$ is generated by a Monte Carlo (MC) importance sampling procedure, in which the probability of a given configuration is proportional to $\exp(-F[\varphi_i]/T)$ with F given by Eq. (7). With $\{\varphi_i\}$ given, the Hamiltonian of Eq. (3) is diagonalized and the single particle density of states $N(\omega, T, \{\varphi_i\})$ is calculated. Further MC $\{\varphi_i\}$ configurations are generated and an average density of states $N(\omega, T) = \langle N(\omega, T, \{\varphi_i\}) \rangle$ at a given temperature is determined.

As stressed already above, our point of view is that the XY action, used in the MC simulations is not an *ad hoc* assumption but, in principle, arises from integrating out the shorter wavelength fermion degrees of freedom [22] up to the scale of the Cooper-pair size, so that only the *center of mass* phase fluctuations are dominant. Thus, the scale of the lattice spacing for $F[\varphi_i]$ is set by the pair size coherence length $\xi_0 \sim v_F/\pi\Delta_0$ and is of order 3 to 4 times the basic Cu-Cu lattice spacing of the fermion Hamiltonian Eq. (3).

The computationally intensive part of the calculation is the diagonalization and in order to get meaningful results as T approaches T_{KT}, we found it necessary to average over a large number ($\sim 10^4$) of MC $\{\varphi_i\}$ configurations. This required that some compromise had to be made with respect to the lattice size. The results we will present are for a 32×32 Hamiltonian lattice. However, if we were to take $\xi_0 \sim 4$ lattice spacings, this would lead to only an 8×8 lattice for the φ_i simulations. This would not allow a sufficient range for the Kosterlitz-Thouless phase coherence length to grow as T approaches T_{KT}. Thus, we have chosen to set $\Delta = 1.0t$ giving $\xi_0 \sim 1$ so that the φ_i simulation can be carried out on the same $L \times L$ lattice that is used for the diagonalization of H. The important physical point is that this procedure effectively cuts off phase fluctuations on a scale less than the Cooper-pair size ξ_0.

Results for $N(\omega, T)$ are shown in Fig. 1. For each temperature we have generated up to $25,000$ independent MC $\{\varphi_i\}$ configurations, diagonalized H for each of these configurations, and computed $\langle N(\omega, T, \{\varphi_i\}) \rangle$. In these calculations, as discussed above, we have set $\Delta = 1.0t$ corresponding to $T_c^{MF} \simeq 0.5t$ and selected J so that $T_{KT} = 0.1t$ [25]. In order to reduce

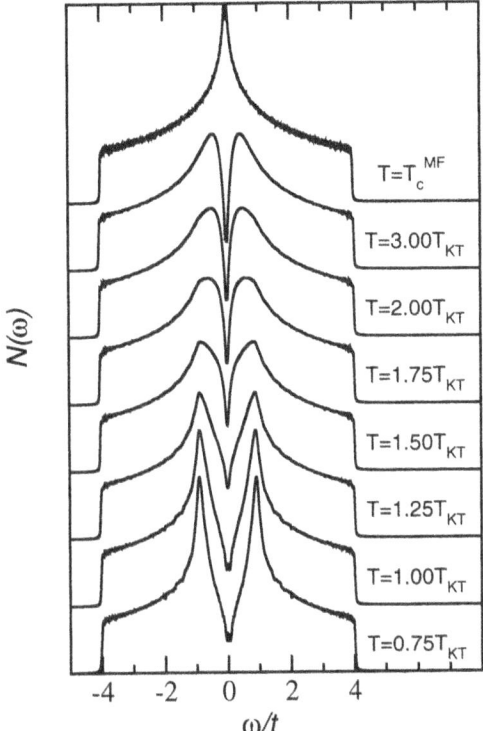

Fig. 1. Single particle density of states $N(\omega)$ for different temperatures T for a 32×32 lattice with $\Delta = 1.0t$ and $T_{KT} = 0.1t$. A pseudogap appears below $T_c^{MF} \simeq 0.5t$ and coherence peaks develop as T approaches T_{KT}.

finite-size effects, we employ a very effective scheme recently suggested by F. F. Assaad [26].

For $T > T_c^{MF}$, the gap amplitude vanishes and the density of states exhibits the usual Van Hove peak at $\omega = 0$. For $T < T_c^{MF}$, the presence of a finite gap amplitude gives rise to a pseudogap whose size is set by 2Δ. Then, as T approaches T_{KT} and the XY phase correlation length rapidly increases, coherence peaks evolve, the separation of which is determined by 2Δ. An important point is that the scale in temperature over which the evolution of the coherence peaks occurs, is set by some fraction of T_{KT} which means that it appears suddenly on a scale set by T_c^{MF}.

An *effective* correlation length $\xi(T)$, extracted by fitting an exponential form to the correlation function

$$C(\ell) = \left\langle e^{-i\varphi_{i+\ell}} e^{i\varphi_i} \right\rangle \tag{8}$$

is plotted versus T in Fig. 2 for our 32×32 lattice. The rapid onset of $\xi(T)$ as T_{KT} is approached is clearly seen. It is this sudden increase of $\xi(T)$ that is

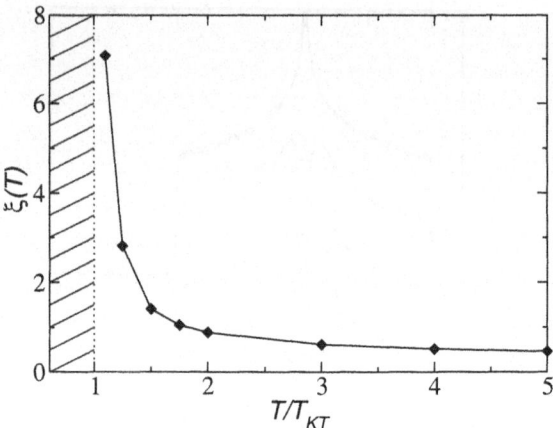

Fig. 2. The *effective* correlation length $\xi(T)$ versus T/T_{KT} for the 32×32 lattice. Here $T_c^{MF}/T_{KT} \simeq 5$ so that the pseudogap regime which extends from $T/T_{KT} \simeq 1.5$ to 5 is large compared to the superconducting region which extends from 0 to $T/T_{KT} = 1$. The pronounced increase of $\xi(T)$ occurs over a narrow temperature region, on a scale set by T_c^{MF}, as T_{KT} is approached.

responsible for the appearance of the coherence peaks as T approaches T_{KT}. This effect is further enhanced by the $2D$ to $3D$ crossover that occurs in the actual materials.

In order to compare these results for $N(\omega, T)$ with scanning tunneling spectra dI/dV, we have calculated $dI(V, T)/dV$ using the standard quasi-particle expression for the tunneling current,

$$\frac{dI(V,T)}{dV} \propto \int N(\omega) \frac{\partial f(\omega - V)}{\partial V} \, d\omega. \tag{9}$$

Here, $f(\omega) = (\exp(\omega/T) + 1)^{-1}$ is the usual Fermi factor. Results for $dI(V, T)/dV$ are displayed in Fig. 3. In accordance with the experiments [17], we have plotted in Fig. 3 the normalized tunneling conductance $dI/dV|_{norm} = (dI(V, T)/dV)/(dI(V, T_c^{MF})/dV)$ which is unbiased from band structure effects. The effect of the Fermi factors is to provide a thermal smoothing of the quasi-particle density of states over a region of order $2T$. This becomes significant at the higher temperatures and the prominent pseudogap dependence of $N(\omega, T)$ seen in Fig. 1 is smoothed out in dI/dV. One sees that the size of the pseudogap scales with the spacing between the coherence peaks and evolves continuously out of the superconducting state. The pseudogap persists over a large temperature range measured in units of T_{KT}, becoming smoothed out by the thermal effects as T approaches T_c^{MF} and vanishing above T_c^{MF}.

These results for $dI(V, T)/dV$ are similar to recent scanning tunneling measurements of Bi2212 and Bi2201 [4, 5, 17, 18]. The superconducting gap

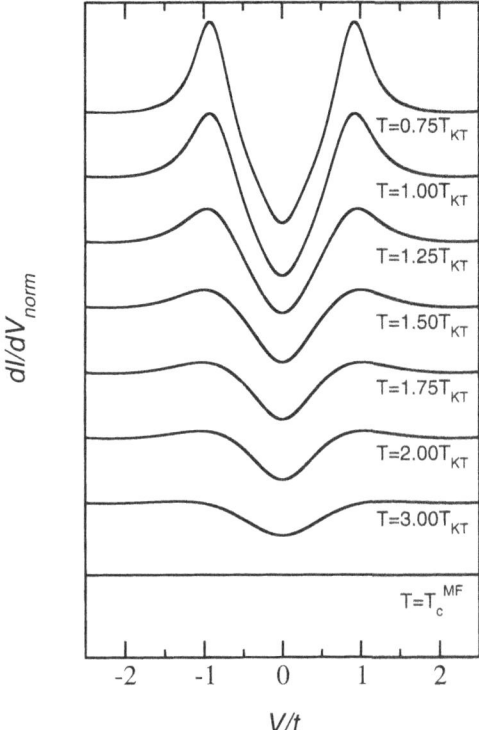

Fig. 3. Tunneling conductance, $\frac{dI}{dV}$, normalized to its value at T_c^{MF} for different temperatures.

for $T < T_{KT}$ evolves continuously into the pseudogap regime which extends up to $T = T_c^{MF}$. The coherence peaks appear suddenly as T_{KT} is approached. At higher temperatures, the pseudogap fills in rather than closing and the temperature range associated with the pseudogap regime can be large compared with the size of the superconducting regime.

2.2 Kinetic Energy Loss and Optical Sum Rule

With the Hamiltonian given in Eq. (3), it is straight forward to show that the optical sum rule yields [33]:

$$W_{low+D} = \int_0^\Omega Re\ \sigma_{xx}(\omega)\ d\omega = -e^2\pi\langle k_x\rangle/2, \qquad (10)$$

in units where $\hbar = c = 1$, with $\langle k_x\rangle$ being the expectation value of the local next neighbor hopping (kinetic energy), i. e.

$$k_x = -t\sum_\sigma (c_{i\,\sigma}^\dagger c_{i+x\,\sigma} + c_{i+x\,\sigma}^\dagger c_{i\,\sigma}). \qquad (11)$$

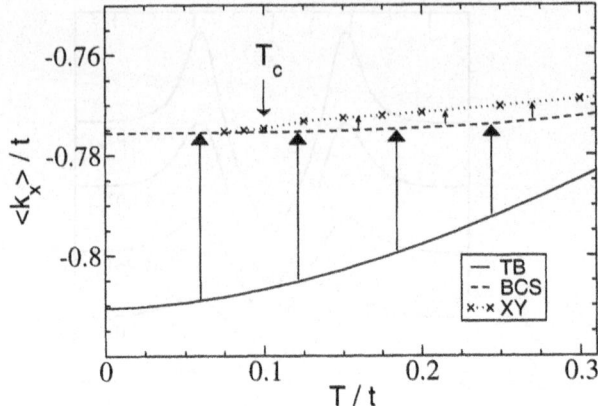

Fig. 4. Kinetic energy per bond $\langle k_x \rangle$ as a function of temperature for the *free* tight-binding electrons (TB), the BCS solution (BCS) and our phase-fluctuation model (XY). The large vertical arrows indicate the increase in kinetic energy upon pairing relative to the free tight-binding model and the small arrows indicate the additional increase due to phase fluctuations. This additional *phase fluctuation energy* goes to zero near $T_c \equiv T_{KT}$, which causes the significant change in the optical integral upon entering the superconducting state.

A violation of the optical sum rule at high-frequencies with a loss in high-frequency spectral weight W_{high} upon entering the superconducting state, would therefore show up as an increase of the low frequency spectral weight (including the superfluid condensation peak) W_{low+D} with $W_{low+D} \sim -\langle k_x \rangle$.

In these calculations, where we assume a BCS temperature dependence of the pairing gap $\Delta(T)$, we have set $\Delta(T = 0) = 1.0t$ corresponding to $T_c^{MF} \simeq 0.42t$ and selected J so that $T_{KT} = 0.1t$ [25]. The condition that $\xi > \xi_0$ is thus always fulfilled, if we are not too close to T_c^{MF}. The calculation of the kinetic energy for an $L \times L$ ($L = 32$) periodic lattice now proceeds along the same lines as in Sect. 2.1: for a given set of phases $\{\varphi_i\}$, the Hamiltonian of Eq. (3) is diagonalized and the kinetic energy $E_{kin}(T, \{\varphi_i\}) = \langle k_x \rangle_{\{\varphi_i\}}$ is calculated. Further Monte-Carlo $\{\varphi_i\}$ configurations are generated and an average kinetic energy $E_{kin}(T) = \langle k_x \rangle$ at a given temperature is determined.

Figure 4 displays the kinetic energy $\langle k_x \rangle$ as a function of temperature for non-interacting tight-binding electrons, BCS electrons, and our phase fluctuation model (termed XY in Fig. 4). We can clearly see, that pairing produces an overall increased kinetic energy (indicated as vertical arrows) with respect to the tight-binding model (full curve in Fig. 4). In the phase fluctuation model, the kinetic energy is further increased (small vertical arrows) due to the incoherent movement of the paired electrons. We can further see that the kinetic energy is a smoothly decreasing function of temperature for $T \to 0$. This is in agreement with the experimental results [28, 30] where the

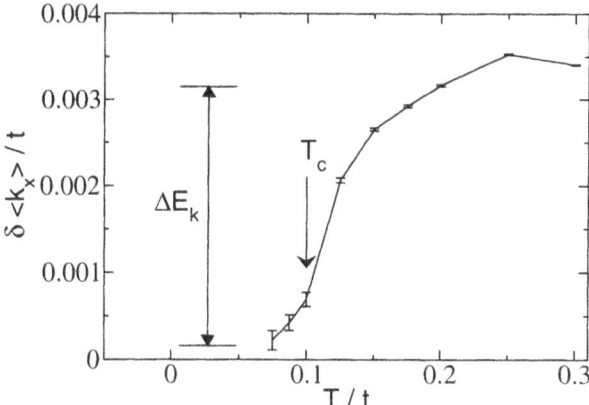

Fig. 5. Kinetic energy contribution from phase fluctuations $\delta\langle k_x\rangle = \langle k_x\rangle_{XY} - \langle k_x\rangle_{BCS}$. One can clearly see the sharp decrease of the kinetic energy near the Kosterlitz-Thouless transition at $T = 0.1t \equiv T_c$. ΔE_k gives a rough estimation of the condensation energy.

corresponding low frequency spectral weight $W_{low+D} \sim -\langle k_x\rangle$ is smoothly increasing for $T \to 0$. What we are especially interested in is the rather pronounced (magnified by using a different scale for the kinetic energy in Fig. 5) change near $T_c \equiv T_{KT}$, where the kinetic energy of our phase fluctuation model suddenly reduces to the BCS value. This sudden deviation from the $T \gtrsim T_c$ behavior can also be seen experimentally, where there is a kink in the temperature dependence of the low frequency spectral weight W_{low+D} at T_c [28] and it can be related to the kinetic condensation energy upon entering the superconducting state.

In order to make this rapid change of in plane kinetic energy more explicit, we plot in Fig. 5 the difference between the BCS kinetic energy and our phase fluctuation model $\delta\langle k_x\rangle = \langle k_x\rangle_{XY} - \langle k_x\rangle_{BCS}$. Since we assume in our model that we already have *preformed* pairs with a fixed gap Δ, but only local phase coherence, the overall kinetic energy within the phase fluctuation scenario his higher than the BCS value. But on the other hand *we loose this additional kinetic energy* at $T_c \equiv T_{KT}$, where the pairs are now allowed to move coherently. The *jump* of the kinetic energy at T_{KT} would be even sharper for a larger lattice. It is due to the appearance of sharp coherence peaks in our single particle spectral function. In Sect. 2.1, we have shown that these coherence peaks appear when long ranged phase correlations develop. In this situation the Cooper pairs can move easier and thereby decrease the kinetic energy.

In order to get a rough estimate of the kinetic condensation energy, we calculate the loss in kinetic energy near T_c

$$\Delta E_k = -\frac{2}{e^2\pi} \int_0^\Omega \left(Re\ \sigma_{xx}^{sc}(\omega) - Re\ \sigma_{xx}^n(\omega) \right) d\omega \qquad (12)$$

as indicated in Fig. 5. Assuming that $t \simeq 250\,meV$, we get a condensation energy of $1.5\,meV$ per Copper site, which is in order of magnitude agreement with the experimental results.

3 Electron Phonon Coupling

In order to gain further insight into the strongly electron-phonon coupling in the correlated systems, we calculate exactly the electron-phonon vertex function by employing a new idea and using Grand Canonical Monte Carlo (GCMC) [47] in the one-band Hubbard model. The central issue is how the vertex function is influenced by correlation effects, i. e. the Hubbard U and, additionally, also the doping density. In the limit of weak coupling, the Feynman diagram calculation is used to help us understand the physical processes entering the vertex function.

The Hamiltonian of the one-band Hubbard model reads,

$$H = -t \sum_{\langle ij \rangle, \sigma} (c_{i\sigma}^\dagger c_{j\sigma} + c_{j\sigma}^\dagger c_{i\sigma}) + U \sum_i n_{i\uparrow} n_{i\downarrow}, \qquad (13)$$

where $\langle ij \rangle$ denotes nearest neighbor lattice sites i and j. The operators $c_{i\sigma}^\dagger$ and $c_{i\sigma}$ create and destroy an electron with spin σ at site i. $n_{i\sigma} \equiv c_{i\sigma}^\dagger c_{i\sigma}$ is the number operator for an electron with spin σ at lattice site i.

In our simulations, we have used two techniques to extract the vertex function. The first is the so-called *frozen phonon technique* [48]. Here, we are in the adiabatic limit of the electron-phonon coupling, where a static, i. e. *frozen* lattice displacement acts as a perturbation on the electronic system. In this case, we add a frozen phonon term $\Delta_e \sum_{k\sigma} g_{kQ} c_{k+Q\sigma}^\dagger c_{k\sigma}$ to the above Hamiltonian. Then the vertex function $\Gamma(p, i\omega_n, Q, 0)$ is extracted as (shown in Fig. 6),

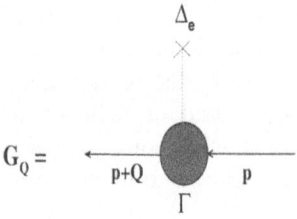

Fig. 6. Feynman diagram of the vertex function.

$$\Gamma(p, i\omega_n, Q, 0) = \lim_{\Delta_e \to 0} \frac{1}{\Delta_e} \frac{G_Q(p, i\omega_n)}{G(p + Q, i\omega_n)G(p, i\omega_n)}. \tag{14}$$

where $G(p, i\omega_n)$ and $G(p + Q, i\omega_n)$ are the usual single particle Green's functions; p, $p + Q$ and Q are momenta of the incoming electron, the outgoing electron and the phonon, respectively. $\omega_n = (2n + 1)\pi T$ is the fermionic Matsubara frequency. Because the frozen phonon term breaks the translational invariance, there is an additional non-vanishing electron Green's function $G_Q(p, i\omega_n)$, in which an electron begins with momentum p, and ends with momentum $p + Q$. This anomalous Green's function is defined as,

$$G_Q(p, i\omega_n) = -\int_0^\infty d\tau e^{i\omega_n \tau} < T_\tau c_{p+Q}(\tau)c_p^\dagger(0) > . \tag{15}$$

The second method to extract the vertex is the linear response technique, in which we can directly obtain the anomalous Green's function by linear response theory, i. e.

$$G_Q(p, i\omega_n) = \Delta_e \int_0^\beta d\tau e^{i\omega_n \tau} \int_0^\beta d\tau' \sum_{kq} g_{kq}*$$

$$*\langle T_\tau C_{k+q}^\dagger(\tau')C_k(\tau')C_{p+Q}(\tau)C_p^\dagger(0)\rangle. \tag{16}$$

The linear response form allows, in principle, to account for dynamic vertex corrections, i. e. where dynamic lattice displacements couple to the electronic system. As to the electron-phonon coupling, there are two kinds of forms: One is a so-called *ionic* coupling where the lattice distortions interact with the local total charge density; the other is a *covalent* coupling, where the hopping matrix couples to the lattice distance between the sites. In this paper, we will concentrate on the ionic coupling, and the bare electron-phonon matrix element is defined as $g_{kq} = 1$ for all phonon modes. Because we are interested in the low frequency processes, i.e., $\omega_n \to 0$, in principle, the numerical simulations should be done at the low temperature regime, where we can extrapolate the vertex function to zero frequency correctly. However, in the two-dimensional system and at finite doping densities, the *sign* problem prevents us from simulating at *very cold* temperatures. We have checked some special cases at low temperatures, and found that the real part of the vertex function $\Gamma(p, i\pi T, Q, 0)$ depends only slightly on temperature, whereas the imaginary part decreases to zero for $T \to 0$. Therefore, we will present our numerical results only for the lowest frequency $\omega_n = \pi T$.

In Fig. 7, the Feynman diagrams are displayed contributing to the vertex function. The thick wavy line represents the screened phonon field, which is expanded in diagrams, shown at the bottom. Here, besides the dielectric screening, we consider the diagrams 2, 3, 4 dominating the proper vertex corrections at small U, where the proper vertex corrections mean, that the Feynman diagram cannot be separated by cutting a single Coulomb line. The summation of all diagrams gives the vertex function, i. e.

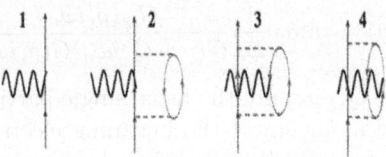

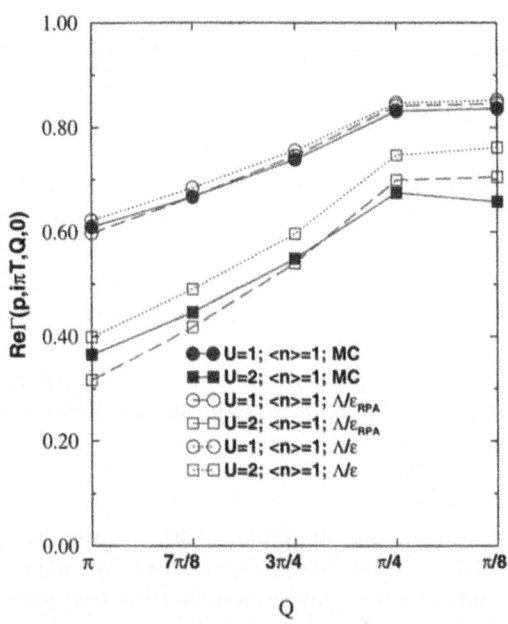

Fig. 7. Feynman diagrams of the vertex function.

Fig. 8. Vertex function versus phonon momentum Q for the phonon ionic coupling to the electrons. $1D$ system, $L = 32$, $\beta = 6$, $p = -\pi/2$.

$$\Gamma(p, i\omega_n, Q, 0) = \Lambda \cdot \frac{1}{\epsilon(Q, 0)}, \qquad (17)$$

with $\Lambda = \Lambda_1 + \Lambda_2 + \Lambda_3 + \Lambda_4$ and the screening $\frac{1}{\epsilon} = 1 - U\Pi(Q, 0)$. Here, the density response function $\Pi(Q, 0)$ is defined as $\Pi(Q, 0) = \frac{1}{2} \int_0^\beta d\tau \langle T_\tau \rho(Q, \tau) \rho^+(Q, 0) \rangle$.

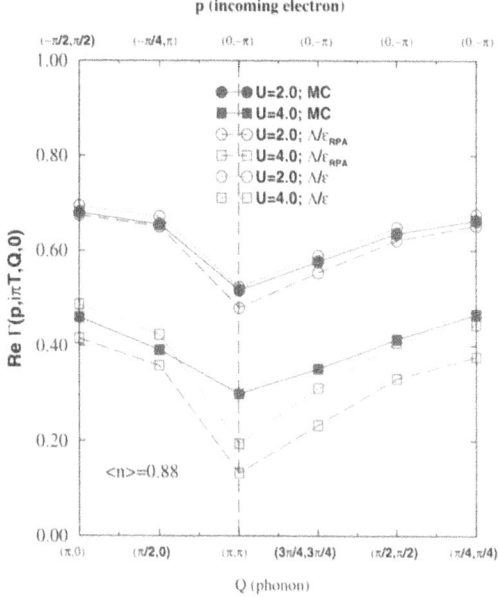

Fig. 9. Real part of vertex function vs phonon momentum Q. 2D system, $L = 8 \times 8$, $< n >= 0.88$, $\beta = 3$.

Within the random-phase- approximation, the dielectric function is given by $\epsilon_{RPA} = 1 + U X_0(Q, 0)$, where $X_0(Q, 0) = \frac{1}{N} \sum_k \frac{f(\epsilon_{k+Q}) - f(\epsilon_k)}{\epsilon_k - \epsilon_{k+Q}}$.

In Fig. 8, we compare the vertex function calculated from Monte Carlo simulations and diagram expansions in the 1D system. We observe, that the numerical results are in reasonable agreement with the analytical results. The general trend is that the backward scattering with large phonon momenta Q is suppressed more than the forward scattering. The effect of Hubbard U is to suppress the electron-phonon coupling at all momenta. The comparison also indicates that the RPA dielectric screening is indeed a good approximation for the weak-coupling Hubbard system, which is considered in Fig. 8 ($U = 1t, 2t$).

In the 2D system, we observe similar behaviors to the 1D system. (However, here we have so far simulated on a relatively *hot* system with $\beta = 3$ or $\beta = 4$.) As shown in Fig. 9, the weak-coupling $U = 2$ diagram expansion with the RPA dielectric function or exact dielectric function gives consistent results with the MC simulation. In the diagonal direction from $(0, 0)$ to (π, π), the vertex function decreases with phonon momentum Q, suggesting that the backward scattering is suppressed much more than the forward scattering. However, our numerical results indicate that the electron-phonon coupling at $Q = (\pi, 0)$ is comparable in amplitude to the case of small phonon momentum. This result is in clear contrast to what previous works ($1/N$, slave-boson calculations) have obtained. The consequences of this finding, which yields to a *violation* of the usually assumed small ratio of the transport to electron-phonon coupling constant λ_{tr} will be investigated in our group in the next step. From the Feynman diagram expansions, there are two important features: (1) the dielectric screening is slightly changed by Hubbard U, and at $U = 4$, the dependence on the momentum Q is rather weak; (2) the proper vertex corrections increase to order U^2. As a result, the structure of the vertex function at large Coulomb interaction is mainly attributed to the contributions of proper vertex corrections, while the dielectric screening reduces the electron-phonon interaction at all momenta, overall.

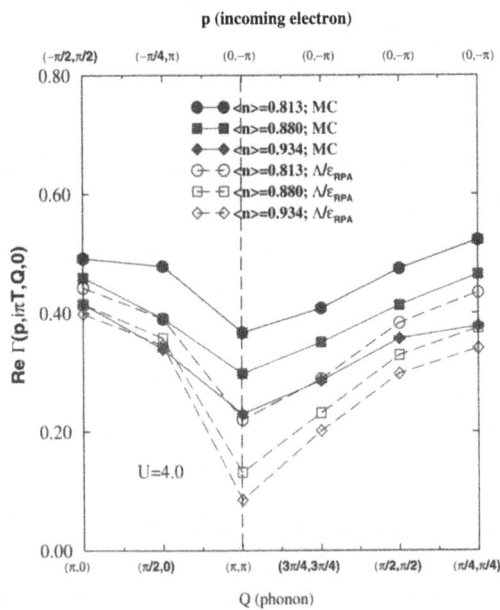

Fig. 10. Real part of vertex function at different dopings. 2D system, $L = 8 \times 8$, $U = 4$, $\beta = 3$.

The doping dependence of the ionic electron-phonon coupling in 2D is shown in Fig. 10. Also illustrated is the diagram expansion with RPA dielectric screening. It is clearly seen that with decreasing the doping density, the vertex functions at all phonon momenta are reduced monotonously, and the reduction depends slightly on the phonon momentum. The RPA calculations agree qualitatively with the MC simulations, but overestimate the vertex corrections, overall. Especially, we note that the RPA calculations give much more reduction at $Q = (\pi, \pi)$ than other phonon momenta. Comparison with Figures 5 and 6 in the reference [46] shows that the dependence of the vertex function on phonon momentum Q and doping density agrees rather well in the diagonal direction for our numerical simulation and Zeyher's *et al.* X-method. However, as already emphasized, a dramatic difference exists at $Q = (\pi, 0)$, where we find that the electron-phonon coupling is still effective even in the intermediate Coulomb interaction, whereas Zeyher *et al.*'s $(1/N)$ results indicate a strong suppression as the phonon at $Q = (\pi, \pi)$.

4 The Need for High-Performance Computing

As already said above, the phase-fluctuation program is a mixture of a Monte-Carlo and a diagonalization program. For the Monte-Carlo part of the calculation, we need a massive (MPI) parallel environment. On the other hand, the exact diagonalization part requires simultaneously a high degree of vectorization, a large memory and a large bandwidth for memory access. Therefore, the program was run in the inter-node MPI mode on 64 nodes of the Hitachi SR8000, using COMPAS for intra-node parallelization. A typical run lasted for about 18 hours with a sustained performance of 3.6 GFlop/s per node for a 30 × 30 lattice (∼ 1.0 GFlop/s per node for a 32 × 32 lattice). The total amount of storage required was about 64 GByte (i. e. ∼ 1.0 GByte per node). The diagonalization of a typical $N \times N$ matrix needs $\mathcal{O}(N^3)$ steps and a memory of $\mathcal{O}(N^2)$. In our problem we have $N = 2L^2$, with L being the size of our physical lattice. Therefore the computing time scales with $\mathcal{O}(L^6)$ and the memory with $\mathcal{O}(L^4)$. In order to get meaningful results, we need $L \gg \xi$. Since the physical interesting regime, is the region where the phase-correlation length ξ diverges ($\xi \to \infty$), we need a very large amount of computing time and memory close to the critical region, which only the Hitachi SR8000 can provide.

   In the electron-phonon vertex function part of our study, the massive (MPI) parallel Monte Carlo program was run in the intra-node mode on the Hitachi SR8000 with a sustained performance of 0.5 GFlop/s per processor and required a storage of 10-100 MB per processor. The total computing time of a complete run on 32 nodes (i. e. 256 processors) was about 16 hours. As already said above, this project aims to study the electron-phonon interactions in the strongly correlated electronic systems based on Quantum Monte Carlo (QMC) techniques. But in order to understand how the electron-phonon coupling depends on the electron and phonon momenta in the physically relevant low energy region, we must perform simulations on as large lattices as possible (which provides us with a dense mesh of **k**-points and, thus, momenta) and at very low temperatures (which is required because of the low-energy scale $E \simeq k_B T_c$, where T_c is the superconducting transition temperature). In general, the computing time scales with the lattice size $N = L \times L$ and inverse temperature β in the form: *time* $\propto N^3 * \beta$ (Frozen-Phonon) and *time* $\propto N^4 * \beta^2$ (Linear-Response), thus the computing time increases dramatically when the lattice size changes from 8 × 8 to 10 × 10, or even larger.

Typically, if we would run our program on a workstation for one fixed set of input-parameters $(\mathbf{p}, \mathbf{Q}_{Phonon}, \text{Doping})$, we would have to wait for one month or even longer.

5 Summary and Conclusion

Summarizing, in order to develop a quantitative understanding for the role of phase fluctuations, we have presented in the first part of this paper a numerically exact solution of the BCS model with a $d_{x^2-y^2}$-gap and fluctuating phases. This is the simplest model with as few approximations as possible, whose exact solution should establish a guide for future, more detailed studies. Nevertheless, already in its simple form, the BCS d-wave model with center of mass pair-phase fluctuations, which are consistently extracted from an XY action used in the Monte Carlo simulation, reproduces salient features of recent STM spectra of high-T_c cuprates. Based upon these results, it appears that the pseudogap behavior observed in the scanning tunneling spectra of Bi2212 and Bi2201 can be understood as arising from phase fluctuations of a $d_{x^2-y^2}$-pairing gap.

We have further shown, that this phase fluctuation model shows a significant loss of in-plane kinetic energy at T_c. This lowering in kinetic energy is due to the sharpening the of quasiparticle peaks at the superconducting transition temperature $T_c \equiv T_{KT}$, where the phase correlation length ξ diverges. For a tight-binding lattice model, this change in in-plane kinetic energy can directly be related to the optical integral, where it appears as a violation of the low frequency sum rule. This sum rule violation should, therefore, also appear in other superconductors with low charge carrier density (phase stiffness), such as the organic superconductors.

In the second part of this paper, based on Quantum Monte Carlo simulations, we study the electron-phonon vertex function in the one-band Hubbard model. In contrast to earlier studies, this allows, at least in principle, for a numerically exact solution for the vertex, which is especially important for strongly correlated systems. We find that the effects of the Hubbard interaction is to suppress the electron-phonon coupling at all phonon momenta, and the reduction of doping density makes the electron-phonon coupling become less effective. As a function of phonon momentum, the forward scattering is rather stronger than the scattering close to $Q = (\pi, \pi)$. The technique will now be applied to the more important *covalent* coupling, where preliminary results call for a substantial revision of the so-far assumed picture of a small $\lambda_{tr}/\lambda_{el-ph}$ ratio.

Acknowledgement. Many stimulating discussions with D. J. Scalapino and S. A. Kivelson are acknowledged. We also want to thank the Leibniz-Rechenzentrum (LRZ) München for computational support. This work was supported by the DFG under Grant No. Ha 1537/16-2, by a Heisenberg fellowship (AR 324/3-1) and the KONWHIR projects OOPCV and CUHE.

References

1. A. G. Loeser, Z. X. Shen *et al.*, Science **273**, 325 (1996).
2. H. Ding *et al.*, Nature (London) **382**, 51 (1996); and, Phys. Rev. Lett. **78**, 2628 (1997).
3. J. W. Loram *et al.*, Phys. Rev. Lett. **71**, 1740 (1993).
4. Ch. Renner *et al.*, Phys. Rev. Lett. **80**, 149 (1998).
5. N. Miyakawa *et al.*, Phys. Rev. Lett. **83**, 1018 (1999).
6. M. Takigawa *et al.*, Phys. Rev. B **43**, 247 (1991); H. Alloul *et al.*, Phys. Rev. Lett. **70**, 1171 (1993).
7. J. Orenstein *et al.*, Phys. Rev. B **42**, 6342 (1990).
8. S. Chakravarty *et al.*, Phys. Rev. B **63**, 94503 (2001).
9. J. R. Schrieffer and A. P. Kampf, J. Phys. Chem. Solids **56**, 1673 (1995).
10. D. Pines, Z. Phys. B **103**, 129 (1997); A. V. Chubukov and J. Schmalian, Phys. Rev. B **57**, R11085 (1998).
11. V. J. Emery and S. A. Kivelson, Nature (London) **374**, 434 (1995).
12. M. Randeria *et al.*, Phys. Rev. Lett. **69**, 2001 (1992).
13. M. Franz and A. J. Millis, Phys. Rev. B **58**, 14572 (1998).
14. H.-J. Kwon and A. T. Dorsey, Phys. Rev. B **59**, 6438 (1999).
15. B. Kyung, S. Allen and A.-M. S. Tremblay, Phys. Rev. B **64**, 75116 (2001).
16. Y. Wang *et al.*, cond-mat/0205299 (unpublished).
17. M. Kugler *et al.*, Phys. Rev. Lett. **86**, 4911 (2001).
18. T. Eckl *et al.*, cond-mat/0110377 (unpublished).
19. J. M. Kosterlitz and D. J. Thouless, J. Phys. C **6**, 1181 (1973).
20. E. W. Carlson *et al.*, Phys. Rev. Lett. **83**, 612 (1999) and refs. therein.
21. V. J. Emery and S. A. Kivelson, Phys. Rev. Lett. **74**, 3253 (1995).
22. A. Paramekanti *et al.*, Phys. Rev. B **62**, 6786 (2000).
23. A more detailed discussion of this static field Monte Carlo approximation for the pair-field is given in T. Eckl *et al.* (to be published); see also N. E. Bickers and D. J. Scalapino (cond-mat/0010480) and P. Monthoux and D. J. Scalapino (cond-mat/0111534).
24. E. Dagotto *et al.*, Phys. Rev. B **58**, 6414 (1998) have used Monte Carlo simulations to study a model of classical spins interacting with electrons.
25. $T_{KT} \simeq 0.89J$, see, for example, J. F. Fernández, M. F. Ferreira, and J. Stankiewicz, Phys. Rev. B **34**, 292 (1986).
26. F. F. Assaad, Phys. Rev. B **65**, 115104 (2002).
27. K. A. Moler *et al.*, Science **279**, 1193 (1998).
28. H. J. A. Molegraaf *et al.*, Science **295**, 2239 (2002).
29. J. E. Hirsch, Science **295**, 2226 (2002).
30. A. F. Santander-Syro *et al.*, cond-mat/0111539 (unpublished).
31. M. R. Norman *et al.*, cond-mat/0201415 (unpublished).
32. *Introduction to Superconductivity* (McGraw–Hill, New York, 1975)
33. D. J. Scalapino *et al.*, Phys. Rev. B **47**, 7995 (1993).
34. M. L. Kulic, Physics Reports **338**, 1-264 (2000).
35. D. J. Scalapino, Physics Reports **250**, 329-365 (1995).

36. V. G. Hadjiev, X. J. Zhou, T. Strohm, M. Cardona, Q. M. Lin, and C. W. Chu, Phys. Rev. B**58**, 1043 (1998).
37. J. P. Frank, S. Harker, and J. H. Brewer, Phys. Rev. Lett. **71**, 283 (1993).
38. G. M. Zhao, H. Keller, and K. Conder, cond-mat/0204447,0204448.
39. D. Shimada, Y. Shiina, A. Mottate, Y. Ohyagi, and N. Tsuda, Phys. Rev. B**51**, R16495 (1995).
40. R. S. Gonnelli, G. A. Ummarino, V. A. Stepanov, Physica C**275**, 162 (1997).
41. Z. X. Shen, A. Lanzara, Ishihara, and N. Nagaosa, cond-mat/0108381.
42. A. Lanzara, *et al.*, Nature **412**, 510 (2001).
43. R. J. Mcqueeney *et al.*, Phy. Rev. Lett. **82**, 628 (1999); Y. Petrov, T. Egami, R. J. McQueeney, M. Yethiraj, H. A. Mook, and F. Dogan, cond-mat/0003414.
44. J. H. Kim, K. Levin, R. Wentzconvitch, and Auerbach, Phys. Rev. B**44**, 5148 (1991); J. H. Kim, and Z. Tesanovic, Phys. Rev. Lett. **71**, 4218 (1993).
45. M. Grilli, and C. Castellani, Phys. Rev. B**50**, 16880 (1994).
46. R. Zeyher, and M. Kulic, Phys. Rev. B**53**, 2850 (1996).
47. R. Blankenbecker, D. J. Scalapino, and R. L. Sugar, Phys. Rev. D**24**, 2278 (1981).
48. R. M. Noack, Ph. D. thesis.

The Cluster-Perturbation-Theory and its Application to Strongly-Correlated Materials

Christopher Dahnken, Enrico Arrigoni, Werner Hanke, Marc G. Zacher, and Robert Eder

[1] Institut für Theoretische Physik
Universität Würzburg
97074 Würzburg, Germany

Abstract. We present a technique whereby the standard restriction of quantum-mechanic many-body simulations to small clusters of 10 to 1000 sites can be overcome by the systematic use of a so-called cluster perturbation theory. In principle, this facilitates the accurate numerical solution of the macroscopic, i.e. infinite-size system. These calculations have become feasible only with the recent advance of high-performance computers. Angular-resolved photoemission data on various strongly-correlated materials are analyzed as a specific example, via a method that extends exact-diagonalization analysis to the infinite lattice by means of a perturbation in the intercluster hopping. We apply this method to relate photoemission data to different phases of high-T_c materials, namely (i) the so-called stripe phase of LaSrCuO and LaNdSrCuO, in which doped holes organize themselves in quasi-one-dimensional structures, (ii) the insulating phase of SrCuOCl, (iii) and the Fermi surface geometry of the overdoped $Bi_2Sr_2CaCu_2O_{8+\delta}$ compound. The computational effort, which is necessary to address this problem and the performance that we have achieved, is detailed.

1 Introduction

A variety of numerical techniques exist which deal with strongly-correlated electron systems, such as the high-temperature superconductors and other transition-metal compounds, on the basis of accurate simulations such as quantum monte carlo (QMC) and exact diagonalization (ED) within relatively small-size (10-1000 sites) cluster. In this project cluster-perturbation techniques and the corresponding algorithms have been systematically developed, in order to be able to extend the so-obtained "local", i.e. cluster many-body physics to the, in principle, infinite-sized (i.e. macroscopic) solid. First applications of this idea are also presented.

The study of high-T_c materials has produced major advances in techniques for the treatment of so-called *strongly correlated systems* [6,7]. In fact, high-T_c materials can be seen as a paradigm for strong correlation. It is widely believed that is is precisely the physics of the strongly interacting electrons, which is the reason for the richness of their phase diagram. Strong

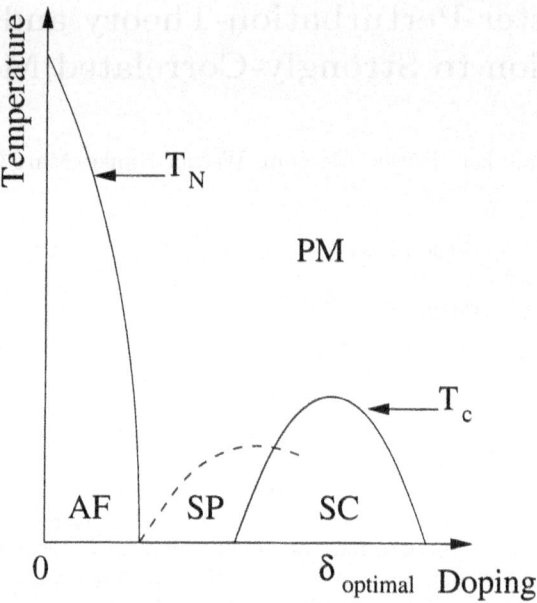

Fig. 1. Generic temperature-versus-doping phase diagram of the high temperature superconductors. The full lines indicate the Neél-temperature T_N in the antiferromagnetic (AF) and the critical temperature T_c in the superconducting (SC) phase. Between the two, at low temperature and partially overlapping with the SC phase, one finds a phase of one-dimensional ordered charge carriers, the stripe phase (SP). At finite doping and higher temperatures the HTSC are generally paramagnetic metals (PM). The label $\delta_{optimal}$ marks the point of the highest critical temperature. In the text we denote the regions left and right of $\delta_{optimal}$ as underdoped and overdoped, respectively.

correlation establishes a fundamental, essentially unsolved problem of theoretical physics, concerning in particular solid-state and statistical physics. In these systems, the electron interaction energy is typically comparable or larger than the kinetic energy. Therefore, one can not apply the usual perturbative schemes, which live from a "small" parameter, here the ratio of kinetic over Coulomb energy.

Consider a metal, which can be seen as a matrix of ions in which electrons move, subject to the interaction with the former. As a first approximation, one can take the ions as fixed at their places and consider only the electron motion. Then, we still have to deal with a large number of electrons, of the order of 10^{23} per cubic cm. The forces between electrons and with the ions are well-known, therefore, in contrast to particle physics, one does not have to worry about the nature of the interactions. Here, the problem consists in the solution of the dynamic equations for such a large number of electrons, in order to make predictions and understand their physical behavior. An approach, which works quite well for most solids, consists in treating the electrons as "independent" from each other: each electron is affected by the other particles in a "mean-field" way only. Strongly-correlated systems are, by definition, those materials where such an "independent"-particle approach fails.

For this reason, a number of numerical and analytical techniques have been developed in order to deal with this "strong-correlation" problem. A central quantity in strongly-correlated systems is the one-particle Green's function, which, in particular, gives information about the behavior of single-particle excitations of the system. The importance of this quantity is related to the fact that its imaginary part – the single-particle spectral function – can be (almost) directly measured by angular-resolved photoemission spectroscopy (ARPES). It is, thus, essential – for a given many-body model – to have an efficient algorithm for calculating the Green's function with sufficient accuracy in order to assess the validity of the chosen model by quantitative comparison with experiment. Due to the failure of "mean-field" methods in strongly correlated materials, as discussed above, few numerical techniques are able to provide controlled results on this quantity. This is due to the fact that numerical methods are either restricted to very small system sizes or resort to statistical methods, which result in noisy data.

Recently, a method has been proposed to approximately solve the Green's function for the infinite lattice starting from the exact solution for smaller clusters [8, 9]. Here, we present first applications of this method (termed cluster perturbation theory (CPT)) to address various issues arising from experimental studies of the high-T_c Superconductors (HTSC).

The paper is structured as follows: We first introduce the CPT technique in Sect. 2.Particular attention is paid here to the relevance of high-performance computing for this new branch of solid-state many-body physics, which aims at going from the microscopic "local" physics to the macroscopic system. We then apply it to an effective single-band model to describe a mixed, so-called stripe phase of two high-T_c materials, namely LaSrCuO and LaNdSrCuO in Sect. 3. As a further application, we investigate the photoemission of the insulating and so-called "overdoped" region of some high-T_c materials by a numerical analysis of the spectral function of the more general three-band model. This allows us to explain inconsistencies recently observed in the Fermi-surface analysis of the high-T_c material $Bi_2Sr_2CaCu_2O_{8+\delta}$ (Sect. 4). Finally, we summarize our results in Sect. 5.

2 Numerical technique

In this Section, we describe the cluster-perturbation algorithm and the step from the microscopic to the macroscopic length scale. An application of this method to the analysis of different phases and models for the high-T_c superconductors will be summarized in Sect. 3 and 4.

Approximate numerical solution

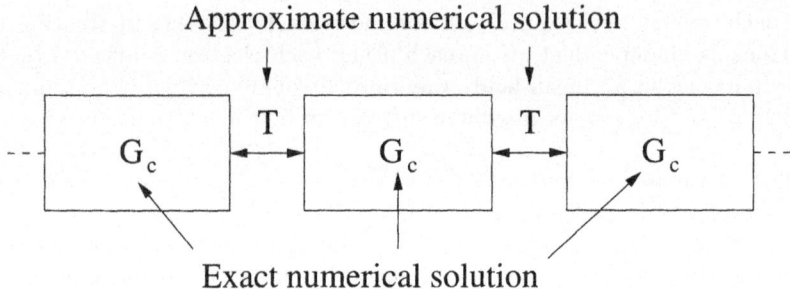

Exact numerical solution

Fig. 2. Visualization of the basic idea of cluster perturbation theory. The perturbation expansion explained in this section allows the subdivision of the infinite lattice into cluster, each of them being a large, but numerically solvable problem. The resulting cluster Green's function G_c again numerically coupled in an approximative way, allowing the transition to the infinite system.

Central piece of the CPT method is the Green's function of the cluster G_c, which can be obtained, e. g. from exact diagonalization (ED). Alternatively, one can derive G_c from other "Cluster" techniques, such as Quantum Monte Carlo (QMC) or the so-called Density-Matrix Renormalization-Group (DMRG) method.

The CPT method tells us how to carry out the step from the cluster and it Green's function "G_c" to the infinite-sized system and its Green's function G^∞ (schematically illustrated in Fig. 2). In an infinite lattice, the single clusters are not isolated but coupled by intercluster terms, T, which favor tunneling processes of electrons between the clusters. Both the clusters Green's function, and the tunneling terms T can be written in terms of large (see below) matrices. The method to take into account all possible couplings between the clusters is then sketched in Fig. 2. Here, G^∞ is obtained from the exactly summed infinite geometrical series

$$G^\infty = G_c + G_c T G_c + G_c T G_c T G_c \ldots = \left(G_c^{-1} - T\right)^{-1} \tag{1}$$

which contains one cluster, then two clusters coupled by T, then three clusters coupled by two T's, etc. Although this method is quite general and can be applied to several fields of physics, in what follows, we will specify the CPT technique to the application of strongly interacting electrons on a lattice, such as the Hubbard model.

The cluster Green's function G_c can be considered as a matrix $G_{c,\mathbf{r}_1,\mathbf{r}_2}$ in the N_0 cluster-site indices $\mathbf{r}_1, \mathbf{r}_2$. The same holds true for the tunneling term $T \equiv T_{\mathbf{r}_1,\mathbf{r}_2}(\mathbf{\Delta})$, which, however also couples two different clusters, and, thus, additionally depends on the relative position vector $\mathbf{\Delta}$ between the two clusters. The sum (1) can be readily carried out (at least symbolically) if one goes over to the Fourier transformation of T with respect to $\mathbf{\Delta}$. We thus introduce the matrix

$$V(\mathbf{q})_{\mathbf{r}_1,\mathbf{r}_2} = \sum_{\mathbf{\Delta}} e^{-i\mathbf{q}\cdot\mathbf{\Delta}} \, T_{\mathbf{r}_1,\mathbf{r}_2}(\mathbf{\Delta}) \, .$$

The CPT expression for the Green's function of the infinite lattice now also depends on the intercluster wavevector $\mathbf{q}$, as well as on the cluster-site indices:

$$G^{\infty}_{\mathbf{r}_1,\mathbf{r}_2}(\mathbf{q}) = \left[G_c^{-1} - V(\mathbf{q}) \right]^{-1}_{\mathbf{r}_1,\mathbf{r}_2} \, . \tag{2}$$

This is the CPT expression. The lattice Green's function can be expressed completely in momentum space by Fourier transforming (2). Its expression, however, is non-diagonal in the momenta, as the partition into clusters produces a smaller "superlattice" Brillouin zone (BZ). Finally, one has

$$G^{\infty}_{\mathbf{q}+\mathbf{Q}_1,\mathbf{q}+\mathbf{Q}_2} = \frac{1}{N_0} \sum_{\mathbf{r}_1,\mathbf{r}_2}^{cluster} e^{i\mathbf{q}\cdot(\mathbf{r}_1-\mathbf{r}_2)} e^{i\mathbf{Q}_1\cdot\mathbf{r}_1 - i\mathbf{Q}_2\cdot\mathbf{r}_2} G^{\infty}_{\mathbf{r}_1,\mathbf{r}_2}(\mathbf{q}) \, , \tag{3}$$

where $\mathbf{Q}_i$ are vectors of the reciprocal superlattice. The non-diagonal terms $\mathbf{Q}_1 \neq \mathbf{Q}_2$ in (3) are produced by the different treatment of the intra- and intercluster hopping terms. Within the CPT, these terms are neglected, and one takes only the $\mathbf{G}_1 = \mathbf{G}_2$ part of the Green's function to evaluate the spectral properties. If we consider more than one electronic degree of freedom such as in the three-band Hubbard model (see Sect. 4), the Green's function is additionally a 3×3 matrix in the orbitals of a single cell.

An interesting aspect is the fact that the lowest-order CPT becomes exact in (i) the trivial case of vanishing intercluster hopping and (ii) for non-interacting systems. The latter observation is due to the fact that corrections to (2) are given by higher-order cumulants, which are vanishing when Wick's theorem holds, i. e. for non-interacting electrons. This fact makes the CPT an appealing interpolation between the strong- and the weak-coupling limits.

In order to address different phases and models for high-T_c materials, the method has to be adapted and generalized in order to deal with different cluster geometries. Below, we we will consider clusters in the form of "stripe" configurations (Sect. 3), or clusters containing multi-orbital structures, such as the so-called three-band- Hubbard model (Sect. 4).

2.1 Relation to high performance computing and KONWIHR

The method outlined above presents a significant advance, in the sense that it aims at extracting important physical properties, which are *relevant for the large scale, macroscopic behaviour of strongly correlated systems from smaller cluster sizes*. Simultaneously it aims at going beyond the established knowledge in cluster simulations by using high-performance methods. This justifies the inclusion in projects at the *Leibniz-Rechenzentrum München* (LRZ) and the *Kompetenznetzwerk für Technisch-Wissenschaftliches Hoch- und Höchstleistungsrechnen in Bayern* (KONWIHR).

The accurate numerical treatment of strongly correlated electron systems is a major problem in even high-performance computational physics. Due to the large dimension of the Hilbert space, which increases exponentially with the number of lattice sites N_0, one is usually restricted to very small systems. In the case of a full diagonalization of the Hubbard model, one can have a maximum of $N_0 \approx 10$ lattice sites, so the Hamilton matrix has $4^{N_0} \times 4^{N_0} \approx 10^{12}$ elements. By exploiting symmetries, such as spin and translational symmetry, and conserved quantities, such as the particle number, one can reduce the Hilbert space dimension drastically. Unfortunately, even with this reduction, the system sizes that can be treated are still quite small ($N_0 \approx 20$ sites in the Hubbard and $N_0 \approx 30$ sites in the t-J model).

The CPT partially overcomes this problem since, as discussed above, it allows for an approximate evaluation of the Green's function, in principle for the infinite lattice. Despite this advantage, CPT requires additional computational effort since the subdivision of the plane in discrete clusters breaks translational symmetry and therefore enlarges the Hilbert space dimension significantly. In addition, the algorithm requires the calculation of *all* the $N_0 \times N_0$ Green's functions in a cluster.

For this reason, CPT calculations require computational tasks that can only be solved on most powerful computers with large memory. Programming and optimization must be carried out with great care, individually adapting each problem to the machine-dependent parameters.

The method we use to diagonalize the Hamilton matrix is the well known Lanczos technique [10], one of the most powerful in large scale matrix computing, which is an iterative scheme suitable for large, especially sparse matrices. This method is extremely efficient in finding the extremal eigenvalues in only a few (≈ 100) iterations while requiring to memorize only three vectors of length N_{States} (which still easily exceeds 1GByte). We especially make use of vectorization capabilities, resulting in excellent performances from 0.6 to 3.8 GFlop per node, dependent on the symmetry of the problem. Memory requirements are naturally large, ranging from 0.5 to 4 GByte per node.

It is especially this need for vectorization and memory that confines CPT calculations of physically relevant system sizes to the domain of high-performance computers such as the *Höchstleistungsrechner in Bayern* (HLRB).

3 Application to microscopically mixed phases, such as the stripe phase in high-temperature superconductors

The wide-ranging applicability of our general scheme for complex questions of up-to-date material science physics can be nicely demonstrated in the case of the so-called *stripe phase*. This mixed phase has been detected recently in the high-T_c compounds in the most relevant (for high-T_c superconductivity) doping regime. It consists of microscopically alternating (Fig. 3) magnetic (antiferromagnetically ordered) regions and metallic, quasi one-dimensional stripes. The imminent need for a theoretical study of the stripe phase is that its very existence has never been established conclusively from experiment. Only indirectly, salient features of various experiments (such as photoemission, to be discussed below) point strongly to its existence. On the other hand, the understanding of this phase is clearly a prerequisite for a theory of high-T_c superconductivity, as it must depend "on where the wholes go" as a function of doping. Our numerical study of the single-electron (photoemission) excitations in a striped phase is carried out via an extension of the above cluster algorithm. The application of the idea to stripes is indicated in Fig. 3: it is based on dividing the 2D plane into alternating clusters of

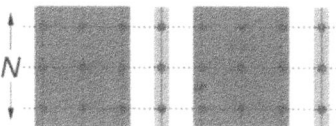

Fig. 3. Visualization of the cluster perturbation approach for stripes: the ground states for the half-filled three-leg ladder $(3 \times N)$ and the quarter-filled 1-leg chain $(1 \times N)$ are calculated exactly via exact diagonalization. The alternating clusters are then coupled via the inter-cluster hopping which is treated perturbatively.

metallic (with hole density $n_h = 0.5$) stripes and antiferromagnetic domains. The local many-body physics within the stripes, including strong correlations, is treated exactly via exact diagonalization (ED). On the other hand, the inter-stripe hopping is incorporated perturbatively. Thus, the CPT idea is translated to the stripe physics: the theory can be used to incorporate long-distance effects into ED data, which already contain short-distance effects – in particular correlations – exactly. A study of the properties of experimentally observed stripe phases *solely* by ED [11] is precluded by the prohibitively large unit cells. The manageable clusters for ED are simply too small to accommodate even a single such unit cell.

3.1 Analysis of the stripe phase

The high-T_c superconducting materials (as well as a number of other so-called strongly-correlated systems) are characterized by the presence of different competing phases in their phase diagram (Fig. 1). Besides the well known insulating antiferromagnetic (AF) and superconducting (SC) states, a number of new unconventional phases have been discovered in the past years. Among those, one might mention a spin-glass phase between the AF and the SC regions, the pseudogap phase at higher temperatures [12,13], and different structural phases obtained by decreasing temperatures. Some years ago, elastic neutron scattering showed that one of these incommensurate magnetic phases becomes static upon replacing some La with Nd in the "classical" high-T_c superconductor $La_{2-x}Sr_xCuO_4$ (LSCO) yielding $La_{1.48}Nd_{0.4}Sr_{0.12}CuO_4$

(Nd-LSCO). More interestingly, these magnetic structures are accompanied by charge-density-waves of half the wavelength of the spin-density waves and appear to be one-dimensional [14, 15].

The following physical picture is quite appealing: in order to minimize the disturbance of the underlying AF structure, the doped charge carriers remain essentially confined in one-dimensional "stripes" separating the AF domains [16, 17]. These domain walls introduce a phase shift between the AF regions producing a charge-density wave with a wavelength equal to the distance between them and a spin-density wave of twice this wavelength.

In this Section, we provide arguments, based on the numerical cluster-perturbation algorithm outlined in Sect. 2, showing that angle-resolved photoemission spectroscopy (ARPES) provide strong support for the existence of stripes both in Nd-LSCO as well as in LSCO [18, 19]. In particular, we will show that the experimental spectra of LSCO and Nd-LSCO can be almost quantitatively reproduced by different stripe geometries for a wide variety of dopings. In other words, we can tell precisely "where the holes, i.e. charge carriers go as a function of doping".

If (static or dynamic) stripes are present in HTSC, it is clear that low-energy excitations should be considerably affected. As a matter of fact, indications of such effects on electronic spectra have recently been accumulated by angle-resolved photoemission spectroscopy (ARPES) both on the static stripes in the Nd-LSCO system [19] and on possible dynamic domain walls in the LSCO compound [18]. Here, we want to show that the electronic structure revealed by ARPES, both in Nd-LSCO as well as LSCO, contains features which can be only explained in terms of a quasi-one dimensional stripe structure.

3.2 Numerical results

The results of our CPT calculations can be summarized as follows: (i) close to the wavevector $\mathbf{k} = (\pi, 0)$ we see, like in the experiments (Fig. 4a), a two-component electronic feature (Fig. 4b): a sharp low-energy feature close to E_F and a more broad feature at higher binding energies. Both features can be explained by the mixing of metallic and antiferromagnetic bands at this $\mathbf{k}$-point, as discussed below. (ii) the excitation near $(\pi/2, \pi/2)$ is at higher binding energies than the low-energy excitation at $(\pi, 0)$ and of reduced weight. (iii) the integrated spectral weight of the cluster-stripe calculation resembles the quasi-one-dimensional segments in momentum space (see Fig. 5c) as seen in the Nd-LSCO experiment (Fig. 5a). Also in agreement with the Nd-LSCO experiment (Fig. 5b), our calculation finds the low-energy excitations near $(\pm\pi, 0)$ and $(0, \pm\pi)$ (Fig. 5d).

The advantage of the CPT technique is that it allows us to resolve for each excitation, whether its main origin is from the insulating (AF) or the metallic part of the stripe configuration: In Fig. 6 we compare the spectra for the unperturbed stripe configuration with inter-cluster hopping set equal

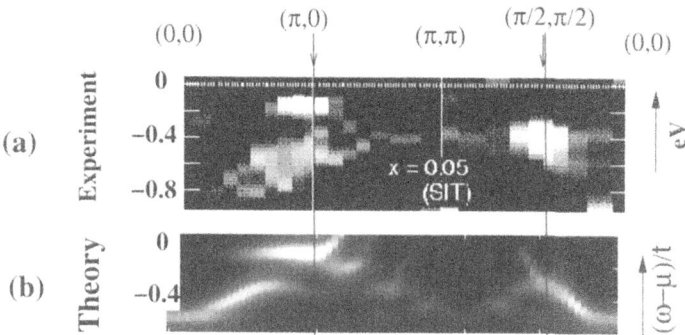

Fig. 4. Comparison of experimental (a) and theoretical (b) ARPES spectra for La$_{2-x}$Sr$_x$CuO$_4$. (a) displays ARPES results for the single-particle spectral function $A(\mathbf{k}, \omega)$: by Ino *et al.* [18]. The gray scale corresponds to the second derivative of the original measured data. Flat regions are black, regions with high curvature (i.e. peaks) are white. (b) shows the result of the stripe CPT calculation. Here, $A(\mathbf{k}, \omega)$ is plotted directly with maximum intensity corresponding to white.

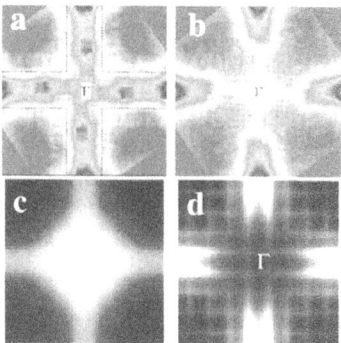

Fig. 5. Integrated spectral weight of stripe configurations. Comparison of experimental (a,b) and CPT (c,d) results. (a,c) show the total integrated weight in photoemission (equal to the momentum distribution), while (b,d) show low-energy excitations only, i. e. only the spectral weight within a window of range $0.2t$ (b) or $100meV$ (d) from the Fermi energy. Regions of high spectral weight correspond to dark areas in the experimental plots (a,b) and to white areas in the theoretical plots (c,d).

to zero (in Fig. 6a: the solid curve stands for the AF domains, the shaded curve for the 1D metal) with the result of our CPT calculation (solid line in Fig. 6b) with inter-cluster hopping t. The stripes are oriented along the y-direction. Therefore, we can conclude that peaks (in the calculation with $t = 0$), that show a dispersion along $(0, 0)$ to $(\pi, 0)$ direction stem from the AF domains (solid line in Fig. 6a) whereas the metallic excitations prior to the mixing (shaded curve in Fig. 6a) are dispersionless. By comparing the

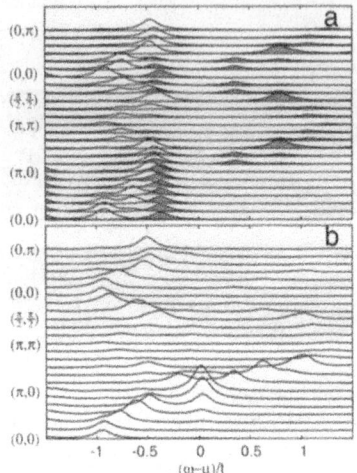

Fig. 6. Single-particle spectral weight prior (a) and after (b) application of the CPT: In (a) the shaded curve gives the spectrum of the 1D metallic chains and the solid line corresponds to the spectrum of the uncoupled AF domains (see Fig. 3). In (b) the result of the CPT is plotted.

three curves, we therefore conclude that the sharp quasiparticle peak near $(\pi, 0)$ results from the mixing of the (dispersionless) $(k_x, 0)$ minimum of the metallic band and the top of the insulating valence band situated at $(\pi, 0)$. Going from $(\pi, 0)$ to (π, π), the metallic band becomes dispersive and crosses, in agreement with experiments [19, 18], the Fermi surface at $k_y = \pi/4$ (since it is quarter-filled). The dispersion of the insulating band, however, is in the opposite direction. For this reason, in the final spectrum, we observe that the sharp quasiparticle peak at $(\pi, 0)$ becomes dispersive going into $(0, \pi)$ direction and eventually crosses the Fermi surface, however, with diminishing weight due to the absence of mixing with the insulating band. This effect is best visible in the gray scale plot of Fig. 4b. This finding may serve to clarify questions raised in the experimental ref. [19], concerning the origin of the quasiparticle peak at $(\pi, 0)$.

4 Three-band Hubbard model

In the previous section, we used the so-called $t-J$ model, which is an effective model used for microscopic descriptions of the high-T_c materials. It is the simplest model possible which just considers the hopping t of one electronic degree of freedom (e.g. the $Cud_{x^2-y^2}$ orbitals) and the magnetic exchange $J \propto \frac{t^2}{U}$, where U denotes the local Coulomb repulsion. A more complete model to describe the physics of the CuO_2 layers in the high-T_c materials should contain all relevant orbitals involved in the low-energy excitations. There is

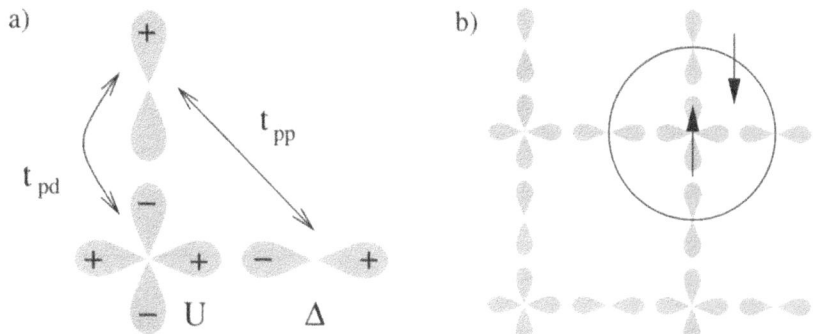

Fig. 7. a): Unit cell of the three-band Hubbard model with the relevant orbitals $Cu3d_{x^2-y^2}$,$O2p_x$ and $O2p_y$. Arrows indicate the hoppings between Cu and $O_{x,y}$ (t_{pd}) and O_x and O_y (t_{pp}). U and Δ denote the Coulomb repulsion and the charge transfer energy, respectively. b): 12-site cluster considered in the CPT calculation.

a general agreement [20] that the pure in-plane characteristics is sufficiently described by a three-orbital model, containing the orbitals $Cu3d_{x^2-y^2}$,$O2p_x$ and $O2p_y$, i. e. the so-called three-band model:

$$H_{3b} = -t_{pd} \sum_{\langle ij\rangle,\sigma} \alpha_{ij} \left(d^{\dagger}_{i,\sigma}p_{j,\sigma} + p^{\dagger}_{j,\sigma}d_{i,\sigma}\right)$$
$$-t_{pp} \sum_{\langle jj'\rangle,\sigma} \alpha'_{jj'} \left(p^{\dagger}_{j,\sigma}p_{j',\sigma} + p^{\dagger}_{j',\sigma}p_{j,\sigma}\right)$$
$$+\frac{U_d}{2} \sum_{i,\sigma,\sigma'} d^{\dagger}_{i,\sigma}d_{i,\sigma}d^{\dagger}_{i,\sigma'}d_{i,\sigma'}$$
$$+\Delta\sum_{j,\sigma} p^{\dagger}_{j,\sigma}p_{j,\sigma} . \tag{4}$$

Here, the operators $d^{\dagger}_{i\sigma}$ and $p^{\dagger}_{j,\sigma}$ create holes in the Cu 3d and in the O 2p orbitals, respectively, and α_{ij} and $\alpha'_{jj'}$ give the usual orbital phase factors. $\langle...\rangle$ denotes summation over nearest neighbors. The geometry of the model is shown in Fig. 7.

Because of the larger orbital space, exact diagonalizations of H_{3b} are usually limited to 4 unit cells, allowing for just three inequivalent wave vectors, which are not enough to address questions of quasiparticle dispersion, let alone of the Fermi surface. Nevertheless, since the important correlations of these materials should be relatively short ranged (a celebrated example is the so-called Zhang-Rice singlet [21]), one can still envisage to capture these short-range correlations (resulting in the singlet formation) by exact diagonalization of a small cluster. Then one continues the cluster properties to the infinite lattice within a perturbation in the inter-cluster hopping elements. This makes the three-band model an ideal candidate for CPT calculations.

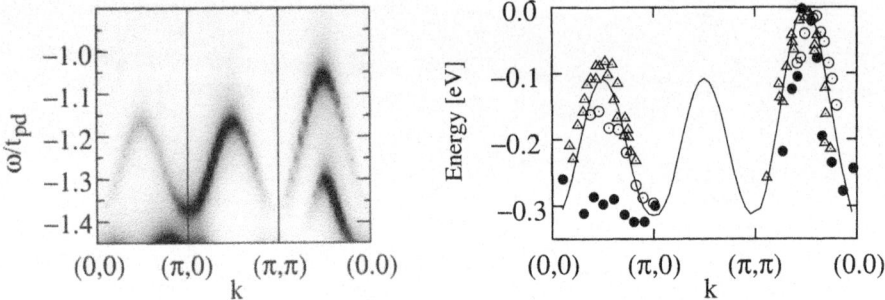

Fig. 8. Grayscale plot of the spectral density of the half filled three-band Hubbard model, (4), with $t_{pp} = 0.4$, $U_d = 6$ and $\Delta = 4$ (in units of $t_{pd} = 1.3eV$), as obtained from CPT (left). The points mark the peak positions. The right panel shows the band dispersion obtained from different experiments (circles [31], filled circles [37] and triangles [32]).

Here, we present results on the three-band model obtained by ED and CPT in the insulating and overdoped region (see the phase diagram in Fig. 1). As we will show below, our results show excellent agreement with recent experimental and theoretical results and, thus, provide a consistent picture of the high-T_c materials. Our main interest concerns recent high-resolution ARPES studies of the Fermi surface of BSSCO [22,23,24,25,26,27,28]. These experiments detected the so-called bilayer splitting between the bonding and antibonding band in this material, induced by a small, but relevant coupling between two CuO$_2$ layers. The general picture arising suggests that the different Fermi-surface geometries observed so far in ARPES experiments [22,23,24,25,26,27,28] can be attributed to this effect [29].

First, we discuss the dispersion of the low-energy excitations on the basis of the most recent experimental studies of the half-filled cuprate Sr$_2$CuO$_2$Cl$_2$. Up to few exceptions [30], these studies provide a consistent picture of the quasiparticle dispersion of the HTSC parent compounds [31, 32], showing two parabola centered at $k = (\pi/2, 0)$ and $k = (\pi/2, \pi/2)$. A characteristic energy difference $\Delta_a \approx 100 - 150meV$ between the maxima of the parabola is observed. The total width W of the first band is about $300 - 400meV$ (compare Fig. 8 for experimental data). In accordance with earlier numerical studies [33, 34, 35, 36] of the three-band model, we find that the parameter set $t_{pp} = 0.4t_{pd}$, $U = 6t_{pd}$ and $\Delta = 4t_{pd}$ gives the best agreement with the experimental band structure. With the value of the copper-copper hopping matrix element [36] taken from the literature, $t_{pd} = 1.3eV$, our calculations yield $\Delta_a = 150meV$ and $W = 400meV$. All these quantities are well within margins of the experimental constraints. Therefore, these parameters should give a rather good description of the materials.

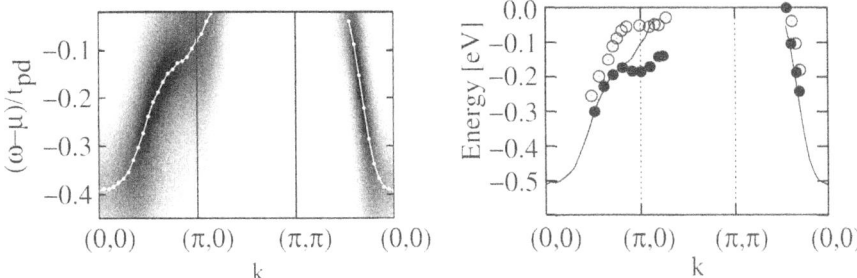

Fig. 9. Spectral density of the overdoped ($\delta = 0.25$) three-band Hubbard model as obtained by CPT (left panel). Parameters are as in Fig. 8. The white dots mark the peak positions. Corresponding experimental results from Ref. [37] are included for comparison (right panel). Empty circles denote the overdoped, filled circles the underdoped case. The solid line marks the CPT result.

Figure 8 shows a density plot of the theoretical single-particle spectral function $A(k,\omega)$ along the high-symmetry directions of the Brillouin zone (left panel). On the right panel we show corresponding experimental ARPES data for comparison [30, 31, 32]. The dispersion of the band is reproduced quite well by our calculation. Note that the scale of the binding energy is not absolute, since we are dealing with insulators.

We next turn to the doped systems. Here, we assume for simplicity that the parameters are not significantly dependent on doping and on the particular material, and keep them unchanged. The minimum doping, that can be achieved with a 12-site cluster by adding an additional electron, is $\delta = 0.25$. This undoubtedly "high" doping is, nevertheless, well suited to be compared to the detailed experimental results on dispersion [37] and Fermi surface [26, 38, 39, 24, 27, 25] of the overdoped cuprates. The results of our calculation are plotted in Fig. 9. A similar dispersion is observed in the experimental data for $Bi_2Sr_2CaCu_2O_{8+\delta}$ [37], shown for comparison in the right panel of Fig. 9. The good qualitative agreement of our numerical results with the experimental ones suggests that our approach of taking the same parameter values in the undoped and in the overdoped systems is quite reasonable. This was also found earlier by intensive QMC simulation [34, 35].

The magnitude of the spectral weight at the Fermi energy is displayed as a grayscale plot in Fig. 10 for the whole Brillouin zone. The darkest regions, thus, give an estimate of the location of the Fermi surface. The data clearly show a hole-like "barrel" closed around $k = (\pi, \pi)$, as usually observed in ARPES experiments on the overdoped cuprates [26, 24, 22, 40], mostly at a photon energy of $E_\nu \approx 22eV$. On the other hand, experiments carried out at a higher energy [38, 25, 28] $E_\nu = 33eV$ disagree with these results, and report an electron-like Fermi surface. This is quite puzzling, and there is not yet a full consensus about this issue. Some recent results suggest that the observation of two different Fermi surfaces in BSSCO is actually due to the splitting between the bonding and antibonding band originating from the coupling between the bilayers [25, 39]. Nevertheless, our data suggest that, if the bilayer splitting is not relevant, the Fermi surface should be hole-like.

The coupling between the two layers can be treated without further difficulties within our method. Again, we diagonalize a 2×2 unit cell exactly and include an interlayer hopping term $t_\perp$ within the CPT treatment. Since the effective hopping $t_\perp$ under consideration is relatively small and CPT amounts to a perturbation in the intercluster hopping, we believe that this is a good procedure.

The interplane hopping has been analyzed in studies based on LDA calculations and has been mapped onto effective multi-orbital models by integrating out high-energy degrees of freedom [41]. This approach provides results which are consistent with the experimentally observed bilayer splitting [39]. In our calculation, we adopt this idea by treating the orbital phase factors of the various interplane hoppings accordingly. We take a value of $t_\perp \approx 0.1 t_{pd}$, which is sufficient to produce an observable splitting of the Fermi surface.

Figure 10 (right panel) displays our results obtained for the bilayer. Different experimental results are shown in the last panel for comparison. Two branches can be distinguished, one being almost a square closed around $k = (\pi, \pi)$, The other one is closing around $k = (0, 0)$, forming an electron-like Fermi surface as observed in [25]. Similar results have also been found by local density-functional band calculations of ARPES in Bi2212 [42]. These results suggests, that a possible reason for the crucial differences in the experimental observations might be found in interlayer processes. An explanation of the apparent dependence of the Fermi surface on the photon energy would require the consideration of matrix element effects which have not been addressed within this paper, but are treated in detail elsewhere [42, 43].

5 Summary and Conclusions

Summarizing, we have shown that the Cluster-Perturbation-Theory (CPT) is an appropriate method to treat infinite-size lattice systems starting from numerical solutions of finite clusters. In particular, we have applied this method to investigate photoemission spectra in different regions of the phase diagram of the high-T_c superconducting materials. Detailed comparison establishes the "stripe phase" as a real phase which is built up out of microscopic AF and metallic (SC) stripes. In particular, this establishes where the charge carriers "go as a function of doping", a question which is clearly a prerequisite for a microscopic theory of high temperature superconductivity.

In particular, we would like to stress that the use of high-performance computing was absolutely necessary to achieve such a detailed understanding

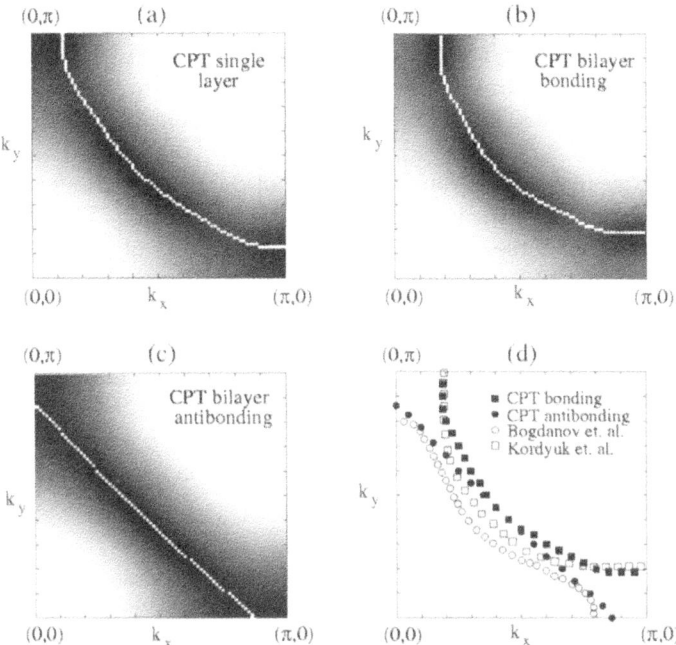

Fig. 10. Spectral weight at the Fermi energy giving an estimate of the position of the Fermi surface. Results are displayed for a single CuO_2 layer (a) and for the bonding (b) and antibonding (c) band of a bilayer with interlayer hopping $t_\perp = 0.1 t_{pd}$. The white dots indicate the maxima of the spectral weight on a given branch. Panel (d) shows the comparison with experimental results on overdoped (Bi,Pb)-2212 at 22eV photon energy [22] (empty squares) and and slightly over-doped (Bi,Pb)-2212 at 55eV photon energy [25](empty circles). Filled squares and circles denote the results of the CPT calculation for the bonding and antibonding branch, respectively.

of material science. This concerns essentially all the facets of our simulations, ranging from the storage required, the sustained performance reached to the total computing time. This most important computational issue is detailed in Sect. 2.1 of the present paper.

We have explained in detail the theoretical basis of CPT in Sect. 2 and how it is applied to the problems considered in this paper. We pointed out the relevance of our work for the research in computational solid-state physics and explained the computational effort necessary to carry out the computations.

In Sect. 3, we have compared in detail the single-particle spectral weight $A(\mathbf{k}, \omega)$ in the underdoped region with recent ARPES experiments. Salient features observed in experiments such as the two-peak structure around $(\pi, 0)$ with a sharp excitation close to the Fermi-energy and a broader feature at higher binding energies, the quasi-1D distribution of spectral weight, and low energy excitations located around the $(\pi, 0)$-points in the Brillouin zone are reproduced by the $A(\mathbf{k}, \omega)$ result for the site-centered stripe-configuration. The origin of these features can be naturally explained by hybridization effects of the metallic and insulating bands coming from the different stripe domains.

In Sect. 4, we reported a systematic study of the single-particle properties of the three-band model in the half-filled and doped region. This study was undertaken as an example to demonstrate the feasibility of including crucial orbital degrees of freedom. This step is essential in our thrust to replace the so far entirely empirical search for improved material properties (higher T_c , etc) by a systematic search. This analysis allows us to determine a parameter set, which appropriately describes both the insulating as well as the overdoped region and gives good agreement with ARPES data. The Fermi surface obtained at high dopings is hole like, although an interlayer coupling of about $\approx 0.1 t_{pd}$ produces a splitting into an electron and a hole like Fermi surface, in agreement with ARPES experiments.

Acknowledgement. We would like to acknowledge useful discussions and comments by Z.-X. Shen (Stanford), and S. A. Kivelson (UCLA). This work was supported by the DFG under Grant No. Ha 1537/16-2 and by a Heisenberg fellowship (AR 324/3-1), by the Bavaria California Technology Center (BaCaTeC), the KONWIHR projects OOPCV and CUHE. The calculations were carried out at the high-performance computing centers LRZ (München) and HLRS (Stuttgart). We would like to gratefully acknowledge the help of the staff at both institutions.

References

1. E. Arrigoni *et al.*, proceedings of the SNS 2001 Conference, to appear in Journal of Physics and Chemistry of Solids (unpublished).
2. E. Arrigoni *et al.*, invited contribution to the Proceedings of the 2002 Fruehjahrstagung der Deutschen Physikalischen Gesellschaft. To appear in Advances in Solid State Physics (unpublished).
3. M. G. Zacher, R. Eder, E. Arrigoni, and W. Hanke, Phys. Rev. B **65**, 045109 (2002).
4. M. G. Zacher, R. Eder, E. Arrigoni, and W. Hanke, Phys. Rev. Lett. **85**, 2585 (2000).
5. M. G. Zacher, R. Eder, E. Arrigoni, and W. Hanke, Int. J. Mod. Phys. **14**, 3783 (2000).
6. D. Vollhardt, in *Perspectives in Many-Particle Physics, Proceedings of the International School of Physics "Enrico Fermi", Course CXXI, Varenna, 1992*, edited by R. A. Broglia and J. R. Schrieffer (North-Holland, Amsterdam, 1994), p. 37.
7. P. Fulde, *Electron Correlations in Molecules and Solids* (Springer, Berlin, 1995).
8. C. Gros and R. Valenti, Phys. Rev. B **48**, 418 (1993).

9. D. Senechal, D. Perez, and M. Pioro-Ladriere, Phys. Rev. Lett. **84**, 522 (2000).
10. C. Lanczos, J. Res. Nat. Bur. Stand. **45**, 255 (1950).
11. T. Tohyama, S. Nagai, Y. Shibata, and S. Maekawa, Phys. Rev. Lett. **82**, 4910 (1999).
12. H. Ding *et al.*, Nature (London) **382**, 51 (1996).
13. A. G. Loeser *et al.*, Science **273**, 325 (1996).
14. J. M. Tranquada *et al.*, Nature (London) **375**, 561 (1995).
15. V. J. Emery, S. A. Kivelson, and J. M. Tranquada, Proc. Natl. Acad. Sci. USA **96**, 8814 (1999).
16. J. Zaanen and O. Gunnarsson, Phys. Rev. B **40**, 7391 (1989).
17. J. Zaanen, Science **286**, 251 (1999).
18. A. Ino *et al.*, Phys. Rev. B **62**, 4137 (2000).
19. X. J. Zhou *et al.*, Science **286**, 268 (1999).
20. Z.-X. Shen and D. S. Dessau, Phys. Rep. **253**, 1 (1995).
21. F. C. Zhang and T. M. Rice, Phys. Rev. B **37**, 3759 (1988).
22. A. A. Kordyuk *et al.*, Phys. Rev. B **66**, 014502 (2002).
23. S. V. Borisenko *et al.*, cond-mat/0102323 (unpublished).
24. J. Mesot *et al.*, Phys. Rev. B **63**, 224516 (2001).
25. P. V. Bogdanov *et al.*, cond-mat/0005394 (unpublished).
26. H. Fretwell *et al.*, Phys. Rev. Lett. **84**, 4449 (2000).
27. M. S. Golden *et al.*, Physica C **341–348**, 2099 (2000).
28. Y.-D. Chuang *et al.*, Phys. Rev. Lett. **83**, 3717 (1999).
29. C. Dahnken, E. Arrigoni, and W. Hanke, Journal of Low Temperature Physics **126**, 949 (2002).
30. B. O. Wells *et al.*, Phys. Rev. Lett. **74**, 964 (1995).
31. C. Dürr *et al.*, Phys. Rev. B **63**, 014505 (2001).
32. S. LaRosa *et al.*, Phys. Rev. B **56**, R525 (1997).
33. G. Dopf, A. Muramatsu, and W. Hanke, Phys. Rev. B **41**, 9264 (1990).
34. G. Dopf, A. Muramatsu, and W. Hanke, Phys. Rev. Lett. **68**, 353 (1992).
35. G. Dopf *et al.*, Phys. Rev. Lett. **68**, 2082 (1992).
36. W. Brenig, Phys. Rep. **251**, 154 (1995).
37. D. S. Marshall *et al.*, Phys. Rev. Lett. **76**, 4841 (1996).
38. A. D. Gromko *et al.*, cond-mat/0003017 (unpublished).
39. D. L. Feng *et al.*, Phys. Rev. Lett. **86**, 5550 (2001).
40. S. Legner *et al.*, Phys. Rev. B **62**, 154 (2000).
41. O. K. Andersen, A. I. Liechtenstein, O. Jepsen, and F. Paulsen, J. Phys. Chem. Solids **56**, 1573 (1995).
42. A. Bansil and M. Lindroos, Phys. Rev. Lett. **83**, 5154 (1999).
43. C. Dahnken and R. Eder, cond-mat/0109036 (unpublished).

9. ...

10. ...

11. ...

Object-Oriented C++ Class Library for Many Body Physics on Finite Lattices and a First Application to High-Temperature Superconductivity

Ansgar Dorneich[1], Martin Jöstingmeier[1], Enrico Arrigoni[1], Christopher Dahnken[1], Thomas Eckl[1], Werner Hanke[1], Shou Cheng Zhang[2], and Matthias Troyer[3]

[1] Institute for Theoretical Physics and Astrophysics, University of Würzburg
Am Hubland, 97074 Würzburg, Germany
ardornei@physik.uni-wuerzburg.de
[2] Department of Physics, Stanford University
Stanford, 94305 California, USA
sczhang@stanford.edu
[3] Institute for Theoretical Physics, ETH Zürich, 8093 Zürich, Switzerland
troyer@itp.phys.ethz.ch

Abstract. We present the design and implementation details for an object-oriented C++ class library for many-body physics on finite lattices. We divide the simulation in five modules which are strictly separated and interact via well defined interfaces. Special emphasis is put on the simulation algorithms, where we review the stochastic series expansion and the loop-operator update, both used in our Quantum-Monte-Carlo simulations. The second part of the paper is dedicated to an application from solid-state physics: the SO(5) model in two dimensions as a model for high-temperature-superconductivity. We demonstrate that this microscopic model, which aims at unifying antiferromagnetism and superconductivity, reproduces salient features of the temperature versus doping phase diagram.

Introduction

Strongly correlated electron systems rank amongst the most intensively studied topics in modern theoretical solid state physics [1, 2]. The reason for this interest is that strongly interacting electrons are the key ingredient for many unusual optical, electrical and magnetic properties of technologically promising materials such as high-temperature superconductors (HTSC) [3] or colossal magneto-resistance materials (CMR) [4]. Amongst others, possible future applications for these classes of materials may range from magnetic data recording to significantly faster computers, loss-free transport and storage of electrical energy, new means of transport based on magnetic levitation, medical imaging, better speech quality in mobile communication, and hyperfine detectors in scientific research and material testing.

The fundamental property of strongly correlated electron systems is an extraordinarily large entanglement of the many-body wave function of a "macroscopic" number, about 10^{23}, of electrons within typically a cubic centimetre. This essentially "quantum mechanical behavior on a macroscopic level" causes unusual material properties as mentioned above. On the other hand, the same behavior makes a theoretical description extremely difficult. The standard analytical approach for the description of interacting many-particle systems is to start from the corresponding noninteracting systems and to add the interaction effects a posteriori as (hopefully) small perturbations or effective fields and potentials. Concepts of this kind have been applied successfully throughout the modern theory of simple metals, insulators, or semiconductors [5,6]. More generally, traditional theories in solid state physics rely on the existence of a small parameter such as the ratio between Coulomb interactions of the charge carriers and the kinetic energy, or bandwidth, in a typical metal. However, this approach, which implicitly assumes that the properties of the interacting system evolve 'smoothly' from those of the noninteracting one, must fail if the effect to be studied relies on the strong interaction of many particles. In the case of the HTSC, for example, this fundamental problem is expressed by the fact that the Coulomb interaction is equal to or even larger than the bandwidth.

For this reason, numerical simulations have become important tools within solid state physics, and a substantial part of the literature and ongoing works in this area rely on these 'computer experiments' [2, 1, 7]. Numerical simulation allows measurements of the properties of well defined microscopic model Hamiltonians, within a certain finite cluster, up to essentially arbitrary accuracy, thereby taking into account all quantum mechanical interactions without any approximation or perturbation expansion. In this paper, we present a new concept of high performance computing for strongly correlated electron systems on finite lattices. In section 1, we will present our concept of an object-oriented class library and discuss, in two examples, two different techniques of simulating a many-body system. Section 2 is dedicated to an up-to-date application of the developed program package: the projected-SO(5) model in two dimensions, which is a currently much-discussed theory for HTSC. In section 3, we summarize our work.

1 Concept of the C++ class library

In many fields, researchers can resort to powerful libraries, helping them quickly to achieve a solution of their particular problems. The situation in computational solid-state physics is, until today, radically different: Here each scientist or research group develops their own programs, spending large amounts of time with programming, etc.. Furthermore, the way of designing the programs is mostly still the same as two decades ago. To a certain extent this is due to the still massive use of Fortran, being related to its advantages:

– standard support of a big number of different data types (e.g. complex numbers)

– easy optimization of the program, caused by its limited number of instructions and forbidden pointer-aliasing [8].

On the other hand, programming in Fortran 77 brings also definite disadvantages: Fortran 77 programs are very difficult to maintain. Also debugging can be quite cumbersome as elements ensuring data encapsulation are missing, at least if one compares the abilities of Fortran with modern languages as C++. The performance penalty, that had to be payed for object-oriented programming belongs to the past, as nowadays C++ compilers generating efficient binaries are available. Also the programming techniques, which allow to generate 'performing binaries' from object-oriented codes have been developed. Details can be found in [9, 10].

Based on these general observations, we decided to build an object-oriented library for many-body physics on finite lattices using C++. In the object-oriented software design of the new library codes, much attention has been spent on the creation of a flexible program, which can easily be applied to a wide range of classes of lattice geometries and Hamiltonians. This has been achieved by strictly separating the numerical simulation routines from the program modules describing the crystal lattice and from the parts modeling the particles and their energetic interactions. A specific application then requires nothing more than selecting a certain lattice geometry and a certain Hamiltonian and combining them with the suitable simulation routines. The "raw" structure of these simulations is shown in Fig. 1:

The module "finite lattice". The module "finite lattice" represents the geometric properties of the lattice and the relations between the lattice points. For the simulation one needs typically to consider:

– Which lattice points are nearest neighbours to a given lattice point?

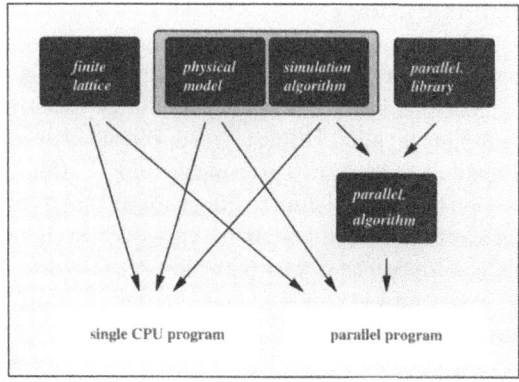

Fig. 1. Assembly of the basic modules for a lattice simulation

- Which lattice points can interact with a given lattice point?
- Which boundary conditions are given? Open? Periodic? Aperiodic?
- What is the distance between two given lattice points?
- Are there topologically equivalent lattice points?
- What are the symmetries of the lattice? Translation? Inversion? Rotation?
- Which point is mapped by which symmetry operation on a given lattice point?

Furthermore, the lattice module also has to connect real space with momentum space and establish functions that

- inform about allowed wave-vectors **k** in reciprocal (i.e. momentum) -space.
- distinguish between topologically equivalent and different wave-vectors.
- calculate Bloch-factors
- Fourier transform between the Green's functions $g(\mathbf{r})$ and $G(\mathbf{k})$ in real and momentum-space.

The module 'physical model'. The module 'physical model' describes the particles and their interactions, i.e. specifies the Hamiltonian. This module includes different types of particles (fermions, bosons, spin, charge, orbital) and their statistics, e.g. how many particles per lattice site are allowed. Any interaction, characterized by the range and the mechanism between the particles can be simulated. Typical examples of such interactions are Coulomb-interaction, spin-spin coupling, particle hopping, or external potentials like magnetic or electric fields. The implementation of the module 'physical model' depends strongly on the desired simulation algorithm. To demonstrate the different needs, we will give a short introduction in two of the simulation algorithms we implemented during this project and which are typically used in solid-state many-body physics, i.e., a standard technique the so called 'exact diagonalisation' (ED) and a more sophisticated one, the 'stochastic series expansion' (SSE):

- **The exact diagonalisation**
  The ED is the most elementary numerical simulation technique in solid state physics. Systems of quantum mechanical particles on discrete points of a lattice, e.g. electrons in a crystal lattice, can be described by a finite-dimensional linear space, called Hilbert space; physically measurable quantities become linear operators in this space, and – after the choice of a basis – they adopt the form of finite-dimensional matrices. The physical properties of the system studied are then expressed by the eigenvalues and eigenvectors of these matrices, which can be obtained from diagonalizing the matrices. This method is called 'exact diagonalization' (ED), because it solves the quantum mechanical eigenvalue problem exactly, without the approximations and systematic or statistical errors inevitably present in other methods like quantum Monte-Carlo techniques (QMC) [11, 12, 13], or density-matrix renormalization (DMRG). [14, 15] In practice, the range

of system sizes which are accessible via ED is severely limited by the exponential increase of Hilbert space dimension with the number of orbitals in the system. A complete and exact solution of the eigenvalue problem is impossible with currently available methods and computers for systems with more than 10 or 12 orbitals (or cluster sites, with one orbital per site). Two "tricks" help to push this limit to 20 or 22 orbitals: first, spatial symmetries of the system's geometry – translational, rotational and inversion *symmetries* – can be used to split up the full Hamiltonian into a series of smaller blocks, which can then be treated independently [16, 1, 17]. Second, instead of calculating *all* eigenvalues and eigenstates exactly, one can refer to much faster iterative methods returning good approximations for the lowest eigenvalues only. Since the latter determine the physical properties of the system, at least at lower temperatures, – the ground state and the low-energy excitations – the information gained in this way is, in most cases, sufficient to describe the system. [18, 1]. Another advantage of ED is its ability to calculate dynamical-response functions describing the system's reaction to external perturbations. Formally, these dynamical-response functions are expressed by the imaginary part of Green functions [19] of the form

$$A(\mathbf{k}, \omega) = \Im \frac{1}{\pi} \langle E_n \, |\hat{O}^\dagger(\mathbf{k}) \frac{1}{\omega - (H - E_n) - i0_+} \hat{O}(\mathbf{k})| \, E_n \rangle, \qquad (1)$$

where $\langle E_n \, | \ldots | \, E_n \rangle$ denotes a quantum-mechanical expectation value over the energy eigenstate $| \, E_n \rangle$, $\mathbf{k}$ is the wave number and ω the angular frequency of the perturbation. Here $\hbar = 1$ is taken for convenience, so that $\mathbf{k}$ becomes the momentum and ω the energy of the excitation. $\hat{O}(\mathbf{k})$ stands for a $\mathbf{k}$-dependent operator.

The ED method usually involves no temperature statistics (no thermodynamic partition sum), hence it describes the system's physical properties at zero temperature. Therefore, ED is a powerful complementary approach (however limited to typically 10-20 "orbitals" or lattices sites as discussed above) to other lattice-simulation methods in many-particle physics, particularly the important finite-temperature QMC methods. The latter techniques work efficiently at high temperatures (and can be extended to significantly larger lattices, i.e. $\mathcal{O}(100)$ for fermions, $\mathcal{O}(10.000)$ for bosons), but run into severe numerical problems if T tends to zero. Next we will present as a member of the QMC class of techniques, the so called "SSE" method.

– The Stochastic Series Expansion

QMC techniques are currently the most important simulation tools of strongly-correlated solid-state physics. Many different variants have been developed, and we refer to [20] for an overview. The basic idea of all these variants is not to use *all* basis states of the Hilbert space for exactly diagonalizing the measurement operators, but to calculate statistical mean values of the measurable quantity based on a relatively small ensemble of the 'statistically most relevant' states of the system. These states are sampled starting from an arbitrarily chosen initial state and using a statistical sampling process. The statistical process has to be organized in such a way that the correct thermodynamics of the system studied is captured: i.e. each state s should be sampled with a probability, which is proportional to its Boltzmann factor $e^{-E(s)/k_B T}$ [21, 22]. It can be shown [21] that this is guaranteed if the process fulfills the *detailed balance* criterion. Unlike ED and DMRG (apart from some rather involved extensions to $T > 0$), it also works at finite temperature, thus providing access to thermodynamical properties of the system. The central quantity to be sampled in a QMC simulation is the partition function, i.e.

$$Z = \text{Tr}(e^{-\beta H}), \qquad (2)$$

where H is the system's Hamiltonian and $\beta = 1/T$ the inverse temperature. Standard QMC techniques [20] split up the exponential into a product of many 'imaginary time slices' $e^{-\Delta\tau H}$ and truncate the Taylor expansion of this expression after a certain order in $\Delta\tau$, thereby introducing a discretization error of order $(\Delta\tau)^n$.

In SSE, however, one chooses a convenient Hilbert base $\{|\alpha\rangle\}$ (for example the S^z eigenbase $\{|\alpha\rangle\} = \{|S_1^z, S_2^z, ..., S_N^z\rangle\}$ of the z-spin component of the particles) and expands Z into the power series

$$Z = \sum_{\alpha} \sum_{n=0}^{\infty} \frac{(-\beta)^n}{n!} \langle\alpha|H^n|\alpha\rangle. \qquad (3)$$

It can be shown that the statistically relevant exponents of this power series are centered around

$$\langle n\rangle \propto N_s\beta, \qquad (4)$$

where N_s is the number of sites (or orbitals) in the system [23]. We can, thus, truncate the infinite sum over n at a finite cut-off length $L \propto N_s\beta$ without introducing any systematic error for practical computations. The best value for L can be determined and adjusted during an initial thermalization phase of the QMC simulation: beginning with a relatively small value of L, one can start the QMC update process, stop it whenever the cut-off L is exceeded and restart with L increased by 10 to 20%. Now let H be composed of a certain number of elementary interactions involving one or two sites (such as on-site potentials, nearest neighbor hopping etc.). In order to obtain a uniform notation, we combine those interactions, affecting only one site to new 'bond' interactions. One can, for example, take two chemical potential terms $\mu \cdot \hat{n}(\text{site}1)$ and $\mu \cdot \hat{n}(\text{site}2)$ and form the bond term $\frac{1}{C}\mu(\hat{n}(\text{site}1) + \hat{n}(\text{site}2))$ with the constant C assuring that the sum over all new bond terms equals the sum over all initial on-site terms. We can thus assume in the following that H is a finite sum of bond terms H_b and that the operator strings H^n in (3) can be split into terms of the form

$$\prod_{i=1}^{n} H_{b_i}^{(a_i)}, \tag{5}$$

where b_i labels the bond on which the elementary interaction term operates and a_i the operator type (e.g. density–density interaction or hopping). By introducing "empty" unit operators $H^{(0)} = id$, one can artificially grow all operator strings to length L and then obtains [23]

$$Z = \sum_{\alpha} \sum_{\{S_L\}} \frac{\beta^n (L-n)!}{L!} \langle \alpha \mid \prod_{i=0}^{L} (-H_{b_i}^{(a_i)}) \mid \alpha \rangle. \tag{6}$$

Here, $\{S_L\}$ denotes the set of all concatenations of L bond operators $H_b^{(a)}$ and n is the number of non-unit operators in S_L.

If we want to sample the (α, S_L) configurations according to their relative weights with a Monte-Carlo procedure, we have to make sure that the energy of each bond operator is zero or negative since, in order to fulfill detailed balance, we choose the acceptance probability p of a bond interaction to be proportional to its negative matrix element. This requires, however, that all matrix elements are non-positive.

Update mechanism. Having outlined the basic idea of SSE, we next review the non-local updating scheme. The basic idea was proposed by Sandvik [24]. However, we added a couple of extensions. In particular, we formulated the access and computability of time-dependent Green's functions, as in (1) [25]. These Green's functions are the tools to extract the experimentally accessible information, such as the response to an external perturbation. A world-line representation is used, in which the x-axis represents the spatial dimensions and the y-axis the propagation level $l = 1...L$. We symbolize type-1 bosons by single solid lines, type-2 bosons by double lines and empty sites by dotted lines (see Fig. 2, left).

Following Sandvik, we separate the set of all bond operators into three classes: empty operators $H^{(0)} = \mathbb{1}$, diagonal operators $H^{(d)}$ and non-diagonal operators $H^{(nd)}$. The QMC process starts with an arbitrarily chosen initial state $\mid \alpha \rangle$ and an empty operator string: in Fig. 2 (left), for example, three sites are occupied with type-1 bosons, two sites are empty and site 2 is occupied by a type-2 particle. Now two different update steps are performed in alternating order: a diagonal update, exchanging empty and diagonal bond operators, and an operator loop update, transforming and exchanging diagonal and non-diagonal operators.

In the diagonal update step, the operator string positions $l = 1...L$ are traversed in ascending order. If the current bond operator is a non-diagonal one, it is left unchanged; if it is an empty or diagonal operator, it is replaced by a diagonal or empty one, with a certain probability satisfying detailed balance, i.e. an operator with lower energy is more likely to be maintained

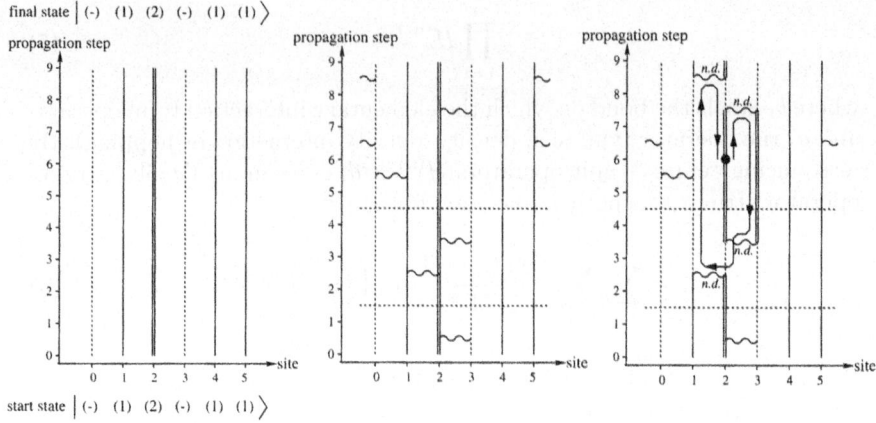

Fig. 2. Left: *world-line representation of an arbitrarily chosen start state for a physical system with three allowed occupations per site: empty (dashed line), particle 1 (solid line) or particle 2 (double line). The initial cut-off length L has been set to L = 9, and the initial bond-operator string consists only of "empty" operators.* **Middle:** *in the diagonal update step a certain number of empty bond operators is replaced by diagonal ones (and vice versa). In this example 7 of the initial 9 empty operators have been replaced.* **Right:** *The loop update closes if the initial insertion point – here the propagation level 6 at world line 2 – is reached again and if the inserted world-line discontinuity is removed in this step.*

or inserted than an operator with higher energy (Fig. 2 middle).

Non-diagonal bond operators cannot simply be inserted into the world-line configuration as diagonal operators can: their insertion and modification requires local changes of the world-line occupations. Sandvik proposed the following method to construct such a loop: a certain world-line and a propagation level l on it is chosen arbitrarily; at the chosen point one disturbs the world-line by a local change – for example the creation or annihilation of a particle. Then one chooses a direction (up or down in propagation direction) and starts moving the disturbance in this direction. The aim is to move this disturbance (we'll call it 'loop head') through the network of world-lines and interaction vertices until the initial discontinuity is reached again and healed. Whenever the loop head reaches an interaction vertex, we must decide how to go on, satisfying detailed balance.

Using this scheme a wide range of quantum mechanical observables are easily accessible by this operator loop update:

– All observables $H^{(a)}$, appearing as elementary interactions in the Hamiltonian, can be measured very easily by counting the corresponding interaction vertices in the bond-operator string S_L: if S_L contains on average $\langle N(a) \rangle$ of such vertices, the measured value of $H^{(a)}$ is simply

$$\langle H^{(a)} \rangle = -\frac{1}{\beta} \langle N(a) \rangle. \qquad (7)$$

- Summing over all elementary terms $H^{(a)}$ gives an estimator for the system's internal energy (which, since $\langle E \rangle \propto N_s$ explains (4)):

$$\langle E \rangle = -\frac{1}{\beta} \langle n \rangle, \tag{8}$$

where n is the number of non-empty interaction vertices in S_L. (This equation can be derived very easily from $\langle E \rangle = \frac{\partial}{\partial \beta} \ln Z$.)

- For the heat capacity C_V, we additionally have to measure the fluctuations of n:

$$C_V = \langle n^2 \rangle - \langle n \rangle^2 - \langle n \rangle. \tag{9}$$

- Equal-time correlations of two diagonal operators D_1 and D_2 can be measured via

$$\langle D_1 D_2 \rangle = \left\langle \frac{1}{n} \sum_{l=0}^{n-1} d_2[l]\, d_1[l] \right\rangle, \tag{10}$$

where $d_i[l]$ is the eigenvalue of D_i acting on the state $|\alpha(l)\rangle$. It results from applying the first l non-empty bond-operators $H^{(b)}$ to the initial state $|\alpha\rangle$.

Are there equally efficient estimators for time-dependent observables? In SSE the propagation index l describes the evolution of an initial state, when a series of elementary terms of the Hamiltonian is acting on it; thus l plays a role analogous to imaginary time in a standard path integral. More detailed calculations [26,27] show that an imaginary time separation τ corresponds to a binomial distribution of propagation distances Δl; the time-dependent correlation $\langle D_2(\tau)D_1(0)\rangle$, for example, is related to the correlator

$$C_{12}(\Delta l) = \frac{1}{n} \sum_{l=0}^{n-1} d_2[l + \Delta l]\, d_1[l] \tag{11}$$

via

$$\langle D_2(\tau)D_1(0)\rangle = \left\langle \sum_{\Delta l=0}^{n} \binom{n}{\Delta l} \left(\frac{\tau}{\beta}\right)^{\Delta l} \left(1 - \frac{\tau}{\beta}\right)^{n-\Delta l} C_{12}(\Delta l) \right\rangle. \tag{12}$$

The corresponding generalized susceptibilities which contain, for example, the response of the many-body system to an external perturbation, can be calculated straightforward by numerically integrating $\langle D_2(\tau)D_1(0)\rangle$ over τ, i.e.

$$\chi_{12} = \int_0^\beta \langle D_2(\tau)D_1(0)\rangle \, d\tau. \tag{13}$$

Another method, which is numerically more stable, artificially adds an external field $h_2 D_2$ to the Hamiltonian and calculates the response function

$$\chi_{12} = \left. \frac{\partial \langle D_1 \rangle}{\partial h_2} \right|_{h_2=0}. \tag{14}$$

The result is [26,27]

$$\chi_{12} = \left\langle \frac{\beta}{n(n+1)} \left(\sum_{l=0}^{n-1} d_2[l] \right) \left(\sum_{l=0}^{n-1} d_1[l] \right) + \frac{\beta}{n(n+1)} \sum_{l=0}^{n-1} d_2[l]\, d_1[l] \right\rangle.$$

(15)

The observables listed above serve to access important static thermodynamic properties of the system studied. However, properties such as photoemission, described by the Green's function $\langle a^\dagger(\mathbf{k},\omega)\, a(0,0) \rangle$, where $a(\mathbf{k},\omega)$ $\left(a^\dagger(\mathbf{k},\omega) \right)$ creates a hole (electron) with momentum $\mathbf{k}$ end energy ω, or spin flip response functions $\langle S^-(\mathbf{k},\omega)\, S^+(0,0) \rangle$, where S^- and S^+ are the spin flip operators known from standard quantum-mechanics, are often even more interesting as they provide insights into the system's dynamics. Within the framework of SSE, measuring these Green's functions $G(\mathbf{k},\omega)$ requires the insertion of local changes on certain world-lines (such as removing a particle at propagation level l_1 on world-line w_1 and re-inserting it at propagation level l_2 on world-line w_2). Performing these insertions is a highly non-trivial task, since on the one hand detailed balance must be assured. On the other hand, the whole process has to sample all distances $r = w_2 - w_1$ and all propagation differences $\Delta l = l_2 - l_1$ efficiently. Both requirements also exist for the QMC update steps between different measurements, and within SSE they are fulfilled by introducing the non-local operator loop-update mechanism. Since this update inserts and moves local changes on the network of world-lines and connecting interaction vertices, it can be used to record the corresponding Green's functions $G(\mathbf{r}, \Delta l)$ 'on the run', while constructing the closed loop.

Having measured and recorded the quantities $G(\mathbf{r}, \Delta l)$ (or a correlation function $C(r, \Delta l) = \langle D_2(r,\tau) D_1(0,0) \rangle$), we still have to perform a couple of non-trivial transformation steps to obtain the desired quantities $G(\mathbf{k},\omega)$ and $C(\mathbf{k},\omega)$, which describe the dynamical response of the system to external perturbations. First, we have to translate the propagation levels Δl to imaginary times τ, then a Fourier transform brings us from $\mathbf{r}$-space to $\mathbf{k}$-space; finally, we need an inverse Laplace transform to convert imaginary time τ to excitation energy ω.

Efficiently accessing the system's dynamics. In this paragraph, an efficient implementation strategy for recording $G(\mathbf{r}, \Delta l)$ and for the adjacent transformation steps, mentioned above, is presented. The transformation from propagation levels Δl to imaginary time τ requires the same weight factors as discussed earlier for diagonal correlation functions:

$$G(r,\tau) = \sum_{\Delta l=0}^{n} \binom{L}{\Delta l} \left(\frac{\tau}{\beta} \right)^{\Delta l} \left(1 - \frac{\tau}{\beta} \right)^{L-\Delta l} G(r, \Delta l)$$

(16)

$$\equiv \sum_{\Delta l=0}^{L} w(\tau, \Delta l)\, G(r, \Delta l),$$

where

$$w(\tau, \Delta l) = \binom{L}{\Delta l}\left(\frac{\tau}{\beta}\right)^{\Delta l}\left(1 - \frac{\tau}{\beta}\right)^{L - \Delta l}. \tag{17}$$

Working in a fixed string size representation with fixed L instead of varying n is more convenient because the binomial weight prefactors are fixed during the entire simulation and can easily be calculated once, at the beginning of the simulation.

There are several possible ways to implement the recording of $G(\mathbf{r}, \Delta l)$ measurements and the following transformation to $G(\mathbf{r}, \tau)$. The easiest and, at first glance, fastest way simply writes all recorded $G(\mathbf{r}, \Delta l)$ data into a two-dimensional array with dimensions N_s and $L \propto N_s \beta$. The transformation to $G(\mathbf{r}, \tau)$ can then be performed at the end of the simulation. However, this method has two problems. First a separate measurement has to be recorded each time the loop head steps up or down by one level on a world-line and, whenever it traverses an interaction vertex. Recording all these measurements drastically slows down the loop update process. Second, for large systems ($N_s \approx 5000$) and low temperatures ($\beta \approx 40$), the two-dimensional array needed to store $G(\mathbf{r}, \Delta l)$ contains about one billion elements and needs more memory than available on many computer systems.

In our code we have adopted a different strategy to overcome these problems: we perform *all* possible $G(\mathbf{r}, \Delta l)$ measurements (thereby exploiting the fact that $G(\mathbf{r}, \Delta l)$ is constant on the entire world-line fragment between two adjacent vertices) and directly transform these into $G(\mathbf{r}, \tau)$ at the end of each loop update step. On the one hand, the transformation introduced after each QMC update step is necessary to keep memory requirements manageable. On the other hand the number of floating-point operations grows significantly. Simply applying (16) with its computationally expensive operations (divisions,powers,binomial coefficients,large sums), would now cost by far too much computation time. Instead, we remember that $G(\mathbf{r}, \Delta l)$ is composed out of a relatively small number of intervals $I =]\Delta l_1(I), \Delta l_2(I)]$ with constant function value (Fig. 3b)). Therefore, we can compute the contribution of an entire Δl-interval to $G(\mathbf{r}, \tau)$ in one step:

$$G(r, \tau) = \sum_I G(r, I)\big(W(\tau, \Delta l_2(I)) - W(\tau, \Delta l_1(I))\big), \tag{18}$$

where W is the "integrated weight function"

$$W(\tau, \Delta l) = \sum_{m=0}^{\Delta l} w(\tau, m). \tag{19}$$

The Δl-range in which $W(\tau, \Delta l)$ considerably differs from 0 and 1 is determined by the mean value and the standard deviation of the binomial distribution $w(\tau, \Delta l)$:

$$\langle \Delta l \rangle = L \frac{\tau}{\beta} \tag{20}$$

$$\sigma_{\Delta l} = \sqrt{L \frac{\tau}{\beta} \left(1 - \frac{\tau}{\beta}\right)}. \tag{21}$$

Below $(\langle \Delta l \rangle - 5\,\sigma_{\Delta l})$ the integrated weight is zero, above $(\langle \Delta l \rangle + 5\,\sigma_{\Delta l})$ it is 1 (up to an error of less than 10^{-7}). The remaining interval rarely contains more than fifty or hundred Δl-points (see Fig. 3d); these values can easily be stored after having been computed once for each τ. Thus, $W(\tau, \Delta l)$ can be calculated very rapidly with nothing but a couple of "cheap" elementary operations. For very large systems and very low temperatures

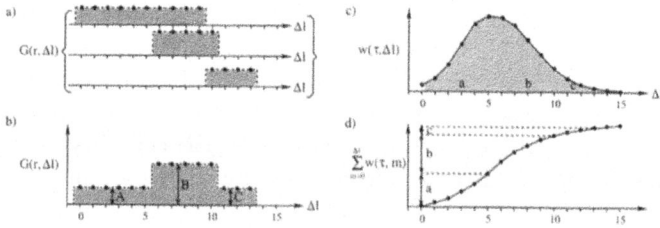

Fig. 3. Transformation of Green's functions measurements from propagation level Δl to imaginary time τ: the raw measurements recorded during loop update on different world-line segments (a) are combined into a single function $G(\mathbf{r}, \Delta l)$ (b). For a given τ $G(\mathbf{r}, \tau)$ could be computed by summing up all $G(\mathbf{r}, \Delta l)$ weighted with $w(\tau, \Delta l) = \left(\frac{L}{\Delta l}\right)\left(\frac{\tau}{\beta}\right)^{\Delta l}\left(1 - \frac{\tau}{\beta}\right)^{L - \Delta l}$ (c). A much more efficient way uses the "integrated weight function" $W(\tau, \Delta l) = \sum_{m=0}^{\Delta l} w(\tau, m)$ (d) to get the total contribution of each range $]\Delta l_1, \Delta l_2]$ in which $G(\mathbf{r}, \Delta l)$ is constant. In the example shown here $G(\mathbf{r}, \tau)$ is then simply $aA + bB + cC$.

the "relevant" Δl-ranges might become so large that it is unfavorable to store all needed $W(\tau, \Delta l)$ values – for example, because accessing the large array $W[\tau_i, \Delta l]$ would cause too many cache misses. In this case one can store the coefficients of some interpolation functions for $W(\tau, \Delta l)$ instead of the function values themselves. Practical tests have shown that dividing the relevant interval $[\langle \Delta l \rangle - 5\sigma_{\Delta l}, \langle \Delta l \rangle + 5\sigma_{\Delta l}]$ into six sub-intervals with boundaries $\langle \Delta l \rangle - 5\,\sigma_{\Delta l}$, $\langle \Delta l \rangle - 2.8\,\sigma_{\Delta l}$, $\langle \Delta l \rangle - 1.3\,\sigma_{\Delta l}$, $\langle \Delta l \rangle$, $\langle \Delta l \rangle + 1.3\,\sigma_{\Delta l}$, $\langle \Delta l \rangle + 2.8\,\sigma_{\Delta l}$ and $\langle \Delta l \rangle + 5\,\sigma_{\Delta l}$ and interpolating W in each sub-interval by a fifth-order polynomial is a good compromise between evaluation speed (about 15 elementary operations), storage requirements (36 floating point numbers for each τ) and interpolation accuracy (better than $2..3 \times 10^{-7}$). The next transformation step, Fourier transform from $G(\mathbf{r})$ to $G(\mathbf{k})$, is a well-known standard method that does not impose any fundamental problems. However, standard Fast Fourier Transform (FFT) algorithms perform best if *all* $G(\mathbf{k})$ values are to be calculated, whereas in practice one rarely needs all $\mathbf{k}$-values and is interested only in one $\mathbf{k}$-point or in some special points of the Brillouin zone, e.g. the point $\mathbf{k} = (\pi, \pi)$ and its immediate neighborhood, as in our application to super-conductivity blow. Then, one can save a lot of computation time by not recurring to FFT but using optimized algorithms designed particularly for these cases. If we are interested in only one or a few $\mathbf{k}$-points, we can use a simple Fourier transform to get $\{G(\mathbf{k}, \tau)\}$ from $\{G(\mathbf{r}, \tau)\}$ in $\mathcal{O}(N_s \cdot n_k)$ operations (n_k is the number of $\mathbf{k}$-points). Correlation functions $C(\mathbf{k}, \tau)$ can even be measured directly in $\mathbf{k}$-space, which also can be done in $\mathcal{O}(N_s \cdot n_k)$ operations. For the case $1 \ll n_k \ll N_s$, we have implemented a new Fourier transform algorithm performing much better than FFT, in this situation. [28] Unlike a Fourier transform, a Laplace transform in general cannot be inverted . Therefore, the last transition step from τ to ω is by far more complicated than the previous one from r to k. We use efficient *Maximum Entropy* techniques developed within the last years and refer to earlier publications. [29]

2 First Application: The Two-Dimensional SO(5) Model for high-temperature Superconductivity

In recent years, more and more novel materials with exciting prospects of application have been detected, which share one common feature: They display a rich phase diagram with competing orders, such as insulating and metallic phases as well as anti-ferro magnetic (AF) and superconducting (SC) phases. It is becoming more evident, with the amazing developments of both, very accurate experimental techniques and high- performance computing, that the underlying universal reason for this rich phase diagram is strong electronic correlations.

Examples are the high-temperature superconductors, organic materials, quantum-hall systems, heavy-fermion compounds and the manganites with the latter being, for example, of substantial importance in technical applications, such as magnetic data recording. Depending on temperature and doping - in the high-T_c compounds- an insulating (AF) phase, a so called "striped" phase (where AF and SC are "phase separated" on a microscopic scale) and a superconducting phase with unusual $\left(d_{x^2-y^2}\right)$ symmetry are found in experiment.

All these competing orders appear at very low temperature and very low energy. They are nearly degenerate and have to develop out of one and the same microscopic Hamiltonian, comprised out of electrons, ions and Coulomb interactions, which corresponds to the bare, i.e. "unrenormalized" energy scale at high temperature and energy. A central, if not the most important, issue in present solid-state theory aims at developing – as much as possible – a universal picture of how these competing orders and emerging phenomena are driven by strong electronic correlations.

In the SO(5)-theory of high-T_c superconductivity, which was first proposed on mathematical grounds by S. C. Zhang [30] and then further developed and physically motivated by our group [31,32,33], such a unification is proposed on the grounds of rather general symmetry arguments.

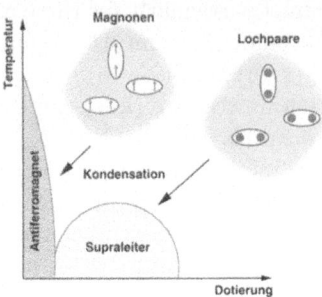

Fig. 4. Generic Temperature versus phase diagram of the cuprate HTSC. In real HTSC crystals, the chemical potential μ can be varied by various hole doping concentrations.

This theory embeds the SO(3) rotational symmetry of spin systems and the U(1) symmetry of charge conservation – both are also obvious symmetries of the so-called t-J and Hubbard model – into a larger five dimensional rotation symmetry, SO(5). SO(5) is the smallest group which contains both SO(3) and U(1) as subgroups. The five-dimensional order parameter of this new symmetry group, the so-called superspin, is a combination of the three components of the AF order parameter – the Néel vector (N_x, N_y, N_z) – and the two components forming the order parameter of d-wave SC, namely the real and imaginary part of the superconducting gap function Δ. The interpretation of AF and SC phase of the HTSC within SO(5) theory is as follows: in a completely SO(5)-symmetric system the superspin vector can rotate within a five-dimensional sphere, and we would expect mixed states of coexisting AF and SC order. In reality, however, the chemical potential (which controls the hole doping) induces an anisotropy between AF and d-wave SC and explicitly breaks SO(5) symmetry. The chemical potential in SO(5) theory plays the same role as magnetic fields do in SO(3) angular momentum multiplets: a magnetic field splits up these multiplets by introducing small energy offsets between states with different z-spin or L_z quantum numbers (Zeeman effect). Therefore, at half-filling all superspin directions within the 3D 'easy sphere' of zero SC and finite AF order parameter values are energetically favorable, and upon doping the superspin flips into the 'easy plane' of zero AF and finite SC order parameter values [30,34] (see Fig. 5). The SO(5) theory of HTSC provides an elegant explanation for many features of the cuprates such as the close vicinity of an AF and a SC phase or the neutron resonance peak at $\mathbf{k} = (\pi, \pi)$ [30,35,36,37] and makes a number of experimental

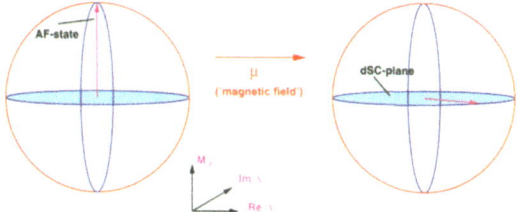

Fig. 5. The chemical potential forces the superspin into the 'easy sphere' of zero SC order parameter at half-filling and into the 'easy plane' of zero AF order parameter at finite hole doping.

predictions [38, 39, 40, 41, 42]. Fingerprints of approximate SO(5) symmetry have also been detected in some widely studied microscopic effective models for the CuO_2 planes, e.g. the t-J [43] or the Hubbard model [44]. However, the cuprates' Mott insulating behavior at half-filling severely challenges the validity of SO(5) theory [45, 44, 46, 47]: SO(5) symmetry requires collective charge pair excitations to have the same (vanishing) mass as collective spin-wave excitations. The real cuprates, on the contrary, are Mott insulators at half-filling and possess a large energy gap U of several eV due to electron-electron interaction. Therefore, in physically realistic models the 'upper half' of all SO(5) multiplets, i.e. the states with $Q > 0$, should be separated from the lower part of the multiplet by the large energy difference U (see Fig. 6).

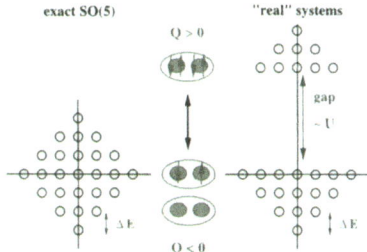

Fig. 6. In exactly SO(5)-symmetric systems collective charge and spin excitations within an SO(5) multiplet should both have vanishing energy. If exact SO(5)-symmetry is broken by a chemical potential, there are constant energy steps ΔE between neighbored Q levels (left). In physically realistic systems, however, charge excitations above half-filling must pay the large energy U due to the strong on-site repulsion of two electrons in the same orbital.

The only way to overcome these problems is to incorporate additional SO(5)-symmetry breaking terms into the SO(5) description of the HTSC. Most important is to attain a correct description of the Mott gap at half-filling. This can be done by projecting out all states containing doubly occupied sites (see Fig. 6). These states are separated from the states without

double occupancies by the energy scale $2U$ of more than 10 eV, which is
by orders of magnitude higher than the low-energy scales T_N (Néel temperature) and T_c (superconducting transition temperature). Therefore, states
with double occupancies should not even be important as intermediate states
for scattering processes and, therefore should be projected out. The resulting models, which exactly implement the so-called Gutzwiller constraint of
no-double-occupancy, are called 'projected SO(5)' or 'pSO(5)' models.

The simplest bosonic projected SO(5) Hamiltonian takes the form [48]

$$H = \Delta_s \sum_{x,\alpha=2,3,4} t_\alpha^\dagger(x) t_\alpha(x) + (\Delta_c - 2\mu) \sum_x t_h^\dagger(x) t_h(x) \qquad (22)$$
$$- J_s \sum_{<xx'>,\alpha=2,3,4} n_\alpha(x) n_\alpha(x') - J_c \sum_{<xx'>} (t_h^\dagger(x) t_h(x') + \text{h.c.}),$$

where $n_\alpha = (t_\alpha + t_\alpha^\dagger)/\sqrt{2}$ are the three components of the Néel order
parameter. Δ_s and $\Delta_c \sim U$ are the energies to create a magnon and a hole-
pair excitation, respectively, at vanishing chemical potential $\mu = 0$. As one
can see, the excitation energy for hole pairs can be compensated by μ in order
to have equal energies for spin and hole-pair excitations. (SO(5) condition).

Phase separation and coexistence of AF and SC We start our analysis with numerical simulations on an isotropic 2D square lattice, which is
believed to be a good model for the physics of the cuprates' CuO_2 planes.
We choose $J_s = J_c/2$, and define $J := J_s$ as our unit of energy. The value of
Δ_s turns out not to be crucial for the dynamics of the model or the general
structure of the phase diagram, as long as $\Delta_s \lesssim 4J$. We, thus, choose $\Delta_s = J$,
and shift the chemical potential so that $\Delta_c = \Delta_s$.
The nature of the phase transition is displayed in Fig. 7. We find a phase
separation into a hole-rich and almost hole-free phases. Such a phase separation into coexisting phases can also be observed in real cuprates, for example
in the HTSC compound La_2CuO_{4+y} [49]. Figure 8 shows the crystal struc-

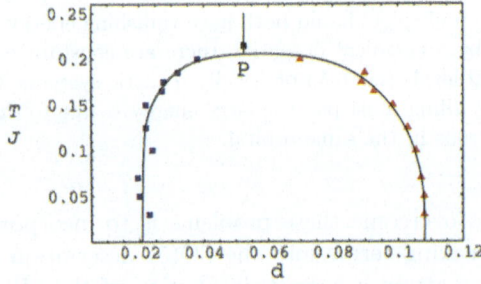

Fig. 7. Hole densities of the coexisting phases on the first order transition line from
(almost) zero to finite hole density at $\mu = \mu_c$ as a function of temperature.

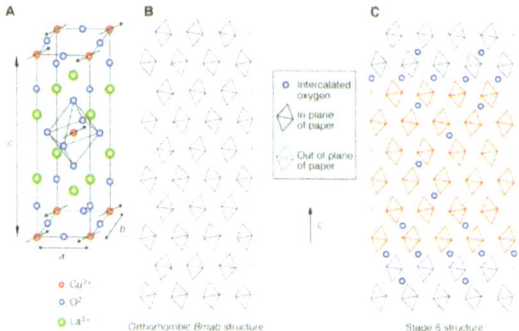

Fig. 8. (A) The tetragonal unit cell of La_2CuO_{4+y}. **(B)** Schematic crystal structure of undoped La_2CuO_4: the CuO_6 octahedra arrange in a tilted structure. **(C)** Schematic structure of the "stage 6" phase of La_2CuO_{4+y} at oxygen doping $y \approx 0.06$. (Picture by B.O. Wells *et al.*, **Science 277**, p. 1068;).

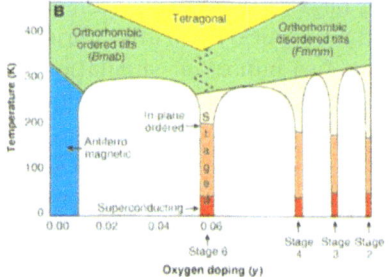

Fig. 9. Phase diagram of La_2CuO_{4+y} as a function of oxygen doping y. There are several doping ranges in which the material separates into two phases with different doping concentrations, e.g. the doping range $y = 0.01, ..., 0.06$. (Picture by B.O. Wells *et al.*, **Science 277**, p. 1068;).

ture and Fig. 9 the phase diagram of La_2CuO_{4+y}. For doping concentrations $y = 0.01$ to 0.055 the material displays a mixture of two phases with different oxygen concentrations: there are alternating AF regions with very low doping concentration $(y = 0.01)$ and hole-rich 'stage 6' regions $(y = 0.055)$ which become superconducting at low temperatures. This is in remarkable accordance with the pSO(5) results in Fig. 7. Due to the fact that each additional oxygen attracts and immobilizes 2 electrons in the CUO_2 planes, thereby introducing 2 holes, the agreement between theory and experiment is perfect even on a quantitative level: the doping densities of $y = 0.01$ and $y = 0.055$ of the two phases exactly correspond to the values $\delta = 2y = 0.02$ and $\delta = 0.11$ obtained from the pSO(5) model.

A variety of other experimental facts, known to be salient fingerprints of the high-T_c cuprates are reproduced by the pSO(5)-theory. Among them are an unusual doping dependence of the chemical potential, which is a direct consequence of the above microscopic phase separation and a resonance found in the spin-response function in the superconducting state. The latter is, in fact the fingerprint of the AF state in the SC state and is quite naturally explained by the SO(5)-"rotation". The interested reader is referred to our recent publication [33].

3 Summary

This report for the KONWIHR project OOPCV consist of two main parts. In the first section, we present the design of a new C++ class library for strongly correlated electron systems on finite lattices. One of the main strategies when designing the library was to separate the simulation technique from the physical specification of the studied system. To achieve this goal, the simulation was split into 5 independently usable object-oriented modules (see Fig. 1).These software modules interact via well-defined interfaces, so that a new application of SSE on a certain lattice geometry and a certain physical model can be composed very quickly. We give a brief description of the different modules, and review two important simulation algorithms: the ED and the SSE. Unlike many 'traditional' QMC techniques, the Stochastic Series Expansion (SSE) is an exact method without any systematic error such as the Trotter discretization error. Stochastic Series Expansion together with the implementation "tricks" and Green's-functions measurements described in this paper, is a high performance quantum Monte-Carlo simulation technique. It allows (in the bosonic case) to access both static and dynamical properties of very large systems of thousands of sites and at very low temperatures. Compared to the loop-algorithm, which is slightly faster on large systems for some specific Hamiltonians, SSE has the advantages of not suffering from exponential slowing down in external fields; furthermore, SSE is more easily applicable to wide classes of Hamiltonians.

In the second part of this paper, the SSE codes created during this project were applied to a very pertinent problem from current solid state physics: to a numerical analysis of the projected SO(5)-symmetric model of high-temperature superconductivity. This model maps the low-energy physics of the CuO_2 planes in the cuprate superconductors onto a model consisting of four bosonic quasiparticles: three triplet excitations and a hole-pair excitation. We started the study with a short review on the basic principles and ideas of SO(5)-theory and projected SO(5)-theory. The latter is the simplest bosonic model containing two key ingredients for the HTSC: the Mott gap and the vicinity of AF and SC phases. Then, we demonstrated that the projected SO(5) model gives a semiquantitative description of many properties of the HTSC in a consistent way. In particular, an AF and a SC phase were identified whose phase boundaries look similar to the ones of real cuprate materials. Furthermore, the pSO(5) model was shown to reproduce effects like phase separation or AF-SC coexistence, which appear in a variety of novel superconducting compounds.

Future work will, among other things, be directed to studying the scaling, i.e. critical properties of the quantum projected SO(5) model in three dimensions. This will become for the first time possible due to high-performance computing in combination with the highly accurate SSE QMC analysis. Here the central aim is to show that the scaling behavior is consistent with an SO(5)-symmetric critical behavior. The 2D-study reported here is a necessary, but not yet sufficient, first step, because of the "limited",.i.e. Kosterlitz-Thoules nature of the 2D phase transition, in general.

High performance computing is an absolutely necessary tool to treat the upper problems, because studying phase transitions causes extreme numerical difficulties:

- The determination of the phase boundaries requires an extensive finite-size scaling of the order parameters.
- Close to phase boundaries, the autocorrelation time τ grows from values smaller than one to sizes of over several thousand.

This work was supported by KONWIHR OOPCV in close cooperation with KONWIHR CUHE , DFG HA1537/16-1,2 and the Swiss National Science Foundation. High performance calculations per performed at HLRZ Jülich and LRZ Munich.

References

1. E. Dagotto, Rev. Mod. Phys. **66**, 763 (1994).
2. R. Eder, Numerical studies of strongly correlated electrons, Habilitationsschrift, Würzburg, 1998.
3. J. G. Bednorz and K. A. Müller, Zeit. Phys. B **64**, 189 (1986).
4. M. Imada, A. Fujimori, and Y. Tokura, Rev. Mod. Phys. **70**, 1039 (1998).
5. N. W. Ashcroft and N. D. Mermin, *Solid State Physics* (Saunders College, Philadelphia, 1976).
6. C. Kittel, *Introduction to solid state physics* (Wiley, New York, 1971).
7. W. Hanke, A. Muramatsu, and G. Dopf, Phys. Blätter **47**, 1061 (1991).
8. S. Haney, Computers in Physics **10/6**, 552 (1906).
9. A. Dorneich, New computational techniques for strongly correlated electron systems, PhD Thesis, Universität Würzburg, 2001.
10. T. Veldhuizen, Techniques for Scientific C++, http://www.osl.iu.edu/ tveldhui/papers/techniques/, 2000.
11. M. Suzuki, Prog. Theor. Phys. **65**, 1454 (1976).
12. J. E. Hirsch, R. L. Sugar, D. J. Scalapino, and R. Blankenbecler, Phys. Rev. B **26**, 5033 (1982).
13. J. E. Hirsch, Phys. Rev. B **38**, 12023 (1988).
14. S. White, Phys. Rev. B **46**, 5678 (1992).
15. S. R. White, Phys. Rev. B **48**, 10345 (1993).
16. H. Q. Lin and J. E. Gubernatis, Computers in Physics **7**, 400 (1993).
17. A. Dorneich, Exakte-Diagonalisierungs-Studien zur SO(5)-Symmetrie an Leitermodellen, Diplomarbeit, Universität Würzburg, (available from ftp.physik.uni-wuerzburg.de/pub/diplom/dorneich.ps.gz), 1998.

18. C. Lanczos, J. Res. Nat. Bur. Stand. **45**, 255 (1950).
19. A. L. Fetter and J. D. Walecka, *Quantum Theory of Many-Particle Systems* (Mc Graw-Hill, New York, 1971).
20. W. von der Linden, Physics Reports **220**, 53 (1992).
21. W. Kinzel and G. Reents, *Physik per Computer* (Spektrum Verlag, Heidelberb, Berlin, Oxford, 1996).
22. R. Landau and M. Páez, *Computational Physics* (John Wiley & Sons, New York, 1997).
23. A. W. Sandvik, Phys. Rev. B **56**, 11678 (1997).
24. A. W. Sandvik, Phys. Rev. B **59**, R14157 (1999).
25. A. Dorneich and M. Troyer, Phys. Rev. E **64**, 066701 (2001).
26. A. W. Sandvik and J. Kurkijärvi, Phys. Rev. B **43**, 5950 (1991).
27. A. W. Sandvik, J. Phys. A **25**, 3667 (1992).
28. A. Dorneich, submitted to Comp. Phys. Comm. (unpublished).
29. W. von der Linden, r. Preuß, and W. Hanke, J. Phys. Condens. Matter **8**, 3881 (1996).
30. S.-C. Zhang, Science **275**, 1089 (1997).
31. R. Eder, A. Dorneich, M. G. Zacher, W. Hanke, and S.-C. Zhang, Phys. Rev. B **59**, 561 (1999).
32. M. G. Zacher, W. Hanke, E. Arrigoni, and S.-C. Zhang, Phys. Rev. Lett. **85**, 824 (2000).
33. A. Dorneich, W. Hanke, E. Arrigoni, M. Troyer, and S. C. Zhang, Phys. Rev. Lett. **88**, 057003 (2002).
34. M. G. Zacher, From one to two dimensions: Numerical and analytical studies of strongly correlated electron systems, PhD thesis, Universität Würzburg, 1999.
35. C. P. Burgess and C. A. Lutken, Phys. Rev. B **57**, 8642 (1998).
36. X. Hu, T. Koyama, and M. Tachiki, Phys. Rev. Lett. **82**, 2568 (1999).
37. E. Demler, H. Kohno, and S.-C. Zhang, Phys. Rev. B **58**, 5719 (1998).
38. Y. Bazaliy, E. Demler, and S.-C. Zhang, Phys. Rev. Lett. **79**, 1921 (1997).
39. D. P. Arovas, A. J. Berlinsky, C. Kallin, and S.-C. Zhang, Phys. Rev. Lett. **79**, 2871 (1997).
40. E. Demler, A. J. Berlinsky, C. Kallin, G. B. Arnold, and M. R. Beasley, Phys. Rev. Lett. **80**, 2917 (1998).
41. D. E. Sheehy and P. M. Goldbart, Phys. Rev. B **57**, 8131 (1998).
42. D. Goldhaber-Gordon, H. Shtrikman, D. Mahalu, D. Abusch-Magder, U. Meirav, and M. A. Kastner, Nature (London) **391**, 156 (1998).
43. R. Eder, W. Hanke, and S.-C. Zhang, Phys. Rev. B **57**, 13781 (1998).
44. S. Meixner, W. Hanke, E. Demler, and S.-C. Zhang, Phys. Rev. Lett. **79**, 4902 (1997).
45. M. Greiter, Phys. Rev. Lett. **79**, 4898 (1997).
46. G. Baskaran and P. W. Anderson, J. Phys. and Chem. Solids **59**, 1780 (1998).
47. S. cheng Zhang, Journal of Physics and Chemistry of Solids **59**, 1774 (1998).
48. S.-C. Zhang, J.-P. Hu, E. Arrigoni, W. Hanke, and A. Auerbach, Phys. Rev. B **60**, 13070 (1999).
49. B. O. Wells, Y. S. Lee, M. A. Kastner, R. J. Christianson, R. J. Birgeneau, K. Yamada, Y. Endoh, and G. Shirane, Science **277**, 1067 (1997).

From Fermi Liquid to Non-Fermi Liquid Physics – Influence of Non-Local Fluctuations in Low-Dimensional Fermion Systems

Thomas Pruschke[1], Thomas Maier[2], and Mark Jarrell[2]

[1] Center for Electronic Correlations and Magnetism
   Theoretical Physics III, Institute for Physics, University of Augsburg
   86135 Augsburg, Germany
[2] Department of Physics, University of Cincinnati
   Cincinnati OH 45221, USA

Abstract. We discuss single particle dynamics of the 2D Hubbard model in the intermediate coupling regime calculated with the dynamical cluster approximation, which is a non-perturbative approach. We find a crossover from a normal Fermi liquid with a Fermi surface satisfiying Luttinger's theorem at large doping to a non-Fermi liquid for small doping with clear violation of Luttinger's theorem.

1 Introduction

The rich phenomenology of high-T_c superconductors [1] has stimulated strong experimental and theoretical interest in the field of strongly correlated electron systems. Apart from the anomalously high transition temperatures, these compounds are also of interest due to their unusual normal state properties. Most of these anomalous properties are found in spectra and transport quantities, i.e. are intimately linked to the dynamics of the electronic degrees of freedom. Thus, much of the experimental and theoretical effort has concentrated on the development of an understanding of the single-particle dynamics. Among the fundamental and controversial questions are whether the cuprates can be described as a Fermi liquid or not and what shape and volume a possible Fermi surface will have.

Early in the theoretical investigation of the high-T_c cuprates it was realized that the 2D Hubbard model

$$H = \sum_{i,j,\sigma} t_{ij} c_{i\sigma}^\dagger c_{j\sigma} + \frac{U}{2} \sum_{i\sigma} c_{i\sigma}^\dagger c_{i\sigma} c_{i\bar\sigma}^\dagger c_{i\bar\sigma} \tag{1}$$

in the intermediate coupling regime, or closely related models like the t-J model probably capture the essential physics [2]. In the wake of this conjecture, a huge effort has been directed to the study of these models [3]. There is now a general consensus that the appropriate parameter regime for the cuprates is the intermediate coupling regime where the Coulomb parameter U is roughly equal to the bandwidth. However, this is the most complicated regime of the model since both weak and strong coupling perturbative approaches fail. Exact diagonalization of small clusters [3] suffers from strong finite-size effects, often ruling out the reliable extraction of low-energy dynamics. Conventional Quantum Monte Carlo for finite sized systems suffers from a severe minus sign problem in this parameter regime. The resulting data is of insufficient quality to allow for reliable calculations of dynamic quantities at low enough temperatures. High-temperature series has provided some of the most informative results for the Fermi surface topology, but it does not yield spectra, and, so far, only results for the t-J model are available [4].

Thus, also from a theoretical point of view, the question of whether a Fermi surface does actually exist and what its topology is, still is a matter of debate. In that connection it is of special interest that some experiments indicate a violation of Luttinger's theorem; if true, any theory, such as FLEX [5], based on a weak-coupling expansion around the non-interacting limit would be inadequate.

Thus a treatment within a non-perturbative scheme going beyond conventional finite-size calculations clearly is desirable. In this paper we therefore use the recently developed dynamical cluster approximation (DCA) [6,7,8,9,10] to study the low-energy behavior of the $2D$ Hubbard model in the intermediate coupling regime with nearest-neighbor hopping t and on-site correlation U equal to the band width W. The DCA systematically incorporates non-local corrections to local approximations like the dynamical mean field, by mapping the lattice onto a self-consistently embedded cluster. We solve the cluster problem using a combination of quantum Monte Carlo (QMC) and the maximum entropy method to obtain dynamics. This technique produces results in the thermodynamic limit and has a mild minus-sign problem [10].

The paper is organized as follows. The next section contains a brief introduction to the DCA. The numerical results will be presented in the third section followed by a discussion and summary.

2 Formalism

2.1 Theoretical background

A detailed discussion of the DCA formalism was already given in previous publications [6,7,8,9,10]. The main assumption underlying the DCA is that the single-particle self-energy $\Sigma(\mathbf{k}, z)$ is a slowly varying function of the momentum $\mathbf{k}$ and can be approximated by a constant within each of a set of cells centered at a corresponding set of momenta $\mathbf{K}$ in the first Brillouin zone [6]. This prescription is sketched in Fig. 1 for the cluster size $N_c = 16$. The set of cluster $\mathbf{K}$-points is given by $K_\alpha^n = \pi(n_\alpha/2 - 1)$ with the spatial index $\alpha = x, y$ and $1 \leq n_\alpha \leq 4$ and the self-energy is assumed to be constant within the

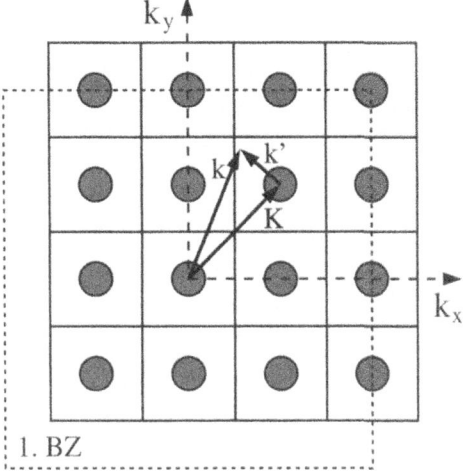

Fig. 1. The DCA coarse graining cells (shaded regions) for the cluster size $N_c = 16$. Each cell is represented by a cluster **K**-point (filled circles).

shaded region around each **K**, i.e. $\Sigma(\mathbf{K}+\mathbf{k}', z) = \Sigma(\mathbf{K}, z)$. The single-particle lattice Green functions are then coarse-grained or averaged within these cells, $\bar{G}(\mathbf{K}, z) = N_c/N \sum_{\mathbf{k}'} G(\mathbf{K} + \mathbf{k}', z)$, and used to calculate the lattice self energy and other irreducible quantities. Within this approximation, one can set up a self-consistency cycle similar to the one in the dynamical mean-field theory (DMFT) [11]. However, in contrast to the DMFT, where only local correlations are taken into account, the DCA includes non-local dynamical correlations. The length scales of these non-local correlations can be varied systematically from short ranged to long ranged by increasing the number of coarse-graining cells. The DCA collapses to the DMFT if one represents the Brillouin zone by one cell only, thus setting the characteristic length scale to zero.

By construction, the DCA preserves the translational and point group symmetries of the lattice. Comparisons to other extensions of the DMFT developed during the past years [12, 13] either show that these are identical to the DCA or converge more slowly as function of cluster size [14].

For the impurity problem of the DMFT a large set of reliable numerical techniques has been developed over the past ten years [15,11,16]. We have employed Quantum Monte Carlo (QMC) and the Non-Crossing Approximation (NCA) [7] as non-perturbative (in U) methods to solve the cluster problem of the DCA. For cluster sizes larger than $N_c = 4$ however, only the QMC technique is presently available for the DCA. From the QMC data, the spectra are obtained by analytic continuation with the maximum entropy method [17]. An interesting side-effect of the coupling of the cluster to a dynamic bath is a rather mild minus-sign problem, even for comparatively large values of U

and small temperatures [10]. Finally, the self energy is interpolated on to the full Brillouin zone using Akima splines, which is a sensible step as long as the assumption of a slow variation in **k**-space is valid. Note that it is very important to interpolate irreducible quantities like the self energy and *not* for example the cluster Green function itself.

2.2 Computational requirements

The embedded cluster problem arising in the DCA is solved using the Hirsch-Fye impurity algorithm [18] modified to simulate an embedded cluster [10]. In contrast to standard finite-size QMC algorithms, where locality in space-time allows an efficient reduction of the computational effort [19], the coupling to a dynamic bath in the DCA leads to a manifest non-locality of the effective action in euclidian time. This non-locality of the Hirsch-Fye algorithm leads to an extremely bad scaling with the number of space-time points N like N^3. For a typical calculation with $N_c = 16$, intermediate coupling U and low temperature this means computation times of the order of *four to five months* on a modern workstation *per DCA iteration step*. Since one typically needs $5 \ldots 10$ DCA iterations, this scaling would render the method unpracticable.

On the other hand, the QMC method is based on averaging statistically independent measurments. It has been realised early, that this series of measurments can be broken up into an in principle arbitrary number of independent processes. This observation is the basis for the high parallelizability of the QMC, which is thus almost perfect for the use on high-performance computers like the Hitachi SR8000: The communication between the different processes is reduced to spreading the initial data among the processes when starting the calculation and collecting them in the end, leading to an almost linear speed-up with the number of processes. The aforementioned calculation for example needs approximately 11 hours on the Hitachi SR8000 , using 512 processors, and has a peak performance of approximately 0.5TFlop/s, i.e. ≈ 1GFlop/s per processor.

Only due to this perfomance on modern massively parallel computer systems the application of the DCA to physical problems is possible at all.

3 Results

For a proper description of the CuO_2 planes of the high-T_c cuprates within the Hubbard model (1) it is generally accepted that the tight-binding dispersion has the form

$$t_{\mathbf{k}} = -2t \left(\cos(k_x) + \cos(k_y) \right) \tag{2}$$

$$-4t' \cos(k_x) \cos(k_y)$$

with a nearest neighbor hopping amplitude $t > 0$ and a next-nearest neighbor hopping amplitude t', which in principle can have any sign. From bandstructure calculations and the general form of the measured Fermi surface, especially in the overdoped regime, conventionally a negative t' is inferred [20].

The case for nearest-neighbor hopping only, i.e. $t' = 0$, was discussed in [21]. The more realistic situation with $t' \neq 0$ is the subject of the present contribution. In the following we set $t = 1/4\text{eV}$ and $t' = -0.2t$ in accordance with typical values extracted from experiment and bandstructure calculations and choose $U = W = 2\text{eV}$. This value of U is sufficiently large that for $N_c \geq 4$ a Mott gap is present in the half-filled model [22]. We performed our simulations at a range of temperatures, but will present results for $T = 0.033\text{eV}$ only, which is roughly room temperature. A pseudogap due to short-ranged spin correlations is also present in the weakly doped model for slightly lower temperatures [10].

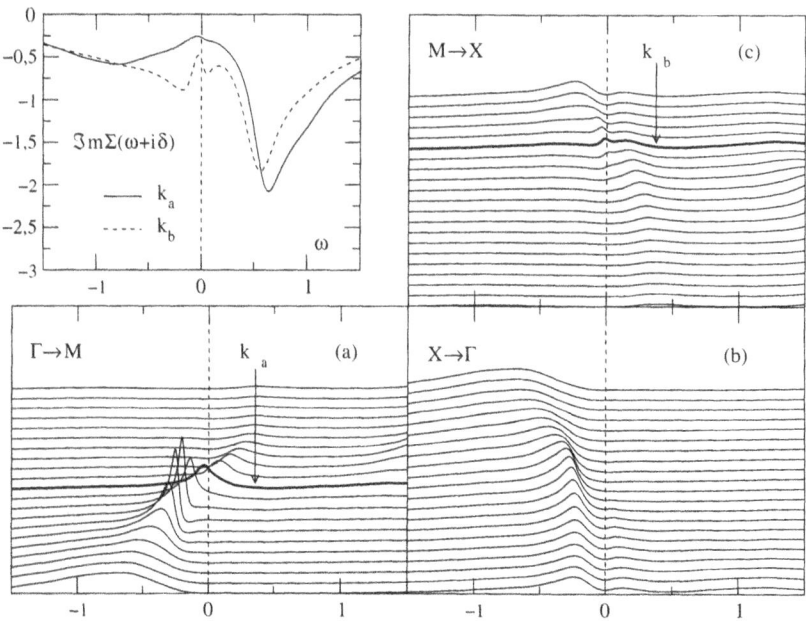

Fig. 2. (a)–(c) The single-particle spectrum $A(\mathbf{k}, \omega)$ for $U = 2\text{eV}$, $T = \frac{1}{30}\text{eV}$, $\delta = 0.05$, $t' = -0.2t$ and $N_c = 16$ along certain high symmetry directions (spectra for different $\mathbf{k}$ are shifted along the y-axis). The thick lines in Figs. (a) and (c) indicate the spectra which cross the Fermi energy with a peak closest to $\omega = 0$. In (b), no such peak is found which crosses the Fermi energy. (d) the imaginary part of the self energy versus frequency at the Fermi surface crossing found in (a) and (c).

The single-particle spectra for certain high symmetry directions are plotted in Figs. 2 and 3 for $n = 0.95$ and $n = 0.80$, respectively. We use the standard convention to identify the high symmetry points in the first Brillouin zone, viz $\Gamma = (0,0)$, $M = (\pi, \pi)$ and $X = (\pi, 0)$. Peaks in the spectrum cross the Fermi energy along the $\Gamma \to M$ and $M \to X$ directions, indicated by the colored lines and the blue arrows. Note that due to a finite, negative t' the Fermi surface is hole like for all dopings studied, in contrast to the case $t' = 0$, where a crossover from a hole-like Fermi surface at small doping to an electron-like at large doping is observed [21]. The imaginary part of the self energy at these crossing points is plotted versus frequency in panel (d) of Figs. 2 and 3.

One very interesting feature of the spectra at low doping, Fig. 2, is that the peak near $(\pi/2, \pi/2)$ broadens dramatically before crossing the Fermi energy. Near X, on the other hand, one does not observe any dramatic change in the spectrum when crossing the Fermi energy. This indicates that near $(\pi/2, \pi/2)$ hole-like quasi-particle (QP) excitations with $k < k_F$ appear to have longer lifetimes than electronic excitations with $k > k_F$. This asymmetry between particles and holes near the Fermi surface is a strong indication of non-Fermi liquid (NFL) behavior, at least along the Γ-M direction. Note also, that the spectrum near X in Fig. 2 is strongly damped and one cannot truely speak of a quasiparticle crossing the Fermi energy.

It is also quite instructive to look at the imaginary part of the self-energy at the **k**-points where the peak in the spectrum crosses the Fermi energy (Fig. 2(d)). In particular at the crossing point $\mathbf{k}_b$ close to $(\pi, 0)$, $\Im m \Sigma(\mathbf{k}_b, \omega)$ shows a striking asymmetry in the low-frequency regime as compared to $\Im m \Sigma(\mathbf{k}_a, \omega)$, where $\mathbf{k}_a$ is the crossing point close to $(\pi/2, \pi/2)$. In fact, $\Im m \Sigma(\mathbf{k}_b, \omega)$ starts to develop an additional feature which eventually leads to the formation of a pseudo gap in the spectra along the $M \to X$ direction at lower temperatures [10]. In addition, one observes a rather large residual scattering rate for both momenta $\mathbf{k}_a$ and $\mathbf{k}_b$. This can either be taken as further evidence for NFL behavior or as signal for the occurrence of a new very small low-energy scale [23]. Obviously, we cannot decide this question on the basis of the present data, but would have to look at much lower temperatures. Unfortunately, this is not possible at present. Note that in the low-energy regime the self energy displays significant k-dependence. This clearly renders theories based on a local approximation like the DMFA inadequate at least for small doping.

At high doping, Fig. 3, the peaks in the spectrum close to the Fermi energy are far sharper. Here it makes sense to speak of a conventional Fermi liquid and quasi particles again. As already mentioned, the Fermi energy crossings can be found along $\Gamma \to M$ and $M \to X$. Again, this is in qualitative accordance with ARPES experiments for strongly overdoped LSCO [25, 24], although these experiments still find rather broad structures even in heavily overdoped samples. In addition, there is no evidence for particle-hole asym-

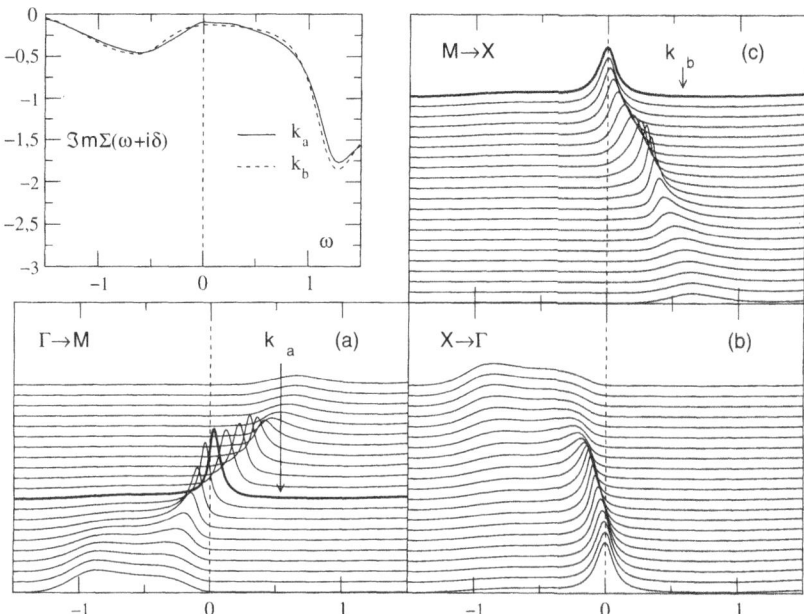

Fig. 3. (a)–(c) The single-particle spectrum $A(\mathbf{k}, \omega)$ $\delta = 0.20$ along certain high symmetry directions. Other parameters as in Fig. 2. The thick lines in Figs. (a) and (c) indicate the spectra which cross the Fermi energy with a peak closest to $\omega = 0$. In (c), no such peak is found which crosses the Fermi energy. (d) the imaginary part of the self energy versus frequency at the Fermi surface crossing found in (a) and (b).

metry in our data. Especially at $(\pi/2, \pi/2)$ the structure crossing the Fermi level appears to be rather symmetric with respect to the crossing point.

The imaginary part of the self energy, shown in Fig. 3(d), has a broad maximum at $\omega = 0$ with a very small residual scattering rate and changes little as $\mathbf{k}$ moves along the Fermi surface. This weak dependence on $\mathbf{k}$ is an indication that approximations like the DMFA should be accurate here, i.e. that there is little effect of non-local correlations. All indications are that for this doping regime standard Fermi-liquid behavior has returned.

More evidence for NFL behavior can be seen in the shape of the Fermi surface. An analysis of the Fermi surface for the t-J model has been performed recently within a high-temperature expansion [4] and a physically motivated decoupling scheme [27]. The results have been considered as clear evidence for a violation of Luttinger's theorem and the possible formation of a non-Fermi liquid at small doping.

We map the Fermi surface from a constant energy plot of the single-particle spectra at the Fermi energy $A(\mathbf{k}, \omega = 0)$. This method is equivalent to mapping out the regions in $\mathbf{k}$-space where peaks in the spectral function

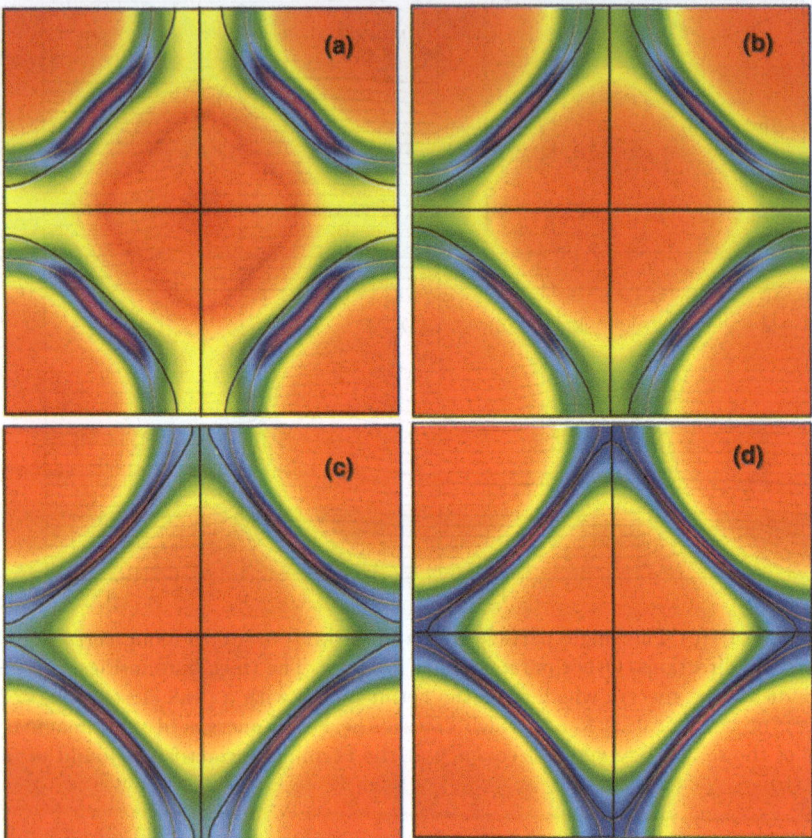

Fig. 4. Constant energy scans of $A(\mathbf{k}, \omega = 0)$ for $U = 2\text{eV}$, $T = \frac{1}{30}\text{eV}$, $t' = -0.2t$ and $N_c = 16$ for $n = 0.95$ (a), $n = 0.90$ (b), $n = 0.85$ (c) and $n = 0.80$ (d). The violet (red) represents regions of high (low) electronic density. The blue and violet regions map out the Fermi surface. The non-interacting Fermi surface is represented by a black line. The solid red line indicates the points where the real part of the denominator of the Green function vanishes. At low doping, the interacting Fermi surface encloses significantly more volume than the non-interacting Fermi surface. This indicates a violation of Luttinger's theorem.

$A(\mathbf{k}, \omega)$ cross the Fermi energy (see Figs. 2 and 3) and thus allows more direct contact with ARPES experiments. The constant energy plots $A(\mathbf{k}, \omega = 0)$ are shown in Fig. 4a-d for $n = 0.95$, $n = 0.90$ and $n = 0.80$, respectively, for the first Brillouin zone. The regions of high density are colored in violet and low density in red. The solid black lines as before represent the non-interacting Fermi surface. To distinguish between structures resulting from either quasiparticle excitations or incoherent background, we also plot the points where the real part of the denominator of the Green functions vanishes at zero frequency. These points which solve the quasiparticle equation $-\epsilon_{\mathbf{k}} - \Re\Sigma(\mathbf{k}, \omega = 0) = 0$ are given by the solid red line.

The Fermi surface resulting from our calculations is hole-like, centered around (π, π), and encloses a volume larger than the non-interacting Fermi surface for $n = 0.95$ and $n = 0.90$. This indicates a violation of Luttinger's theorem. In addition, the shape of the Fermi surface close to $(\pi/2, \pi/2)$, especially the clear shift above $(\pi/2, \pi/2)$ cannot be interpreted neither in terms of a simple tight-binding band structure nor a weak-coupling theory [26]. At least for $n = 0.95$ this structure rather is compatible with the notion of a small Fermi surface, i.e. pockets around $\mathbf{k} = (\pi/2, \pi/2)$, predicted by studies of the t-J model in the limit doping $\delta \to 0$ [4, 27].

With increasing doping the structures around $(\pi/2, \pi/2)$ start to follow more or less the noninteracting Fermi surface again. This points towards a restoration of Luttinger's theorem. For $n = 0.80$, the Fermi surface has essentially the same volume and shape as the non-interacting surface, indicating a return to Fermi-liquid like behavior. Note, however, that at the X point the interacting Fermi surface is still hole-like, while the non-interacting already is closed around Γ. This feature, i.e. the tendency towards a hole-like Fermi surface of the interacting system even i nthe overdoped regime, is also in qualitative accordance with experiment [28].

4 Summary and conclusions

The increasing precision and quality of experimental ARPES spectra in recent years has led to a number of new results on the single-particle dynamics of the high-T_c cuprates, both partially resolving long-standing issues and posing new questions and problems. Motivated by especially the interesting observations concerning the changes of Fermi surface topology with doping, we have studied the two-dimensional (2D) Hubbard model in the intermediate coupling regime (on-site correlation U equal to the bandwidth) at the temperature $T = 0.033$eV. To study the model in this most problematic parameter regime we used quantum Monte Carlo within the dynamical cluster approximation for the cluster size $N_c = 16$. Since this method allows for controlled and reliable calculations of low-energy features within a non-perturbative scheme and in the thermodynamic limit, fundamental problems in this field can be addressed. These include the single-particle spectral properties at low energies, especially possible deviations from Luttinger's theorem or the formation of non Fermi liquid states, and the resulting topology of the Fermi surface.

From the constant energy scans of the spectrum at the Fermi energy we find that the there is clear evidence for a violation of Luttinger's theorem at small doping. For $\delta = 5\%$ the only sharp structures appear at $\mathbf{k} = (\pi/2, \pi/2)$ and strongly resemble the notion of a small Fermi surface, i.e. hole pockets.

Non Fermi liquid behavior is also evidenced by the spectrum. A strong particle-hole asymmetry is found at the Fermi surface crossing near $(\pi/2, \pi/2)$ with hole-like excitations having much longer life-times than electronic excitations. Furthermore the corresponding self-energy shows strong **k**-dependence rendering local approximations like the DMFA irrelevant. Additional structures in the self energy and a rather large residual scattering rate can also be interpreted as signs for non Fermi liquid behavior. However, to really distinguish a non Fermi liquid from a Fermi liquid with a possibly extremely small energy scale much lower temperatures must be studied.

With increasing doping the Fermi surface changes its topology around $(\pi/2, \pi/2)$, and beyond $\delta \approx 0.10$ one can infer a tendency towards restoration of Luttinger's theorem and conventional Fermi liquid behavior. Finally when $\delta = 0.20$ the Fermi surface coincides with the non-interacting Fermi surface. The corresponding spectra show well defined and sharp quasi-particle peaks around the Fermi energy with particle-hole symmetry being recovered in the low-energy region.

Although the Hubbard model surely presents an oversimplification of the real cuprates, the common belief is that at least the essential qualitative features of the low energy dynamics should be reproduced. In particular, the results presented here show some interesting qualitative agreement with experiments regarding the behavior of the different structures observed in the spectra. Moreover, these features could be related to at least an apparent violation of Luttinger's theorem or possibly even non Fermi liquid behavior at low doping. From our results it becomes very clear that neither weak-coupling treatments nor local theories like the DMFT are able to capture the essentials of the physics of the cuprate in the weakly doped regime. There are, of course, a variety of further questions to be addressed. For example, the effects of a bilayer coupling on the topology of the Fermi surface or more generally on the low-energy single particle dynamics have to be investigated. We believe that such an investigation can also address some of the still mysterious features in behavior of the Fermi surface topology of the high-T_c cuprates. Moreover it is very important to find new methods to solve the DCA self consistency cycle at lower temperatures or preferably at $T = 0$. This would enable us to clearly distinguish between a strong coupling Fermi liquid and genuine non Fermi liquid behavior at low doping.

We acknowledge useful conversations with M. Hettler, C. Huscroft, H.R. Krishnamurthy, M. Sigrist and T.M. Rice. This work was supported by NSF grant DMR-0073308 and by the Deutsche Forschungsgemeinschaft through the Graduiertenkolleg "Komplexität in Festkörpern" and SFB 484 "Kooperative Phänomene im Festkörper". We acknowledge supercomputer support by the Leibniz Rechenzentrum in Munich under grant h0301.

References

1. For a review, see M.B. Maple, J. Mag. Mat. **177-181** pp18-30 (1998) ; J.L. Tallon and J.W. Loram, Physica C: Superconductivity **349**1-2 pp53-68 (2001).
2. P.W. Anderson, **The Theory of Superconductivity in the High-T_c Cuprates**, Princeton University Press, Princeton, NJ (1997).
3. E. Dagotto, Reviews of Mod. Physics **66**, 763(1994).
4. W.O. Putikka, M.U. Luchini and R.R.P. Singh, Phys. Rev. Letters **81**, 2966(1998).
5. N. E. Bickers, D. J. Scalapino, S. R. White, Phys. Rev. Lett. **62**, 961 (1989).
6. M.H. Hettler, A.N. Tahvildar-Zadeh, M. Jarrell, T. Pruschke and H. R. Krishnamurthy, Phys. Rev. B **58**, 7475 (1998); M.H. Hettler, M. Mukherjee, M. Jarrell and H. R. Krishnamurthy, Phys. Rev. B **61**, 12739 (2000).
7. Th. Maier *et al.*, Eur. Phys. J. B **13**, 613 (2000); Th. Maier, M. Jarrell, Th. Pruschke, and J. Keller, Phys. Rev. Lett. **85**, 1524 (2000).
8. C. Huscroft, M. Jarrell, Th. Maier, S. Moukouri, and A.N. Tahvildarzadeh, Phys. Rev. Lett. **86**, 139 (2001).
9. S. Moukouri and M. Jarrell, to appear in Computer Simulations in Condensed Matter Physics VII, Eds. D.P. Landau, K. K. Mon, and H. B. Schuttler (Springer-Verlang, Heidelberg, Berlin, 2000).
10. M. Jarrell, Th. Maier, C. Huscroft, S. Moukouri, Phys. Rev. B, to appear, cond-mat/0108140.
11. T. Pruschke *et al.*, Adv. in Phys. **42**, 187 (1995); A. Georges *et al.*, Rev. Mod. Phys. **68**, 13 (1996).
12. A.I. Lichtenstein and M.I. Katsnelson, Phys. Rev. B **62**, R9283 (2000).
13. G. Kotliar, S.Y. Savrasov, G. Palsson, cond-mat/0010328.
14. Th. Maier and M. Jarrell, Phys. Rev. B 65, 041104 (2002).
15. M. Jarrell, Phys. Rev. Lett. **69**, 168-71 (1992).
16. R. Bulla, Th. Pruschke and A.C Hewson, J. Phys. – Condens. Matter **10**, 8365(1998).
17. M. Jarrell and J.E. Gubernatis, Phys. Rep. **269**, 135 (1996).
18. J.E. Hirsch and R.M. Fye, Phys. Rev. Lett. **56**, 2521 (1986).
19. R. Blankenbecler, D.J. Scalapino and R.L. Sugar, Phys. Rev. D**24**, 2278(1981).
20. M.S. Hybertsen, E.B. Stechel, M. Schluter and D.R. Jennison, Phys. Rev. B41, 11068(1990); S.B. Bacci, E.R. Gagliano, R.M. Martin and J.F. Annett, Phys. Rev. B44, 7504(1991).
21. Th. Maier, Th. Pruschke and M. Jarrell, Phys. Rev. B, in press.
22. S. Moukouri and M. Jarrell , unpublished.
23. J. Altmann, W. Brenig and A.P. Kampf, Eur. Phys. J. B**18**, 429(2000).
24. A. Ino, C. Kim, T. Mizokawa, Z.-X. Shen, A. Fujimori, M. Takaba, K. Tamasaku, H. Eisaki and S. Uchida, J. Phys. Soc. Jap. **68**, 1496(1999).
25. A. Damascelli, D.H. Lu and Z.-X. Shen, Journal of Electron Spectroscopy **117-118**, 165(2001).
26. V. Zlatić, B. Horvatić, B. Dolički, S. Grabowski, P. Entel, and K.-D. Schotte, Phys. Rev. B**63**, 35104(2001) and references therein.
27. P. Prelovšek and A. Ramšak, cond-mat/0109209 (2001).
28. A.A. Korduyk, S.V. Borisenko, M.S. Golden, S. Legner, K.A Nenkov, M. Knupfer and J. Fink, cond-mat/0104294.

One-Dimensional Electron-Phonon Systems: Mott- Versus Peierls-Insulators

Holger Fehske[1,2], Gerhard Wellein[3], Arno P. Kampf[4], Michael Sekania[4], Georg Hager[3], Alexander Weiße[2], Helmut Büttner[2], and Alan R. Bishop[5]

[1] Institut für Physik, Universität Greifswald
   17487 Greifswald, Germany
[2] Physikalisches Institut, Universität Bayreuth
   95440 Bayreuth, Germany
[3] Regionales Rechenzentrum Erlangen, Universität Erlangen
   91058 Erlangen, Germany
[4] Institut für Physik, Universität Augsburg
   86135 Augsburg, Germany
[5] Theoretical Division and Center for Nonlinear Studies
   Los Alamos National Laboratory
   Los Alamos, New Mexico 87545, USA

Abstract. We analyze ground state and spectral properties of the one-dimensional half-filled Holstein Hubbard model with respect to the Peierls-insulator to Mott-insulator transition, exploiting Lanczos diagonalization, density matrix renormalization, kernel polynomial expansion, and maximum entropy methods on the Hitachi SR8000-F1 supercomputer.

1 Introduction

Quasi-one-dimensional (1D) strongly coupled electron-phonon systems like MX-chain compounds or conjugated polymers are particularly rewarding to study for a number of reasons. They exhibit a remarkable wide range of strengths of competing forces and, as a result, physical properties. These systems share fundamental features with higher dimensional novel materials, such as high-temperature superconductors, charge-ordered nickelates or colossal magneto-resistance manganites, i.e., they are complex enough to investigate the interplay of charge, spin, and lattice degrees of freedom which is important for strongly correlated electronic systems in two and three dimensions as well. Nevertheless they are simple enough to allow for a nearly microscopic modeling. Thus they are suited modern systems to develop and test new theoretical methods by bringing together techniques from quantum chemistry, electronic band structure investigations, and many-body physics.

Two properties of quasi-1D materials are crucial for their unusual electronic, magnetic and optical properties: first they have broken-symmetry ground states, and second, in the insulating phases, they have gap states. The first feature arises, because the itinerancy of the electrons strongly competes with electron-electron and electron-phonon (EP) interactions, which tend to localize the charge carriers by establishing spin-density-wave (SDW) and charge-density-wave (CDW) ground states, respectively. Hence, at half-filling, Mott (MI) or Peierls (PI) insulating phases are energetically favored over the metallic state. An interesting and still controversial question is whether or not only one quantum critical point separates the PI and MI phases at temperature $T = 0$ [2]. The second feature, i.e. the existence of gap states, may be caused by donating electrons to or accepting electrons from the half-filled host material. Those charged gap states are related to local lattice distortions (polarons or bipolarons). Another possibility is "photo-doping", i.e. the creation of neutral excitations (e.g., excitons). More recently, the existence and stability of intrinsically localized vibrational modes has been demonstrated both experimentally and theoretically in clean strong-CDW MX-materials [3]. These multi-phonon bound states occur inside the CDW gap when the effective lattice potential, dynamically self-generated in the process of carrier localization, exhibits a significant nonlinearity as a consequence of a non-adiabatic electron-phonon interaction of intermediate strength.

Significant progress has been made towards understanding the physics underlying these various effects by numerical investigations of generic model Hamiltonians. Neglecting in a first step the lattice dynamics, a frequently used starting point has been the *adiabatic Holstein-Hubbard model* (AHHM):

$$H_{AHHM} = H_{t-U} - \sum_{i,\sigma} \Delta_i n_{i\sigma} + \frac{\kappa}{2} \sum_i \Delta_i^2 , \tag{1}$$

$$H_{t-U} = -t \sum_{i,\sigma} (c_{i\sigma}^\dagger c_{i+1\sigma} + \text{H.c.}) + U \sum_i n_{i\uparrow} n_{i\downarrow} . \tag{2}$$

Here, H_{t-U} constitutes the conventional Hubbard Hamiltonian with hopping amplitude t and on-site Coulomb repulsion strength U; $c_{i\sigma}^\dagger$ creates a spin-σ electron at Wannier site i and $n_{i\sigma} = c_{i\sigma}^\dagger c_{i\sigma}$. In addition, H_{AHHM} includes the elastic energy of a harmonic lattice with a "spring constant" κ. Within this so-called frozen phonon approach, $\Delta_i = (-1)^i \Delta$ is a measure of the static, staggered density modulations of the PI phase. Equation (1) with $\kappa = 0$ and fixed Δ is known as the ionic Hubbard model (IHM) for which a band insulator (BI) to MI transition has been established previously [2]. The BI-MI transition of the IHM on finite lattices was shown to be connected to a ground state level crossing with a site-parity change, where the site inversion symmetry operator P is defined by $P c_{i\sigma}^\dagger P^\dagger = c_{N-i\sigma}^\dagger$ with $N = 4n$ [4].

Of course, dynamical phonon effects are known to be particularly important in quasi-1D materials, where the lattice zero-point motion is usually comparable to the Peierls distortion [5]. By any means quantum phonon effects should be included in a theoretical analysis of transport and optical properties of electronically 1D compounds. Introducing phonon creation $b_i^\dagger$ and destruction operators b_i, the general *Holstein-Hubbard Hamiltonian*

(HHM) takes the form

$$H = H_{t-U} - g\omega_0 \sum_{i,\sigma}(b_i^\dagger + b_i)n_{i\sigma} + \omega_0 \sum_i b_i^\dagger b_i \,, \qquad (3)$$

where $g = \sqrt{\varepsilon_p/\omega_0}$ is a dimensionless EP coupling constant and ω_0 denotes the frequency of the optical phonon mode. The physics of the HHM is governed by three competing effects: the itinerancy of the electrons, their Coulomb repulsion and the local EP interaction. There are two dimensionless energy ratios, U/t and ε_p/t, which determine the tendency of the itinerant quantum mechanical system to establish a magnetic or charge ordered state, respectively. Since the EP coupling is retarded, the phonon frequency ω_0 defines a third relevant energy scale of the problem. At $U = 0$, the ground state of the pure half-filled Holstein model is a Peierls distorted state with staggered charge order in the adiabatic limit $\omega_0 \to 0$ for any finite ε_p. However, as in the HM of spinless fermions, quantum phonon fluctuations destroy the Peierls state for small EP interaction strength [6]. Above a critical threshold $g_c(\omega_0)$, the HM describes a PI with equal spin (Δ_s) and charge (Δ_c) excitation gaps – the characteristic feature of a band insulator.

In what follows, exact numerical methods [7] are used to diagonalize the HHM on finite chains, preserving the full dynamics of the phonons, and the density matrix renormalization group (DMRG) technique is applied to the AHHM and IHM.

2 Peierls- to Mott-insulator transition

In order to draw conclusions about the phase diagram in the *adiabatic limit*, we plot in Fig. 1 $\Delta(U,\kappa)$ obtained from DMRG on an open chain of length $N = 64$, where the value for the stiffness constant is fixed at $\kappa = 0.74$. In contrast to the behavior of the 8-site chain $\Delta(U,\kappa)$ decreases more smoothly with increasing U and vanishes discontinuously near $U/2\varepsilon_p \approx 0.75$. Also

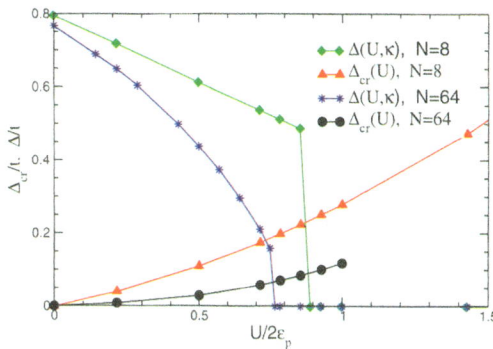

Fig. 1. Level crossing line $\Delta_{cr}(U)$ of the IHM for an 8-site ring (red triangles) and from finite-size scaling (black circles). In addition: ionic potential strength $\Delta(U,\kappa)$ of the AHHM for an 8-site ring (green diamonds) and on an open 64-site chain (blue stars).

shown in Fig. 1 is the level crossing line $\Delta_{cr}(U)$ of the IHM obtained from finite-size scaling of Lanczos results for rings of up to 14 sites; $\Delta_{cr}(U)$ remains finite in the infinite chain length limit. Importantly, $\Delta(U, \kappa)$ and $\Delta_{cr}(U)$ do not intercept because $\Delta(U, \kappa)$ jumps to zero before reaching the level crossing point of the IHM. The discontinuous nature of the PI-MI transition in the AHHM is easily verified in the atomic limit $t = 0$ where $\Delta = 1/\kappa$ for $U < U_c = 1/\kappa$ and $\Delta = 0$ for $U > U_c$. The first order nature persists for finite small t, i.e. in the strong coupling regime U, $\kappa^{-1} \gg t$. However, as we have explicitly verified by diagonalization of a periodic AHHM ring of length $N = 14$, the transition is second order in the weak coupling regime U, $\kappa^{-1} \ll t$. This implies a continuous decrease of $\Delta(U)$ and therefore $\Delta(U)$ necessarily intercepts the $\Delta_{cr}(U)$ line of the IHM (cf. Fig. 3). This intercept marks the point U_{opt} when the site-parity sectors become degenerate and the optical absorption gap Δ_{opt} disappears. For weak coupling the PI-MI transition therefore evolves across *two* critical points. We summarize these findings in the phase diagram shown in Fig. 2. In the Peierls BI phase for $U < U_{opt}$ the spin and charge excitation gaps are equal and finite, and remarkably $\Delta_{opt} \neq \Delta_c$ [8]. When the site-parity sectors become degenerate at $U = U_{opt}$, $\Delta_{opt} = 0$ but $\Delta_c = \Delta_s > 0$. For $U \geq U_s$ the usual MI phase with $\Delta_{opt} = \Delta_c > \Delta_s = 0$ is realized. For strong coupling $U_{opt} = U_s$ holds. In weak coupling there exists an intermediate region $U_{opt} < U < U_s$ in which all excitation gaps are finite. The CDW persists for all $U < U_s$. The site-parity eigenvalue is $P = +1$ in the PI and $P = -1$ in the MI phase. It is natural

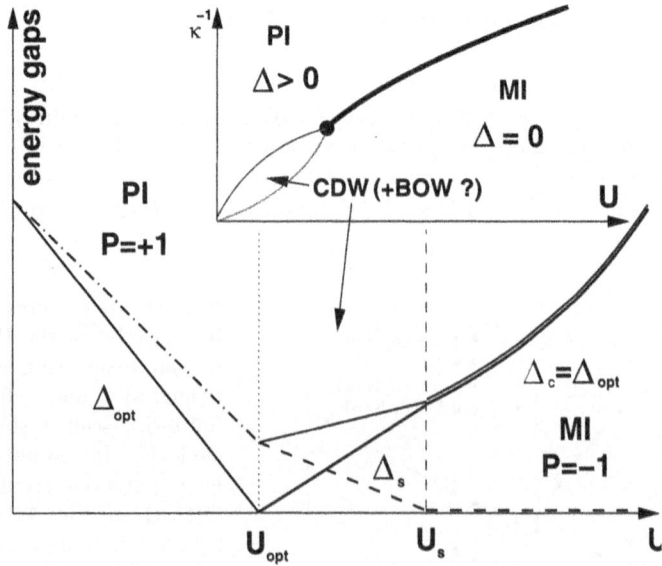

Fig. 2. Qualitative phase diagram of the AHHM. Inset: Transition scenario in the weak coupling regime $U, \kappa^{-1} \ll t$.

to expect an additional ordering phenomenon in the window $U_{opt} < U < U_s$. A bond order wave (BOW) with a finite expectation value of the staggered bond charge $B = \frac{1}{N}\sum_{i\sigma}(-1)^i\langle c_{i\sigma}^\dagger c_{i+1\,\sigma} + \text{H.c.}\rangle$ is the natural candidate.

In order to discuss the PI-MI transition at finite phonon frequencies, i.e. in the *non-adiabatic regime*, we have calculated the regular part of the optical conductivity at $T = 0$,

$$\sigma^{reg}(\omega) = \frac{\pi}{N}\sum_{m\neq 0}\frac{|\langle\psi_0|\hat{j}|\psi_m\rangle|^2}{E_m - E_0}\,\delta(\omega - E_m + E_0)\,. \tag{4}$$

Here $|\psi_0\rangle$ and $|\psi_m\rangle$ denote the ground state and excited states, respectively, with corresponding energies E_0 and E_m. Note that the current operator $\hat{j} = -\mathrm{i}et\sum_{i\sigma}(c_{i\sigma}^\dagger c_{i+1\,\sigma} - c_{i+1\,\sigma}^\dagger c_{i\sigma})$ has finite matrix elements between states of different site-parity only.

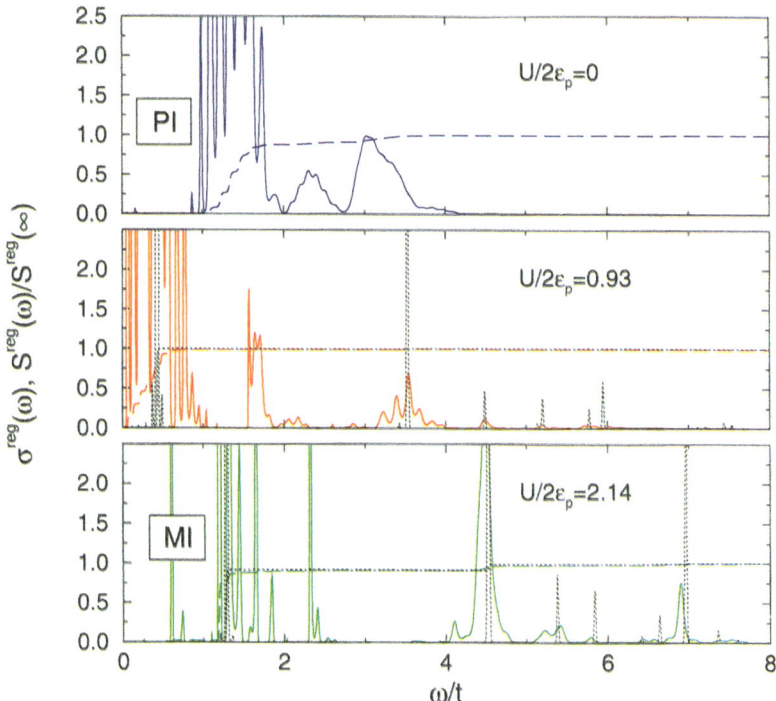

Fig. 3. Optical conductivity in the 8-site HHM ring for $\omega_0 = 0.1t$ and $g^2 = 7$. Top panel: PI phase for $U = 0$; middle panel: near criticality $U \sim U_{opt}$; lower panel: MI phase for $U = 3t$. The lower two panels include σ^{reg} for $g = 0$ (dotted lines), i.e. for the pure Hubbard chain. The dashed lines give the normalized integrated spectral weights $S^{reg}(\omega)$. $S^{reg}(\omega)/S^{reg}(\infty)$ with $S^{reg}(\omega) = \int_0^\omega \sigma^{reg}(\omega')\,\mathrm{d}\omega'$ is a natural measure for the relative weight of the different optical absorption processes.

The evolution of the frequency dependence of $\sigma^{reg}(\omega)$ from the PI to the MI phase with increasing U is illustrated in Fig. 3. In the PI regime the electronic excitations are gapped due to the pronounced CDW correlations. The broad optical absorption band for $U = 0$ results from particle-hole excitations across the BI gap which are accompanied by multi-phonon absorption and emission processes. The shape of the absorption band reflects the phonon distribution function in the ground state. Excitonic gap states may occur in the process of structural relaxation. In the MI phase the optical gap is by its nature a correlation gap. The lower panel in Fig. 3 shows clearly that $\sigma(\omega)$ of the HHM in the MI phase is dominated by excitations which can be related to those of the pure Hubbard model. In addition, phononic sidebands appear. More interesting, we found phonon-induced states with low spectral weight within the Mott-Hubbard gap. These states can be viewed as a "fingerprint" of the lower Hubbard band and will be discussed in more details elsewhere [9]. Most notably, in-between the PI and MI phases the optical gap closes at U_{opt} and, due to the selection rules for optical transitions, this necessarily implies a ground state level crossing with a site-parity change. We have explicitly verified that the ground state site parity in the PI phase is $P = +1$ and $P = -1$ in the MI phase. For the HHM on finite rings U_{opt} is identical to the critical point where $S_c(\pi)$ sharply drops. From our conductivity data we found evidence for *only one* critical point in the non-adiabatic region.

This finding is corroborated by the results obtained for the spectral density of single-particle excitations associated with the injection of a spin-σ electron with wave number K (inverse photoemission (IPE))

$$A_{K\sigma}^+(\omega) = \sum_m |\langle \psi_m^{(N_{el}+1)} | c_{K\sigma}^\dagger | \psi_0^{(N_{el})} \rangle|^2 \, \delta[\omega - (E_m^{(N_{el}+1)} - E_0^{(N_{el})})], \qquad (5)$$

and the corresponding quantity for the emission of an electron (photoemission (PE))

$$A_{K\sigma}^-(\omega) = \sum_m |\langle \psi_m^{(N_{el}-1)} | c_{K\sigma} | \psi_0^{(N_{el})} \rangle|^2 \, \delta[\omega + (E_m^{(N_{el}-1)} - E_0^{(N_{el})})]. \qquad (6)$$

Here $|\psi_0^{(N_{el})}\rangle$ is the ground state of the system with N_{el} electrons and $|\psi_m^{(N_{el}\pm1)}\rangle$ are eigenstates of the $(N_{el}\pm1)$-particle system. $E_0^{(N_{el})}$ and $E_m^{(N_{el}\pm1)}$ are the corresponding energies.

For the half-filled band case ($N_{el} = N$, total $S^z=0$ sector), the single-particle spectral function $A_{K\sigma}(\omega) = A_{K\sigma}^+(\omega) + A_{K\sigma}^-(\omega)$ of the interacting system is gapped at $E_F \, \forall \, K$, indicating massive charge excitations across the CDW ($U < U_{opt}$) and Mott-Hubbard ($U > U_{opt}$) gaps in the PI and MI phases [9]. When U approaches the critical value U_{opt} from both below and above, the gap feature vanishes for the Fermi momenta $K_F = \pm\pi/2$. Figure 2 displays the K-resolved IPE and PE spectra at $U = U_{opt}$. One recognizes that the spectral function $A_{K\sigma}(\omega)$ obeys various sum rules. The simplest one,

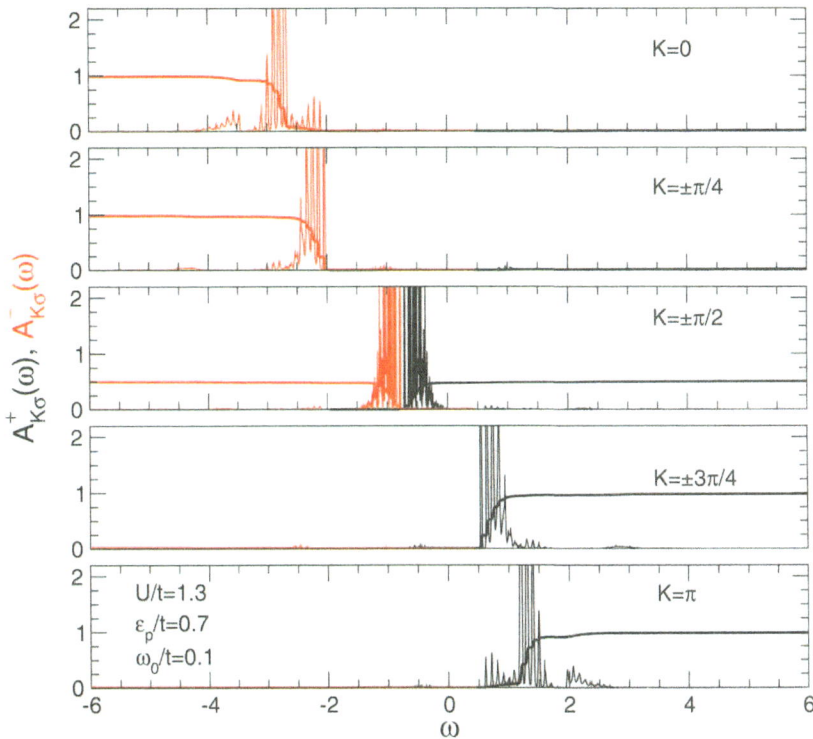

Fig. 4. K-resolved spectral densities (black lines for IPE, red lines for PE) at $U/2\epsilon_p = 0.93$, i.e. in the vicinity of the PI-MI transition point. Included are the integrated spectral weights. Measuring all energies relative to the Fermi energy we note the "particle-hole" symmetry: $A^-_{K,\sigma}(-\bar{\omega}) = A^+_{\pi-K,\sigma}(\bar{\omega})$.

$\int^{\infty}_{-\infty} A_{K\sigma}(\omega)d\omega = 1$, is not useful for (angle-resolved) photoemission spectroscopy (ARPES) since it involves both occupied and unoccupied states [10]. The one-particle density of states is given by $N_\sigma(\omega) = \sum_K A_{K\sigma}(\omega)$. The important sum rule for ARPES, however, is $\sum_\sigma \int^{\infty}_{-\infty} n_F(\omega)A_{K\sigma}(\omega)d\omega = n(K)$ ($n_F(\omega)$ is the Fermi function), which relates the ARPES intensity to the number of electrons in a momentum state K: $n(K) = \sum_\sigma \langle c^\dagger_{K\sigma} c_{K\sigma}\rangle$. The momentum distribution $n(K)$ is one for K's far below K_F and shows a sharp drop for K near K_F.

3 Performance analysis on the Hitachi SR8000-F1

Although there has been a tremendous increase in the computational power during the past decade, the numerical simulation of the interacting quantum systems discussed in the preceding section remains a "grand-challenge application" for modern supercomputers. Even for small clusters, exact diagonalization (ED) studies of microscopic electron-phonon models like the Holstein Hubbard model involve very large sparse matrices. In order to achieve high accuracy within our numerical calculations and to ensure an efficient use of supercomputers DMRG algorithms [11] and *one–step* ED methods (Lanczos [12], Jacobi-Davidson [13], Maximum-Entropy [14]) have been used. The DMRG algorithm which is an iterative method with hundreds of ED steps on a reduced subspace can be run on workstations or multi-processor systems. On the contrary, *one–step* ED techniques require matrix vector multiplications (MVM) involving a sparse matrix representation of the total Hilbert space but provide to date the only method free of any approximations. Since the MVM step determines the computational time as well as the memory requirements for these algorithms, two supplementary MVM strategies have been implemented.

The *in–core* implementation stores the non-zero matrix elements using sparse storage formats, such as *Compressed Row Storage* (CRS) [15] or *Jagged Diagonals Storage* (JDS) [15] yielding high performance at the cost of additional storage. It has been demonstrated [17], that a combination of JDS format (which achieves best performance on vector computers [16]) and an hybrid programming approach is best suited for the Hitachi SR8000-F1, where the 100 GFlop/s barrier can be exceeded on 128 nodes. Furthermore we have shown that, although based on RISC technology, one Hitachi SR8000-F1 node can exceed the performance of comparable present-day vector processors (NEC SX5e) even if vector-gather operations are involved [17]. The *in–core* JDS implementation is used in combination with the Jacobi-Davidson method to compute several low lying eigenstates (10–100) where thousands of MVM steps are required.

A memory-saving, scalable, and parallel algorithm that recomputes the non-zero matrix elements in each MVM step is used for the *out-of-core* MVM implementation (cf. [7, 18]). In combination with Lanczos (Maximum Entropy) algorithms - which typically need 50–250 MVM steps - the total memory requirements for our ED studies of ground state (spectral) properties can be reduced to approximately 3–4 arrays of the matrix dimension (D_{mat}). Here the available main memory sets the only limit for the matrix dimension accessible for ED and thus determines the quality of our numerical results.

For more than four years the CRAY T3E systems at HLR Stuttgart and NIC Jülich have provided the largest amount of aggregate main memory with a maximum of 128 GB available for production runs. Including the final upgrade at the beginning of 2002 the Hitachi SR8000-F1 system allows us to increase the matrix dimensions by a factor of roughly 7 with approximately 900 GB main memory available for batch jobs on 152 nodes. A brief summary of the development of supercomputers and *out-of-core* MVM implementations used in our ED projects during the past decade is given in Table 1. Note that programming language did not change all along, while the parallelization

Table 1. Development of supercomputer resources and programming techniques used for ED in the past decade. System specifications, programming languages and parallelization techniques used in the MVM are given in the first five rows. Maximum matrix sizes achieved on each system and the corresponding time per MVM step are given in the last two rows.

System	TM CM5 GMD St. Augustin	CRAY T3E NIC Jülich	Hitachi SR8000-F1 LRZ Munich
Util. time	1993/1994	1998-	2001-
#CPUs	64	256	1216
Memory	2 GB	128 GB	900 GB
Language	FORTRAN	FORTRAN	FORTRAN
Parallelization	CMFortran	MPI/CRAY-shmem	MPI+OpenMP
$D_{\mathrm{mat}}^{\max}$	5.6×10^7	4.4×10^9	3.3×10^{10}
MVM [s]	156	33	63

strategy of the MVM step had to be adapted several times to ensure the best use of the supercomputer architecture.

Performance and scalability of different *out-of-core* MVM parallelization strategies on the Hitachi SR8000-F1 are depicted in Fig. 5 for a fixed matrix size ($D_{\mathrm{mat}}^{\max} = 1.6 \times 10^9$). Running a pure MPI parallelization (MPP-mode) and

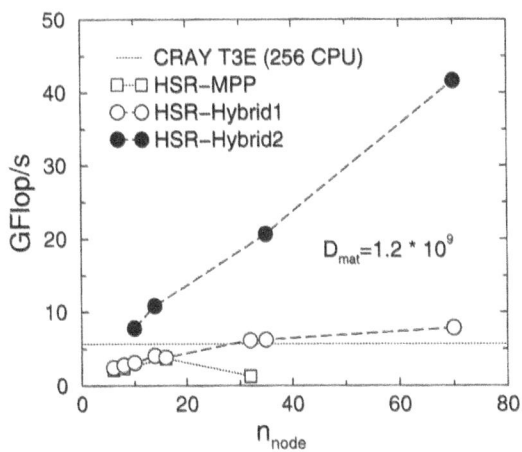

Fig. 5. MVM performance on Hitachi SR8000-F1 (HSR): (i) HSR-MPP: pure MPI mode. (ii) HSR-Hybrid1: Hybrid mode with no additional communication buffer. (iii) HSR-Hybrid2: Hybrid mode with additional communication buffer space. For comparison also MVM performance of a 256 processor CRAY T3E-1200 system is depicted (red line).

assigning one process to each processor, satisfying performance numbers cannot be achieved. At larger processor counts the total performance even drops with increasing number of processors, because a large number of outstanding MPI communication requests over-floods internal message queues. The impact of this bottleneck can be substantially reduced by applying an hybrid programming strategy, where MPI has been used for inter-node communication while shared memory parallelization is carried out within one node (HSR-Hybrid1). As a result, the total number of MPI processes is reduced by a factor of 8 (when compared to the MPP-mode) resulting in performance numbers comparable to CRAY T3E-1200 systems. When spending two additional arrays (of matrix dimension D_{mat}) for buffering of messages and separating communication from computation (HSR-Hybrid2) a performance boost by a factor of 3-4 is obtained. At the same time, however, the maximum matrix dimension achievable by the HSR-Hybrid2 implementation is reduced by a factor of roughly 2 when compared to the HSR-Hybrid1 implementation (cf. Table 1).

In conclusion, for the *one–step* ED techniques applied in our project, the hybrid programming approach is the method of choice to achieve high performance and high scalability on the Hitachi SR8000-F1.

4 Conclusions

The work presented in this report is an example for the predictive power of large-scale numerical many-body calculations performed on modern supercomputers. The implementation of the various optimized program packages on the Hitachi SR8000-F1 at the LRZ München allows us to obtain new and exciting insights into the complex interplay of charge, spin, and lattice degrees of freedom in the currently intensively studied quasi-1D materials. More precisely, using quasi-exact numerical techniques to determine the ground state and spectral properties of the one-dimensional half-filled Holstein-Hubbard model, we have revealed the physics behind the crossover from a Peierls band insulator to a correlated Mott-Hubbard insulator. The transition results from a ground state level crossing with a change in the ground state site-parity eigenvalue. In the adiabatic limit, where the quantum phase transition is connected to the band- to Mott-insulator transition of the ionic Hubbard model, two scenarios emerge: a discontinuous PI-MI transition in the strong-coupling regime, and two continuous transitions for weak interactions with an intermediate phase of possible coexistence of charge-density-wave and bond-order-wave. In the non-adiabatic regime, optical conductivity and charge structure factor data indicate that the PI-MI transition proceeds continuously.

Acknowledgement. This work was supported by the *Competence Network for Technical and Scientific High Performance Computing in Bavaria* (KONWIHR, project HQS@HPC) and the Deutsche Forschungsgemeinschaft (project Fe 398-1/2). A. P. K.

and M. S. acknowledge support by SFB 484. A. R. B. appreciates the partial support of a Senior Humboldt Fellowship at the University of Bayreuth. Work at Los Alamos is supported by the U.S. D.O.E.

References

1. A. R. Bishop and B. I. Swanson, Los Alamos Science **21**, 133 (1993).
2. M. Fabrizio, A. O. Gogolin, and A. A. Nersesyan, Phys. Rev. Lett. **83**, 2014 (1999). P. Brune, G.I. Japaridze, and A.P. Kampf, arXiv:cond-mat/0106007. H. Fehske *et al.*, Physica B **312-313**, 562 (2002).
3. B. I. Swanson *et al.*, Phys. Rev. Lett. **82**, 3288 (1999); Phys. Rev. Lett. **82**, 3288 (1999); H. Fehske, M. Kinateder, G. Wellein, and A. R. Bishop, Phys. Rev. B **63**, 245121 (2001).
4. N. Gidopoulos, S. Sorella, and E. Tosatti, Eur. Phys. J. B **14**, 217 (2000).
5. R. Peierls, *Quantum Theory of Solids* (Oxford University Press, Oxford 1955); R. H. McKenzie and J. W. Wilkins, Phys. Rev. Lett. **69**, 1085 (1993); H. Fehske, M. Holicki, and A. Weiße, Adv. Sol. State Phys. **40**, 235 (2000).
6. R. J. Bursill, R. H. McKenzie, and C. J. Hamer, Phys. Rev. Lett. **80**, 5607 (1998); A. Weiße and H. Fehske, Phys. Rev. B **58**, 13526 (1998); E. Jeckelmann, C. Zhang, and S. R. White, Phys. Rev. B **60**, 7950 (1999).
7. B. Bäuml, G. Wellein, and H. Fehske, Phys. Rev. B **58**, 3663 (1998); A. Weiße, H. Fehske, G. Wellein, and A. R. Bishop, Phys. Rev. B **62**, R747 (2000).
8. For a similar conclusion in the IHM see S. Qin, J. Lou, Z. Su, G.-S. Tian, arXiv:cond-mat/0004162v2.
9. H. Fehske *et al.*, to be published.
10. M. Randeria *et al.*, Phys. Rev. Lett. **74**, 4951 (1995).
11. S. R. White, Phys. Rev. B **48**, 1993 (1993).
12. J. K. Cullum, R. A. Willoughby, *Lanczos Algorithms for Large Symmetric Eigenvalue Computations*, Volumes I & II, Birkhäuser, Boston (1985).
13. G.L.G. Sleijpen and H.A. van der Vorst, J. Matrix Anal. Appl. **17**, 401 (1996).
14. R. N. Silver and H. Röder, Phys. Rev. E **56**, 4822 (1997).
15. R. Barrett *et al.*, *Templates for the Solution of Linear Systems: Building Blocks for Iterative Methods*, SIAM, Philadelphia (1993).
16. M. Kinateder, G. Wellein, A. Basermann, and H. Fehske, in *High Performance Computing in Science and Engineering '00* edited by E. Krause and W. Jäger, Springer-Verlag, Berlin Heidelberg (2001), pp. 188–204.
17. G. Wellein, G. Hager, A. Basermann, and H. Fehske, Proceedings of *VEC-PAR2002*, Porto (2002).
18. G. Wellein and H. Fehske, in *High Performance Computing in Science and Engineering '99* edited by E. Krause and W. Jäger, Springer-Verlag, Berlin Heidelberg (2000), pp 112-129.

Part VI

Geophysics

Werner Hanke

Institut für Theoretische Physik und Astrophysik
Universität Würzburg
Am Hubland
97074 Würzburg, Germany

The project of H. Igel and his group from the University in Munich is concerned with the computational study of seismology. This topic has, in the last few years, experienced a dramatic increase of excitement. This is due to the fact that, with the current possibilities of high-performance computing, seismic waves which are located in a frequency band following regional or global earthquakes, can be simulated numerically for realistic three-dimensional earth models, for the first time. The corresponding simulations lead to considerable improvements in the understanding of structure properties, e. g. of the earth mantle, the inside of a sedimentary basin or of a volcano. Also in forecasting strong ground motion for realistic earthquake scenarios, the corresponding calculations are of importance. During the initial phase of the above project, some very high resolution simulations have been performed. These calculations have important implications for future directions in computational seismology. As an example, the low seismic velocities inside active faults, e. g. the famous San Andreas Fault in California, may act as an amplifier for ground motion. This has e. g. implications for buildings in the vicinity of those faults. In the first phase of this project, which is reported here, (1) parallelization and implementation of algorithms for numerical wave propagation on the Hitachi SR8000-F1; (2) verification of the codes and analysis of the efficiency and (3) first applications to realistic problems, as discussed above, were achieved.

3-D Seismic Wave Propagation on a Global and Regional Scale: Earthquakes, Fault Zones, Volcanoes

Heiner Igel, Gilbert Brietzke, Michael Ewald, Miko Fohrmann, Gunnar Jahnke, Tarje Nissen-Meyer, Johannes Ripperger, Max Strasser, Markus Treml, and Guoquan Wang

Department of Earth and Environmental Sciences, Geophysics Section
Ludwig-Maximilians-University
Theresienstrasse 41
Munich, Germany
Heiner.Igel@lmu.de

Abstract. For computational seismology the present years are extremely exciting. The reason is, that with the current supercomputer technology, the frequency band in which seismic waves are observed following regional or global earthquakes, can be simulated numerically for realistic 3D earth models for the first time. Depending on the spatial scales under consideration (whole planet, a sedimentary basin at risk from local earthquakes, a volcano with high risk for future eruptions) this will lead to considerable improvement (1) in the understanding of the structural properties (e.g. the Earth's mantle, the inside of a sedimentary basin or a volcano) and (2) in forecasting strong ground motion for realistic earthquake scenarios. The latter point may have considerable long-term societal benefits, as the short-term prediction of large earthquakes seems out of sight. During the first phase of this project some of the highest-resolution simulations ever done were carried out with important implications for future directions in computational seismology. The most important scientific results can be summarized as: (1) 3D Simulations of several earthquakes in the Cologne Basin in Germany demonstrate that the main characteristics of ground motion (e.g. peak motion amplitude, shaking duration) are successfully predicted through numerical simulations; (2) The low seismic velocities inside active faults (e.g. San Andreas Fault, California) may act as an amplifier for ground motion. This has implications for buildings in the vicinity of faults; (3) Large scale simulations of strong earthquakes in subduction zones show that the local 3D structure at depth strongly influences the waves propagating to the surface. Ignoring this will lead to severe misinterpretations. (4) Including topography to understand wave propagation inside volcanoes is crucial. Our simulations demonstrate the scattering effects due to topography. If we want to understand the state of a volcanic system prior to eruptions from seismic waves these effects have to be taken into account.

1 Introduction

The accurate simulation of seismic wave propagation through realistic 3-D Earth models plays a fundamental role in several areas of geophysics: (1) in global seismology knowledge of the structure of the Earth's deep interior is crucial to understand the dynamic behaviour of our planet such as mantle convection, slab subduction or hot spot activity. Accurate synthetic 3-D seismograms which can be compared with globally recorded data require a numerical approach. The structural resolution of today's tomographic models can only be improved by exploiting the 3-D wave effects of the geodynamically important regions inside the Earth. (2) As deterministic short-term earthquake fore-casting seems out of sight, the accurate prediction of likely ground motion following earthquakes in seismically active regions is a major goal which will allow measures (e.g. applying strict building codes) to be taken before major events. 3-D modelling will allow local (e.g. amplifying) effects such as low-velocity zones or topography to be studied. These so-called site effects are being investigated for several areas at risk (e.g. Cologne Basin, Germany; Los Angeles Basin, USA; Beijing Basin, China). (3) Active volcanic areas show very characteristic complex ground motion which is usually recorded on local networks monitoring the activity and risk of eruption. The origin of the seismically recorded signals are poorly understood. One of the reasons is the structural complexity of volcanic areas with strong 3-D heterogeneities, topography and sources in the summit region.

The first phase of this project was dedicated to (1) Parallelization and implementation of algorithms for numerical wave propagation on the Hitachi SR8000-F1; (2) Verification of the codes and analysis of their efficiency; and (3) first applications to realistic problems.

2 Numerical simulation of seismic wave propagation

The algorithms implemented to date constitute numerical solutions to the (visco-)elastic wave equations in Cartesian and spherical coordinates. The time-dependent partial differential equations are solved numerically using high-order finite-difference methods (e.g. [6, 7]). This implies that - no matter the particular problem or coordinate system - the space-dependent fields are defined on a 3-D grid and the time extrapolation is carried out using a Taylor expansion or a Runge-Kutta method. The space derivatives are calculated by explicit high-order finite-difference schemes which do not necessitate the use of matrix inversion techniques. This approach leads to naturally parallel problems where communication is only needed at the boundaries of the decomposed domains, when calculating space derivatives.

Before complex 3D models were run on the Hitachi Sr8000 all algorithms were verified by comparing the numerical solutions to analytical solutions for simple (layered) model geometries. Thereby the parameter space for stable and accurate numerical solutions for the given problems could be identified. The final programming model uses (1) the automatic parallelization for the (intra-node) shared-memory part and (2) the MPI model across the nodes. This seems to be the optimal approach for explicit FD algorithms.

In the following we review the results from the first production runs of these algorithms.

3 Scientific and technical results

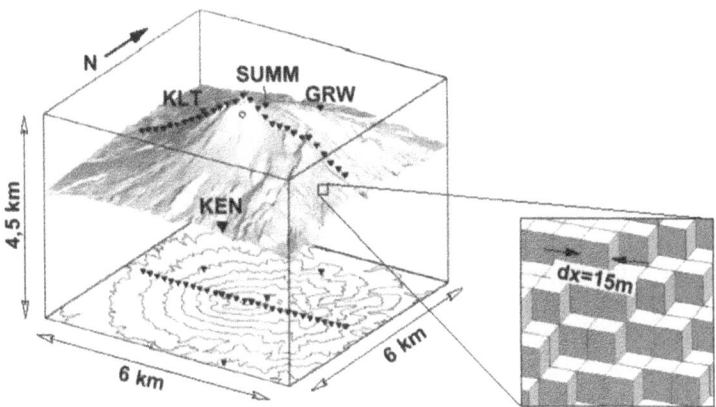

Fig. 1. Topography of volcano Merapi, Indonesia. The free surface boundary condition is implemented on the (blocky) surface of the regular finite-difference grid. Other options for the irregular free surface are being investigated.

3.1 Volcano topography in 3-D seismic wave propagation

How does the (usually strong) topography affect the seismic wavefield observed on volcanoes? How can we model the topographic effects and the scattering effects inside the medium? Two methods which allow the modelling of wave propagation for strong topography were implemented and tested for a realistic topography (digital elevation map of the Merapi volcano, Indonesia). The first method is based on stretching a regular Cartesian grid in a way that the surface follows the topography. An important technical result is that while this approach is applicable to moderate topography it becomes unstable for strong topography [12], a situation frequently encountered around volcanoes.

Fig. 2. Snapshot of seismic wave propagation for a source near the summit of volcano Merapi. Such events frequently occur during active periods of volcanoes and they are associated with magma movement inside the volcano.

An alternative method uses a modified free surface boundary condition which is applied to a blocky representation of the real topography. While this approach needs a large number of grid points per wavelength, it leads to a stable algorithm and was our method of choice for the first 3D simulation of wave propagation through a real volcano model [12, 13].

Figure 1 shows the principles of the latter technique applied to the topography of the high-risk volcano Merapi in Indonesia. On this volcano a seismic network permanently records the ground motion. This project aims at modelling these data using 3D simulation techniques. We intend to use this code to investigate (1) the seismic signature of pyroclastic flows; (2) seismic sources inside magma chambers and volcanic dykes; (3) scattering vs. topographic effects as observed on Merapi. In Fig 2 a snapshot of wave propagation a few seconds after a source emitted seismic energy near the summit of the volcano is shown. Such seismic sources may be indicative of magma movements inside the volcano.

3.2 Simulation of earthquake scenarios

As reliable deterministic earthquake prediction is not in sight, the simulation of realistic earthquake scenarios is one of the most important tools to assess the seismic hazard of active regions. The ground motion observed at the surface after large earthquakes predominantly depends on (1) the source depth; (2) the magnitude; (3) the source mechanism (the orientation and size of the slip on the fault plane); and (4) the structure of the Earth's crust. One of the key factors in shaking hazard is the local seismic velocity structure. Low seismic velocities near the surface (e.g. sedimentary basins as in Los Angeles, Mexico City, the Cologne Basin area) may amplify the ground motion up to ten-fold and thereby increase the hazard even for moderate earthquake magnitudes or distant large earthquakes. This is an inherent 3D effect and can only be properly modelled with the use of 3D modelling techniques.

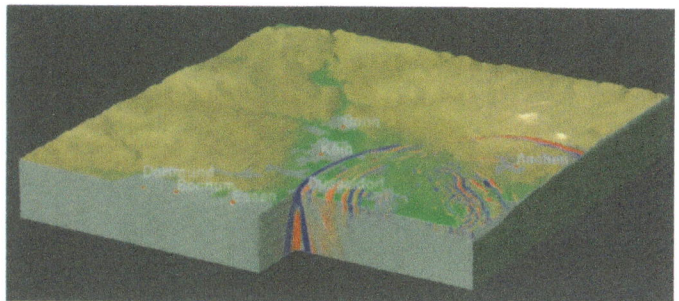

Fig. 3. Snapshot of ground motion for a 3D simulation of the M5.9 Roermond earthquake that shook the Cologne Basin in 1992 [2]. The complex structure of the local sedimentary basin leads to strong lateral variations in peak ground acceleration and shaking duration. These parameters are crucial for the estimation of structural stability.

In this project the first high-resolution 3D calculations for the area in Germany with the highest seismic risk - the Cologne Basin - were undertaken. The simulations were carried out on a 3D spatial grid with approx. 150 million grid points. This allowed us to carry out a simulation with a dominant frequency of approx. 1 Hz. It is important to note that the data available for the simulated M5.9 earthquake near the village of Roermond in 1992, which shook the area of the Cologne Basin, are in a similar frequency range (see Fig. 3).

Even though the simulations are based on a relatively simple 3D velocity model of the sedimentary structure, they show good agreement with observed data as far as the general characteristics of the ground motion are concerned. This tells us that we may be on the right way to be able to calculate the possible ground motion amplification due to 3D structure for this (and other) areas. The numerical algorithms provide us with the ground motion at the Earth's surface at all times of the simulation. These data enable us to extract crucial information on the maximum amplitudes of motion or the shaking duration and convert them into so-called shaking hazard maps (see Fig. 4), a measure of (expected) damage in a region. One of the main goals of this study is to compare and calibrate simulated results with observed ground motion (or damage observations). At present our study areas are the Cologne Basin, the Beijing area and the Los Angeles Basin.

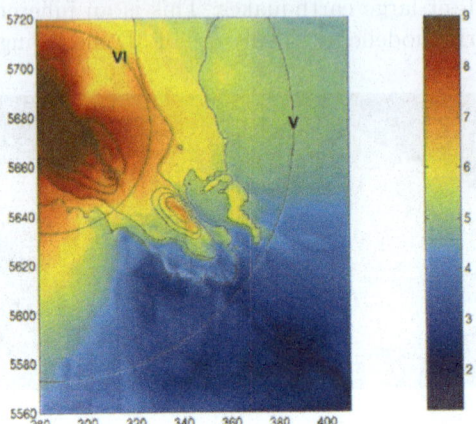

Fig. 4. Shaking intensity map calculated from 3D simulations of the M5.9 Roermond earthquake in 1992. The graphic shows the same geographical area as the previous figure, with the upward direction being North. The colour scaling denotes the modified Mercalli scale, an indicator for expected damage in the region. The theoretical damage predictions can be compared to the observed damage during earthquakes. The most important goal of this area of research is to make the simulated results accurate enough so that local authorities can use them for planning purposes.

3.3 Seismic wave propagation on a planetary scale

To first order the structure of the Earth's interior is spherically symmetric. Yet, the 3D perturbations thereof play a fundamental role in the dynamic behaviour of our planet as lateral variations in temperature and density drive convective processes in the Earth's mantle and core. Therefore, we need to understand how such 3D structural perturbations (e.g. subduction zones, plumes, lateral variations at the core-mantle boundary) affect the seismic wavefield observed at the Earth's surface after large earthquakes.

The problem of wave propagation in spherical geometry is complicated by the occurrence of singularities in the wave equation in spherical coordinates. Therefore, standard numerical methods such as the finite difference method can not directly be applied to the simulation of whole Earth wave propagation. The options are (1) using an axi-symmetric approach, thereby reducing the problem to two dimensions (see Fig. 5); (2) considering a spherical section while centering the physical domain around the equator thus avoiding problems with the singularities. For whole earth wave propagation unstructured grid methods (e.g. finite elements, spectral elements) have to be employed.

Subduction zones contain the largest earthquakes on Earth. Knowledge of their structural details not only is important for hazard assessment but also to understand the dynamics of subduction and mantle convection. In

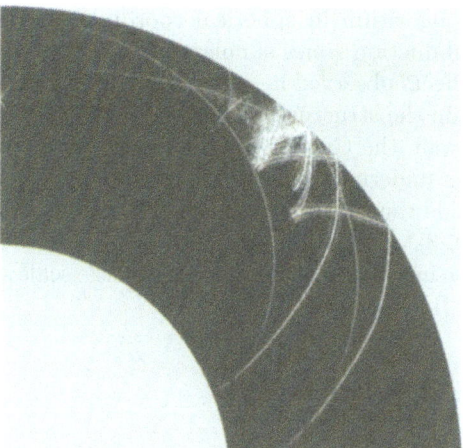

Fig. 5. Snapshot of shear wave propagation inside the Earth 10 minutes after the origin time of an earthquake at the surface. Waves travel along the surface and through the interior. The bright colours denote the energy in the wave field. This simulation was carried out with an axi-symmetric solution of the wave equation, which leads to a 2D computational domain, while allowing the calculation of a 3D wave field. The dominant period of this simulation 5s. This is the actual frequency band which is observed around the globe for shear wave propagation. The grid spacing for such simulations is below 1km [10].

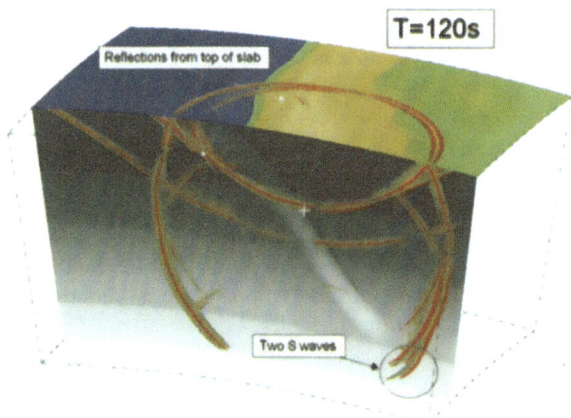

Fig. 6. Snapshot (curl of the wavefield) for a source inside a subduction zone. The calculations are calculated in spherical geometry in order to account for the Earth's spherical surface relevant to large scale wave propagation problems. The complex 3D structure of subduction zones, which host the largest earthquakes observed on Earth, leads to wave effects which need to be accounted for when investigating the rupture mechanisms of deep earthquakes.

this study a 3D algorithm in spherical coordinates was implemented and earthquakes in subduction zones simulated [11, 7]. We were able to simulate particular wave effects observed in nature which - in the future - can be used to further constrain the structure of subduction zones (Fig. 6). Furthermore, hot spots like Hawaii, the Galapagos Islands or Iceland are characterized by large scale plumes underneath them, which are thought to consist of rising material from inside the mantle. However, it is not clear, how deep the roots of the plumes are. 3D simulation for plume models were investigated. These simulations will be important to design future large scale seismic experiments around hot spots [14].

3.4 Fault zone wave propagation

Fig. 7. Snapshot of 3D wave propagation for a source outside a shallow low-seismic velocity zone associated with an active seismic fault (e.g. San-Andreas fault, North-Anatolian fault). An important result of recent simulations is the fact that sources outside faults can generate high-amplitude trapped waves and therefore lead to elevated shaking hazard in the vicinity of faults.

The detailed structure of fault zones (FZs) plays an important role in problems related to fault mechanics, earthquake rupture, wave propagation and seismic hazard. FZs are thought to consist of a O(10-100)m wide region of decreased seismic velocity but structural details such as their depth extent, lateral and vertical variations etc. are elusive. The small spatial scales involved make such structures difficult to image with ray-theoretical methods such as tomography. However, seismic energy trapped inside FZ layers can provide dispersive wave trains that carry information on the FZ structure.

These waves can travel many kilometers inside the FZ before reaching the surface and are therefore strongly altered by its properties. Candidate trapped waves have been observed above several active faults. Inversion algorithms exist that can model these observations in terms of planar fault zone structures. However, at present it is not clear how reliable these estimates are, as the effects of (even small) 3-D variations on trapped waves are not well understood. The goal of this study is to distinguish 3-D structures that do and do not significantly affect FZ waves. To achieve this, we perform numerical calculations of wave propagation in various FZ geometries and analyze the waveforms, spectra and envelopes of the synthetic seismograms. The main results are that (1) moderate changes of the shape of FZ or (2) small-scale heterogeneities or (3) depth-dependent properties do not strongly affect the observed FZ waves. In contrast, strong effects are to be expected from (4) breaks in the continuity of FZ structure (e.g. offsets), which may at some point allow imaging such features at depth [9,8].

Fig. 8. Arial view of the surface manifestation of the San-Andreas fault, California. It is thought that the fault contains a O(100m) wide zone with low seismic velocities down to a depth of several kilometres. This zone acts like a wave guide and leads to characteristic wave phenomena relevant to shaking hazard as well as the understanding of the rupture process. (Courtesy United States Geological Survey)

An important discovery in the course of these 3D simulations was the fact that sources outside shallow FZs are capable of producing considerable trapped wave energy [3,4,5]. This means that a much larger volume of possible earthquake hypocenters may lead to amplified motion at the surface in the vicinity of faults. These numerical results are supported by observations of aftershocks of the large earthquakes in Turkey of 1999 [1].

Acknowledgement. We thank the steering committee for access to the Hitachi supercomputer and the staff of the Leibniz Rechenzentrum for help in optimizing our codes at the initial stages. These projects were partly funded through KON-WIHR (konwihr.in.tum.de) and the German Research Foundation (Ig16/2). We thank Yehuda Ben-Zion for his input in the area of fault zone wave propagation. Thanks to the German Academic Exchange Service for partial support through the International Quality Network: Georisk (www.iqn-georisk.de).

References

1. Ben-Zion, Y., Peng, Z., Okaya, D., Seeber, L., Armbruster, J:G., Ozer, N., Michael, A.J., Baris, S., Aktar, M., (2002): Shallow low-velocity fault zone layer in the KAradere-Duzce Branch of the North Anatolian Fault based on analysis of trapped waves, submitted to Geophysical Journal International.
2. Ewald, M., (2001): Numerical simulation of site effects with applications to the Cologne basin, Diplom thesis, Institute of Geophysics, LMU.
3. Fohrmann, M., Jahnke, G., Igel, H., Ben-Zion, Y., (2001): Guided waves generated by sources outside a low velocity fault zone layer, presented at the Annual Meeting of the American Geophysical Union, 2001.
4. Fohrmann, M., (2002): Trapped waves generated by sources outside a low-velocity layer, Diplom thesis, LMU.
5. Fohrmann, M., Jahnke, G., Igel, H., Ben-Zion, Y., (2002): Trapped waves from sources outside fault zones, submitted to Pure and Applied Geophysics.
6. Igel, H., Riollet, B., Mora, P., (1995): Anisotropic wave propagation through finite difference grids, Geophysics, 60, 1203-1216.
7. Igel, H., Nissen-Meyer, T., Jahnke, G., (2002a): Wave propagation in 3-D spherical sections: effects of subduction zones, Phys. Earth Planet. Int., in print.
8. Igel, H., Jahnke, G., Ben-Zion, Y., (2002b): Numerical simulation of fault zone guided waves: accuracy and 3-D effects, Pure and Applied Geophysics, in print.
9. Jahnke, G., Igel, H., Ben-Zion, Y., (2002): Fault zone guided waves: the effects of 3D structure, Geophys. J. Int., in print.
10. Jahnke, G., Igel, H., Nissen-Meyer, T., (2002): Short-period global wave propagation using a finite-difference method, presented at the Meeting of the German Geophysical Sovciety 2002.
11. Nissen-Meyer, T., (2001): Wave propagation through 3D subduction zones. Diplom thesis, Institute of Geophysics, LMU.
12. Ripperger, J., (2001): Volcano topography in 3-D seismic wave simulation, Diplom thesis, Institute of Geophysics, LMU.
13. Ripperger, J., Igel, H., Wassermann, J., (2002): Seismic simulation in the presence of real volcano topography, submitted to J. Volcanol. Geotherm. Res.
14. Strasser, M., (2001): Numerical Modeling of Wavefield Effects of Mantle Plume Structures, Diplom thesis, Institute of Geophysics, LMU.

Part VII

Fundamental Physics

Bengt Petersson

Fakultät für Physik
Universität Bielefeld
33615 Bielefeld, Germany

High performance computing has had a very big impact on theoretical high energy physics. The forces of Nature, apart from gravity, are described by the so-called Standard Model, which includes the strong, electromagnetic and weak forces. Although the basic laws are thus supposed to be known, there are still important unsolved problems. The Standard Model has more than 20 free parameters, which should be determined from experiment. It is hoped that their determination will help to find a more fundamental theory, perhaps also including quantum gravity. Another difficulty is that there is no known systematic analytic approximation method, which can be used in the case of the strong interaction, apart from special situations, e.g. where a large momentum transfer is involved. The only method, which one can use to determine the properties of the strongly interacting particles, are very large scale numerical simulations. Such simulations are also needed to determine the free parameters of the Standard Model, because the experimentally observed quantities in general are influenced by strong interactions in the final state.

Within the Standard Model, the strong interactions are given by a quantum field theory, Quantum Chromodynamics. The theory is described by a Lagrangian, given by an explicit analytic expression. The basic constituents are quarks and gluons. They have not been directly observed in experiments, but many indirect pieces of evidence have been found, which lend strong support to the theory. In fact the quarks and gluons are supposed to be essentially free at very short distances, but absolutely confined to distances smaller than the size of the hadrons. These particles, formerly denoted "elementary particles" are the ones we observe, like the proton, neutron and pi-meson and many other particles and resonances found at the large accelerators. In Quantum Chromodynamics the quarks and gluons are represented by fields (functions) of the position in space and time. It is possible to perform an analytic continuation in the time variable, to represent the quantum theory as a statistical model in a four dimensional Euclidean space. All observables can be calculated by functional integrals over the fields, with a positive definite Boltzmann weight determined by the Lagrangian.

In 1974 Ken Wilson proposed to approximate the four-dimensional continuum by a four-dimensional regular lattice and introduced a suitable formulation. Then the theory becomes a well defined statistical model, where numerical methods similar to the methods of statistical mechanics can be used. The theory has as parameters only one scale parameter and the quark masses. The simulations have to be performed in a systematic way, so that one can extrapolate to the continuum, and analytically continue the results back to real time. The main difficulty in the numerical simulations is posed by the quark fields. Because the quarks are fermions, the corresponding weight is a determinant of a very big matrix, where the indices are the position in space-time times the spin and internal indices. In realistic simualtions, the matrix has a size of several tenthousand lines and columns.

Since the middle of the 1980's, intensive work has been put into the simulations of the strong interactions. Methods have been developed and mainly quenched simulations, where the quarks are treated as external sources have been successfully performed. However, the real theory with dynamical quarks could only be simulated with a realistic continuum extrapolation, once computers in the Teraflops range have become available. It is therefore of great importance that the Hitachi Teraflops computer has made such a simulation possible. The first results are described in the contribution of G. Schierholz and collaborators. They find a good agreement with the experimental hadron masses and other properties of the hadrons. This is a very important pioneering calculation and certainly confirms the hypothesis that Quantum Chromodynamics is the correct theory. One can expect many more interesting results from this simulation.

An important symmetry of Quantum Chromodynamics in the continuum is the chiral symmetry, in the limit of zero quark mass. The quarks have a small mass, so this symmetry should be nearly fulfilled in the Lagrangian. In the physical ground state it should be spontaneously broken. This leads to the prediction that the physical fermions are massive, while the pion is nearly massless, in comparison. Several other relations follow from this symmetry. It has turned out to be quite difficult to find a formulation, in which the symmetry is manifest in the lattice formulation. In the original formulation of Wilson it is explicitly broken, but should be recovered in the continuum, by appropriate tuning. It would, however, be much more satisfying if the symmetry is explicit already in the lattice formulation. Recently it has been shown that such formulation of the theory is possible. However, the fermionic part of the action becomes considerably more complicated. A thorough investigation of the possibility to use such a formulation, is studied in the contribution of

Gattringer et al. As this formulation is one or two orders of magnitude more demanding in computer time, the quenched approximation, where the quarks do not take part in the dynamics, is studied. It is shown that nevertheless the use of a chirally symmetric discretization is very promising, and leads to a much better behaviour when the quark masses are given small (nearly realistic) values, so that a well controlled extrapolation to the physical value can be performed. This is certainly a very important result, and gives the hope that with computers in the 10–100 Teraflops range one can obtain results, which can fix the parameters of the Standard Model to an accuracy of a few percent.

In nuclear physics it would not be possible to calculate the properties of the nuclei from the fundamental theory. Instead one starts from phenomenological two and three body potentials between the nuclei. To describe the properties of 4He, this becomes nevertheless a very demanding problem. In the contribution of H.M. Hoffmann and G.M. Hale, such a calculation is, however, performed. This the first calculation where one attempts to describe all the main properties of this nucleus by a microscopic calculation. The use of the LRZ Hitachi computer has certainly been paramount for performing such a calculation. It seems to be in the reach of the method that by using this computer also the properties of other light nuclei can be calculated by this method.

A further contribution in fundamental physics is the simulation of the local universe by S. Gottlöber et al. This is also a highly non linear problem, although in the realm of classical physics. It is shown that using large scale constraints and a cosmological model consistent with cosmological observations, the local part of the Universe with its peculiar structures can be well reproduced. The large amount of data for the local neighbourhood of our Galaxy can be compared with the theory of structure and galaxy formation. The use of a supercomupter in the Teraflops range is very important for such studies, which will shed light on the unanswered questions in the field of galaxy formation and distribution.

In conclusion the contributions to the conference mentioned above show the extreme importance of a computer in the Teraflops range for the advancement of fundamental physics. Many questions from the consistency of the theory to the properties of Nature at scales from the subnuclear to the size of the Universe could only be answered by numerical calculations on a computer of this size. It is very encouraging that at the HLRB so many important results have already been achieved in this field. In fact they are pioneering, and have contributed very much to the leading position of German physicists in this branch of physics. It is clear that the German expertise is strong enough, so that a continuation of the HLRB program towards even faster computers would be very rewarding.

Simulation of QCD with Dynamical Quarks

Gerrit Schierholz[*]

[1] John von Neumann-Institut für Computing NIC
Deutsches Elektronen-Synchrotron DESY
Platanenallee 6
D-15738 Zeuthen
[2] Deutsches Elektronen-Synchrotron DESY
Notkestr. 85
D-22603 Hamburg
Gerrit.Schierholz@desy.de

1 Introduction

Quantum chromodynamics (QCD) is considered to be the theory of the strong interactions. It binds quarks and gluons, the smallest building blocks of matter, to nucleons and mesons, and these to nuclei. The forces are so strong, that quarks and gluons are not observed in isolation. This phenomenon is called quark confinement.

However, only after we have quantitatively understood QCD and found agreement with experiment, can we be sure that it is indeed the correct theory of matter. It is the biggest challenge in theoretical particle physics, to solve QCD from first principles and understand the mechanism of quark confinement.

The equations of this theory are so complicated, that they cannot be solved by traditional techniques. The modern way of tackling them is through the use of powerful computers. In order for the computer to solve the theory, one approximates space and time by a finite box divided into a discrete set of points, the lattice. This reduces the problem to a finite system of coupled equations, which can be solved by standard Monte Carlo techniques. Later on one may remove this approximation by letting the lattice spacing go to zero and the box size to infinity.

The precision of the calculation is largely a question of computing power. Previous simulations were mainly done in the so-called quenched approximation, in which contributions of virtual quark-antiquark pairs (also called

---

[*] For the **QCDSF** Collaboration: A. Ali Khan (HU Berlin), V. Bornyakov (IHEP Protvino), S. Capitani (DESY Zeuthen), M. Göckeler (Leipzig), T. Hemmert (TU München), R. Horsley (Edinburgh), B. Klaus (FU Berlin), W. Kürzinger (FU Berlin), H. Perlt (Leipzig), D. Petters (FU Berlin), P. Rakow (Regensburg), S. Schaefer (Regensburg), A. Schäfer (Regensburg), G. Schierholz (DESY Zeuthen), and A. Schiller (Leipzig).

dynamical quarks) are neglected. The reason is that the cost of computing grows inversely to some large power of the (virtual) quark mass m_q, while in the quenched approximation $m_q = \infty$. The physical up and down quark masses are of $O(10)$ MeV.

In this project we will overcome this approximation and perform a more realistic simulation of QCD, including the effect of two flavors ($N_f = 2$) of dynamical quarks. Dynamical quarks account for the pion cloud that surrounds the hadrons, and thus are expected to have a significant effect on masses, form factors and structure functions, to name only a few observables. Though one need not simulate at the physical quark mass, m_q must be small enough so that the lattice results can be safely extrapolated to the physical value.

2 Preliminaries

To reduce finite cut-off effects, and to facilitate the extrapolation to the continuum limit ($a \to 0$, a being the lattice spacing), we use improved Wilson fermions [1]:

$$S_F \to S_F - \frac{a}{4} c_{SW} \, g \sum_x \bar{\psi}(x) \sigma_{\mu\nu} F_{\mu\nu}(x) \psi(x), \tag{1}$$

which have discretization errors of $O(a^2)$ only. Likewise, we shall improve the operators:

$$\mathcal{O} \to (1 + c_0 \, am)\mathcal{O} + a \sum_{i \geq 1} c_i \, \mathcal{O}_i. \tag{2}$$

The price to pay is that the improvement coefficients c_{SW}, c_0, c_1, $\cdots$ have to be computed to non-perturbative precision.

The calculation proceeds in two steps. In step one we generate a statistical ensemble of gauge field configurations using a hybrid Monte Carlo algorithm. This calculation, which is the most (computer time) expensive part, is done on the Hitachi SR8000. Using seven CPUs (out of eight on a board) for calculation and one CPU for remote communication, we achieve a sustained speed of 600 Mflop/s per CPU. In step two we will then compute averages of correlation functions, operator matrix elements, etc. over these configurations to obtain physical answers. We have done simulations for a relatively wide range of quark masses and at several lattice spacings. In Fig. 1 we show our present (parameter) set of gauge field configurations. The lattice sizes are $16^3\,32$ at the heavier quark masses, and $24^3\,48$ at the lighter quark masses.

An important issue is renormalization. The lattice operators are, in general, divergent (in the limit $a \to 0$) and need to be normalized:

$$\mathcal{O}^S(\mu) = Z_{\mathcal{O}}^S \mathcal{O}(a), \quad S : \text{scheme}, \tag{3}$$

where S must match the normalization condition of the corresponding, perturbatively calculated Wilson coefficient. Usually this is the $\overline{MS}$ scheme. As

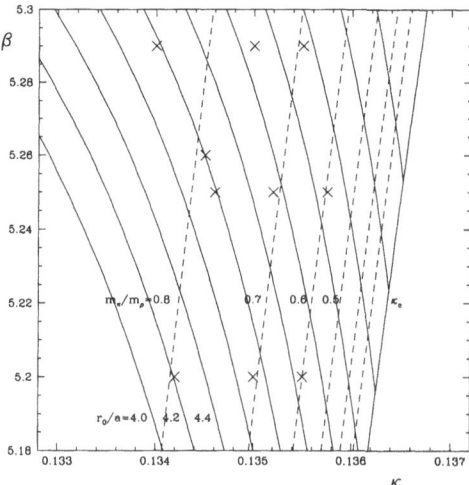

Fig. 1. The parameters of our simulation ($\times$) in the (κ, β) plane, together with lines of constant a (solid lines) and m_π/m_ρ (dashed lines). Also shown is the critical hopping parameter κ_c, corresponding to zero quark mass. The quark mass is given by (5), and $r_0 = 0.5$ fm.

long as there are at most logarithmic divergences present, a perturbative calculation of the renormalization constants is possible, and many results have been reported in recent years [2]. Because the coupling constant is large, perturbative results are not always reliable, and non-perturbative techniques have been developed [3, 4, 5].

In the following we will focus on the more phenomenological aspects of QCD. Due to space limitations, we will be able to highlight a few results only, which are of current interest.

3 Hadron and Quark Masses

The first goal is to compute the masses of mesons and baryons. This provides an important test of QCD and a benchmark calculation of lattice QCD. In Fig. 2 we show results for some of the low-lying meson and baryon masses. The errors are largely due to the extrapolation of our results to the physical up and down quark masses. In the near future we hope to be able to reduce the errors significantly by doing simulations at smaller quark masses.

Among the least known parameters of the Standard Model are the masses of the light (up and down) and strange quarks. The reason is that the connec-

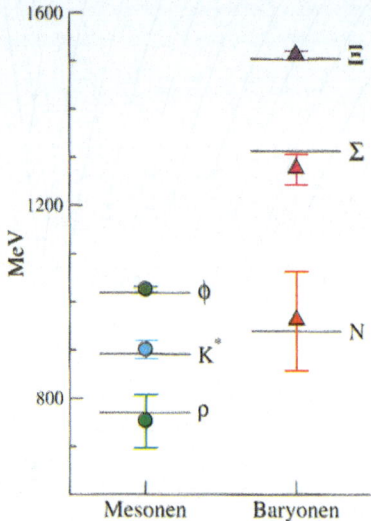

Fig. 2. The spectrum of hadrons, as predicted by QCD, compared with experiment, for a range of meson and baryon states. The experimental values are marked by horizontal lines. The pion and kaon masses were used to fix the light and strange quark masses.

tion between quark masses and observables like hadron masses is highly nonperturbative. Due to confinement, quarks are not eigenstates of the Hamiltonian and are not directly observable. A definition of the quark mass from experiment thus involves prescribing a measurement procedure. This amounts to assigning a renormalization scheme and a scale μ. Conventionally, quark masses are given in a mass independent renormalization scheme, such as $\overline{MS}$. We define the renormalized quark mass by [6,7]

$$m_q^{\overline{MS}} = \frac{(1 + b_A a m_q) Z_A}{(1 + b_P a m_q) Z_P^{\overline{MS}}(\mu)} \tilde{m}_q, \tag{4}$$

where m_q is the bare quark mass

$$a m_q = \frac{1}{2\kappa} - \frac{1}{2\kappa_c}, \tag{5}$$

and $\tilde{m}_q$ is the Ward identity quark mass

$$a \tilde{m}_q = \frac{\langle \partial_4 A_4(t) P(0) \rangle}{2 \langle P(t) P(0) \rangle}, \tag{6}$$

where A_μ is the (improved) partially conserved axial vector current and P the pseudoscalar density. The renormalization constants Z and improvement

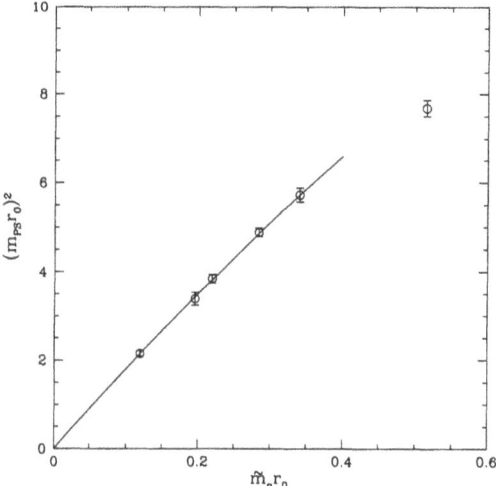

Fig. 3. The pseudoscalar mass m_{PS} against the quark mass $\tilde{m}_q$.

coefficients (b_P, b_A, $\cdots$) are computed in tadpole improved perturbation theory [2]. A fully non-perturbative calculation of the renormalization constants [5] is in progress. The Ward identity quark masses can be computed from the physical pion and kaon masses. In Fig. 3 we plot the pseudoscalar mass as a function of the Ward identity quark mass. This suggests to make the following ansatz:

$$m_\pi^2 = A(m_\ell)2\tilde{m}_\ell + B(m_\ell)(2\tilde{m}_\ell)^2,$$
$$m_K^2 = A(m_\ell)(\tilde{m}_\ell + \tilde{m}_s + B(m_\ell)(\tilde{m}_\ell + \tilde{m}_s)^2, \tag{7}$$

where $m_\ell = (m_u + m_d)/2$. From (7) we can compute $\tilde{m}_\ell$ and $\tilde{m}_s$, and from (4) we obtain the renormalized quark masses

$$m_\ell^{\overline{MS}}(2\,\text{GeV}) = 3.5(2)\,\text{MeV}$$
$$m_s^{\overline{MS}}(2\,\text{GeV}) = 90(5)\,\text{MeV}. \tag{8}$$

4 The scale parameter Λ

The Λ parameter sets the scale in QCD. In the chiral limit ($m_q \to 0$) it is the only parameter of the theory, and hence it is a quantity of fundamental interest. (Note that the real world is very close to the chiral limit.) It is defined by the running of the strong coupling constant α_s at high energies,

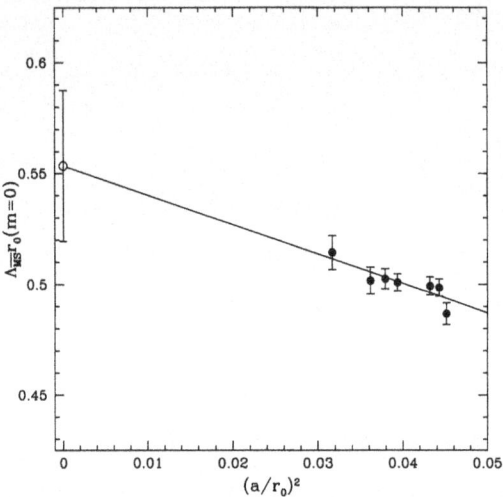

Fig. 4. The scale parameter $\Lambda_{\overline{MS}}$, extrapolated to the chiral limit, against a^2 for $a\mu = 2.5$. The solid line is a linear extrapolation to the continuum limit.

where non-perturbative effects are supposed to become small. The running coupling constant $g(\mu)$ (in the scheme $\mathcal{S}$) is given implicitly by

$$\frac{\mu}{\Lambda_{\mathcal{S}}} = \left(b_0 g_{\mathcal{S}}^2\right)^{\frac{b_1}{2b_0^2}} \exp\left(\frac{1}{2b_0 g_{\mathcal{S}}^2} + \int_0^{g_{\mathcal{S}}} d\xi \left(\frac{1}{\beta^{\mathcal{S}}(\xi)} + \frac{1}{b_0 \xi^3} - \frac{b_1}{b_0^2 \xi}\right)\right), \quad (9)$$

where $\Lambda_{\mathcal{S}}$ appears as an integration constant. To obtain Λ, we need to compute $g_{\mathcal{S}}(\mu)$ at some large scale μ. In the following we shall be interested in $\Lambda_{\overline{MS}}$.

We start from the boosted coupling

$$g_{\square}^2(a) = \frac{g^2(a)}{P}, \quad (10)$$

where P is the average plaquette value, rather than the bare lattice coupling $(g^2(a) = 6/\beta)$, because perturbative expansions in $g_{\square}$ converge more rapidly than expansions in g. The boosted coupling can be calculated easily and precisely. From $g_{\square}$ we compute $g_{\overline{MS}}$ using the three-loop formula [8]

$$\frac{1}{g_{\overline{MS}}^2(\mu)} = \frac{1}{g_{\square}^2(a)} + 2b_0 \ln a\mu - t_1^{\square} + \left(2b_1 \ln a\mu - t_2^{\square}\right) g_{\square}^2(a)$$

$$+ \left(-2b_0 b_1 \ln^2 a\mu + 2(b_2^{\overline{MS}} + b_1 t_1^{\square}) \ln a\mu - t_3^{\square} + p_1 t_2^{\square}\right) g_{\square}^4(a). \quad (11)$$

To convert the lattice spacing to a physical scale we use the force parameter r_0. Our result for $\Lambda_{\overline{MS}}$ is shown in Fig. 4. We have varied μ from 4 to 10

GeV and found practically no difference. The data still show a residual cut-off dependence, depite our using an improved action, and thus have to be extrapolated to the continuum limit. The final result is

$$\Lambda_{\overline{MS}}^{N_f=2} = 217(16)(11)\,\text{MeV}, \tag{12}$$

where the first error is purely statistical, while the second one is an estimate of the systematic uncertainties. In the quenched approximation we find

$$\Lambda_{\overline{MS}}^{N_f=0} = 243(1)(10)\,\text{MeV}. \tag{13}$$

5 Structure Functions

Understanding the structure of hadrons in terms of quark and gluon con-stituents, in particular how quarks and gluons provide the binding and spin of the nucleon, is one of the outstanding problems in particle physics. The nucleon has four structure functions, xF_1, F_2, g_1 and g_2. The operator prod-uct expansion (OPE) relates moments of them to nucleon matrix elements. For the unpolarized structure functions this reads

$$\begin{aligned}
2\int_0^1 dx\,x^{n-1}F_1(x,Q^2) &= c_{1,n}^{(2)}(Q^2,\mu)v_n^{(2)}(\mu) + O\!\left(\frac{1}{Q^2}\right), \\
\int_0^1 dx\,x^{n-2}F_2(x,Q^2) &= c_{2,n}^{(2)}(Q^2,\mu)v_n^{(2)}(\mu) + O\!\left(\frac{1}{Q^2}\right),
\end{aligned} \tag{14}$$

where the Wilson coefficients c are independent of the target, and $v_n^{(2)}$ are the leading, twist-two reduced matrix elements of the nucleon defined by

$$\begin{aligned}
\frac{1}{2}\sum_s \langle p,s|\mathcal{O}_{\{\mu_1\cdots\mu_n\}}|p,s\rangle &= 2v_n^{(2)}(p_{\mu_1}\cdots p_{\mu_n} - \text{traces}), \\
\mathcal{O}_{\mu_1\cdots\mu_n} &= \left(\frac{i}{2}\right)^{n-1}\bar\psi\gamma_{\mu_1}\overleftrightarrow{D}_{\mu_2}\cdots\overleftrightarrow{D}_{\mu_n}\psi - \text{traces},
\end{aligned}$$

and being renormalized at the scale μ. In parton model language

$$v_n^{(2)} =: \langle x^{n-1}\rangle = \int_0^1 dx\,x^{n-1}\Big(q_\uparrow(x,\mu) + q_\downarrow(x,\mu)\Big), \tag{15}$$

where $q_\uparrow(x,\mu)$ $(q_\downarrow(x,\mu))$, the parton distribution function, measures the probability of finding a quark q $(= u,\,d,\,\cdots)$ with fractional momentum x and helicity $+$ $(-)$ inside the nucleon. Similarly, for the polarized structure functions we have

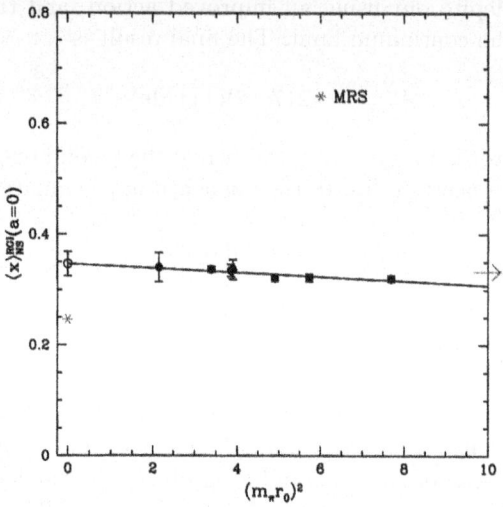

Fig. 5. The lowest non-singlet moment $\langle x \rangle$ against m_π^2, together with the experimental value (∗). The data have been converted to the renormalization group invariant (RGI) renormalization scheme.

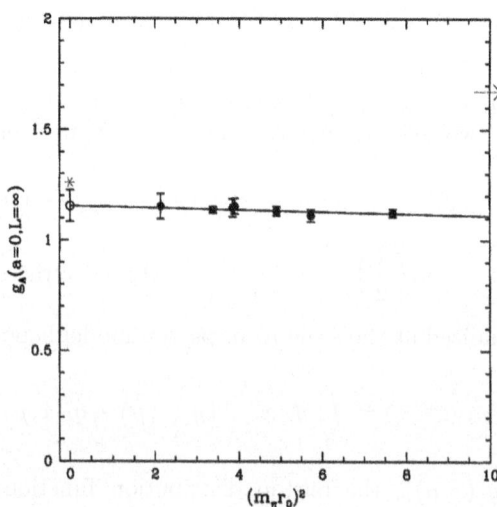

Fig. 6. The axial vector charge g_A against m_π^2, together with the experimental value (∗).

$$2 \int_0^1 dx\, x^n g_1(x, Q^2) = \frac{1}{2} e_{1,n}(Q^2, \mu) a_n(\mu) + O\Big(\frac{1}{Q^2}\Big),$$

$$2 \int_0^1 dx\, x^n g_2(x, Q^2) = \frac{n}{2(n+1)} \Big(e_{2,n}(Q^2, \mu) d_n(\mu) \tag{16}$$

$$- e_{1,n}(Q^2, \mu) a_n(\mu) \Big) + O\Big(\frac{1}{Q^2}\Big)$$

($n \geq 2$ for g_2), where

$$\langle p, s | \mathcal{O}^5_{\{\sigma\mu_1\cdots\mu_n\}} | p, s \rangle = \frac{1}{n+1} a_n (s_\sigma p_{\mu_1} \cdots p_{\mu_n} + \cdots - \text{traces}),$$

$$\langle p, s | \mathcal{O}^5_{[\sigma\{\mu_1]\cdots\mu_n\}} | p, s \rangle = \frac{1}{n+1} d_n \big((s_\sigma p_{\mu_1} - s_{\mu_1} p_\sigma) p_{\mu_2} \cdots p_{\mu_n} + \cdots - \text{traces} \big),$$

$$\mathcal{O}^5_{\sigma\mu_1\cdots\mu_n} = \Big(\frac{i}{2}\Big)^n \bar{\psi} \gamma_\sigma \gamma_5 \overleftrightarrow{D}_{\mu_1} \cdots \overleftrightarrow{D}_{\mu_n} \psi - \text{traces}.$$

In parton model language

$$\frac{1}{2} a_n =: \Delta^n q = \int_0^1 dx\, x^n \Big(q_\uparrow(x, \mu) - q_\downarrow(x, \mu) \Big), \tag{17}$$

where

$$\Delta u - \Delta d = g_A \tag{18}$$

($\Delta := \Delta^0$), while d_n has twist three and no parton model interpretation.

In Figs. 5 and 6 we show the lowest moments of the unpolarized and polarized parton distribution functions [9], respectively. In both cases we focus on the non-singlet part. The lines are linear extrapolations to the chiral limit. The arrow on the right-hand side of the figures points to the result in the non-relativistic limit. In the case of g_A we find good agreement with experiment, while a linear extrapolation of $\langle x \rangle$ to the physical quark mass overestimates the experimental result by approximately 40%. It has been argued [10] that the light pion cloud is not adequately represented by a linear (quadratic) extrapolation in the quark (pion) mass. Indeed, chiral perturbation theory predicts

$$\langle x^n \rangle_{NS} = A \Big(1 - \frac{3g_A^2 + 1}{(4\pi f_\pi)^2} m_\pi^2 \ln \Big(\frac{m_\pi^2}{m_\pi^2 + \Lambda^2} \Big) \Big) + \text{analytic terms}, \tag{19}$$

where Λ is a phenomenological parameter, which results in a large deviation from linearity at small m_π. To test that, we are currently doing simulations at pion masses of 300 MeV (corresponding to quark masses of ≈ 40 MeV) and less in quenched QCD. In Fig. 7 we show some very preliminary results. It appears that the data indeed bend down towards the experimental value. The curve shows a fit of (19) to the lattice data, which reproduces the experimental result. If true, this means that one will have to do simulations at pion masses well below 300 MeV in order to determine the parameters of the non-linear extrapolation reasonably well.

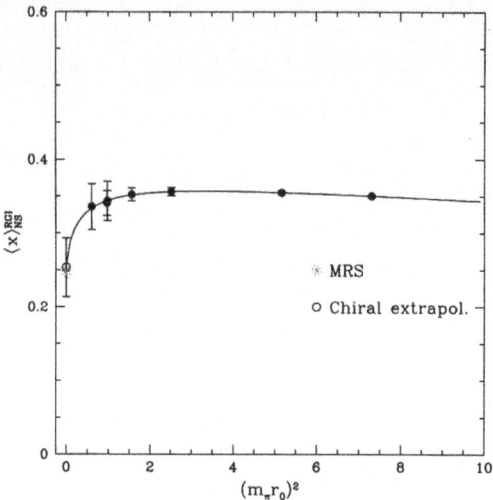

Fig. 7. The moment $\langle x \rangle$ at $\beta = 6.0$ in quenched QCD. The two leftmost data points with larger error bars are obtained on $32^3 \, 48$ lattices, the other points on $24^3 \, 48$ and $16^3 \, 32$ lattices.

6 Conclusions

Serious computations including the effects of dynamical quarks have only just begun. We have obtained results for a host of other observables which we could not show here. By and large we find good agreement with experiment, except for the moments of the unpolarized structure functions. The message here is that important physics is being omitted by a naive extrapolation to the physical quark mass. In the future we will gradually push our simulations towards pion masses of the order of twice the physical pion mass.

We were not able to discuss the more fundamental question of quark confinement. This is a lively subject, and we have seen already that dynamical quarks have a significant effect on the dynamics of the QCD vacuum [11].

Acknowledgement. We like to thank the Leibniz-Rechenzentrum and its committees for their generous award of computer time on the Hitachi SR8000, without which the calculations reported here would not have been possible. We also thank the supporting staff, in particular Dr. M. Brehm and Dr. H.-D. Steinhöfer, for scientific advice and administrative help.

References

1. B. Sheikholeslami and R. Wohlert, Nucl. Phys. B259 (1985) 572.
2. S. Capitani et al., Nucl. Phys. B593 (2001) 183.
3. K. Jansen et al., Phys. Lett. B372 (1996) 275.
4. G. Martinelli et al., Nucl. Phys. B445 (1995) 81.
5. M. Göckeler et al., Nucl. Phys. B544 (1999) 699.
6. M. Göckeler et al., Phys. Rev. D62 (2000) 054504.
7. D. Pleiter, Nucl. Phys. (Proc. Suppl.) 94 (2001) 265; M. Göckeler et al. [QCDSF Collaboration], in preparation.
8. S. Booth et al., Phys. Lett. B519 (2001) 229, Nucl. Phys. (Proc. Suppl.) 106 (2002) 308.
9. S. Capitani et al., Nucl. Phys. (Proc. Suppl.) 106 (2002) 299; M. Göckeler et al. [QCDSF Collaboration], in preparation.
10. W. Detmold et al., Phys. Rev. Lett. 87 (2001) 172001; D. Arndt and M. Savage, Nucl. Phys. A697 (2002) 429; J.-W. Chen and X. Ji, Phys. Lett. B523 (2001) 107.
11. V. Bornyakov et al., Nucl. Phys. (Proc. Suppl.) 106 (2002) 634; R. Horsley et al., Nucl. Phys. (Proc. Suppl.) 106 (2002) 569; V. Bornyakov et al. [QCDSF Collaboration], in preparation.

Quantum Chromodynamics with Chiral Quarks

Christof Gattringer[1], Meinulf Göckeler[2][1], Peter Hasenfratz[3], Simon Hauswirth[3], Kieran Holland[4], Thomas Jörg[3], Christian B. Lang[5], Ferenc Niedermayer[3], Paul E.L. Rakow[1], Stefan Schaefer[1], and Andreas Schäfer[1]

[1] Institut für Theoretische Physik, Universität Regensburg
    93040 Regensburg, Germany
[2] Institut für Theoretische Physik, Universität Leipzig
    04109 Leipzig, Germany
[3] Institut für Theoretische Physik, Universität Bern
    3012 Bern, Switzerland
[4] Department of Physics, University of California at San Diego
    San Diego, USA
[5] Institut für Theoretische Physik, Universität Graz
    8010 Graz, Austria

Abstract. Quantum-Chromodynamics (QCD) is the theory of quarks, gluons and their interaction. It has an important almost exact symmetry, the so-called chiral symmetry. This symmetry plays a major role in all low-energy hadronic processes. For traditional formulations of lattice QCD, CPU-time and memory limitations prevent simulations with light quark masses and this symmetry is seriously violated. During the last few years successful implementations of the chiral symmetry for lattice QCD have been constructed. The new chiral formulations should, among other advantages, allow simulations at much smaller quark masses, thus greatly improving the predictive power. We use two implementations to study the relevance of chiral symmetry for hadron masses and confirm these expectations. [1]

1 Introduction

Beyond presenting some of our results and discussing the methods we used, the aim of this contribution is to communicate the fascination and excitement amongst practitioners of computer aided particle physics. This excitement is induced by the fact that, for the first time ever, this community has controlled results on fundamental problems which were completely beyond reach before.

[1] Supported by the Austrian Academy of Sciences (APART 654), the DFG (Forschergruppe 'Gitter-Hadronen-Phänomenologie'), the European Community's Human Potential Programme under HPRN-CT-2000-00145 Hadrons/Lattice QCD, BBW Nr. 99.0143, the Schweizerischer Nationalfonds and the US Department of Energy under grant DOE-FG03-97ER40546.

The story of research of fundamental physics in the 20th century has a circular character: objects which looked like elementary building blocks turned out to be composite having an underlying structure. This quest for elementarity led from atoms to nuclei, from nuclei to protons and neutrons, which then were found to be composed of unexpected new objects called quarks.

And the quarks? There are indications that nature prepared here something new for us. First, no substructure of quarks has been observed in experiments yet. Second, no experiment could break up the proton, or other hadrons made of quarks into its constituents. It seems, quarks are permanently confined in the hadrons.

The physics of hadrons (their masses, interactions and decays), the 'strong interaction', is described by a special quantum field theory called quantum chromodynamics (QCD). It is believed that QCD describes the strong interaction exactly. Any deviation of the theoretical predictions from experiments would indicate, therefore, new unknown fundamental physics.

The phenomena described by QCD belong to the difficult problems in physics. Such problems are characterized by having a large (in principle infinite) number of strongly interacting relevant degrees of freedom and by involving simultaneously large and small distance scales. Traditional methods, like perturbation theory, break down. Introducing a space-time grid QCD can be brought in a form ('lattice QCD') where these non-perturbative problems can be rigorously formulated and can be answered by stochastic numerical methods.

Lattice QCD is well suited for parallel computing. The two main reasons are that the interactions are local and homogenous since there are no external forces in a fundamental theory. Lattice QCD practitioners have been active in the development of dedicated parallel computers and were among the most efficient users of commercial machines.

The structure of QCD is similar to that of quantum electrodynamics where quarks and gluons (which communicate the interaction between quarks) play the role of electrons and photons, respectively. The physics, however (as confinement illustrates) is very different. Symmetries play an important role in understanding how QCD works. Chiral symmetry, which transforms left and right handed quarks (i.e. quarks for which the spin and momentum point in the opposite or same direction) independently, becomes an exact symmetry of QCD in the limit of massless quarks. In nature, this symmetry is only slightly broken explicitly by the small quark masses. Its spontaneous breaking by dynamical effects produces very characteristic features in hadron spectroscopy like the existence of massless Goldstone bosons.

It has been understood only recently how to define QCD on a lattice without destroying this important symmetry. In QCD the quark-gluon interaction is coded by the Dirac operator D which is a large complex matrix on the lattice whose matrix elements depend on the gluon field. Chiral symmetry is preserved if this matrix satisfies a non-linear relation (Ginsparg-Wilson relation)

$$\gamma_5 \, D \; + \; D \, \gamma_5 \; = \; 2 \, a \, D \, \gamma_5 \, R \, D \, , \tag{1}$$

where γ_5 is a Dirac matrix, a denotes the lattice spacing and R is a local operator. All the solutions of this equation have a complicated structure whose effective handling in numerical simulations is a demanding problem. However, for the reliable extrapolation to realistically small quark masses, which is highly non-trivial for many observables [1] it seems to be very advantageous. For certain Goldstone boson related problems chiral symmetry on the lattice is even mandatory.

In this work we pursue the construction of two different Dirac operators [2, 3] which satisfy eq.(1) to a very good approximation although not exactly (Section 2). These two realizations are based on two different general approaches. The fixed-point action is expected to be closely scale invariant (i.e. independent of the lattice spacing a). We either use it as it is or augment it by a small number of iteration steps towards an exact solution of the Ginsparg-Wilson equation. The second approach pursues a direct solution of an approximate Ginsparg-Wilson equation. Both approaches have characteristic features and in order to compare them we use the two Dirac operators in parallel. In Section 3 we discuss how to handle the corresponding large sparse matrices effectively on a parallel computer and the performance in particularly on the Hitachi SR8000-F1. In Section 4 some of our physics results and the conclusions are presented.

2 Construction of the Dirac operator

As discussed above, the Ginsparg-Wilson equation (1) is the key to chiral symmetry on the lattice. In this section we present two Dirac operators which are approximate solutions of (1). They are known as the fixed-point (FP) [2] and chirally improved (CI) [3] Dirac operator. For both methods the quality of the approximation can be improved by adding more terms to the parameterized Dirac operator. There exists also a systematic procedure (overlap [4]) which can be used to decrease the remaining chiral symmetry violation to any desired level.

2.1 Systematic expansion of a general Dirac operator

For both the FP and the CI operator the first step of the construction is an expansion of the most general Dirac operator on the lattice which is compatible with all the symmetries in a series of simple operators [3, 5].

A general Dirac operator is a large matrix of the form $D(x, y)_{\alpha d, \alpha' d'}$ where $x, y = 1, \ldots, V$ ($V = N_s^3 N_t =$ lattice volume, the number of sites), $\alpha, \alpha' = 1, 2, 3$ (color index), $d, d' = 1, .., 4$ (Dirac index). Due to the locality of the interaction D is a sparse matrix: the relative distances $|x - y|$ are restricted, in practice to the hypercube.

Since the action has to be gauge invariant we write the Dirac operator in the general form

$$D(x,y)_{\alpha d, \alpha' d'} = \sum_{a=1}^{16} (\Gamma^a)_{dd'} \, (U_p(x, x + r_p))_{\alpha\alpha'} \, c_{ap}\delta(x + r_p, y) \,. \qquad (2)$$

Here Γ^a are the generators of the Clifford algebra (4×4 matrices), p denotes a path with offset r_p, i.e. connecting the sites x and $x + r_p$. The 3×3 ("color") matrix U_p is a product of gauge links along the path p. The coefficients c_{ap} are gauge invariant functions of the gauge fields. The invariance under the discrete symmetries (cubic symmetries, charge conjugation) restricts the coefficients c_{ap} for paths related to each other by these transformations to be equal (up to a sign). The freedom left in the choice of c_{ap} for independent paths has been used to satisfy approximately the GW relation (1). For the FP Dirac operator the coefficients c_{ap} are real polynomials of closed gauge loops, while for the chirally improved operator they are real constants. The two approaches differ by the methods used to determine the coefficients.

2.2 The fixed-point Dirac operator

For the FP Dirac operator the basic idea is to determine the coefficients by systematically integrating out degrees of freedom. The corresponding renormalization group (RG) transformation maps the lattice action given on a fine lattice with the lattice spacing a to an action on a coarser lattice with lattice spacing $a' = 2a$. Repeating the procedure many times the action converges to a fixed-point action [2]. The FP action has many attractive features, in particular it satisfies the GW relation (1) with some gauge field dependent R.

The FP equation for the fermionic action is solved approximately with an ansatz containing paths on the hypercube (i.e. with $3^4 = 81$ different offsets). Altogether the parametrized FP Dirac operator used in the simulations contains 82 parameters optimized to satisfy the FP equations as well as possible within the given ansatz. Consequently, the violation of the GW relation will be minimized.

2.3 Chirally improved fermions

For the chirally improved fermions the method for determining the coefficients c_{ap} in (2) is different [3,6]. Here the expanded Dirac operator (2) is inserted into the Ginsparg-Wilson equation (1) with a trivial matrix $2R = \mathbf{1}$ on the right-hand side. The left-hand side of (1) is evaluated easily since it only amounts to a commutator of the Clifford algebra elements Γ_α with γ_5. The evaluation of the right-hand side proceeds in two steps: Firstly, the product of the two involved Γ_α and the γ_5 has to be computed. Since the Clifford algebra is closed this is just another Clifford element Γ_β. The second, more demanding step is the multiplication of the paths p. However, for this operation a simple algebraic representation was found which allows a numerical evaluation of this product [6].

Once the two sides of (1) are evaluated for the expanded Dirac operator (2), one finds that both sides are again series expansions in terms of simple operators. It can be shown that the terms are linearly independent and thus the coefficients in front of the left-hand side contributions have to equal the coefficients on the right-hand side. The former are linear combinations of the c_{ap} while the coefficients on the right-hand side are quadratic in the c_{ap}. Thus one can read off a system of coupled quadratic equations for the coefficients c_{ap}. This system of equations is equivalent to the Ginsparg-Wilson equation.

In a practical application one has to truncate the expansion (2). Here we use only terms on the hypercube plus one extra term of length $\sqrt{5}$. The parametrization of the chirally improved Dirac operator is described by altogether 19 real coefficients c_{ap}. After truncation the corresponding system of quadratic equations plus boundary conditions, ensuring the correct limit for free fermions, can be solved numerically and the resulting coefficients c_{ap} give rise to an approximate solution of (1).

3 QCD on a parallel computer

Lattice QCD simulations of realistic problems such as the scattering of two pions or the decay $K \rightarrow \pi\pi$, turn out to be numerically enormously demanding. This is for several reasons: First, the spatial lattice volume has to be large enough for the wave functions of all involved particles to fit into the box of side length $L_s = N_s a$. The temporal size $L_t = N_t a$ also has to be large, as one is mainly interested in the asymptotic behavior of correlation functions at large Euclidean time. Second, in order to get rid of discretization errors, the continuum limit $a \rightarrow 0$ has to be taken by performing simulations at several lattice spacings a and extrapolating the measured observables to the continuum. The lattice spacings a have to be small enough to ensure a controlled extrapolation, and therefore N_s and N_t get large quickly. Third, the computational effort grows with inverse powers of the quark mass. For this reason, simulations are commonly performed at values larger than the tiny mass of the physical up and down quarks, and another extrapolation from results calculated at several higher quark masses is necessary to obtain physical results.

3.1 Remarks on the implementation

In QCD spectroscopy without dynamical fermions (quenched approximation), most of the CPU time is spent on the calculation of the quark propagator. This requires a (partial) inversion of the Dirac operator, which is a large sparse complex matrix of rank $12V$ on a lattice with volume $V = N_s^3 \times N_t$. Such a matrix inversion can be done very efficiently on massively parallel computers, since it boils down to complex matrix-vector multiplication, vector addition and dot product operations, requiring minimal communication when distributed among several nodes.

The FP and chirally improved Dirac operators are constructed from the gauge fields in a complicated manner (compare (2)), and therefore they have to be precalculated and stored. To minimize memory usage, the matrix is stored in two arrays: A `double complex` array containing the non-zero elements and an `integer` array from which the column index of each matrix element can be determined. Since every matrix row contains the same number of non-zero elements ($\mathcal{O}(1000)$), the row index can be reconstructed from the ordering of the array elements. The matrix R which is non-trivial for the FP operator is handled analogously to D.

The key to efficient lattice simulations is a fast matrix inversion algorithm. As the rank of the Dirac operator is far too large to perform an exact inversion, iterative methods are used, which are commonly variants of Krylov subspace solvers. A significant computational gain can be obtained using multi-mass solvers [7], which return the solutions of the shifted linear system $(D + m)x = b$ for a whole set of values of m at the cost of only one inversion. As a consequence, in QCD the cost of inverting the Dirac operator at a range of quark masses is equivalent to the cost of a single inversion at the smallest quark mass. For our spectroscopy calculations, we worked with the multi-mass BiCGstab algorithm described in [8].

Table 1. Memory requirements for storing the D and R operators, a gauge configuration U and a vector b of size $12V$ at different lattice volumes $V = N_s^3 \times N_t$.

array	$8^3 \times 24$	$12^3 \times 24$	$16^3 \times 24$
D	2.1 GB	7.2 GB	22.8 GB
R	0.15 GB	0.5 GB	1.5 GB
U	6.8 MB	23 MB	72 MB
b	2.3 MB	7.6 MB	24 MB

For all but the smallest lattices, parallel code is required. The memory requirements in Table 1 for various arrays at the lattice volumes used in the simulations show that the shared memory available on most machines is exceeded even at moderate lattice sizes. Fortunately, explicit parallelization of the quark propagator code for n nodes is easily implemented by splitting

Table 2. Time in milliseconds and overhead factor for dot product of two double complex vectors of size $12V$ at different levels of MPI parallelization on the Hitachi SR8000-F1.

# nodes	$8^3 \times 24$		$12^3 \times 24$		$16^3 \times 24$	
	t_{dot}[ms]	ω	t_{dot}[ms]	ω	t_{dot}[ms]	ω
1	0.33	1.0	0.89	1.0	2.60	1.00
2	0.31	1.9	0.59	1.3	1.43	1.10
4	0.27	3.3	0.41	1.8	0.82	1.26
8	0.25	6.1	0.34	3.1	0.54	1.66
16	0.39	19	0.44	7.9	0.44	2.71

Table 3. Time in seconds and overhead factor for one iteration of the matrix inversion algorithm as a function of lattice size and parallelization level for the Dirac operator.

# nodes	$8^3 \times 24$		$12^3 \times 24$		$16^3 \times 32$	
	t_{iter}[s]	ω	t_{iter}[s]	ω	t_{iter}[s]	ω
1	0.564	1.00				
2	0.313	1.11	1.045	1.00		
4	0.185	1.31	0.619	1.18		
8	0.122	1.73	0.405	1.55	1.30	1.00
16					1.05	1.62

the Dirac operator into n parts, each containing $12V/n$ rows. Vectors of size $12V$ are analogously distributed. Vector and matrix operations are then performed in parallel, and communication is only needed for gathering the result vectors.

3.2 Performance

In this section we present some benchmark results for the parallelization of our code and its overall performance on the Hitachi SR8000-F1. To quantify the parallelization overhead, we list in Tables 2, 3 for several computational tasks the wall-clock time and the overhead factor $\omega = nt_n/t_{\text{ref}}$, where t_n is the wall-clock time for the task running on n nodes and t_{ref} is the wall-clock time for the smallest n on which the task could be run due to memory limits. As shown in Table 2, the dot product of two vectors does not really profit from parallelization on the smallest lattice. The situation changes drastically on the larger lattices, where distribution of the vectors is essential.

The crucial quantity in our simulations is the time for one iteration of the matrix inverter, listed in Table 3. This is by far the most time-consuming task, and here the parallelization overhead is considerable. Comparing the results on the different lattice sizes running on 8 nodes shows that the time increases in proportion to the volume. Hence it seems that the performance does not profit anymore from longer loops as they appear for larger volumes.

Table 4. Overall performance measurements for typical quark propagator runs. Shown are the percentage of the overall time spent for the D and R multiplication routines including communication of the resulting vector and the overall speed in MFLOPS per node with theoretical peak speed of 12 GFLOPS. The overall run time is measured from program start to finish and thus includes I/O time, MPI initialization and finalization and precalculation of D and R operators.

# nodes	$8^3 \times 24$			$12^3 \times 24$			$16^3 \times 32$		
	$t_D[\%]$	$t_R[\%]$	MFLOPS	$t_D[\%]$	$t_R[\%]$	MFLOPS	$t_D[\%]$	$t_R[\%]$	MFLOPS
1	56	16	4200						
2	55	18	3800	60	19	4160			
4	53	22	3240	59	23	3570			
8	51	27	2560	57	29	2850	56	29	2770
16							51	34	1700

The overall performance of the quark propagator code is given in Table 4. These values depend on the number of iterations required for the inversion, and thus on the smallest quark mass in a given run. The bottom line is that our code is reasonably efficient, running at an overall rate of around 30% of peak performance in production runs. This number includes the precalculation of the D and R operators and the time for I/O and MPI setup which decrease overall performance. Neglecting the communication, the matrix-vector multiplication runs at 6.3 GFLOPS per node for the Dirac operator and at 8.6 GFLOPS per node for the R operator, which is remarkably fast.

4 Results and Conclusions

While the last sections presented some technical details of our calculation in this section we want to discuss our results in a rather general manner, aiming at the non-expert. Nowadays available computer resources do not allow high precision measurements when including dynamical (sea-) quarks; therefore our study is in the quenched approximation.

For conventional lattice implementations of QCD which strongly violate chiral symmetry on the lattice, the artifacts generated have to be corrected by the non-trivial simultaneous extrapolations to infinitely fine lattices ($a \to 0$), large physical volume and nearly vanishing quark masses. These extrapolations introduce substantial systematic errors. Lattice implementations with exact or nearly-exact chiral symmetry promise to significantly reduce these problems. We have calculated standard QCD quantities, which are well studied with conventional lattice implementations, to test whether this expectation is indeed justified. We want to limit our discussion to only two sets of results, namely the value of several hadron masses as a function of the bare quark mass m_0 and the so-called Edinburgh-plot relating two hadronic mass ratios. See Figs. 1 and 2. The message from these figures and our other results

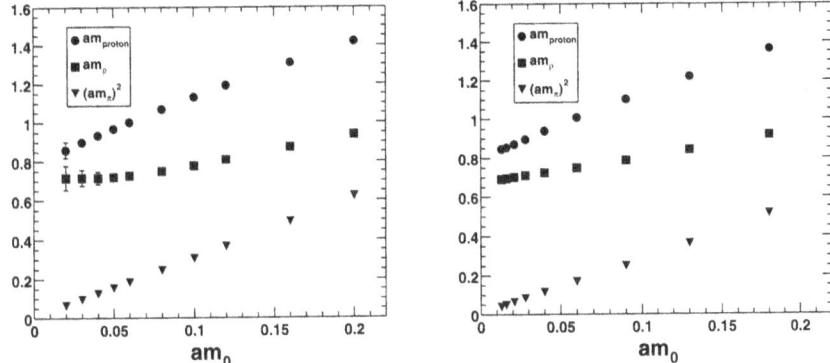

Fig. 1. Hadron masses as a function of the bare quark mass m_0 (everything in lattice units). **Left:** CI fermions; the pion mass for the left-most point is approximately 280 MeV. **Right:** FP fermions. The pion mass for the left-most point is approximately 230 MeV.

can be summarized as follows:

- Comparing results obtained at fixed physical volume but different lattice spacings we find agreement within the statistical errors for most cases studied, i.e. the discretization errors are smaller than our statistical accuracy. This implies that they are substantially smaller than e.g. for Clover-improved Wilson fermions.
- Finite-size effects should be of the same size for a given physical volume for all lattice implementations (up to discretization errors). In fact the observed finite-size effects for our two approaches coincide within our statistical accuracy. We see no finite size effects for box sizes larger than 1.9 fm.
- Within our statistical accuracy our result for the m_p/m_ρ mass ratio extrapolated to the physical pion mass agrees with the experimental value. So the error introduced by the quenched approximation seems to be smaller than our statistical precision for this quantity.
- Most importantly: The smallest pion mass we reach is substantially smaller than for conventional implementations even if the latter use much more computing power. In the mass range of Fig. 1 all the matrix inversions (propagators) converged, i.e. we had no "exceptional configurations". In conventional approaches such configurations occur which have to be treated in a somewhat dubious manner. The improvement is related to the strongly reduced fluctuations of the smallest eigenvalues of the Dirac operator.

A pion mass of $\sim$ 230 MeV overlaps with the region where a powerful analytic technique, chiral perturbation theory, can be used to continue the numerical results to the physical pion mass of $\sim$ 130 MeV. This was one of the missing conditions to make the promise of lattice QCD come true, which is

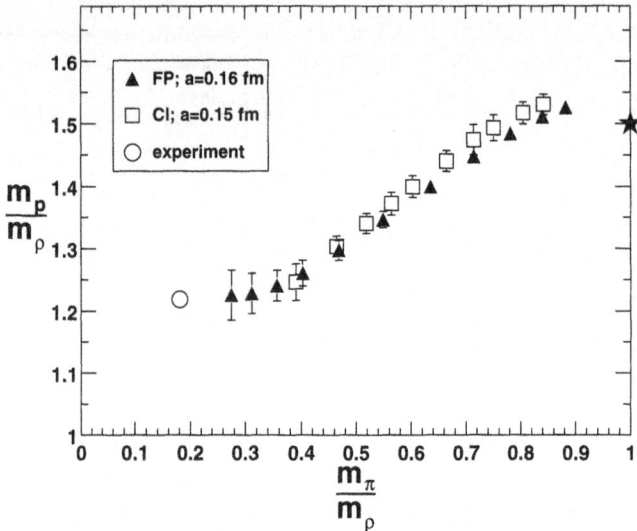

Fig. 2. Results for the physical mass ratios proton mass over ρ mass and pion mass over ρ mass. Even with much higher computing power no calculation with Wilson fermions has up to now reached values of m_π/m_ρ below 0.4.

to give reliable, parameter-free ab-initio QCD-predictions for a large number of measurable hadron properties. Our results suggest that the use of chirally improved lattice implementations could indeed be the best avenue to reach this goal.

References

1. W. Detmold, W. Melnitchouk, and A.W. Thomas, hep-lat/0206001 and references therein.
2. P. Hasenfratz, F. Niedermayer, Nucl. Phys. B 414 (1994) 785; S. Hauswirth, *Light hadron spectroscopy in quenched lattice QCD with chiral fixed-point fermions*, Thesis, Bern University 2002, arXiv:hep-lat/0204015; T. Jörg, *Chiral measurements in quenched lattice QCD with fixed-point fermions*, Thesis, Bern University 2002.
3. C. Gattringer, Phys. Rev. D 63 (2001) 114501.
4. R. Narayanan and H. Neuberger, Phys. Lett. B 302 (1993) 62, Nucl. Phys. B 443 (1995) 305.
5. P. Hasenfratz, S. Hauswirth, K. Holland, Th. Jörg, F. Niedermayer, U. Wenger, Int. J. Mod. Phys. C 12 (2001) 691.
6. C. Gattringer, I. Hip, C.B. Lang, Nucl. Phys. B597 (2001) 451.
7. A. Frommer, B. Nöckel, S. Güsken, T. Lippert and K. Schilling, Int. J. Mod. Phys. C **6** (1995) 627.
8. B. Jegerlehner, arXiv:hep-lat/9612014.

Three-Nucleon Force in the ^{4}He Scattering System

Hartmut M. Hofmann[1] and Gerald M. Hale[2]

[1] Institut für Theoretische Physik III
  Universität Erlangen–Nürnberg
  91058 Erlangen, Germany
  hmh@theorie3.physik.uni-erlangen.de
[2] Theoretical Division Los Alamos National Laboratory
  Los Alamos N.M. 87544, USA
  ghale@lanl.gov

Abstract. We report on a consistent, microscopic calculation of the bound and scattering states in the ^{4}He system employing modern realistic two-nucleon and three-nucleon potentials in the framework of the resonating group model (RGM). We present for comparison with these microscopic RGM calculations the results from a charge-independent, Coulomb-corrected R-matrix analysis of all types of data for reactions in the $A = 4$ system. Comparisons are made for selected examples of phase shifts and measurements from reactions sensitive to three-nucleon force effects.

Introduction

The ^{4}He atomic nucleus is one of the best studied few-body systems, both experimentally and theoretically, as summarized in the recent $A = 4$ compilation [1]. Besides the many textbook examples of gross structure, there are subtle points yielding large effects that are only qualitatively understood. Except for [2] none of the existing calculations aims at a complete understanding of the many features of ^{4}He, which is not surprising in view of the number of different phenomena studied so far [1]. With the recent compilation [1] and the comprehensive R-matrix analysis [3] of a large amount of scattering data below $E_x = 30$ MeV, a new, microscopic calculation for the ^{4}He system in this energy range using modern realistic two- and three-nucleon forces is most desirable.

It is well known that realistic nucleon-nucleon (NN) forces cannot reproduce the ^{3}H, ^{3}He, and ^{4}He binding energies. Three-nucleon interactions (TNIs) are added to give the necessary small corrections but they still fail to reproduce certain properties of the three nucleon system, most notably the A_y analyzing power in Nd scattering [4]. Yet the 30% deviation of A_y can be resolved by tiny changes in the Nd scattering phase shifts (on the order of 0.1 degrees [5,6,7]). Furthermore very many operators can contribute to a

TNI and the lack of stringent conditions in the three-nucleon system on the structure of the TNI makes its application to other systems desirable. In [2] it was shown that although a realistic NN force can generally reproduce the ^{4}He system, there remain differences, most notably in the analyzing powers. Since the intensely studied ^{4}He system [1] is unfortunately very difficult to describe due to the many resonances and the ^{4}He bound state, the much simpler systems $p-^3$He and $n-^3$H where data exist in the energy range of interest were investigated in [8].

The essential findings of this work are that realistic NN interactions describe most of the phase shifts quite well but fail to reproduce the $^3P_2$ and $^3P_0$ phase shifts. The calculated splitting between these two channels is much too small, and neither the Urbana IX (UIX) [9] nor the TLA [10] three-nucleon force is able to improve the splitting significantly. In fact, there it is more important to include in the calculation negative parity states of the three-nucleon subsystem than one of these two TNIs. These findings suggest that new contributions to the NNN force acting on the P-waves should be considered, like an LS type TNI, as proposed in [11] for the $N-d$ analyzing powers, or the V_3^* operators proposed in [12]. Based on these findings we choose as NN force only the AV18 [13] and as TNI the Urbana IX [9] and in addition the V_3^* [12].

We organize the paper in the following way. The next section contains a brief discussion of the Resonating Group Model calculation together with the model spaces used. Then we compare R-matrix and RGM results of a few typical examples of scattering phase shifts for various model spaces and combinations of interactions. Finally we compare with data for examples sensitive to the TNIs.

RGM and model space

We use the Resonating Group Model [14, 15, 16] to compute the scattering in the ^{4}He system using the Kohn-Hulthén variational principle [17]. The main technical problem is the evaluation of the many-body matrix elements in coordinate space. The restriction to a Gaussian basis for the radial dependencies of the wave function allows for a fast and efficient calculation of the individual matrix elements [14, 16]. However, to use these techniques the potentials must also be given in terms of Gaussians. In this work we use suitably parametrized versions of the Argonne AV18 [13] NN potential and the Urbana IX [9] and the V_3^* [12]. The inclusion of an additional TNI requires almost two orders of magnitude more computing power than the realistic NN forces alone.

In the ^{4}He system we use a model space with six two-fragment channels, namely the $p-^3$H, the $n-^3$He, the ^{2}H$-^2$H, the singlet deuteron and deuteron d- ^{2}H, the d- d, and the $(nn)-(pp)$ channels. The last three are an approximation to the three- and four-body breakup channels that cannot in practice be treated within the RGM. The ^{4}He is treated as four clusters in the frame-work of the RGM to allow for the required internal orbital angular momenta of ^{3}He, ^{3}H or ^{2}H.

For the scattering calculation we include the S, P and D wave contri-butions to the $J^\pi = 0^+, 1^+, 2^+, 0^-, 1^-$ and 2^- channels. From the R-matrix analysis these channels are known to give essentially the experimental data. (We discuss cases where this is not the case.) The full wave functions for these channels contain over 200 different spin and orbital angular momentum con-figurations, hence it is too complicated to be given in detail. The simplest wave functions we use for ^{3}He are those described in [8].

This small 29-dimensional model space yields -6.37 MeV binding energy, an rms radius of 1.78 fm and a D state probability of 7.7% for the ^{3}He using AV18. In order to avoid fake effects the relative thresholds in ^{4}He should be reproduced well, therefore we used also a 35-dimensional modelspace, called large,by allowing additional configurations with two orbital angular momenta on the two Jacoby coordinates yielding -6.69 MeV binding energy. This must be compared to -6.92MeV known from Faddeev calculations [18]. For the deuteron we use the structure given in [2], yielding a binding energy of -1.921 MeV, which could be easily improved. But then the relative threshold energies deteriorate, see Table 1. All the Gaussian width parameters were obtained by a non-linear optimization using a genetic algorithm [19] for the combination AV18 and UIX.

Once the fragment wave functions are fixed the scattering problem is solved with our RGM code relying on the Kohn-Hulthén variational principle [17]:

$$\delta\left(\langle \Psi_t | H - E | \Psi_t \rangle - \frac{1}{2} a_{ll}\right) = 0,$$

where a_{ij} denotes the reactance matrix.

The model space described above (consisting of four to ten physical scat-tering channels for each J^π) is by no means sufficient to find reasonable re-sults. So-called distortion or pseudo-inelastic channels [16] have to be added to improve the description of the wave function within the interaction region. Accordingly, the distortion channels have no asymptotic part.

For practical purposes it is obvious to reuse some of the already calculated matrix elements as additional distortion channels. In that way we include all the positive parity states of the three-nucleon subsystems with $J_3^\pi \leq 5/2^+$ in our calculation. However, it was recently pointed out by A. Fonseca [20] that states having a negative parity J_3^- in the three-nucleon fragment increase the $n-^3$H cross section noteably. Therefore we also added the appropriate dis-tortion channels in a similar complexity as in the J_3^+ case to our calculation, thereby roughly doubling the size of the model space. The 20 percent increase of the model space for the 3N bound states from 29 to 35 resulted in almost a factor of two in the computing time. The parameters of the V_3^* TNI were adjusted that the binding energy of triton and ^{3}He did change by less than 10 keV. Therefore we do not give the corresponding energies in Table 1.

Table 1. Comparison of experimental and calculated total binding energies and relative thresholds (in MeV) for the various potential models used

| potential | E_{bin} | | E_{thres} | |
	^{3}H	^{3}He	^{3}He $- p$	$d - d$
av18	-7.068	-6.370	0.698	3.227
av18, large	-7.413	-6.588	0.725	3.572
av18 + UIX	-7.586	-6.875	0.710	3.745
av18 + UIX,large	-8.241	-7.493	0.748	4.400
exp.	-8.481	-7.718	0.763	4.033

Results

Since we are mainly interested in the effects of 3N-forces, we mention the bound state results only briefly. In the large model space AV18 plus UIX yields -27.81 MeV close to the experimental -28.296 MeV, to which the parameters of the UIX are fitted to. Although the parameters of the V_3^* TNI were chosen as to give only minute changes in the three-nucleon system, for ^{4}He it resulted in 650 keV additional binding.

The most detailed comparison between calculation and data is on the level of an energy dependent phase-shift analysis. This is given by the R-matrix analysis as described in [2] in detail. The lowest channel, triton-proton contains the intriguing first exited state of ^{4}He, a 0^+-resonance, sometimes considered a breathing mode, which is clearly seen in the 0^+ phase shift, see Fig. 1, but does not show-up in the angular distributions, see Fig. 3. Neglecting the Coulomb force this resonance is moved just below the p-triton threshold, i.e. it becomes a bound state, bound by less than 50 keV depending on the force and model space used. Therefore all approaches, which neglect the Coulomb force like [20] cannot aim at this energy region. In Fig. 1 the R-matrix results are compared to the pure NN-calculations for various model spaces. For the small model space the calculation is slightly above the R-matrix data. Adding the negative parity distortion channels, we find a small increase due to the enlarged attraction. This effect is much smaller than the one found in triton-neutron scattering [20,8]. Increasing the 3N-model space reduces the phase shifts considerably, due to the better ^{4}He binding and the additional thresholds shifted to higher energy, see Table 1. In Fig. 2 we find strong sensitivity to the additional TNI. This sensitivity in specific partial waves might help to unravel the operator structure of the TNI, especially as all the thresholds are unchanged from UIX to V_3^*.

For 58 and 120 degrees exist measured exitation functions for the p-triton differential cross section. For the forward angle the R-matrix analysis is on top of the data [21], the pure NN-calculation a bit below, with UIX almost on top of the data and with V_3^* a bit above. Since there are only minor differencies, we do not show them. For the backward angle,however, see Fig. 3, even the

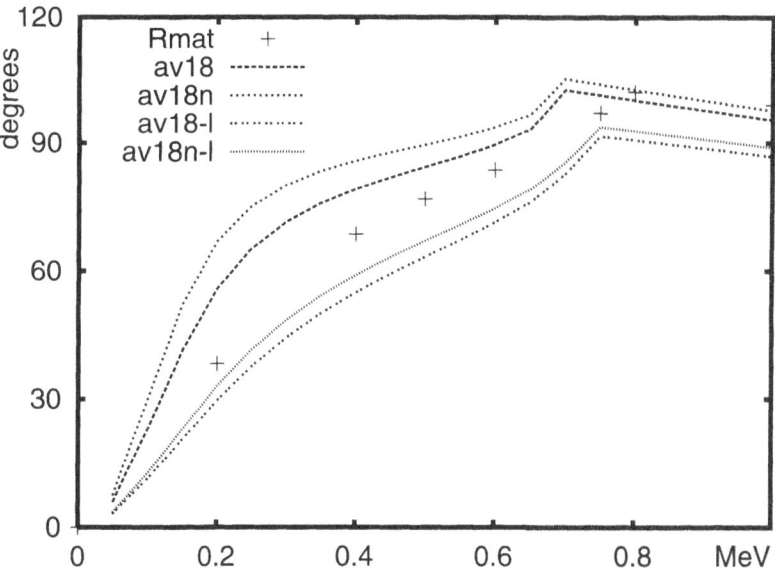

Fig. 1. Comparison of the 0^+ triton-proton phase shifts from the R-matrix analysis (crosses) and calculations employing AV18 in the small model space (av18), adding negative parity distortion channels (av18n), for the large model space (av18-l) and adding negative parity distortion channels (av18n-l)

R-matrix analysis cannot fully reproduce the data in the Coulomb-nuclear interference region. The data from [23] seem to be consistently above those from [21] and [22]. In the Coulomb-nuclear interference the sensitivity to TNI is very large, whereas around the (p,n) threshold data, R-matrix and microscopic calculations agree essentially. Since all the data are very old and not consistent a new measurement is urgently called for. Unfortunately due to radiation hazards of the triton this is not very likely.

A recent measurement of the real parts of the neutron-^{3}He spin-dependent scattering lengths [24] $a_0 = 7.370(58)$ fm and $a_1 = 3.278(53)$ fm is therefore very important. The corresponding R-matrix results are $a_0 = 7.398$ fm and $a_1 = 3.257$ fm. Due to the strong coupling via the 0^+ resonance a_0 has a large imaginary part. Hence, the numerical extraction of this scattering length is not easy. Therefore we give only preliminary numbers in the following. For the large model space the microscopic calculations yield $a_0 = 7.402 \div 7.590$ fm and $a_1 = 3.289 \div 3.424$ fm, depending on the combination of forces. These results are close enough to be used for a further determination of the operator structure of the three-nucleon force.

Out of the many possible data we choose an example, which demonstrates the limitation of our total partial wave model space. The well studied deuteron-deuteron reactions allow a detailed comparison since together

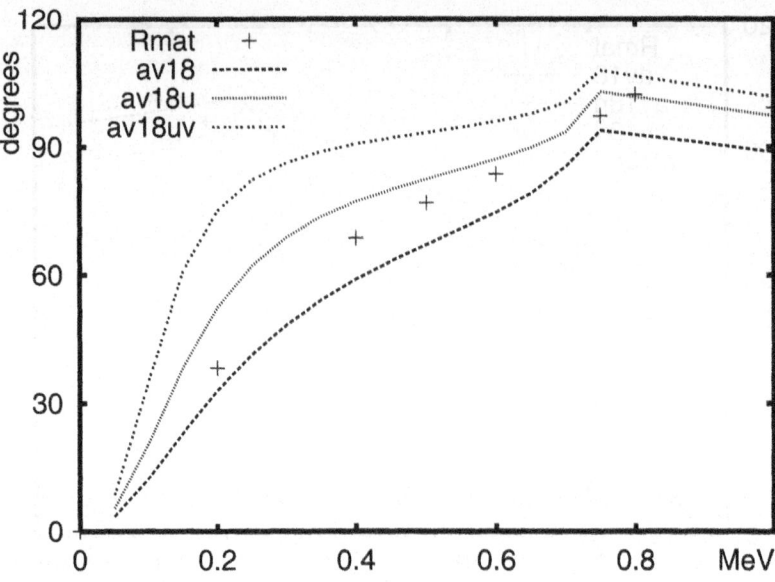

Fig. 2. As Fig. 1, but R-matrix results (crosses) are compared to the full NN-calculation (av18), adding UIX (av18u) and adding V_3^* (av18uv)

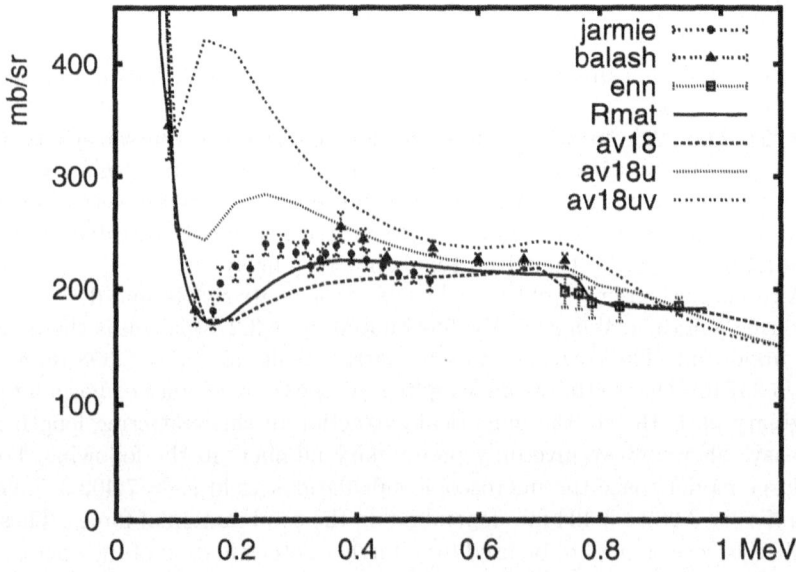

Fig. 3. Differential triton-proton cross section at 120 degrees as function of energy. The data are from Jarmie [21], Balashko [23], and Ennis [22]. The other lines are as in Fig. 2.

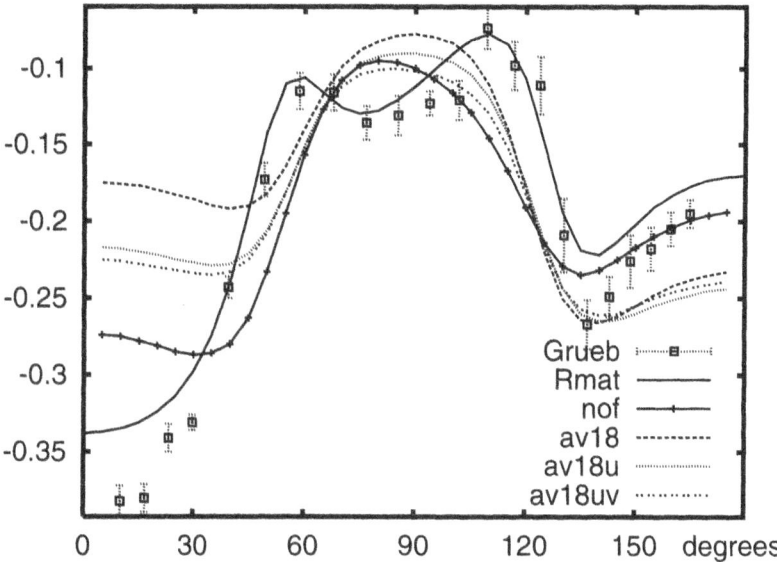

Fig. 4. Tensor analysing power T_{20} for the reaction ^{2}H(d,p)^{3}H at 2.0 MeV center-of-mass energy. The data are from [25], the lines are as in Fig. 2 and additionally the R-matrix results without F-waves (nof).

with differential cross sections for many energies also analysing powers are available. The tensor analysing power T_{20} displays a pronounced angular distribution, see Fig. 4. The full R-matrix analysis nicely reproduces the data, whereas all microscopic calculations fail at forward angles and show no sign of the double-hump structure. Omitting all F-wave S-matrix elements in the R-matrix analysis corresponding to the model space used for the RGM calculation yields an angular dependence quite similar to the microscopic results, see Fig. 4. This demonstrates the need for $L_{rel} = 3$ partial waves in the RGM calculations.

Conclusion

We presented a complete microscopic calculation in the ^{4}He system employing modern realistic two- and three-nucleon forces. We demonstrated that in specific examples the inclusion of NNN-forces yields large effects in phase shifts, differential cross sections and analysing powers. Hence, the ^{4}He-system seems well suited for a detailed study of different NNN-forces, especially since a comprehensive R-matrix analysis exists, which reproduces the vast amount of data for various reactions very well, thus allowing for a comparison on the level of individual partial waves. Therefore a determination of the operator structure of the NNN-force is within reach, provided the microscopic calculations are converged. To aim at this goal the internal triton and ^{3}He wave functions have to be improved, such that the binding energy is within say 50 keV of the experimental value. This can only be achieved by a major increase of the model space. For the deuteron the corresponding modification is trivial. To describe the deuteron-deuteron reactions, we have to allow also for F-waves on all relative coordinates. All these improvements are relative straightforward, but require about a factor 5 in the CPU-time compared to the needs of the work reported here.

Acknowledgement. The work of H.M.H is supported by the BMBF (contract 06ER926) and that of G.M.H. by the Department of Energy. The grant of computer time at the HLRB and the RRZE is gratefully acknowledged. We want to thank G. Wellein and G. Hager at the RRZE for their help.

References

1. Tilley, D.R., Weller, H.R., Hale, G.M. (1992): Energy levels of light nuclei A = 4. Nucl. Phys. **A541** 1–104
2. Hofmann, H.M., Hale, G.M. (1997): Microscopic calculation of the ^{4}He system. Nucl. Phys. **A613** 69–106
3. Hale,G.M., Dodder,D.C., and Witte,K. (unpublished)
4. Witala, H., Hüber, D., Glöckle, W. (1994): Analyzing power puzzle in low energy elastic Nd scattering: Phys. Rev **C49** R14–R16
5. Knutson, L. D., Lamm, L. O., McAninch, J. E. (1993): Determination of the Phase Shifts for p–d Elastic Scattering at E_p = 3 MeV. Phys. Rev. Lett. **71** 3762–3765
6. Kievsky, A., Rosati, S., Tornow, W., Viviani, M. (1996): Critical comparison of experimental data and theoretical predictions for N-d scattering below the breakup threshold. Nucl. Phys. **A607** 402–424
7. Kievsky, A., Tornow, W. (1999): Proton-Deuteron Phase-Shift Analysis above the Deuteron Breakup Threshold. TUNL Progress Report - XXXVIII, Durham, USA
8. Pfitzinger, B., Hofmann, H. M., Hale, G. M. (2001): Elastic p–^{3}He and n–^{3}H scattering with two- and three-body forces. Phys. Rev. **C64** 044003–044008
9. Pudliner, B. S., Pandharipande, V. R., Carlson, J., Pieper, S. C., Wiringa, R. B. (1997): Quantum Monte Carlo calculations of nuclei with $A \leq 7$. Phys. Rev. **C56** 1720–1750
10. Hüber, D., Friar, J. L., Nogga, A., Witala, H., van Kolck, U. (1999): New Three-Nucleon-Force Terms in the Three-Nucleon System. nucl-th/9910034
11. Kievsky, A. (1999): Phenomenological spin-orbit three-body force. Phys. Rev. **C60** 034001–034008
12. Canton, L., Schadow, W. (2000): Why is the three-nucleon force so odd? Phys. Rev. **C62** 044005–044013
13. Wiringa, R. B., Stokes, V. G. J., Schiavilla, R. (1995): Accurate nucleon-nucleon potential with charge-independence breaking. Phys. Rev. **C51** 38–51

14. Hofmann, H. M. (1987): Resonating Group Calculations in Light Nuclear Systems. In: Ferreira, L. S., Fonseca, A. C., Streit, L. (ed) Proceedings of Models and Methods in Few-Body Physics, Lisboa, Portugal 1986. Springer Lecture Notes in Physics **273** 243–282

15. Wildermuth. K., Tang, Y. C. (1977): A Unified Theory of the Nucleus. Vieweg, Braunschweig

16. Tang, Y. C. (1981): Topics in Nuclear Physics. Lecture Notes in Physics **145** Springer, Heidelberg

17. Kohn, W. (1948): Variational methods in nuclear collision problems. Phys. Rev. **74** 1763–1772

18. Nogga, A., Kamada. H., Glöckle, W. (2000): Modern nuclear-force predictions for the α-particle. Phys. Rev. Lett **85** 944–947

19. Winkler, C., Hofmann, H. M. (1997): Determination of bound state wavefunctions by a genetic algorithm. Phys. Rev. **C55** 684–687

20. Fonseca, A. C. (1999): Contribution of Nucleon-Nucleon P Waves to nt-nt, dd -pt and dd-dd Scattering Observables. Phys. Rev. Lett. **83** 4021–4042

21. Jarmie, N., Allen, R. C. (1959): T(p,p)T scattering near the T(p,n)^{3}He threshold. Phys. Rev. **114** 176-178

22. Ennis, M. E., Hemmendinger, A. (1954): Small angle cross section for the scattering of protons by tritons. Phys. Rev. **95** 772–775

23. Balashko, Y. G., Barit, I. Y., Dulcova, L. S., Kurepin, A. B. (1964): Elastic scattering of protons by tritons at energies below the threshold of the (p,n) reaction JETP (Soviet Physics) **19** 1281-1283

24. Zimmer, O., Ehlers, G., Farago, B., Humblot, H., Ketter, W., Scherm, R. (2002): A precise measurement of the spin-dependent neutron scattering length of ^{3}He. EPJdirect **A1** 1-28

25. Grüebler, W., König, V., Risler, R., Schmelzbach, P.A., White, R.E., Marmier, P. (1972): Nucl. Phys. **A193** 149–154

Simulations of the Local Universe

Stefan Gottlöber[1], Anatoly Klypin[2], Andrey Kravtsov[3], Yehuda Hoffman[4], and Andreas Faltenbacher[1]

[1] Astrophysikalisches Institut Potsdam
   An der Sternwarte 16, 14482 Potsdam, Germany
   sgottloeber@aip.de
[2] Astronomy Department
   New Mexico State University, USA
   aklypin@nmsu.edu
[3] Dept. of Astronomy & Astrophysics, CfCP
   The University of Chicago, USA
[4] Racah Institute of Physics
   Hebrew University, Jerusalem, Israel

1 Introduction

One of the greatest challenges of modern astrophysics is understanding how galaxies, such as our Milky Way, form within the framework of the Big Bang cosmology. The current theory of structure formation, the extension of the Big Bang model called the Cold Dark Matter (CDM) scenario, predicts that galaxies form within extended massive dark matter halos built from smaller pieces that collided and merged, resulting in the hierarchy of galaxies, groups, and clusters observed today. The entire sequence of events is thought to be seeded by quantum fluctuations in the very early Universe and governed by mysterious "dark matter" which constitutes about 85% of all matter in the universe. Although the accurate properties of galaxies depend on complicated baryonic processes (radiative cooling, formation and evolution of stars, etc.) operating on small scales, we expect that overall spatial distribution of dark matter halos is closely related to the observed galaxy distribution. Here we present numerical simulations designed to study the formation, evolution and present day properties of such dark matter halos in different cosmological environments.

In all simulations the spatially flat cold dark matter model with a cosmological constant (ΛCDM with $\Omega_{\mathrm{M}} = 0.3$, $\Omega_\Lambda = 0.7$, $\sigma_8 = 0.9$, and $h = 0.7$), favored by most current observations, has been assumed.

2 Numerical simulations

The Adaptive Refinement Tree (ART) N-body code [8,9] was used to run all numerical simulation analyzed in this paper. In some of the simulations described below the code also included eulerian gasdynamics [10]. The code uses Adaptive Mesh Refinement technique to achieve high resolution in the regions of interests. The computational box is covered with a uniform grid which defines the lowest (zeroth) level of resolution. The code then reaches high force resolution by recursively refining all high density regions using an automated refinement algorithm. This creates an hierarchy of refinement meshes of different resolutions, sizes, and geometries covering regions of interest. The refinement data structures and algorithms [4] allow individual cubic cells to be refined. The shape of the refinement mesh can thus effectively match the geometry of the region of interest. This algorithm is well suited for simulations of a selected region within a large computational box, as in the simulations presented below.

During the integration, spatial refinement is accompanied by temporal refinement. Namely, each level of refinement, l, is integrated with its own time step $\Delta a_l = \Delta a_0/2^l$, where Δa_0 is the global time step of the zeroth refinement level. This variable time stepping is very important for accuracy of the results. As the force resolution increases, more steps are needed to integrate the trajectories accurately. In addition to spatial and temporal refinement, simulations described below also use non-adaptive mass refinement to increase the mass (and correspondingly the force) resolution inside a specific region. The multiple mass resolution is implemented in the following way [5]. We first set up a realization of the initial spectrum of perturbations in such a way that initial conditions for a large number (1024^3) of particles can be generated in the simulation box. Initial coordinates and velocities of the particles are then calculated using all waves ranging from the fundamental mode $k = 2\pi/L$ to the Nyquist frequency $k = 2\pi/L \times N^{1/3}/2$, where L is the box size and N is the number of particles in the simulation. Particles outside high-resolution regions are then merged into particles of larger mass and this process can be repeated for merged particles. The larger mass (merged) particle is assigned a velocity and displacement equal to the average velocity and displacement of the smaller-mass particles. Extensive tests of the code and comparisons with other numerical N-body codes can be found in [9, 7].

With increasing number of particles and resolution (i.e., number of refinement cells) the memory as well as the computing time requirement of the code increase. In practice, on nodes with 16 Gb of shared memory and up to 16 CPUs we are limited to simulations with $\leq 256^3$ particles (runtime is CPU-bound). The only way to overcome this problem is to use MPI to distribute computations accross nodes. We have developed two different MPI algorithms to handle larger simulations. In the first approach, we select spatially distinct objects of interest and simulate each of these objects on one node with high mass and force resolution whereas in the remaining part of the simulation box lower resolutions will be used. Each node uses then standard ART with OpenMP on shared memory. The inter-node communication is minimal. In the second approach, one can divide the whole simulation box into a number of sub-boxes which will be handled by different nodes. Each node simulates its own sub-box with high mass and force resolution whereas other sub-boxes have lower mass and force resolution. Again one integration step is done by standard ART with OpenMP on each node. After each integration step nodes exchange communications on positions, velocities and masses of particles that cross sub-box boundaries.

2.1 Constrained simulations of the local universe

Here our goal is to perform simulations that match the observed local universe as well as possible. Namely, we are interested in reproducing the observed structures: the Virgo cluster, the Local Supercluster and the Local Group, in the approximately correct locations and embedded within the observed large-scale configuration dominated by the Great Attractor and Perseus-Pisces superclusters.

An efficient algorithm for reconstructing the density and velocity fields from sparse and noisy data of redshift and velocity surveys is provided by the Wiener filter formalism and constrained realizations of gaussian random fields [14]. Here we use the MARK III catalog [13]. The sample consists of ≈ 3400 galaxies and provides radial velocities and inferred distances with fractional errors $\sim 17 - 21\%$. A detailed analysis of the large scale structure in the Local Universe reconstructed from the MARK III survey was presented in [15].

Several simulations with increasing force and mass resolution in the region around the Virgo Cluster were performed [6, 10]. The initial conditions for these simulations were set using multiple mass resolution technique. Using $z = 0$ output of a low-resolution run, we selected all particles within a sphere of $25h^{-1}$Mpc radius centered on the Virgo cluster. The mass resolution in the Lagrangian region occupied by the selected particles was increased and additional small-scale waves from the initial ΛCDM power spectrum of perturbations were added appropriately [5]. For the two high-resolution simulations, the particle mass in the Local Supercluster region is 8 and 64 times smaller than in the low-resolution simulation. The highest resolution simulation has a particle mass of $2.5 \times 10^9 h^{-1} M_\odot$ and the maximum formal force resolution was $2.4h^{-1}$kpc in the Local Supercluster region. The results of both high-resolution simulations agree well with each other at all resolved scales.

All major structures (the Local Supercluster, Great Attractor, Perseus-Pisces supercluster, and Coma cluster) observed within $100h^{-1}$Mpc around the Milky Way exist in the simulations. The positions and morphology of these structures is, of course, fairly well dictated by the constraints imposed on the initial conditions. The Local Supercluster is an elongated structure which extends over $\sim 40h^{-1}$Mpc along the SGX axis. There is a low-density "bridge" (of overdensity just above the average density), which connects the Local Supercluster with the Perseus-Pisces Super-cluster. There is also an even weaker filament connecting the Local Super-cluster with the Great Attractor.

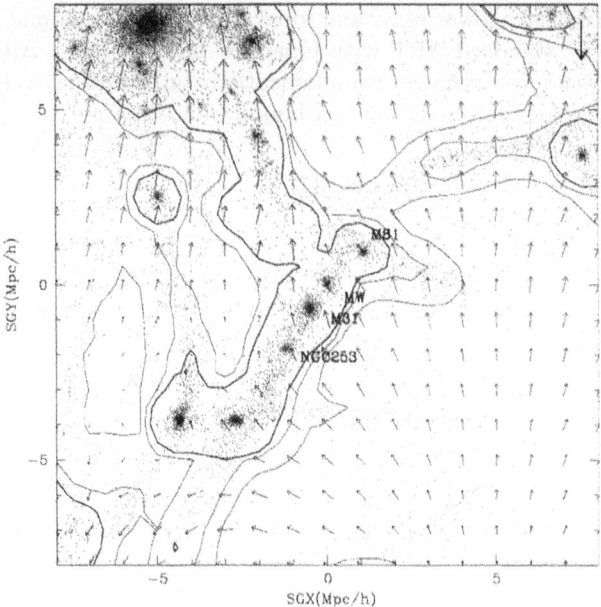

Fig. 1. Density (contours corresponding to overdensities of 1, 2, and 3) and velocity (arrows) fields smoothed with a Gaussian filter of $0.7h^{-1}$Mpc smoothing length around the Local Group. The slice shown has a size and thickness of $15h^{-1}$Mpc and $5h^{-1}$Mpc, respectively, and is centered on the supergalactic plane ($SGZ = 0$). The length of the thick arrow in the top right corner corresponds to a velocity of 500 km/s. The velocities are plotted in the Virgo cluster rest frame.

Just as in the real Universe, the Local Group is located in a weak filament extending between the Virgo and Fornax clusters. This filament is a counterpart of the Coma-Sculptor "cloud" in the distribution of nearby galaxies. Figure 1 shows a zoom-in view of the immediate environment of the simulated Local Group. Note that the structures at these scales are only weakly affected by constraints imposed on the initial conditions. Several possible counterparts to existing objects (e.g., the MW and M31, M51, NGC253) are marked, but their existence is largely fortuitous. As can be seen in this figure, the simulated Local Group is located in a rather weak filament extending to the Virgo cluster. This filament borders an underdense region visible in the right lower corner of Fig. 1, which corresponds to the Local Void in the observed distribution of nearby galaxies. We are now carrying out a series of high-resolution simulations of such voids (see below). The velocity field around the Local Group is rather quiet, in good agreement with observations. The peculiar velocity field in the Local Void exhibits a uniform expansion of matter out of this underdense region, while velocities between the Local Group and the Virgo (upper half of Fig. 1) show a coherent flow onto the main body of the Local Supercluster.

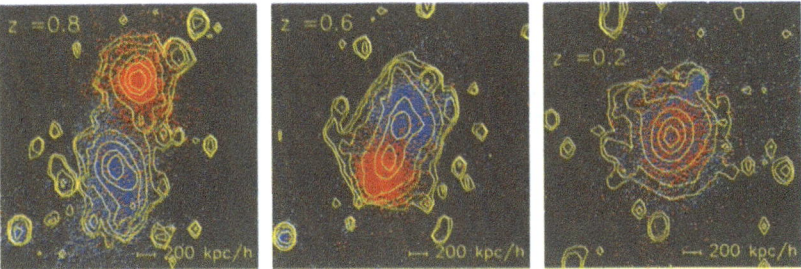

Fig. 2. Merger event between two 6×10^{13} M$_\odot$ clusters. Particles belonging to one of the two clusters before merging are plotted with red and blue dots, respectively. Yellow lines correspond to density contours.

2.2 Galaxy clusters

Within a low mass resolution simulation (128^3 particles, $m_{part} = 2.0 \times 10^{10}h^{-1}M_\odot$) we have identified 15 clusters of galaxies with masses above $1.0 \times 10^{14}h^{-1}M_\odot$. From this set we have selected 8 candidates with different masses and merging histories and added 5 smaller clusters/groups for load balance. We then re-simulated the clusters with higher mass resolution. With particle masses of $3.2 \times 10^8 h^{-1}M_\odot$ a typical cluster and its environment contains more than 1 million particles. The formal force resolution with 9 refinement levels was $1h^{-1}$kpc. Halos with masses above $3.0 \times 10^{10}h^{-1}M_\odot$ are well resolved. A typical cluster contains more than 150 such halos. The simulations were done using an MPI version of the ART code where each of 8 nodes followed the evolution of one or two clusters. We are now carrying out simulations of these clusters including gasdynamics.

The dynamical evolution of a typical major merger between galaxy clusters can be seen in Fig. 2. The figure shows three snapshots of a cluster taken at $z = 0.8$, $z = 0.6$, and $z = 0.2$. Although the cluster does not show significant substructure in its density contours during the merger, it is far from a relaxed state. Most of the energy transfer takes place during the first encounter (between $z = 0.8$ and $z = 0.6$), but the merger remnant does not reach virial equilibrium until it loses all the information about the initial conditions (i.e. the particles are well mixed). For this event, virialisation occurs at $z \sim 0.2$ (~ 4 Gyr after the first encounter). During the first stages of cluster formation, the characteristic time between major mergers can be shorter than the relaxation time. High-redshift clusters may thus be in general far from the virial equilibrium inside their formal virial radius.

2.3 Voids

Cosmological simulations predict many more small DM halos than the observed number of satellites around the Milky Way and Andromeda galaxies.

Do we have the same problem for dwarf galaxies in voids? One naively expects a large number because the Press-Schechter mass function steeply rises with declining mass. In contrast, it seems that observations are failing to find a substantial number of dwarf galaxies inside voids (e.g. [11,3]). However, the situation is complicated because it is very difficult to detect dwarf galaxies, many of them are expected to have low surface brightness.

In order to study the formation of large voids, the simulation box should be sufficiently large; we use a cube of size $80h^{-1}$Mpc. On the other hand, we are interested in the formation of small structure elements inside voids, for which we need highest possible mass resolution. Therefore, we use particles with large masses to follow the evolution of large scale structures and particles with small masses to follow the evolution of structure within one spherical void.

Within a low mass resolution simulation (128^3 particles, $m_{part} = 2.0 \times 10^{10}h^{-1}M_\odot$) we have identified 8387 galactic halos with masses $> 2.0 \times 10^{11}h^{-1}M_\odot$. This corresponds to a mean distance of about $4h^{-1}$Mpc between halos. We then searched voids in the distribution of these galactic halos by constructing the minimal spanning tree using halo positions. The minimal spanning tree was then used to search for the point in the simulation box which has the largest distance r_1 to the set of halos. We identified this point as the center of spherical void with the radius of r_1. Excluding that void we were searching again for the point with the largest distance to the set and thus found the second largest void and so on. The algorithm is similar to that used by [1].

With the algorithm described above we find spherical voids in the halo distribution which do not contain any halo with a mass larger than $2.0 \times 10^{11}h^{-1}M_\odot$; such halos by definition lie on the border of the void. After finding voids in the low-resolution simulation, we re-run the simulation with much higher mass resolution inside the voids ($m_{\text{part}} = 4.0 \times 10^7h^{-1}M_\odot$), which allows identification of objects with masses larger than $10^9h^{-1}M_\odot$. By construction, the voids do not have halos with masses larger than $M_b = 2.0 \times 10^{11}h^{-1}M_\odot$. Five voids have been identified and resimulated, their radii are $r_{void} = 11.6, 10.8, 9.4, 9.1, 9.1$ h^{-1}Mpc. Inside the voids the matter density is typically a factor of 10 smaller than the mean density, but void properties exhibit large differences. Some voids are very isolated: the density within a sphere with radius 30 h^{-1}Mpc centered on one of the void centers reaches only 2/3 of the mean density. On the other hand, the most prominent structure of the simulation, a galaxy cluster of $2 \times 10^{15}h^{-1}$Mpc, is bordering one of the other voids. The density outside of this void rapidly increases. The mass function in voids is about an order of magnitude lower and its shape is different than that of the field galaxies [2].

In Fig. 3 we show the inner 20 h^{-1}Mpc of a spherical void of radius 21.6 h^{-1}Mpc. In this void we found more than 50 halos with circular velocity $v_c > 50$ km/s and more than 600 halos with $v_c > 20$ km/s. There is a certain

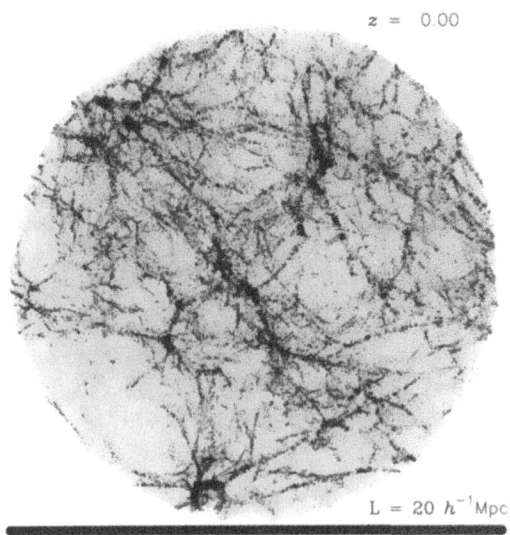

$z = 0.00$

$L = 20 \; h^{-1} Mpc$

Fig. 3. The inner region of a spherical void at $z = 0$.

spatial mass segregation among halos. Typically, more massive halos tend to be situated in the outer part of the void. The largest halos in the plot are actually dwarf-size halos with circular velocities of ~ 50km/s. The void is far from being empty and boring: it has a complex structure with numerous long filaments and small sub-voids. Visually it resembles the large-scale structure of the Universe, but everything in this plot is hundreds and thousands times less massive than in "normal" configurations of interconnected superclusters and filaments.

2.4 Hydro simulations

We have started to repeat [10] many of the simulations described above including eulerian shock-capturing gasdynamics and physical processes such as radiative cooling and stellar feedback. Inclusion of gasdynamics, although significantly increasing computational and memory demands of simulations, will allow us to address a much wider range of questions and more robust comparison with observations. For example, simulations of the Local Supercluster regions with cooling and starformation will allow us to study distribution and properties of galaxies in the simulations as a function of their luminosity and color. High-resolution simulations of the local voids will allow studies of spatial distribution of Ly α absorbers and their connection to galaxies. Simulations of nearby galaxy clusters (e.g., Virgo, Fornax, Coma) formed in realistics large-scale environments will allow for detailed object-to-object

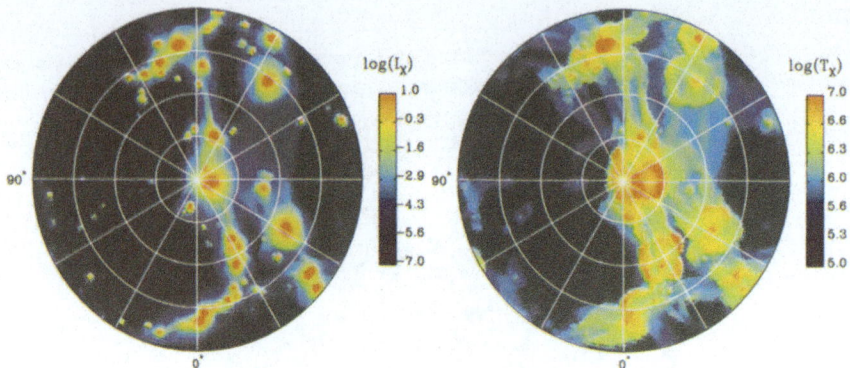

Fig. 4. The projected X-ray intensity and emission-weighted temperature maps of the Local supercluster in galactic coordinates. The Virgo cluster is the brightest object near the center (which corresponds to the North Galactic Pole).

comparison of properties and should give us good insights into physical processes operating within intracluster medium.

As an example, we present here simulations of the Virgo cluster in the context of constrained simulations described in Sect. 2.1. The lagrangian region within five virial radii (at $z = 0$) around the cluster was resimulated with mass resolution 8 times higher (particle mass of $3.1 \times 10^8 h^{-1} M_\odot$) than that of the Local Supercluster simulation and spatial resolution of $\approx 1 h^{-1}$kpc in the central regions of the cluster. Figure 4 shows the sky projection of the X-ray intensity and emission-weighted temperature of gas in the simulations in galactic coordinates (the center of the polar projection corresponds to the North Galactic Pole, NGP). The Virgo cluster is the brightest object located near the NGP, very close to the actual location of the real Virgo cluster. Figure 5 shows radial profiles of dark matter and gas density, as well as profiles of gas temperature and entropy for two simulations: one that included only adiabatic gasdynamics, and the other in which gas was preheated with the energy of 1.5 keV per gas particle at $z = 3$. The preheating very roughly models the possible effect of galactic winds at high-redshifts. The figure shows that preheating increases entropy of the gas thereby lowering its density in the cluster core and increasing the overall gas temperature. It lowers the X-ray luminosity of the cluster by a factor of 8, which brings it in good agreement with the observed luminosity-temperature relation. Note, however, that it does not change the shape of the temperature profile. The proximity of the Virgo cluster makes it one of the best spatially resolved clusters, with the temperature mesurements well outside the virial radius. Figure 6 compares the observed temperature measurements around the center of the Virgo cluster [12] as a function of projected radius to the corresponding measurements in simulations (adiabatic and preheating). We constructed projected map of emission-weighted temperature of the simulated Virgo cluster using $4' \times 4'$

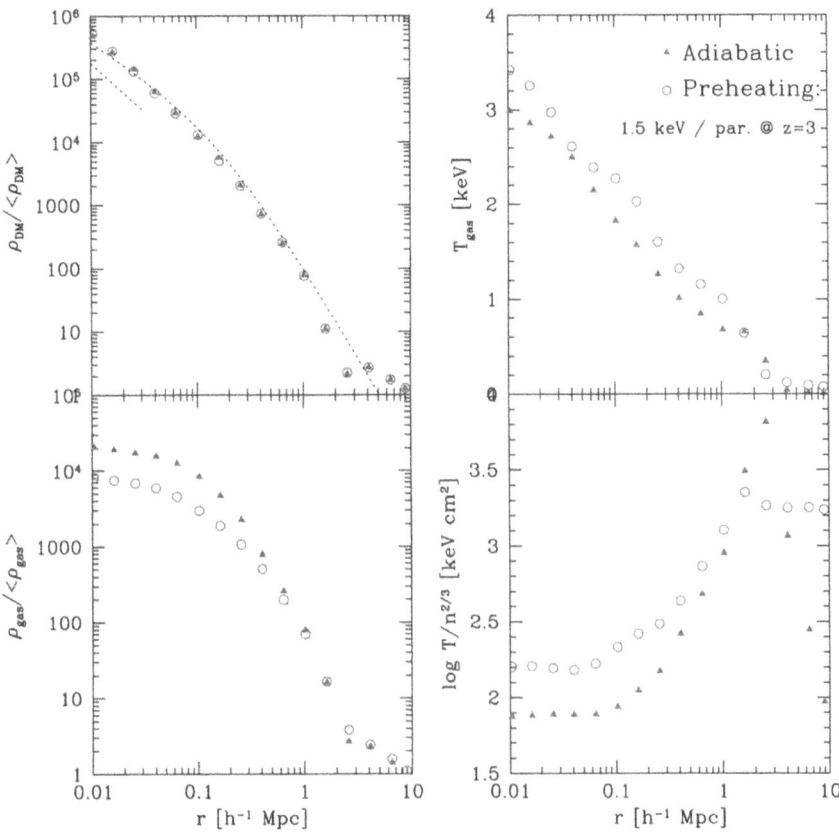

Fig. 5. Radial profiles of dark matter and gas density, gas temperature, and gas entropy of the simulated Virgo cluster. The dotted line in the DM density panel shows the NFW profile with the concentration of 7.

pixels, each pixel represented by a point in the figure. The figure shows that neither simulation can match the much shallower temperature distribution in the central regions of the Virgo cluster. This indicates that preheating alone cannot be the whole story and other processes, such as cooling and/or central heating by AGNs can be important. We will be exploring the effects of these processes in future simulations.

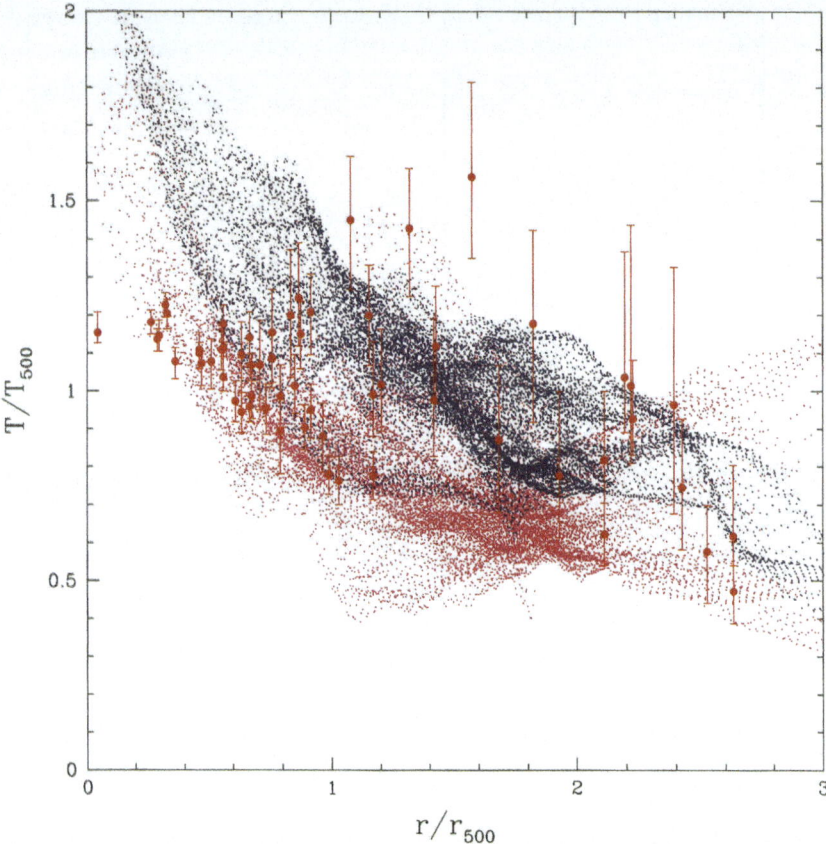

Fig. 6. Projected temperature profiles of the observed (points with errorbars) and simulated (purple and blue points) Virgo cluster. Purple points correspond to the adiabatic simulation while blue points show temperature in the simulation with preheating. The radii are normalized to the radius corresponding to the overdensity of 500 with respect to the critical density.

3 Summary

The adaptive refinement tree code is a useful tool to study cosmological structure formation on different scales and with different resolutions. It runs well on a variety of platforms with shared and distributed memory. The simulations described here have been performed on the Hitachi of LRZ Munich, the small development Hitachi at the AIP, the IBM SP of the Potsdam Institute for Climate Impact Research, and the IBM SP of NERSC Berkeley.

References

1. Einasto, J., Einasto, M., & Gramann, M., 1989, 238, 155
2. Gottlöber, S., Łokas, E.,Klypin, A., astro-ph/0207487
3. Grogin, N.A., Geller, M.J., 2000, Astrophys. J. 119, 32
4. Khokhlov, A.M. 1998, J. Comp. Phys. 143, 519
5. Klypin, A., Kravtsov, A. V., Bullock, J. S., & Primack, J. R. 2001, Astrophys. J. 554, 903
6. Klypin, A., Hoffman, Y., Kravtsov, A. V., & Gottlöber, S. 2001, astro-ph/0107104, submitted to Astrophys. J.
7. Knebe, A., Kravtsov, A. V., Gottlöber, S., & Klypin, A. A. 2000, MNRAS 317, 630
8. Kravtsov, A. V., Klypin, A. A., & Khokhlov, A. M. 1997, Astrophys. J. Suppl. 111, 73
9. Kravtsov, A. V. 1999, PhD thesis, New Mexico State University
10. Kravtsov, A. V., Klypin, A., Hoffman, Y., 2002, Astrophys. J. 571, 563
11. Popescu, C.C., Hopp, U., & Elsässer, H. 1997, A&A 325, 881
12. Shibata, R. et al. 2001, Astrophys. J. 549, 228
13. Willick, J. A., Courteau, S., Faber, S. M., Burstein, D., Dekel, A., & Strauss, M. A. 1997, Astrophys. J. Suppl., 109, 333
14. Zaroubi, S., Hoffman, Y., Fisher, K. B., & Lahav, O. 1995, Astrophys. J., 449, 446
15. Zaroubi, S., Hoffman, Y., & Dekel, A. 1999, Astrophys. J. 520, 413

References

The Free Electron Maser in Pulsar Magnetospheres

Rüdiger Schopper[1], Christoph Nodes[4], Hartmut Ruhl[2], Thomas A. Kunzl[3], and Harald Lesch[4]

[1] Max–Planck–Institut für extraterrestrische Physik
   Giessenbachstraße, 85740 Garching, Germany
[2] General Atomics, San Diego, CA, USA
[3] Max Planck Institut für Quantenoptik, Garching, Germany
[4] Universitäts-Sternwarte München, Centre for Interdisciplinary Plasma Science
   Scheinerstraße 1, 81679 München, Germany,

Abstract. We present the numerical simulations of coherent inverse Compton scattering (CICS) in a highly magnetized plasma process by means of a full three dimensional particle in cell code (PIC), which is mass and energy conservative. We used the parameters of a pusar magnetosphere where CICS is one of the most promising models for the generation of the observed highly coherent radio emission. First we show details of the onset and time evolution of strong Langmuir turbulence driven by a relativistic electron beam penetrating a strongly magnetized background plasma. The Langmuir turbulence acts as self-generated wiggler fields which bunch the beam electrons thereby inducing strong coherent emission of the bunches at frequency γ^2 times the plasma frequency. The emitted power is about $10\,\mathrm{GW}$ in a few nanoseconds. This radiation is interpreted in terms of inverse Compton scattering on nonlinear density fluctuations. CICS is the longitudinal version of a laboratory free electron laser and is applicable in strongly magnetized plasmas like pulsars.

Key words: beam plasma interaction, langmuir turbulence, coherent radio emission, particle in cell simulation, free electron maser

List of Abbreviations and Symbols:

CICS Coherent inverse Compton scattering
FEL Free electron laser
PIC Particle in cell
REB Relativistic electron beam

1 Introduction

A strong source for coherent radiation in laboratory is the free electron laser (FEL) [7]. Its principle is the interaction of a relativistic electron beam (REB) and a magnetic wiggler system. The beam of high energy electrons with energy $\gamma m_e c^2$ passes between permanent magnets of alternating polarity and periodicity λ_W. The energy of the REB is efficiently converted by electron bunching. In the observer's frame the bunches emit the relativistically Doppler-shifted wavelength $\lambda = \lambda_W/\gamma^2$. A comparable effect can be driven by a REB penetrating a strongly magnetized plasma. In phase space such a configuration corresponds to a two-stream-instability. In the plasma version the free energy of the emitted radiation is supplied by the REB, as it is in the case of the FEL, but the positive gradient of the momentum distribution function $df/dp > 0$ is the source of the instability which excites strong density fluctuations. In a strongly magnetized plasma such density oscillations are predominately longitudinal Langmuir waves which grow to nonlinear amplitudes [1, 2, 11]. The interaction of the REB with the electrostatic fluctuations of the Langmuir waves results in very strong coherent electromagnetic waves [2, 4]. In the rest frame of the beam electrons, they experience the Doppler shifted electrostatic wave moving towards them. The nonlinear waves force the REB to wiggle and to emit dipole radiation of the same frequency as the waves. In the laboratory frame this Hertz' dipole radiation is again Doppler boosted and strongly beamed in forward direction due to the relativistic lighthouse effect. This interaction can be viewed as coherent inverse Compton scattering (CICS) off Langmuir waves [1]. Coherence occurs due to the bunch structure of the beam which means a phase coupled electrostatic field and density modulation.

2 Computational Details

Since CICS has been discussed to be promising candidate for the origin of the extremely coherent radio emission from pulsars [3, 6, 5], we use the physical parameters of a pulsar magnetosphere as an instructive application. Radio observations indicate that at about 500 km from the strongly magnetized neutron star the coherent pulsed radio emission is produced. The coherence is proved by pulse substructures on time scales down to a few tens of nanoseconds with increasing flux densities [10]. In these regions of a pulsar magnetosphere typical values are: background electron density $n_e = 10^{12}\,\text{m}^{-3}$, beam electron density equal to the plasma density $n_b = 10^{12}\,\text{m}^{-3}$, beam electron Lorenz factor $\gamma \sim \sqrt{5}$ ($\gamma\beta \sim 2$) and an extremely strong background magnetic field B of about 1000 T. Especially the strong guiding field is important for the efficiency of the CICS-process. It reduces the electron dynamics to the spatial dimension along the magnetic field lines which greatly improves the stability of the excited Langmuir waves and thus leads to strong and stable electrostatic wiggler fields [9]. For the temperature of both the beam– and background electrons values we take the value of $T_e = T_b = 100\,\text{eV}$, typical polar cap temperature of neutron stars derived from their thermal X-ray-emission [8]. The PIC–code used for this numerical simulation is fully three dimensional and it conserves mass and energy. In the simulation the direction of electron beam propagation $\gamma\beta$ and of the magnetic field $\mathbf{B}$ is the positive z–direction, which is also called the longitudinal direction. The simulated box has a numerical extension of $80 \times 80 \times 200$ grid points and a physical extension of $100\,\text{m} \times 100\,\text{m} \times 250\,\text{m}$, which corresponds to the size of the expected features. The resolution of $\delta x = \delta y = \delta z = 1.25\,\text{m}$ is more than sufficient to resolve the expected maximum wavelengths of approximately $\lambda_{\text{Rad}} = 2\pi/k_{\text{Rad}} = 2\pi c/\gamma^2\omega_{\text{pe}} \approx 6.7\,\text{m}$. A particle density of $10^{12}\,\text{m}^{-3}$ is approximated by 10 quasi particles per cell, leading to 16 million background quasi electrons that are accompanied by a varying number of beam quasi electrons of not more than 2.5 million, which eventually enter the computational box. Those numbers are doubled, since for quasi neutrality an equally large number of positive particles are required.

The initial condition is given by a homogeneous background plasma of density n_{e}, temperature T_{e} and a longitudinal magnetic field $B_z = 0.1\,\text{T}$. This value is found to be sufficiently high to represent a much stronger field of 1000 T. Initially there is no beam in the computational box. The beam is injected later at the $z = 0$ plane of the box with a density profile of $n\,(r) = n_{\text{b}}/\left(1 + \exp\left[\frac{r-R}{\Delta R}\right]\right)$. Here r denotes the distance from the x–y center of the box and $R = 20\,\text{m}$ and $\Delta R = 3\,\text{m}$ give the radius and the sharpness of the electron beam. The homogeneous background plasma is disturbed in phase space (see Fig. 1) in order to accelerate the excitation of Langmuir waves by giving the system a hint of the relevant length scales. We used the wave length of the fastest growing Langmuir mode, which follows from an analytic calculation of the linear growth rate. The disturbance is chosen not to induce any finite charge or current densities, thus leaving the background plasma initially homogeneous, neutral and current free.

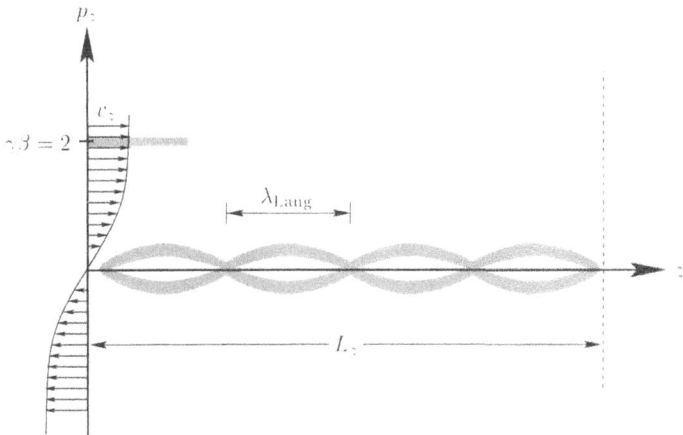

Fig. 1. The initial configuration of electron phase space including the beam. The initial disturbance of the background plasma is heavily exaggerated for reasons of better display.

The boundary conditions during the simulation are quite simple. In the transversal directions x and y we chose periodic boundaries and in the longitudinal direction z the box is open, which means that plasma particles and fields can leave the box unaffected. In addition at the lower z plane ($z = 0$) an electron component is continuously injected with the density profile mentioned above, a mean $\gamma\beta$ of 2 and the temperature T_b.

The simulation runs for 600 timesteps with $\delta t = 2.41$ ns. The total time simulated was $1.44\,\mu$s, after which the growing nonlinear Langmuir waves saturate and a quasi stationary situation had developed as it is shown in the next section.

3 Results

The beam particles entering the computational box interact with the faint traces of fluctuations we introduced in our initial setup and excite the appropriate modes to grow. In Fig. 6 on page 420 the temporal evolution of the elctrons phase space in z–direction is presented by showing phase diagrams for the timesteps 150, 300, 325 and 400. In the first plot (upper left) the penetrating electron beam can be seen together with the background plasma, both are still nearly undisturbed. In the following plots the development of strong Langmuir waves can be observed. After 781 ns, the disturbance in the beam and the background has grown so strong, that both components start to mix, i.e. there is no distinction between beam and background electrons anymore. Beam electrons are strongly decelerated even until $v \simeq 0$, whereas background electrons are accelerated up to a Lorenz factor of 5. This can be seen best in Fig. 7 on page 420, where each of the both compontents are plotted in separate diagrams for the timestep 400. In Fig. 7 left the phase space of the beam electrons is shown. Clearly visible are the ring structures of fully developed Langmuir waves. Some of beam electrons are accelerated up to a $\gamma\beta$ of 4–5 as it is the case for the background electrons, which is shown in Fig. 7 right. On the other hand the beam electrons are also decellerated to zero momentum during their interaction with the Langmuir waves. The extremely strong accelerations are due to the strong electrostatic fields of the excited Langmuir waves. Since the Langmuir turbulence structure in Fig. 6 moves with a much lower speed (which is even continously decreasing throughout the simulation) than the yet unaffected beam electrons, it acts as the wiggler field that stimulates the CICS-emission. The electrostatic fluctuations directly correspond to electron density fluctuations (solid curve in Fig. 6), which reach a value of $\sim +100\%/-50\%$ oriented at $n = n_e + n_b = 2 \cdot 10^{12}\,\mathrm{m}^{-3}$. The structure of the electron density is better seen in Fig. 8 (page 421) and Fig. 9 (page 422), where the temporal evolution of contour surfaces at $n = 1.5 \cdot 10^{12}\,\mathrm{m}^{-3}$ and $n_e = 1.9 \cdot 10^{12}\,\mathrm{m}^{-3}$ and of a cut through the electron beam is shown. The electron density varies between 4 and 1 times $10^{12}\,\mathrm{m}^{-3}$, which implies the strong electrostatic fields. It can be clearly seen, that the beam decays in a series of "pancakes". Such pancakes are absolutely necessary to explain the coherent nature of the pulsed radio emission of neutron stars. Why? Since the emission from relativistic particles is confined to a forward cone with half angle $\sim 1/\gamma$ and the emission is nearly along the magnetic field. Thus, in Fourier space bunching emission corresponds to the component $k_\perp$ to the fields being smaller than the component $k_\parallel$ along the field line by a factor $1/\gamma$. This corresponds to flat pancake shaped bunch with the normal within an angle $1/\gamma$ of B [5]. This is exactly what we observe in our simulations! We briefly note that our simulations are the first ones which show this kind of pancake bunches! The number of particles in such a pancake corresponds to the the the maximum coherence achievable for CICS and it should be noted, that in our simulation the thickness of such a pancake is significantly larger ($\gtrsim 1$ m) than the Debye length which is usually used in analytical models. A typical volume of one bunch is about $10 m^3$ giving a total number of particles per bunch of about 10^{13}. Our simulations prove, that a much higher coherence is achievable than expected from analytical models.

The growth of Langmuir waves can be deduced by the Fourier transform of the electrostatic field $\tilde{E}_z$. In Fig. 2 the energy density of the electrostatic modes is plotted versus the corresponding wavenumber k and time t. The $k \simeq 0.25$ m^{-1} mode is growing by four orders of magnitudes within some ten nanoseconds. It should be noted however, that the fastest growing mode is slightly off compared to the initially given wave number can be seen in Fig. 2 as the small peak at early times. This means, that for our simulation the

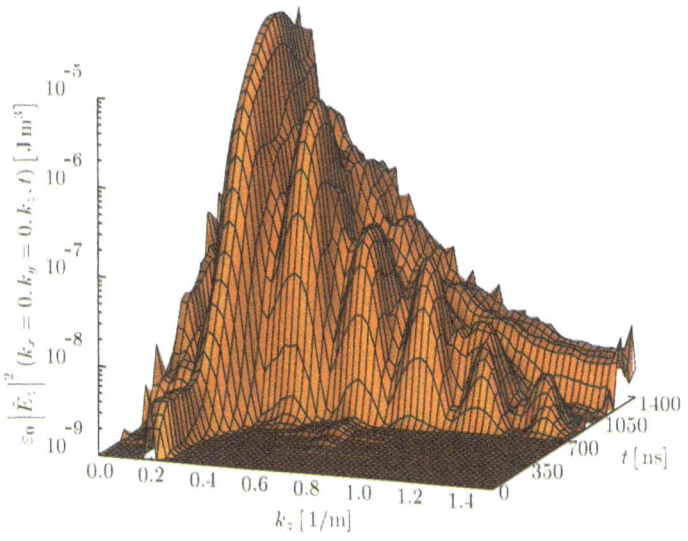

Fig. 2. The electrostatic mode energy density is plotted against wavenumber k and time t.

analytical value for the fastest growing mode is not precisely correct and it proves, that although the initial disturbance has given a hint to the system it has not forced it towards an unphysical solution. We note that the wave energy density is distributed to higher k when it exceeds a value of about $10^{-7}\,\mathrm{J\,m^3}$, driven by nonlinear wave–wave interactions.

It is important to recognize that the generated electrostatic fluctuations react on the beam electrons similar to the wiggler fields in the laboratory FEL, with one major difference. In our case the wave wiggling self-consistently excited is along the propagation direction of the beam, whereas in the FEL-case the external magnetic wiggling is perpendicular to the beam propagation. The strongest emission originates in regions with the largest density gradients Fig. 3. Every de– and acceleration region forces the beam electrons to emit Hertz' dipole radiation, which is beamed in forward direction, as can be seen in Fig. 4 (right). θ_M represents the angle of maximum intensity, expected from theoretical consideration ($\tan\theta_M = 1/2\gamma\beta$) with a $\gamma\beta = 2$. The striking resemblance with a relativistically beamed Hertz' dipole is obvious, but the emission characteristic in our simulation also gives precisely the expected opening angle of the emission cone, which is proposed for a CICS-process. In Fig. 4 (left) the "spectrum" of the emitted radiation is shown, which has a strong peak at $\omega = 5\omega_{\mathrm{pe}}$ and a couple of minor peaks at $4\omega_{\mathrm{pe}}$ and $3\omega_{\mathrm{pe}}$ due to nonlinear wave coupling. Again we find rather precisely

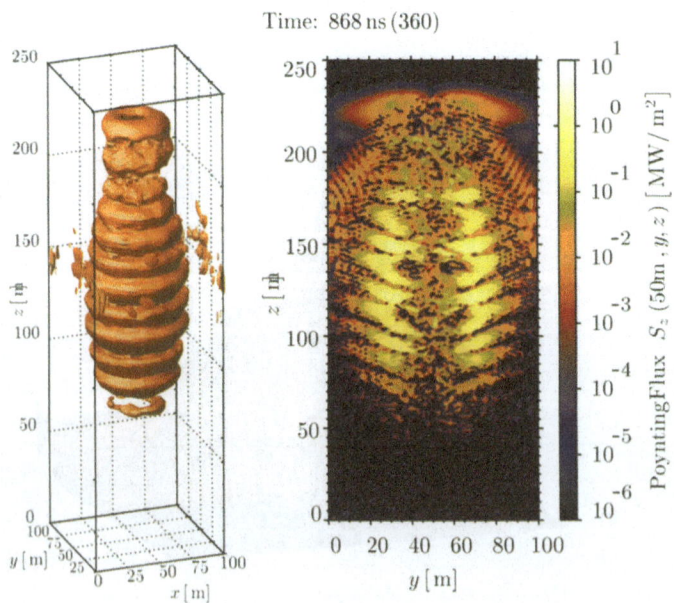

Fig. 3. Structure of the emitted pointing flux as 3D contour surface of the value $2.5 \cdot 10^4\,\mathrm{W/m^2}$ (left) and as a cut through the center of the beam at $x = 50\,\mathrm{m}$ (right).

the theoretically expected value of $w = \gamma^2 \omega_{\mathrm{pe}}$. The total power emitted by the simulated computational box is shown in Fig. 5. It is the escaping poynting flux at the top of the computational box ($z = 250\,\mathrm{m}$) integrated over the top surface. After about 850 ns the emitted power rises within some ten nanoseconds by more than 6 orders of magnitude to a value of 1 GW. Afterwards the total output power increases more slowly by one more order of magnitude and reaches a maximum of 10 GW continuous radiative power at the end of the simulation. This is a significant fraction of the power injected by kinetic energy of the beam particles, shown in Fig. 5 as + symbols at a constant value of 50 GW. The rest of the energy leaves the box as kinetic energy of beam electrons, which would probably be also emitted if the computational would have a greater size in z-direction.

4 Discussion

We have shown that in a strongly magnetized plasma a relativistic electron beam can be forced to emit highly coherent radio emission by self-induced nonlinear density fluctuations. Such slowly moving nonlinear structures oscillate with the local plasma frequency at which the relativistic electrons are scattered. Beam electrons dissipate a significant amount of their kinetic energy by inverse Compton radiation at a frequency of about $\gamma^2 \omega_{\mathrm{pe}}$. Since the beam is sliced into pancake structures which experience the same electric field the inverse Compton scattering is coherent. Such a process is a very promising candidate for the coherent radio emission of pulsars.

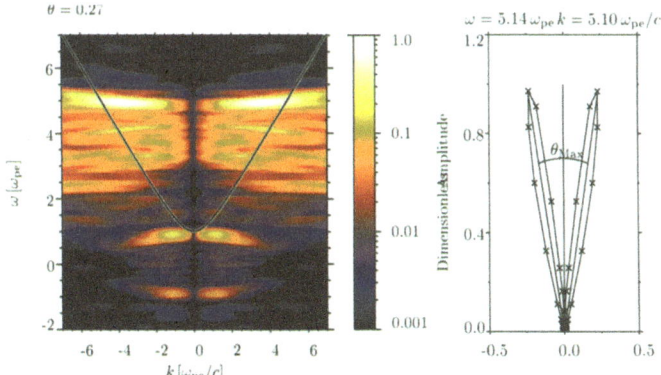

Fig. 4. Dispersion relation (left) and emission characteristic (right) of the emitted radiation. For the left plot the angle θ_{M}, the angle of maximum intensity, has been used. The curve in the left plot gives the dispersion branch of electromagnetic radiation in a plasma. The right diagram presents the intensity of the radiation for a given θ in a polar plot. θ is the angle between beam direction and direction of emission. The values in both diagrams are normalized to a value of one.

5 Why high performance computing

The simulation described above is a small one. For several reasons it is the smallest possible simulation that is physically sensible for the application to pulsars. First, the grid resolution is half the shortest electromagnetic wavelength we expect in our simulation thus leaving no room for the observation of unexpectedly short wavelengths, which might be produced eventually. Second, the Debye length $\lambda_D = 0.07\,\text{m}$, which is the shortest Plasma length, is *not* resolved in the above simulation, which gives rise to artificial heating effects, which are fortunately not important, but nevertheless undesired. Third, the discussed simulation has shown, that within a box of length 250 m only ten percent of the total energy injected is radiated away, indicating, that this might be a geometric (numeric) effect and a higher ratio might be possible in a larger box. Finally, in a pulsar magnetosphere one expects the FEL process acting with Lorentzfactors of 5 to 20, whereas we used a factor of only $\gamma = \sqrt{5}$ as an approximation to 5, which, as it is shown below, makes such a simulation ways *cheaper*.

With a storage requirement of $16 \times 8\,\text{Bytes} + 28 \times 4\,\text{Bytes} = 240\,\text{Bytes}$ per cell for the fields and $9 \times 8\,\text{Byte} = 72\,\text{Bytes}$ per quasi particle the total memory requirement for such a *small* run would be approximately 2.9 GByte. On 16 processing elements such a run of 600 time steps takes three hours or 48 CPU hours total. As already stated such a run is for several reasons unsatisfying, but helps to gain insights into the problem, which is absolutely necessary for further, more expensive studies.

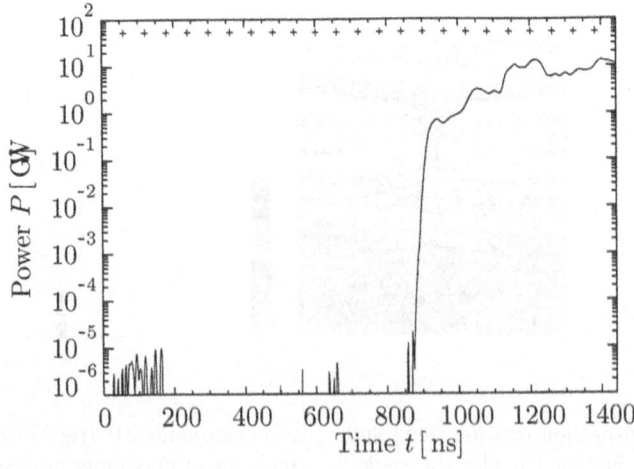

Fig. 5. Total integrated poynting flux leaving the computational box at the top ($z = 250\,\text{m}$) against time. The $+$ symbols show the total power injected as kinetic energy of the beam electrons.

The aim is to do a simulation of the process, which is as close as possible to the real situation and that means to enlarge the box and to go to finer grids at the same time. Enlarging the numerical Box, in order to see if it is possible to convert more energy than what we got is a relatively cheap improvement, since only the z direction is affected and thus the problem grows linearly with the box size in that direction. A refinement of the grid is much worse. All dimensions are equally important due to the radiation and that means, that any refinement has to be done in all directions. Since the largest possible time step δt depends on the spatial resolution ($\delta t \leq \min\left(\delta x, \delta y, \delta z\right)/c$) this means, that the cost grows with a power of four concerning the resolution.

The perfect simulation, with $\delta x = \delta y = \delta z = 0,035\,\mathrm{m} = \lambda_\mathrm{D}/2$ would be by a factor $2 \cdot 10^7$ larger and is just not feasible. Fortunately in our case it is not strictly necessary to fully resolve the small Debye length, so that it is possible to study the FEL process by means of numerical simulations.

One major direction of us is to enlarge the Lorentzfactor γ to more reasonable values. The interesting physical coherent radiation process as such can be investigated with a γ smaller than in a pulsar, however, in order to give quantitative Results one has to close the gap, between the numerical simulation and the real object. Since the wavelength of the expected radiation, which has to be resolved, goes like $\lambda \sim 1/\gamma^2$ the total cost of the simulation grows with $\sim \gamma^8$. So for a run with a Lorentzfactor of 5, which is the lower end of the interesting range, we are talking about 1.9 TByte memory and some 30000 CPU hours, which makes high performance computing for this application absolutely necessary.

References

1. Benford, G., Weatherall, J.C., 1991, ApJ, 378, 543
2. Kato, K., Benford, G., Tzach, D., 1983, Phys. Fluids, 26, 3636
3. Kunzl, T., Lesch, H., Jessner, A., von Hoensbroch, A., 1998, ApJ, 505, L139
4. Levron, P., Benford, G., Tach, D., 1987, PRL, 58, 1336
5. Melrose, D.B., 2000, in Pulsar AstronomyIAU Colloq. 177, ASP. Conf. Series 202, 721
6. Melrose, D.B., Gedalin, M., 1999, ApJ, 521, 351
7. O'Shea, P.G., Freund, H.P., 2001, Science, 292, 1853
8. Pavlov, G.G., Petekhin, A.Y., 1995, ApJ, 450, 883
9. Pelletier, G., Sol, H., Asseo, E. 1988, Phys. Rev. A, 38, 2552
10. Rickett, B.J., 1995, ApJ, 197, 185
11. Weatherall, J.C., 1988, PRL, 60, 1302

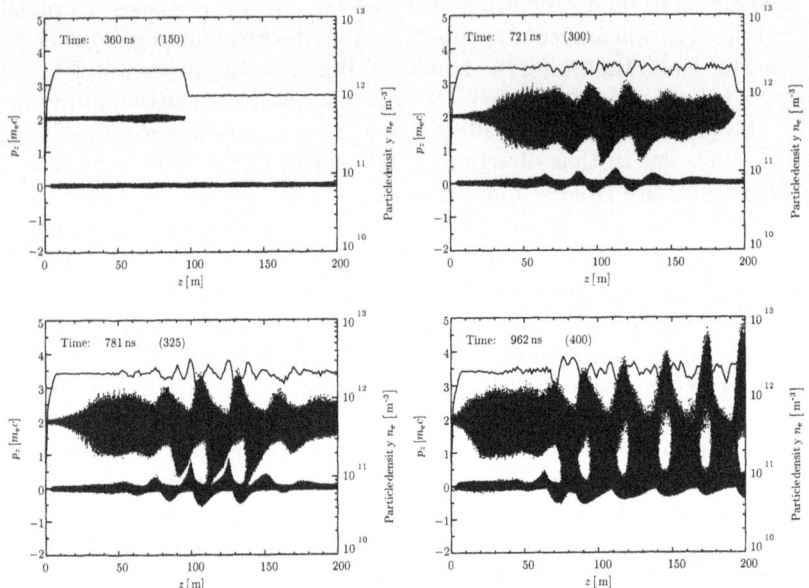

Fig. 6. Electron phase space in z–direction after 150, 300, 325 and 400 timesteps. The quasi particles are plotted as dots according to their corresponding position in z (ordinate) and their momentum p_z (left abscissa). The solid curves present the particle density of the electrons (right abscissa).

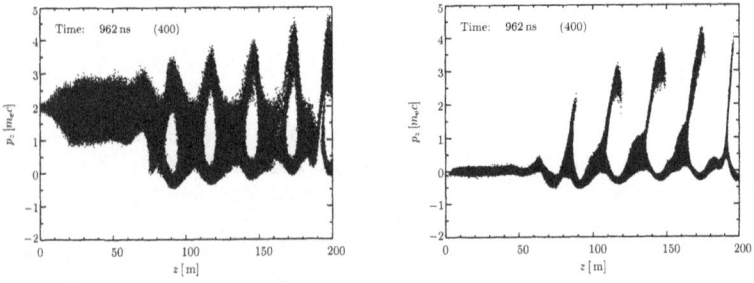

Fig. 7. Electron phase space in z–direction after 400 timesteps separated for beam (left) and background (right) electrons.

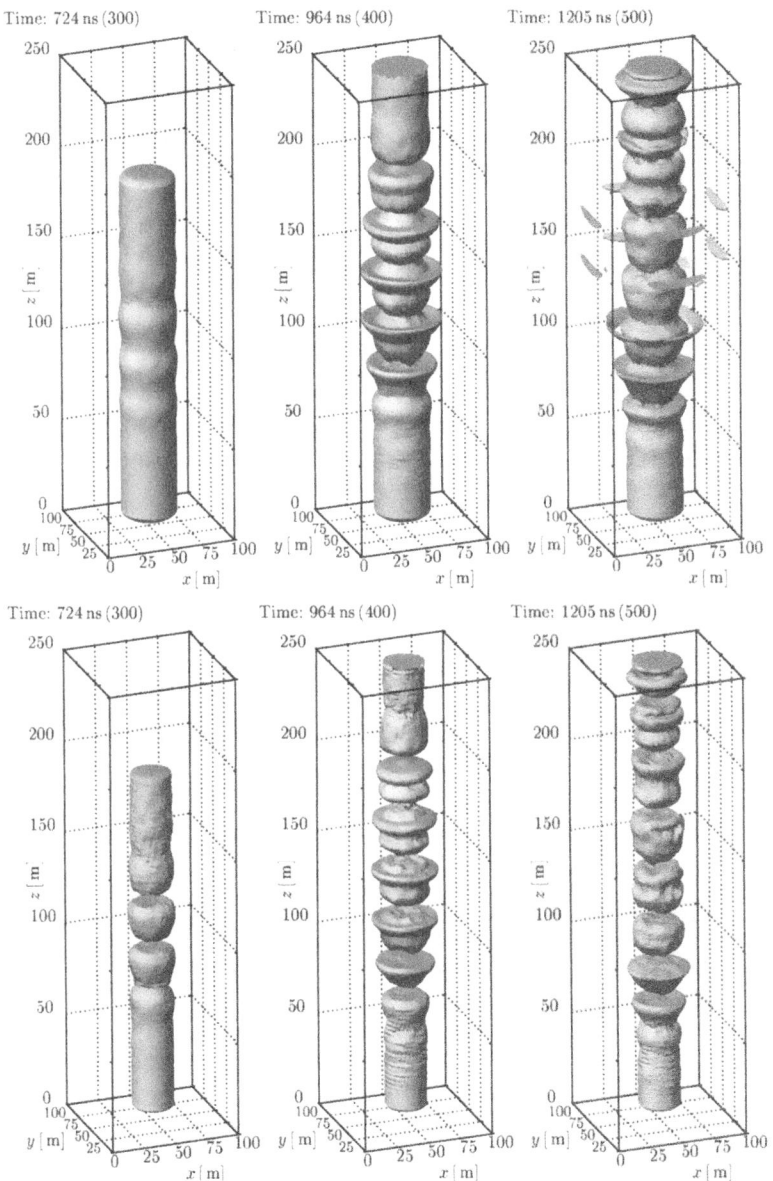

Fig. 8. Three dimensional contour plots of the electron particle density. For the timesteps 300, 400 and 500 (left to right) the surfaces are shown, where the density is $n = 1.5 \cdot 10^{12}\,\mathrm{m}^{-3}$ (top) and $n_{\mathrm{e}} = 1.9 \cdot 10^{12}\,\mathrm{m}^{-3}$ (bottom) respectively.

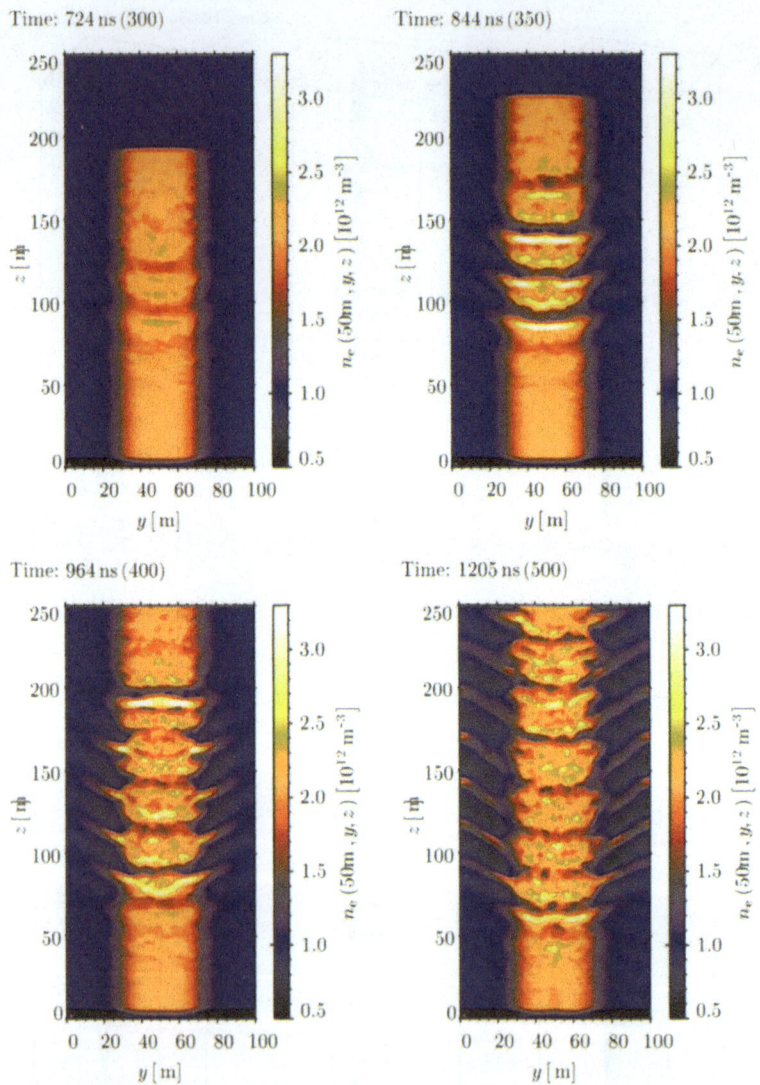

Fig. 9. two dimensional contour plots of the electron particle density at timesteps 300,350,400 and 500. Shown is a cut through the center of the penetrating electron beam at $x = 50$ m.

Part VIII

Computer Science

Christoph Zenger

Fakultät für Informatik
TU München
80290 München, Germany

Computer science plays an important role in many high performance computing projects. However, most of these projects can be found in the sections devoted to specific applications. In contrast, this section comprises a collection of those projects where computer science plays a central role. Programs used in high performance computing almost always have a complex structure which is based on modern numerical methods. They use various program libraries and exploit inherent parallelism in the algorithms. Moreover, they should be well adapted to the underlying architecture of the computer. Therefore, modern software engineering methods and tools are important prerequisites for the construction of reliable and easily maintainable software packages.

The contributions in this section are mainly devoted to issues related to software construction methods (e.g., object orientation), which are well suited for the implementation of modern numerical methods. This allows, for example, the use of multi-level structures and adaptive grids for the solution of partial differential equations. The usage of a design which can be efficiently implemented on the SR8000-F1 supercomputer of the LRZ is also discussed.

Unfortunately, there are almost no contributions in this section covering Computer Science as an application field although several topics in Computer Science exist where high performance computing is strongly needed: data mining and queries on large data bases, just to name a few. The only exception is the paper by Stamatakis et al. where a simulation of a bioinformatics problem is studied. Project proposals dealing with applications in these directions are highly encouraged.

Computer Science

Pseudo-Vectorization and RISC Optimization Techniques for the Hitachi SR8000 Architecture

Georg Hager, Frank Deserno, and Gerhard Wellein

Regionales Rechenzentrum Erlangen
Martensstr. 1
91058 Erlangen, Germany
{*georg.hager, frank.deserno, gerhard.wellein*} *@rrze.uni-erlangen.de*

Abstract. We demonstrate optimization techniques for the Hitachi SR8000 architecture on CPU and SMP level using selected case studies from real-world codes. Special emphasis is given to a comparison with performance characteristics of other modern RISC and vector CPUs.

1 Introduction

Vector supercomputers have been carrying the performance crown in numerical applications for over two decades. Superior memory bandwidth, data parallelism and pipelining abilities have made these systems the premier choice for top notch research. With the advent of powerful superscalar RISC processors in the mid-80s, a change in programming style was required that could not be easily implemented in all numerical codes, so that vector architectures were still the only alternative for many applications. The Hitachi SR8000 system, however, is the first parallel RISC system that has competitive memory bandwidth and internal parallelism, and thus provides a flexible "hybrid" environment with both vector and RISC features. It is generally expected that the hybrid approach will be adopted by more and more supercomputer vendors so that vector systems will continually decrease in significance.

There is a price to be paid for this flexibility, though. Hybrid supercomputer architectures like the SR8000 require a thorough insight into inner workings, compiler abilities and 'weak spots' in order to exploit their unique features in an optimal way. Using selected examples from real-world codes we illustrate SR8000-specific optimization techniques and compare the results with other architectures.

This section gives a coarse overview to the different benchmark platforms used and some of their peculiarities. The second section discusses basic performance issues with vector triads and sparse matrix-vector multiplication on single CPU and node level. In the third section a RISC optimization strategy for loop kernels of a nuclear physics code is given where careful data flow and locality analysis is important. The fourth section deals with the implementa-

Table 1. Benchmark platforms and single processor/node specifications. Peak performance numbers (Peak) are given in GFlops, whereas the memory bandwidth (Memb) per processor is given in GBytes/s. For the RISC based systems the L1 and L2 cache sizes are contained

Platform	Single CPU/node specifications				Remarks
	Peak	Memb	L1 [kB]	L2 [MB]	
Intel-P4 1.5 GHz	1.5	3.2	8	0.256	RD-RAM / pgf90
SGI Origin3400	1.0	1.6	32	8	28 way - ccNUMA
HITACHI SR8000-F1 (1 CPU)	1.5	4.0	128	—	PVP
NEC SX5e (1 CPU)	4.0	32.0	—	—	—
HITACHI SR8000-F1 (1 node)	12.0	32.0	1024	—	PVP+COMPAS

tion and shared memory (SMP) parallelization of a strongly implicit solver which is used in many codes from computational fluid dynamics (CFD).

Benchmark platforms The benchmarks have been performed on the platforms given in Table 1. The Hitachi SR8000 processor is basically an IBM PowerPC design running at 375 MHz with important modifications in hardware and instruction set [2]. These extensions include 160 floating point registers and the possibility of up to 16 outstanding **prefetch** or 128 outstanding **preload** instructions. While the **prefetch** loads one cache line in advance to the L1 cache, **preload** instructions can load single data items directly from the memory to the register set, bypassing the L1 cache. Together with an extensive software pipelining done by the compiler these features are called *Pseudo Vector Processing* (PVP), as they are able to provide a continuous stream of data from main memory to the processor avoiding the penalties of memory latency. Considering the programming style, in particular programs with long inner vectorizable loops benefit from PVP. The performance of programs with high cache locality, however, typically suffers from PVP, especially if **preload** instructions are generated for data already resident in L1 cache. The Hitachi SR8000-F1 node comprises eight compute processors accessing a shared memory. Parallelization within the node can be done either by assigning one MPI process to each processor (MPP mode) or by running shared memory parallelization with eight threads (COMPAS mode). Special emphasis has been put by Hitachi on this **C**ooperative **M**icro**P**rocessors in a single **A**ddress **S**pace mode which is a collection of some unique features:

– Sophisticated auto-parallelization compiler features
– Hardware support for fast collective thread operations

- Memory bandwidth of 32 GBytes/s which matches the aggregated single processor bandwidths (note that one SX5e CPU also has a bandwidth of 32 GBtyes/s).
- 512-way interleaved memory to avoid memory contention and to hide bank busy times

Therefore one Hitachi node can be seen either as an eight-way RISC based SMP node or as a vector-like CPU running one process with eight threads. For all our benchmark tests on the SR8000, the compiler option -Oss was used. For the single-threaded codes, -noparallel was specified additionally.

The NEC SX5e processor runs at a frequency of 250 MHz using eight-track vector pipelines. There are one store and two load pipelines, but loads and stores cannot be performed in the same cycle. Thus effectively, the CPU has one load/store pipe which is twice as fast for loads than for stores.

The MIPS R14000 CPU used in the Origin 3400 implements the MIPS-IV instruction set and runs at 500 MHz. This four-way superscalar processor has the ability to sustain four outstanding memory references, including cache line prefetch from main memory to L2 cache. There are no provisions for preload, though. In the Origin 3000 ccNUMA architecture, the four processors of a node share local memory with an aggregated bandwidth of 3.2 GBytes/s, where the maximum for one CPU is 1.6 Gbytes/s. Links between the nodes can transfer data with up to 1.6 GBytes/s in each direction.

Intel's Pentium 4 is the latest incarnation of the IA32 architecture. The single processor system used for benchmarking was equipped with dual-channel RDRAM memory that has a theoretical bandwidth of 3.2 Gbytes/s. The processor has a hardware prefetch mechanism that tries to detect access patterns and prefetches cachelines into L2. As in the MIPS processor there is no preload instruction. No use was made of the new SSE2 features of the Pentium 4 processor due to the lack of appropriate compilers.

2 Basic loop kernels

As a starting point two basic loop kernels — the vector-triad and a sparse matrix-vector multiplication — are discussed. In principle, they represent the performance characteristics of the large class of scientific applications with performance mainly bounded by the quality of the memory access, e.g. most engineering applications. Of course, for operations that use data from different cache levels, the cache bandwidths (latencies) are decisive. They are usually much higher (lower) than the corresponding numbers for the main memory.

2.1 Vector-triad operation

The vector-triad operation $A(1 : N) = B(1 : N) + C(1 : N) * D(1 : N)$ is widely regarded as the most important operation for engineering applications [3]. With a memory intensity of two memory references per floating point operation (Flop), most processors can not achieve their peak performance even for data already resident in the L1 cache. E.g., the Hitachi processor has a L1-register bandwidth of 32 Bytes (4 Words) per processor cycle and thus can only saturate one of the two multiply-add pipelines, limiting the maximum performance for the vector-triad to 750 MFlops (half of peak performance). Considering that either four load or only two store operations to the L1 cache can be done, the maximum achievable performance is reduced to 600 MFlops. Measuring values of up to 560 MFlops for intermediate loop lengths (N ≈ 1000 − 5000) we come quite close to this maximum value as can be seen from Fig. 1. At longer loop length a drop in performance occurs

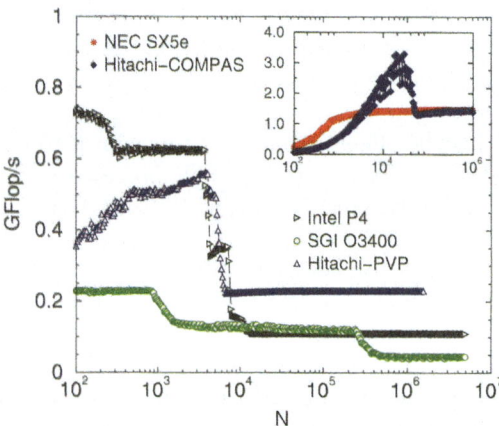

Fig. 1. Performance of vector-triad operation: The main panel depicts cache based systems (PVP enabled for Hitachi SR8000-F1). In the inset performance of one Hitachi SR8000-F1 node (PVP+COMPAS mode) and one NEC SX5e vector processor is shown.

when data has to be loaded from main memory. Since the memory bus can do either two load or two store operations per memory bus cycle (4 ns), basically two (memory bus) cycles are required to perform two Flops for the vector-triad operation. Therefore — ignoring the memory latency — an upper performance limit of 250 MFlops exists for data resident in main memory. Avoiding the penalties from memory latency by using PVP, more than 90 % of this theoretical limit (230 MFlops) can be maintained for long loop lengths (see Fig. 1). The importance of hiding memory latencies even for simple kernel loops like the vector-triads becomes more visible when comparing Intel and Hitachi performance numbers at long loop lengths: Although the Intel system offers 80 % of the Hitachi memory bandwidth it only achieves half of the performance. Of course at short loop length the Intel system outperforms the Hitachi due to the higher clock speed of the processor. As an example for modern RISC processors with large external L2 caches the SGI numbers are given in Fig. 1 for comparison. Please note that the SGI in-cache performance is mainly limited by the fact that the MIPS processor can only issue one memory instruction per cycle which gives an effective cache bandwidth of 4 GByte/s for the vector-triads.

An even more interesting discussion is provided by the inset of Fig. 1 where the performance numbers of combined PVP and COMPAS mode using one Hitachi node and of one NEC vector processor are given. Since the memory bandwidth is the same for both systems, we find the same asymptotic performance of about 1.5 GFlops. This value is about two times CRAY T90 performance and is only slightly below the NEC measurements given in Ref. [3]. At low loop lengths the classical vector processor benefits from fast vector start-up times (when compared to the thread synchronization times of the Hitachi node). However, at intermediate loop lengths the Hitachi outperforms the NEC system by more than a factor of two due to the high aggregate L1 cache bandwidth of the eight processors of the Hitachi node.

2.2 Sparse matrix-vector multiplication

Numerical cores of many scientific applications, e.g. quantum mechanical many body problems in theoretical physics, are iterative sparse matrix algorithms such as sparse eigenvalue solvers. In practice, the performance of such operations is mainly determined by a matrix-vector multiplication (MVM) involving a large sparse matrix. Exploiting the sparsity of the matrix two storage formats are widely used to store the nonzero matrix entries only: The *Compressed Row Storage* (CRS) format and the *Jagged Diagonal Storage* (JDS) scheme [5]. Since the JDS format achieves high performance on vector processors and provides only minor drawbacks on RISC based systems a JDS based implementation has been chosen in what follows [6]:

```
for j= 1,...,max_nz do
   for i= 1,...,jd_ptr(j+1)-jd_ptr(j) do
      y(i) = y(i)+ value(jd_ptr(j)+i-1)* x(col_ind(jd_ptr(j)+i-1))
   end for
end for
```

The MVM is performed along the *Jagged Diagonals*, providing an inner loop length essentially equal to the matrix dimension (D_{Mat}), while the outer loop runs over the maximum number of nonzero entries max_nz per row. A vectorization and/or automatic parallelization of the inner loop is generally done automatically by the compilers. For a detailed description of our JDS implementation we refer to Ref. [4]. Since the JDS MVM involves a vector gather operation on vector x it is expected that cache based systems show poor performance numbers for intermediate and large matrix dimensions, while vector processors can sustain a substantial fraction of their peak performance.

Basic performance characteristics of the systems under consideration are presented in Fig. 2, using a benchmark problem from theoretical physics ($D_{\mathrm{Mat}} \approx 10^3 - 10^7$; max_nz ≈ 20) [4]. For the cache-based SGI Origin and Intel systems the characteristic performance drop at small/intermediate loop

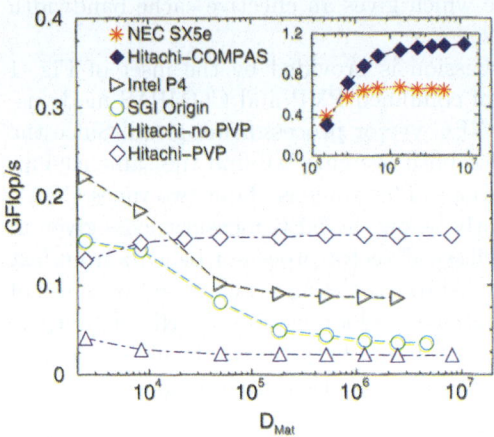

Fig. 2. Performance of sparse MVM kernel: A comparison of cache based systems and single CPU Hitachi SR8000-F1 (PVP enabled/disabled) is depicted in the main panel. The inset shows the performance of one Hitachi SR8000-F1 node (PVP+COMPAS mode) and one NEC SX5e vector processor.

lengths (D_{Mat}) is prominent. Only minor cache effects along with poor performance numbers are obtained for the Hitachi when disabling the PVP feature. However, the performance characteristics change completely if PVP is enabled: The asymptotic performance grows by a factor of nearly 5 to approximately 155 MFlops. Moreover the characteristic performance drop of cache based systems is replaced by a vector processor like behavior, where performance increases with problem size (see also NEC data in inset of Fig. 2) and saturates at a high level. Thus, the memory latency even for the vector-gather operation on vector x can be efficiently hidden by the PVP feature which generates a software pipeline of **prefetch** and **preload** instructions (cf. Fig. 3). The **prefetch** transfers a full cache line (16 or 32 data items depending on the data type) of the index array (**col_ind**) to the L1 cache followed by **load** instructions. For each entry of the cache line non-blocking **preload** instructions are issued. After paying only once the penalty of memory latency, a continuous stream of data from memory to registers is established for non-contiguous access to vector x. To abide the memory flow in the inner loop of the JDS MVM, the compiler performs loop unrolling at a high level (48 times) and builds a long software pipeline (27 stages). Of course, intermediate to long loop lengths are required to make full use of this technique ($D_{Mat} \gtrsim 10^4$).

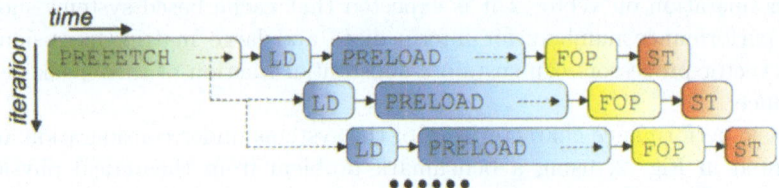

Fig. 3. Diagrammatic view of Pseudo Vector Processing for indirect memory access

Finally, a surprising result is depicted in the inset of Fig. 2: The performance of one Hitachi node using PVP and COMPAS significantly outperforms the NEC vector processor at intermediate to long loop lengths! Since the asymptotic performance of both systems is the same for the vector triads, this difference can not attributed to the memory bandwidth but is mainly determined by the implementation of the vector-gather operation. Obviously the combination of PVP and COMPAS can surpass the hardware implementation of the NEC system, if the loop is long enough. Moreover the single processor performance of combined PVP and COMPAS mode (138 MFlops) shows a decrease of less than 10% when compared to the single processor performance, indicating only minor memory contention problems in the COMPAS mode. However, at small problem sizes the vector processor still outperforms the Hitachi node because of short vector start-up times.

For large scale problems a distributed memory parallelization (using MPI) of the JDS MVM is required. In this context, the Hitachi architecture offers a wide variety of programming approaches (pure MPI, MPI+COMPAS, MPI+OpenMP), which have been discussed in detail in Ref. [4].

3 RISC Optimization

In this section we illustrate special RISC optimizations that are especially suited for the Hitachi CPU. The original code is being used for calculations of scattering problems with three-nucleon forces in nuclear physics [7].

For the benchmark considered, the single subroutine MAT took 98% of the total runtime. The routine achieved only about 26 MFlops, and this could be attributed exclusively to the following two loops:

```
1           DO 40 M=1,IQM
2             DO 30 K=KZHX(M),KZAHL
3    30          F(K)=F(K)*S(MVK(K,M))
4    40      CONTINUE
5    41      DO 50 K=1,KZAHL
6             WERTT(KVK(K)) = WERTT(KVK(K)) + F(K)
7    50      CONTINUE
```

Arrays F(), S() and WERTT() are **double precision**, all other data is **integer**. For any further optimization work one has to consider the following facts:

- S() is short (about 100–200).
- WERTT() is very short (length 1 in the worst case).
- IQM is very small (typically 9).
- The index array KZHX() contains indices in ascending order.
- KZAHL is typically much larger than 1000.

Performance numbers in MFlops are valid for the whole subroutine, unless otherwise noted.

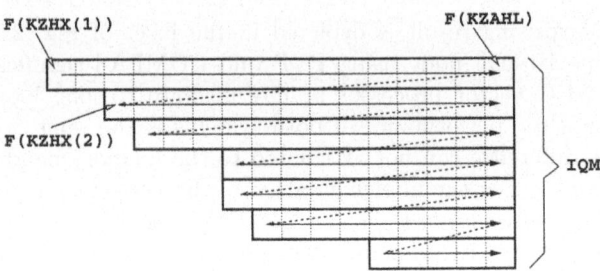

Fig. 4. Original access pattern for F() in the first loop. Outer loop unrolling is impossible.

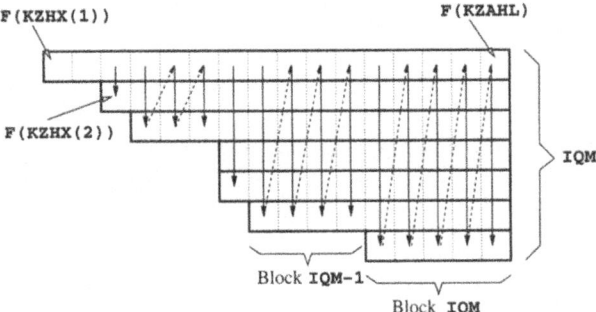

Fig. 5. Naive optimization of the first loop. Registers can be used for F() in an optimal way. A third loop level is required for the blocking pattern.

3.1 First Loop

The Hitachi compiler generates **preload** streams for S() and for WERTT() and **prefetch** streams for F() and MVK(). As S() contains only between 100 and 200 elements and WERTT() is even shorter (only one element in the worst case), PVP is just overhead for those two arrays. Thus the first optimization is to switch off any vectorization for S() by *voption nopreload directives. This already causes a speedup to 69 MFlops. To get even more the ratio of floating-point operations to loads and stores must be improved (cache optimization is of hardly any use because MVK(), the most space-consuming item, is loaded only once). Usually this is achieved by outer loop unrolling, but that cannot be done here in a straightforward manner because of the dependency of the inner loop on M. Here the fourth property from above is of great importance. Figure 4 shows the access pattern for F(). Many of the array elements are loaded more than once, but not in a way that enables register reuse.

Without further knowledge about the CPU architecture one would naively 'transpose' the access pattern, thereby enabling optimal register reuse (Fig. 5). Effectively, the inner and outer loops would be interchanged at the cost of

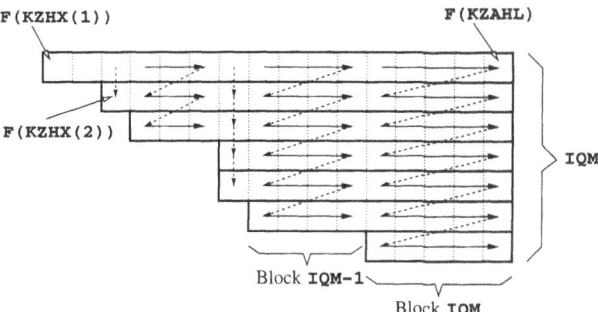

Fig. 6. Optimized version of the first loop. Unrolling of the loop over the vertical dimension is now possible.

introducing a third loop level. One element of F() must then be loaded and stored exactly once instead of up to IQM times, so nearly half of the memory references in the loop can be saved. Access to the index field MVK() is now strided, but this is not a problem due to the smallness of IQM.

The performance of this seemingly optimal solution is unsatisfactory, though. The inner loop length of at most IQM is too short for the Hitachi CPU.

A mixture of both variants is thus in order. The new outer loop (IQM blocks) is kept, but inside one block the access pattern is left as it was in the original code (Fig. 6). Now the compiler should be able to unroll the loop over the 'vertical' dimension, but this is not done automatically and cannot be enforced by compiler directives. Consequently, the loop was unrolled manually by a factor of two:

```
1          do M=1,IQM
2            ISTART=KZHX(M)
3            if(M.NE.IQM) then
4              IEND=KZHX(M+1)-1
5            else
6              IEND=KZAHL
7            endif
8            IS=1
9            if(btest(M,0)) then
10  *voption nopreload(S)
11  *voption noprefetch(S)
12              do K=ISTART,IEND
13                F(K)=F(K)*S(MVK(K,IS))
14              ENDDO
15              IS=IS+1
16            ENDIF
17            do MM=IS,M,2
18  *voption nopreload(S)
19  *voption noprefetch(S)
```

```
20          do K=ISTART,IEND
21             F(K)=F(K)*S(MVK(K,MM))*S(MVK(K,MM+1))
22          ENDDO
23       ENDDO
24     ENDDO
```

With those optimizations the overall performance of the subroutine grows
to 77 MFlops. It is worth noting that the compiler now actually unrolls the
middle loop by three. Further automatic optimizations like modulo variable
expansion and prefetching lead to an overall unrolling factor of 48. Moderate integer register spill occurs, but limiting compiler-induced unrolling to
prevent this does not improve performance any further.

3.2 Second Loop

As mentioned above, disabling pseudo-vectorization for WERTT() is the first
step here. This gives a significant speedup to 87 MFlops, but there is more potential for optimization. Use of functional units in the CPU can be improved
if successive iterations store their results in *different* targets. The compiler
usually does loop unrolling and modulo variable expansion to achieve this,
but here manual intervention is necessary.

Again, unrolling the loop by two is the method of choice. Alternating
iterations write their results into the different arrays WERTT() and WERTT2():

```
1   41     IS=1
2          IM=1
3          if(btest(KZAHL,0)) then
4             WERTT(KVK(IS))=WERTT(KVK(IS)) + F(IS)
5             IM=KVK(IS)
6             IS=IS+1
7          ENDIF
8          IN=IM
9  *voption noprefetch(WERTT,WERTT2)
10 *voption nopreload(WERTT,WERTT2)
11         DO 50 K=IS,KZAHL,2
12            WERTT(KVK(K)) = WERTT(KVK(K)) + F(K)
13 *soption predicate
14            if(IM.lt.KVK(K)) IM=KVK(K)
15            WERTT2(KVK(K+1)) = WERTT2(KVK(K+1)) + F(K+1)
16 *soption predicate
17            if(IN.lt.KVK(K+1)) IN=KVK(K+1)
18 50     CONTINUE
19         IQ=MAX(IM,IN)
20 *voption noprefetch(WERTT,WERTT2)
21 *voption nopreload(WERTT,WERTT2)
22         do k=1,IQ
23            WERTT(K)=WERTT(K)+WERTT2(K)
24         enddo
```

Table 2. Performance improvements on the Hitachi system due to the optimizations described in the text.

Version	MFlops MAT()	MFlops appl.
vanilla	26.0	26.0
1st loop S() opt.	69.4	67.8
1st loop perfect	76.7	74.6
2nd loop WERTT() opt.	86.6	83.6
2nd loop perfect	93.2	89.6

In lines 12 and 15 the two arrays are updated. After the loop a reduction operation collects the results into WERTT() (starting at line 22). To do this it is required to calculate the largest index stored in KVK(). The FORTRAN intrinsic function MAX0 is, however, unsuitable performance-wise. The maximum is calculated 'on the fly' instead (lines 5, 14, 17 and 19) by predicated assignments. This version now achieves a performance of 93 MFlops.

Of course the reduction increases the number of floating point operations in the code, but due to the smallness of IQ this effect is negligible.

3.3 Remarks

Table 2 gives an overview to the effects of the different optimizations. The performance of the whole program could be raised to 90 MFlops. The lesson to be learned is that the pseudo-vectorization features of the Hitachi CPU are not always beneficial and can sometimes even hurt performance. By pinpointing such problems and introducing some simple compiler directives the situation can be improved significantly, however.

It should be stressed that the achieved performance increase is based heavily on the peculiarities of the data structures in the code. If, for instance, only a few small values and a lot of large values (close to KZAHL) were contained in KZHX(), the inner level of the first loop would be very short in the majority of cases which would lead to bad performance on the SR8000. Increasing the length of the array S() to a size comparable with the cache size would also render the prevention of preloads useless.

This code is actually part of a larger MPI-parallel program which typically runs on eight nodes of the SR8000-F1 at LRZ. The performance of the complete application with real-world data sets increased about twofold due to the implemented optimizations.

3.4 Comparison with SGI Origin 3000 architecture

On a CPU with less highly developed pseudo-vectorizing abilities it can be expected that the described optimizations yield smaller performance improvements. This is indeed the case, as a comparison with the SGI architecture reveals. Interestingly, the optimal code for this CPU is the naive approach with

Table 3. Performance improvement with the different code versions on an SGI Origin 3400 (described in Sect. 1. 'Perfect' refers to the fastest version on the SR8000.

Version	MFlops appl.
vanilla	97
1st loop naive	132
1st loop naive, 2nd perfect	149
1st loop perfect, 2nd perfect	123

maximum register reuse. Due to the lack of `preload` functionality, the compiler does not generate pseudo-vectorized data streams for `S()` and `WERTT()` from the start. It is, however, able to unroll the middle loop over the horizontal direction in Fig. 5 to make the loop body fatter. Performance data is as indicated in Table 3. Hardware counter analysis yields an execution rate of two instructions per cycle with the best code version, a value that can not realistically be improved any further.

4 SMP Parallelization of a Strongly Implicit Solver

4.1 Incomplete LU-decomposition – SIP-solver

In this section a strongly implicit solver according to Stone [8] is taken as an example to demonstrate the performance impact of different implementations on the Hitachi SR8000 and other architectures. The algorithm consists of 4 parts, namely the LU-decomposition, which has to be done once, calculation of the residual, forward and backward substitution. The last three steps are performed until the residual is small enough.

All parts of the algorithm except for the calculation of the residual, contain certain data dependencies. As an example, the straightforward way to do the backward substitution looks as follows:

```
1  do k=2,kMaxM
2     do j=2,jMaxM
3        do i=2,iMaxM
4           RES(i,j,k)=(RES(i,j,k)-LB(i,j,k)*RES(i,j,k-1)-
5      *                LW(i,j,k)*RES(i-1,j,k)-
6      *                LS(i,j,k)*RES(i,j-1,k))*LP(i,j,k)
7        enddo
8     enddo
9  enddo
```

Three do-loops iterate over a couple of arrays containing the required data for the update of `RES`. Obviously, before point (i, j, k) is calculated, the points $(i - 1, j, k)$, $(i, j - 1, k)$ and $(i, j, k - 1)$ have to be processed in advance. As a result there is no parallelization possible initially.

4.2 Hyperplane Version

A closer look at the data dependencies leads to a well-known vectorizable version [8]. Given a point (i, j, k) in the so-called "hyperplane" $L = i + j + k = \text{const.}$, only accesses to points in planes $L \pm 1$ are necessary. Having this in mind, one can proceed through all points of a hyperplane, creating long loops with no dependencies. This makes this kind of implementation especially suitable for vector machines. However, by using hyperplanes each point implies an indirect and non-contiguous memory access (see Fig. 7).

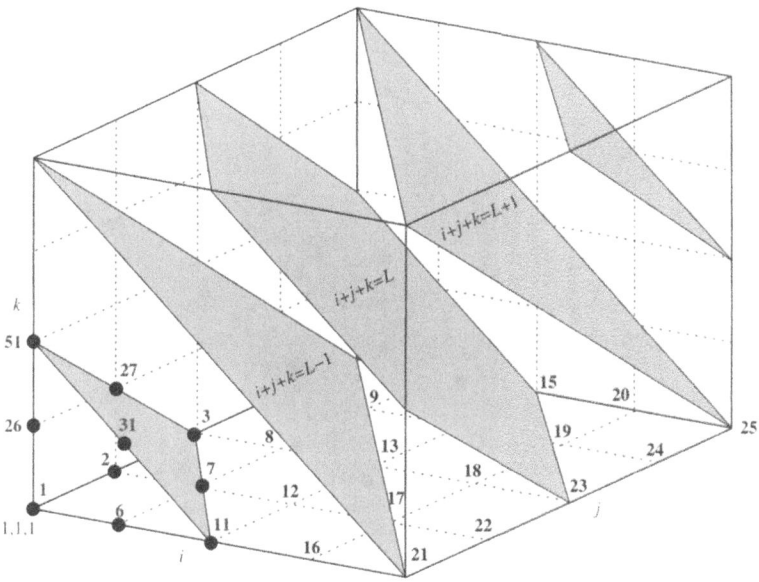

Fig. 7. Hyperplanes with $L = i + j + k = \text{const.}$ in a cube.

In addition, this version allows for a parallelization of the calculation of all points *within* one hyperplane. However, the hyperplanes have to be updated one after another. The following code fraction shows the forward substitution:

```
1   do l=1,hyperplanes
2
3       n=ICL(l)
4
5       do m=n+1,n+LM(l)
6           ijk=IJKV(m)
7           RES(ijk)=(RES(ijk)-LB(ijk)*RES(ijk-ijMax)-LW(ijk)*
8               *           RES(ijk-1)-LS(ijk)*RES(ijk-iMax))*LP(ijk)
9       enddo
10  enddo
```

The inner loop is parallelized using appropriate compiler/OpenMP directives.

4.3 Hyperline Version

One could as well define the hyperplanes by $L = i + k =$ const., making them parallel to the j direction. The term "hyperline" was coined to illustrate this strategy. Again, for constant L only points in $L \pm 1$ are accessed, enabling simple shared memory parallelization within the hyperline. This is shown in Fig. 8. The forward substitution looks similar to the hyperplane version:

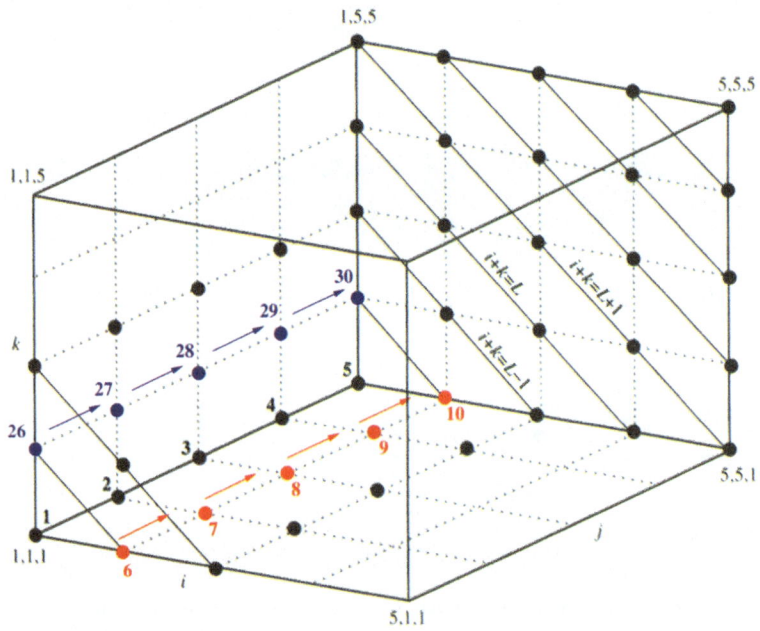

Fig. 8. Hyperlines with $L = i + k =$ const. in a cube.

```
1   do l=1,hyperlines
2
3           do diag = startStopDiag(1,l),startStopDiag(2,l)
4
5           k=coordDiag(1,diag)
6           i=coordDiag(2,diag)
7
8           do j=2,jMaxM
9               ijk=(k-1)*ijMax+(i-1)*jMax+j
10              RES(ijk)=(RES(ijk)-LB(ijk)*RES(ijk-ijMax)-
11          *               LW(ijk)*RES(ijk-1)-LS(ijk)*
```

```
12          *              RES(ijk-jMax))*LP(ijk)
13            enddo
14        enddo
15    enddo
```

The advantage is that now all points in j-direction are accessed continuously, leading to increased data locality. The full content of a cache line that is loaded from memory can be used at least once before the next one is accessed. This is usually not the case for the hyperplane version, where every `load` from memory is punished with latency.

4.4 Pipeline Parallel Processing

Going back to the straightforward version described in Sect. 4.1, Hitachi's Fortran 90 compiler is capable of so-called *pipeline parallel processing* (Fig. 9). Parallelization is done for the loop along the i direction (the middle loop), but calculation of any chunk is delayed by a barrier until the chunk left of it is processed. Consequently, blocks with equal color in Fig. 9 are calculated concurrently. This leads to load imbalance in the "windup" and "winddown" phases of this pipeline, but this effect is negligible for a large enough lattice.

The difference to the hyperline version is that the parallelization is done for the (3D) middle loop instead of a hyperline. As indicated by the arrows inside the rightmost 'C' block in Fig. 9, outer loop unrolling is possible and reduces the load-to-flop ratio.

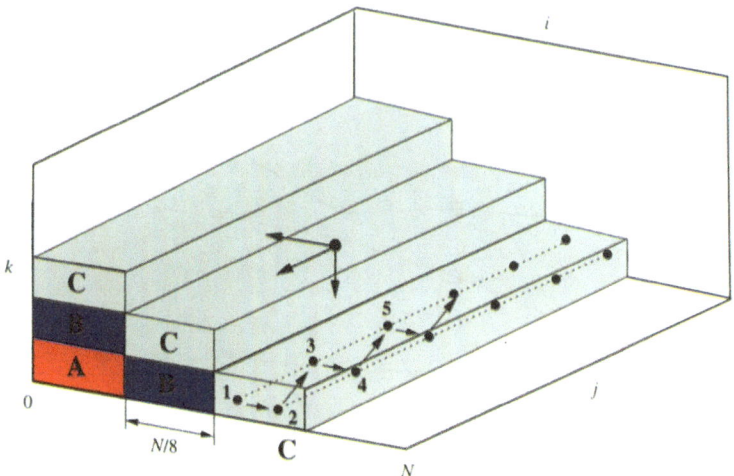

Fig. 9. Schematic view of pipeline parallel processing. Blocks with equal colors (or letters) are calculated simultaneously. One CPU always processes the points inside a certain range of the middle (i) loop.

4.5 Results and Comparison

All experiments were performed with a benchmark code whose basic core was obtained from [9]. It decomposes a given matrix once and performs a fixed number of iterations depending on problem size (10000–500 for size 31^3–91^3). Considering runtime, the LU-decomposition is therefore insignificant. Memory consumption for problem size 91^3 is about 100 MB.

Fig. 10 shows the performance on the Hitachi SR8000 for one CPU and one node for different problem sizes. Obviously the pipeline parallel processing version (3D) performs best and the vectorized hyperplane version is worst.

Fig. 11 contains performance data for all the platforms under consideration; the fastest code variant was chosen for each. The hyperplane version shows outstanding performance on the NEC SX5 for a single CPU run. This version is highly suited for vector architectures as it provides long vector lengths within hyperplanes (there are periodic degradations of the performance which might emerge from disadvantageous access patterns and have not been investigated further). Nevertheless, the 3D version on one Hitachi node outperforms all other versions by far.

4.6 Summary

A straightforward port of the hyperplane version to the Hitachi, exploiting the PVP-feature, yields lower performance than the pipeline parallel processing

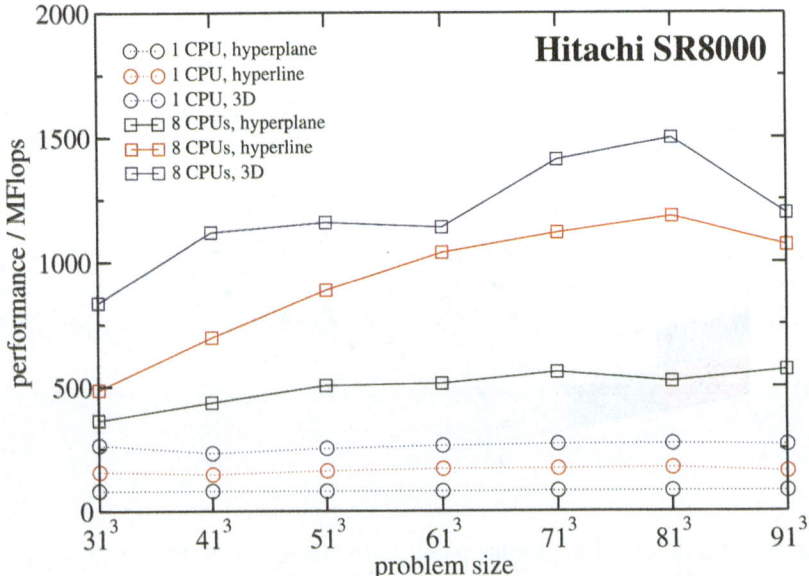

Fig. 10. Hitachi SR8000: performance for SIP-solver variants.

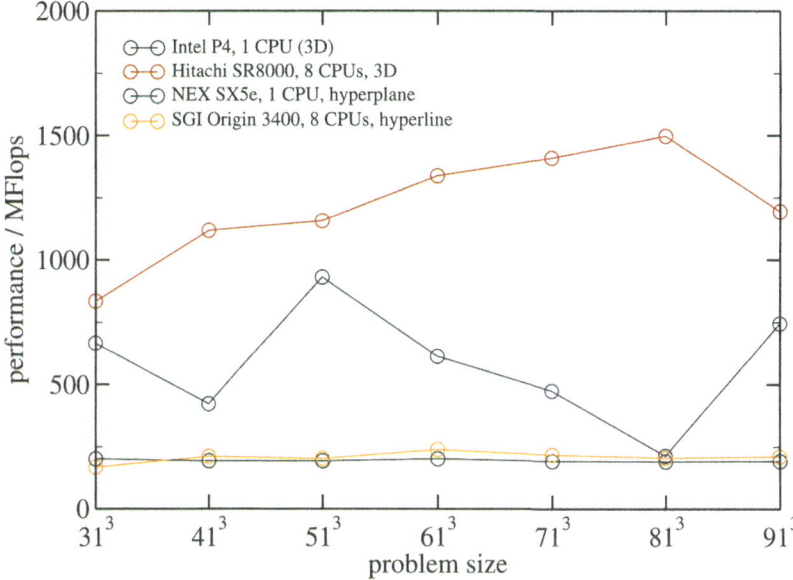

Fig. 11. SIP-solver benchmark: performance on different architectures. The fastest code was chosen on each machine.

version. In spite of the `preload` mechanism, indirect memory access and varying loop lengths make for slower code. Pipelined parallelism, on the other hand, provides the possibility for outer loop unrolling and cache reuse and is thus the best choice on this machine.

5 Conclusions and Acknowledgments

The Hitachi SR8000 system shows performance characteristics which are similar to comparable vector architectures. A vector-like programming style is thus usually a good guess at first, but when cache locality and classic RISC optimization strategies are applicable, it is by all means advisable to use those features. Sometimes, however, one must rely on more uncommon techniques to obtain really outstanding performance.

We would like to thank the HPC team at the LRZ for ongoing support and fruitful discussions. This work was supported by the Competence Network for Scientific High Performance Computing in Bavaria (KONWIHR).

References

1. Available at http://www.top500.org/
2. K. Shimada, T. Kawashimo, M. Hanawa, R. Yamagata and E. Kamada: *A Superscalar RISC Processor with 160 FPRs for Large Scale Scientific Processing.* Proc. of International Conference on Computer Design (1999), pp. 279-280
3. W. Schönauer: *Architecture and Use of Shared and Distributed Memory Parallel Computers,* eds.: W. Schönauer, ISBN 3-00-005484-7.
4. G. Wellein, G. Hager, A. Basermann, and H. Fehske, in *Proceedings of VEC-PAR2002,* Porto (2002).
5. R. Barrett et al.: *Templates for the Solution of Linear Systems: Building Blocks for Iterative Methods,* SIAM, Philadelphia (1993).
6. M. Kinateder, G. Wellein, A. Basermann, and H. Fehske, in *High Performance Computing in Science and Engineering '00,* edited by E. Krause and W. Jäger, Springer-Verlag, Berlin Heidelberg (2001), pp. 188–204.
7. B. Pfitzinger, H. M. Hofmann and G. M. Hale: *Elastic p-^3He and n-^3H scattering with two- and three-body forces.* Phys. Rev. C **64** (2001) 044003
8. J. H. Ferziger, M. Perić: *Computational Methods for Fluid Dynamics.* Springer Verlag, 1999
9. Basic code examples for the algorithms in [8] can be obtained from ftp://ftp.springer.de/pub/technik/peric/

Automatic Performance Analysis on Hitachi SR8000

Michael Gerndt, Andreas Schmidt, Martin Schulz, and Roland Wismüller

Institut für Informatik, LRR
Technische Universität München
85748 Garching, Germany
{*gerndt, andreas.schmidt, schulzm, wismuell*} *@in.tum.de*

Abstract. Performance analysis for applications on teraflop computers requires a new combination of concepts: online processing, automation, and distribution. This article presents the design of a new analysis system that performs an automatic search for performance problems. This search is guided by a specification of performance properties based on the APART Specification Language. The analysis system is currently being implemented for the Hitachi SR8000 teraflop computer within the Peridot project.[1]

1 Introduction

Performance analysis of applications is crucial for teraflop computers. On such machines performance analysis cannot be done in the classical way, i.e. generating massive amounts of performance data, dumping those data to the file system, transferring the data to a workstation, and using parts of the data in the analysis with visual performance tools. Neither the data handling nor the visualization tools are scalable to the level of thousands of processors.

A new combination of concepts has to be applied to do performance analysis on such machines: online processing, automation, and distribution.

The analysis system has to automatically search for performance problems at runtime. This search has to be guided by a specification of performance properties that are typical for applications on such a machine. Taking into account this specification, the amount of performance data to be collected can be reduced to those data that are really necessary.

Although automatic search will already reduce the required data, a distributed implementation is required to deduce performance problems from the raw data locally where the data have been measured.

---

[1] Part of this work is funded by the Compentence Network for High-Performance Computing in Bavaria (KONWIHR, http://konwihr.in.tum.de) and by the European Commission via the working group on Automatic Performance Analysis: Real Tools (APART), http://www.fz-juelich.de/apart

Our approach adds to those concepts support for common design patterns of parallel applications. This information, such as the master/slave pattern, will be utilized to make the analysis process more effective as well as to search for pattern-related performance problems.

Before we present our design in detail we discuss related work in Sect. 2. A big impact on our design had the results of the APART working group on *Automatic Performance Analysis: Resources and Tools.* We outline the concepts of the APART Specification Language (ASL) in Sect. 3 and give examples of performance properties in Sect. 4. The next section introduces the main features of our target machine, the Hitachi SR8000. Sect. 6 introduces parallel program design patterns. The actual design is described in Sect. 7.

2 Related Work

The most relevant project for our work is the Paradyn project at University of Wisconsin-Madison [6]. Paradyn performs an automatic online analysis using dynamic instrumentation for monitoring. The *Performance Consultant* (PC) searches for performance bottlenecks according to the W^3 Search Model. Instrumentation is dynamically inserted taking into account the current set of bottleneck hypotheses and the instrumentation overhead.

Our design differs from Paradyn in several aspects. The analysis is done in a distributed fashion, the potential performance bottlenecks are specified in a high-level notation, and program design patterns will be taken into account. The monitoring will be dynamic but will not require to patch the executable.

The European working group on APART defined a specification language for performance properties of parallel programs. This language and the catalogues of relevant performance properties collected for message passing and shared memory programs will be the foundation for our analysis system.

There are a number of other projects working on automatic performance analysis. KAPPA-PI [1] and Earl/Expert [9] are offline-tools searching for performance properties in trace files of message passing programs. Although these tools apply an automatic search, they will suffer from the well known problems in tracing performance data.

Autopilot [7] is an online tool that consists of sensors and actuators collecting performance data and dynamically tuning the application. Agents are used in the JAMM project [8] to collect performance data for resource management tasks. Both environments are not designed for searching performance problems in parallel application but provide distributed analysis services.

3 APART Specification Language

The *APART Specification Language* (ASL) allows to specify typical performance properties of programs in a formal notation [2, 4]. The specification is based on an object-oriented model of static and dynamic data related to the application's performance. The notation of the object model is similar to Java while a special syntax is used for the specification of properties.

Specifications with ASL are based on the following terminology:

Performance-related Data: Performance-related data, i.e. static program information as well as performance data measured during execution, are required for the evaluation of a program's performance.

Performance Property: A performance property (e.g. load imbalance, communication, cache misses, redundant computations, etc.) characterizes a specific performance behavior of a program and can be checked by a set of *conditions*. Conditions are associated with a *confidence value* (between 0 and 1) indicating the degree of confidence about the existence of a performance property. In addition, for every performance property a *severity figure* is provided that specifies the importance of the property.

Performance Problem: A performance property is a performance problem, iff its severity is greater than a user- or tool-defined threshold.

The ASL provides property templates for specifying similar performance properties in a compact way. In addition, property templates allow to define *metaproperties*. Metaproperties allow to specify new properties by combining alread defined properties. These techniques lead to extremely compact specifications.

4 Performance Properties

The following example demonstrates the specification of performance properties based on property templates.

```
PROPERTY TEMPLATE CostPerProcess  <float CostFunc(MPISummary)>
       (Region r, Experiment e, Process p, Region RankBasis){
LET
  cost = CostFunc(summary(r,e,p))
IN
  CONDITION: cost > 0;
  CONFIDENCE:1;
  SEVERITY:  cost/duration(RankBasis,e);
}

float IoCostFunc(MPISummary rs) =   rs.IoTime;
float SyncCostFunc(MPISummary rs) = rs.SyncTime;
float CommCostFunc(MPISummary rs) = rs.CommTime;
```

```
PROPERTY CostPerProcess <IoCostFunc>   IoCostPerProcess;
PROPERTY CostPerProcess <SyncCostFunc>SyncCostPerProcess;
PROPERTY CostPerProcess <CommCostFunc>CommCostPerProcess;
```

Template *CostPerProcess* depends on a function and defines the context of the property, i.e. the relevant region, the performed experiment, the process, and a region to which the performance data are compared.

The condition for the existence of the property, the confidence into the condition, and the severity are specified in separate sections. All the expressions access data in the performance-related data model not shown here, e.g. *summary (r,e,p)* returns the summary data for region *r* in experiment *e* for process *p*. The severity is determined by comparing the overhead to the total execution time of the ranking basis, usually the entire program.

Based on this template, different properties for IO, synchronization, and communication can be defined based on appropriate overhead functions. These functions access the measured IO overhead, synchronization overhead, and the communication overhead in the summary data of the given program region.

Metaproperties combine allready defined properties. In the following example we define a property template *PropertyOnAllProcesses* that specifies that the property *PropertyPerProcess* holds across the set of all processes. Its condition is fulfilled if the condition of property *PropertyPerProcess* and the corresponding severity holds across all processes involved in an experiment.

```
PROPERTY TEMPLATE PropertyOnAllProcesses
    <PROPERTY x(Region, Experiment, Process, Region)>
    (Region r, Experiment e, Region RankBasis) {

  CONDITION:
    Forall p IN e.processors SUCH THAT
        condition(x(r,e,p,RankBasis)) AND
        confidence(x(r,e,p,RankBasis))==1;
  CONFIDENCE:   1;
  SEVERITY:
    max(severity(x(r,e,p,RankBasis)) WHERE p in e.processes)
}


PROPERTY PropertyOnAllProcesses <IoCostPerProcess> IoCost;
PROPERTY PropertyOnAllProcesses <SynCostPerProcess> SyncCost;
PROPERTY PropertyOnAllProcesses <CommCostPerProcess> CommCost;
```

The properties resulting from this property template combine information from multiple processes. This demonstrates how the ASL can be used to structure properties into a hierarchy according to the processes taken into account.

5 Hitachi SR8000

The target machine for our performance analysis system is the Hitachi SR8000 at Leibniz-Rechenzentrum in Munich. This machine is the first teraflop computer in Europe. The system has a peak performance of 2.2 TFlops.

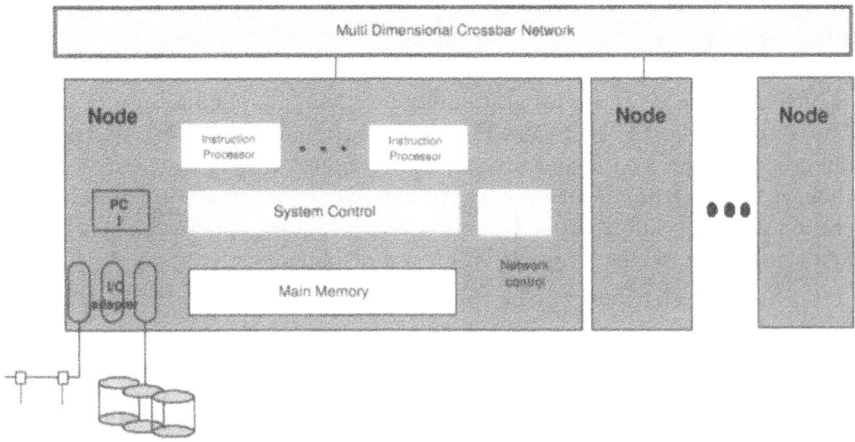

Fig. 1. The Hitachi SR8000 is a clustered SMP system. Each node is a shared memory multiprocessor with nine CPUs. The nodes are connected via multiple crossbar switches in a three dimensional cube topology.

The architecture of the Hitachi is shown in Fig. 1. It is a clustered SMP design, each node consists of eight application processors and one system processor sharing a common memory via a crossbar switch. The processor is based on the RS6000 design which was extended by a large register set and prefetch and preload operations that allow to hide memory latency. The memory bandwidth of a processor is 4 GByte/s, which is quite high for a processor performing up to 1.5 GFlops. A storage controller takes care of the cache coherence and supports efficient shared memory execution. Parallel loops can be set up as well as synchronized via hardware.

The current machine consists of 168 SMP nodes. The nodes are connected via a three dimensional crossbar network, i.e. the nodes in a row, column, or vertical line are connected via a crossbar. The crossbars are connected in a cube topology. Thus, the distance between two arbitrary nodes is no more than three hops.

The processors can be utilized either via pure message passing or via hybrid programming. While in the first model each processor runs one MPI process, in the second model, a node runs an MPI process which itself is parallelized with OpenMP or COMPASS, the Hitachi proprietary shared memory programming interface.

6 Parallel Program Design Patterns

Parallel programs designed for such a machine will be based on a small standard set of design patterns. The most common patterns are: SPMD, Master/Slave, Work Pool, Pipeline, and Embarrasingly Parallel. In addition, typical application-level patterns can be identified, e.g. a simulation code consists of a setup phase, a time-step loop, and a postprocessing phase. These application-level patterns can be specialized for application areas, e.g. special loops in the program computing the forces among the particles.

Currently, there are no programming environments making use of those patterns in standard Fortran and C/C++ programs since these patterns are hidden in the source code and programmers are not willing to use new environments for program development. We will allow the programmer to incrementally add this information to an existing program and thus, we will be able to make use of this information for performance analysis.

7 Design

The distributed performance analysis system consists of a set of *analysis agents*. These agents are autonomous and cooperate in the detection of performance properties.

The agents are logically arranged into a hierarchical structure. This structure is set up according to the hierarchical structure in the ASL specification. Each property is evaluated for a specific set of processes (or threads), e.g. an individual process, all processes on an SMP-node, on a set of nodes, or on the entire collection of nodes assigned to the application program. Examples for sets of processes are the processes being slaves and the process being the master in the master/slave pattern.

Agents are responsible for the evaluation of those properties relevant to processes assigned to them. For example, node level agents check properties for the processes in their node, while the machine level agent checks global properties like *for all processes property xyz holds*.

Higher level agents either combine properties detected by lower level agents or they assign subtasks in the evaluation of a global property to the appropriate lower level agents. The main point is, that agents do not exchange a lot of performance data, but exchange higher level information, such as the set of detected properties.

The agent hierarchy does not imply that each of the logical agents shown must be implemented by a separate physical agent. A physical agent may as well play the role of several logical agents simultaneously, which are located on different levels of the hierarchy. The actual structure at runtime will depend on the size of the application, the amount of resources a user is willing to assign to the agent hierarchy, and the amount of work to be done on the individual levels.

On the Hitachi, we can take advantage of the system processor on each of the nodes. Thus we will assign at least one agent to each node. It will be responsible for all properties on that node, individual properties of the processes as well as node-global properties. Depending on the characteristics mentioned in the previous paragraph the user might want to assign additional processors to the agent hierarchy. Although those processors will not be available for the computation, an efficient performance analysis system will be the only way to analyze large application with thousands of processors on a teraflop machine. The assignment of additional processors will reduce the influence of the analysis on the execution behavior of the application.

The interaction of the application programmer or external tools with the agent hierarchy will be realized via an *analysis portal* executed on a front end machine, e.g. a Web server. It will allow to access information and to send commands from different devices such as workstations and mobile devices.

7.1 Analysis Agent

The analysis agents perform the following tasks:

- Detection of performance problems related to the agent's level in the hierarchy.
- Leaf agents request and receive performance data from application processes. They take into account the instrumentation overhead as well as instrumentation constraints, e.g. limited number of hardware counters.
- Leaf agents receive requests from higher level agents for the partial evaluation of properties. They evaluate those properties with respect to their local information.
- Agents report performance problems and partially evaluated properties back to the next higher-level agents either to allow those agents to deduce more global problems or to make the problems available to the programmer or other clients. The problems will be accompanied by information summarizing the performance data that lead to the detection of the problem.

An agent performs the analysis with respect to program phases. A *program phase* is a portion of the program execution and is related to the execution of a specific code region. It is best determined by the application programmer via the specification of program patterns. Typical phases are the initialization phase, the iterations of a time loop, and the output phase. Phases have to be characterized as being repetitively executed or executed only once. If the programmer did not specify the phases, the fall back position is to either determine phase boundaries based on global synchronization or by the invocation and termination of functions with large granularity.

When a phase starts, the agent creates a set of hypotheses about performance problems based on its internal state, the ASL specification, and the program structure relevant for the phase. It determines the required performance data and sends a request to the application. At the end of the phase the performance data are received from the application and the hypotheses are evaluated. Detected performance problems are reported to the next higher agent level.

The agent consists of the following entities:

ASL Property Specification Database: It contains the ASL specifications of performance properties to be evaluated on that level of the hierarchy. The database can be dynamically updated by higher level agents, e.g. if new properties result from a program pattern specified by the programmer which was not known when the program was started.

Application Structure Database: It contains information of the program structure. This database can be dynamically updated, too.

Performance Hypotheses Database: The hypotheses that have to be evaluated are stored in this database. It can be augmented by evaluation requests from higher level agents. The hypotheses can be also prioritized by higher level agents.

Performance Problem Database: Detected performance problems are stored in this database.

Performance Data Database: It stores the performance data received from the application as long as it is needed for the evaluation of hpotheses.

Control Unit: It performs the analysis algorithm.

Agent Interface: It connects the agents in the hierarchy. The following information is received and sent via this interface:
  - ASL property specifications (from parent)
  - Program structure information (from parent)
  - Priorities for hypotheses (from parent)
  - Evaluation requests (from parent)
  - Partially evaluated hypotheses (from children)
  - Performance problems (from children)

Application Process Interface: It connects leaf agents with the application processes. It allows to:
  - Send instrumentation requests
  - Receive performance data

7.2 Selective Instrumentation

Performance analysis will be enabled via compile time instrumentation. The Fortran and C/C++ compilers on the Hitachi SR8000 can insert calls to a monitoring library at function entry and exit as well as at parallel program regions. This support allows to attach a new monitoring library. The monitoring library will support selective performance monitoring. Which data are gathered at runtime and thus also how much overhead results from monitoring can be determined dynamically. When the program is started, the instrumentation is executed but no monitoring is switched on. During the analysis process, the leaf agents request performance data resulting in monitoring actions at the application processes.

The interface between the leaf agents and the monitor will be designed to be applicable to other environments. For example, the Grid Monitoring Architecture (GMA) developed by the Global Grid Forum is very similar to our approach. It consists of *information producers* that can measure certain metrics and *information consumers* that request measurements. A similar approach is also used in the Windows 2000 operating system, where the sytem monitor can request information from services running on the system.

7.3 Program Pattern Specification

Although the performance analysis system will be able to detect performance problems without additional pattern information, this information will improve the effectiveness and the user friendlieness considerably.

Pattern information will enable to determine pattern specific performance properties. These properties will lead to performance problems related to a more abstract model of the application. For example, performance problems such as *master is overloaded* can be reported for the master/slave pattern. For a multi particle simulation, the problem *load imbalance due to unequal number of particles assigned to the processes* could be detected.

Pattern information will also increase the effectiveness of the analysis since a more precise knowledge of program phases will certainly allow a more precise selection of hypotheses and a higher confidence in the relevance of detected problems.

The pattern specification will be supported via a graphical tool as well as a text oriented implementation for mobile devices. The mapping of pattern entities to program regions will be done based on line number information that can be compiled into the executable. If the patterns include also data structures, the mapping can only be implemented if additional symbol information is available to the specification tools as well as to the monitoring library.

The pattern-specific ASL properties will be extracted from a specification database at the portal and sent to the agent.

8 Future Work

The performance analysis system outlined in this article is currently being implemented in the Peridot project. It is based on the design developed in the last two years by the APART working group www.fz-juelich.de/apart. Although the APART design describes an offline analysis system, parts of the design have been integrated into our approach.

The described system is based on the performance property catalogues developed within APART for message passing and shared memory programs

[4, 3]. These catalogues have been combined and extended within Peridot for the hybrid programming model.

Our approach avoids the most important problem in performance analysis, the creation of massive amounts of performance data. These amounts will be much larger on teraflop machines than they are on current machines. First, only the information required for detecting performance problems is generated, and second, the data do not have to be dumped to external disks and be visualized with difficult to handle environments.

The integration of programm patterns into the analysis system is a new approach and is very important for the effectiveness of the system and the acceptance by the users. Our approach does not force programmers to invest a lot of time into new languages but allows to integrate the pattern information incrementally while the program and the performance analysis is running. After the user provided the information he will see immediate benefits, thus, we hope, that the system's application will not only lead to correctly detected performance problems but will also steer the user's curiosity.

References

1. A. Espinosa, T. Margalef, E. Luque: *Automatic Performance Evaluation of Parallel Programs*, Sixth Euromicro Workshop on Parallel and Distribued Processing, 1998
2. T. Fahringer, M. Gerndt, G. Riley, J.L. Träff, *Specification of Performance Problems in MPI-Programs with ASL*, International Conference on Parallel Processing (ICPP'00), pp. 51-58, 2000
3. T. Fahringer, M. Gerndt, G. Riley, J.L. Träff, *Formalizing OpenMP Performance Properties with the APART Specification Language (ASL)*, International Workshop on OpenMP: Experiences and Implementation, Lecture Notes in Computer Science, Springer Verlag, Tokyo, Japan, pp. 428-439, October, 2000
4. T. Fahringer, M. Gerndt, G. Riley, J.L. Träff, *Knowledge Specification for Automatic Performance Analysis*, APART Technical Report, www.fz-juelich.de/apart, 2001
5. K. Karavanic, B.P. Miller, *Improving Online Performance Diagnosis by the Use of Historical Performance Data* Supercomputing '99, Portland, Oregon (USA), November 1999
6. B.P. Miller, M.D. Callaghan, J.M. Cargille, J.K. Hollingsworth, R.B. Irvin, K.L. Karavanic, K. Kunchithapadam, T. Newhall, *The Paradyn Parallel Performance Measurement Tool*, IEEE Computer, Vol. 28, No. 11, pp. 37-46, 1995
7. R. Ribler, J. Vetter, H. Simitci, D. A. Reed, *Autopilot: Adaptive Control of Distributed Applications*, 7th IEEE Symposium on High-Performance Distributed Computing, 1998
8. B. Tierney, B. Crowley, D. Gunter, M. Holding, J. Lee, M. Thompson, *A Monitoring Sensor Management System for Grid Environments*,IEEE High Performance Distributed Computing conference (HPDC-9), August 2000, LBNL-45260.
9. F. Wolf, B. Mohr, *Automatic Performance Analysis of MPI Applications Based on Event Traces*, EuroPar 2000, LNCS 1900, pp. 123 - 132, August 2000

Adapting PAxML to the Hitachi SR8000-F1 Supercomputer

Alexandros P. Stamatakis[1], Thomas Ludwig[2], and Harald Meier[1]

[1] Technical University of Munich, Department of Computer Science
  Boltzmannstr. 3, 85748 Garching, Germany
  Alexandros.Stamatakis@in.tum.de, Harald.Meier@in.tum.de
[2] Ruprecht-Karls-University, Department of Computer Science
  Im Neuenheimer Feld 348, 69120 Heidelberg, Germany
  thomas.ludwig@informatik.uni-heidelberg.de

Abstract. Heuristics for the NP-complete problem of calculating the optimal phylogenetic tree for a set of aligned rRNA sequences based on the maximum likelihood method are computationally expensive. Thus, supercomputers and appropriately optimized and adapted parallel programs are required in order to perform computations of large phylogenetic tress.

The core of most parallel algorithms is the tree evaluation function, calculating the likelihood value for each tree topology. It accounts for the greatest part of overall computation time. This paper introduces a general method for significantly accelerating the computation of the maximum likelihood value for a given tree topology by using SEVs (Subtree Equality Vectors) to significantly reduce the number of floating point operations during each topology evaluation. Furthermore, we present the parallel implementation of our method in a program called **PAxML** derived from **parallel fastDNAml** and describe its adaptation to the Hitachi SR8000-F1 architecture.

Tests performed with various sequential and parallel phylogeny programs on SUN workstations show global run time improvements of 35% to 47% induced by our method, whereas analogous initial tests on the Hitachi SR8000-F1, with an appropriately adapted version of **PAxML** show global run time improvements of 24% to 35% over **parallel fastDNAml**, as well as a promising floating point performance per processor ranging between 106 Mflops/s and 129 Mflops/s.[1]

1 Introduction

At the **ParBaum** project at the TUM (Technische Universität München) work is conducted to facilitate large-scale parallel phylogenetic tree computations on trees of at least 1000 taxa on the Hitachi SR8000-F1 supercomputer [1] (rank 14 in the top 500 supercomputers list, June 2002 [8]) installed

---

[1] This work is sponsored under the project ID **ParBaum**, within the framework of the "Competence Network for Technical, Scientific High Performance Computing in Bavaria": Kompetenznetzwerk für Technisch-Wissenschaftliches Hoch- und Höchstleistungsrechnen in Bayern (KONWIHR). KONWIHR is funded by means of "High-Tech-Offensive Bayern".

at the LRZ (Leibniz-Rechenzentrum) in Munich. Our work relies on sequence data provided by the ARB [7] rRNA-sequence database, developed jointly by the LRR (Lehrstuhl für Rechnertechnik und Rechnerorganisation) and the Department of Microbiology of the TUM. The ARB database provides a huge amount of sequence data with excellent alignment quality.

Like many problems associated with genome analysis, the perfect phylogeny problem is NP-complete. Thus, the introduction of heuristics for reducing the search space in terms of potential tree topologies evaluated becomes inevitable. Heuristics for phylogenetic tree calculations still remain computationally expensive, mainly due to the high cost of the tree likelihood function, which is invoked repeatedly for each tree topology analyzed.

Thus, only relatively small trees ($\approx$ 500 taxa [5] [6]), compared to the huge amount of data available ($\approx$ 20000 sequences in the ARB database), have been calculated so far.

We focus on *three* key areas to attain our goal of producing large, high quality evolutionary trees:

1. *Improvement of the existing algorithms* by introduction of new heuristics and algorithmic optimizations.
2. Adaptation of the existing algorithms to *hybrid supercomputer architectures.*
3. Integration of *empirical biological knowledge* into algorithms.

This paper describes work concerning point 1 and 2.

Firstly, we present a new algorithmic optimization for accelerating the computation of the topology evaluation function.

We implemented the optimizations in **AxML** [3] (A(x)ccelerated Maximum Likelihood) and **PAxML** (Parallel AxML) based on the latest sequential and parallel releases of **fastDNAml** (v.1.2.2). Furthermore, we have demonstrated the generality of our approach by incorporating our optimization into **TrExML**, a program with a more extensive tree space exploration strategy than **fastDNAml**. We call the resulting program **ATrExML** (Accelerated **TrExML**).

Our experiments obtained total run time reductions ranging from 24% to 47% for various data sets and all three programs mentioned above, i.e. **AxML, PAxML, ATrExML.**

Secondly, after having demonstrated that our approach scales well to the parallel program on conventional processor architectures, we describe the necessary modifications of **PAxML** for an initial adaptation to the Hitachi SR8000-F1 processor architecture. Within this context we present initial experimental results for test runs conducted on the SR8000-F1, including promising Mflops/s/processor performance and global run time improvements over **parallel fastDNAml.**

These results are promising first steps toward efficient determination of large, high quality evolutionary trees using supercomputers, since we have significantly accelerated a program that has already been used for large scale phylogenetic tree computations on supercomputers [5].

2 Subtree Column Equalities

In general the cost of the likelihood function and the branch length optimization function, which accounts for the greatest portion of execution time (95% in the sequential version of **fastDNAml**), can be reduced in two ways:

Firstly, by reducing the size of the search space using some additional heuristics, i.e. reducing the number of topologies evaluated and thus reducing the number of likelihood function invocations. This approach might, however, over look high quality trees.

Secondly, by reducing the number of sequence positions taken into account during computation and thus reducing the number of computations at each inner node during each tree's evaluation.

We consider the second possibility through a detailed analysis of column equalities. Two columns in an alignment are equal and belong to the same *column class* if, on a sequence by sequence basis, the base is the same. A homogeneous column consists of the same base, whereas a heterogeneous column consists of different bases.

More formally, let $s_1, ..., s_n$ be the set of aligned input sequences as depicted in the upper matrix of Fig. 1.

Let m be the number of sequence positions of the alignment. We say, that two columns of the input data set i and j are equal if $\forall s_k, k = 1, ..., n : s_{ki} = s_{kj}$, where s_{kj} is the j-th position of sequence k. One can now calculate the number of equivalent columns for each column class of the input data set.

After calculating column classes, one can compress the input data set by keeping a single representative column for each column class, removing the equivalent columns of the specific class and assigning a count of the number of columns the selected column represents, as depicted in Fig. 1.

Since a necessary prerequisite for a phylogenetic tree calculation is a high-quality multiple alignment of the input sequences one might expect quite a large number of column equalities on a global level. In fact, this kind of global data compression is already performed by most programs. Unfortunately, as the number of aligned sequences grows, the probability of finding two globally equal columns decreases. However, it is reasonable to expect more equalities on the subtree, or local, level.

The fundamental idea of this paper is to extend this compression mechanism to the subtree level, since a large number of column equalities might be expected on the subtree level. Depending on the size of the subtree, fewer sequences have to be compared for column equality and, thus, the probability of finding equal columns is higher.

None the less, we restrain the analysis of subtree column equality to homogeneous columns for the following reason:

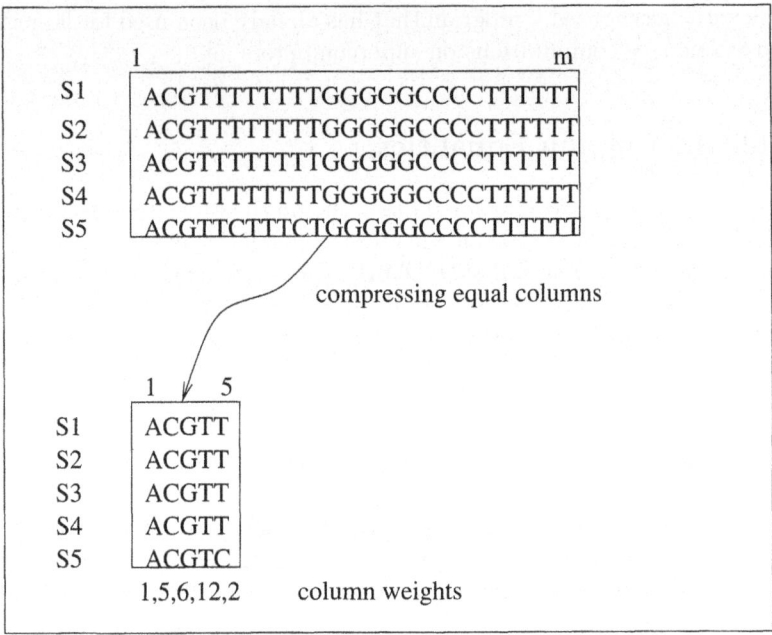

Fig. 1. Global compression of equal columns, all column weights are 1 in the uncompressed matrix

The calculation of heterogeneous equality vectors at an inner node p is complex and requires the search for c^k different column equality classes, where k is the number of tips (sequences) in the subtree of p and c is the number of distinct values the characters of the sequence alignment are mapped to. (E.g., **fastDNAml** uses 15 different values.) This overhead would not amortize well over the additional column equalities we would obtain, especially when $c^k > m'$ where m' is the length of the compressed global sequences.

We now describe an efficient and easy way for recursively calculating subtree column equalities using SEVs (Subtree Equality Vectors).

Let s be the virtual root placed in an unrooted tree for the calculation of its likelihood value. Let p be the root of a subtree with children q and r, relative to s. Let ev_p (ev_q, ev_r) be the equality vector of p (q, r, respectively), with size m'. The value of the equality vector for node p at position i, where $i = 1, ..., m'$ can be calculated by the following function (see example in Fig. 2):

$$ev\_p(i) := \begin{cases} ev\_q(i) \; if \quad ev\_q(i) = ev\_r(i) \\ -1 \qquad else \end{cases} \tag{1}$$

If p is a leaf, we set $ev\_p(i) := map(sequence\_p(i))$, where, $map()$ is a function that maps the character representation of the aligned input sequence $sequence\_p$, at leaf p to values $0, 1, ..., c$. Thus, the values of an inner SEV

ev_p, at position i, range from $-1, 0, ..., c$, i.e. -1 if column i is heterogeneous and from $0, ..., c$ in the case of an homogeneous column.

For SEV values $0, ..., c$ a pointer array $ref\_p(c)$ is maintained, which is initialized with $NULL$ pointers, for storing the references to the first occurrence of the respective column equality class in the likelihood vector of the current node p.

Thus, if the value of the equality vector $ev\_p(j) > -1$ and $ref\_p(ev\_p(j)) \neq NULL$ for an index j of the likelihood vector $lv\_p(j)$ of p, the value for the specific homogeneous column equality class $ev\_p(j)$ has already been calculated for an index $i < j$ and a large block of floating point operations can be replaced by a simple value assignment $lv\_p(j) := lv\_p(i)$. If $ev\_p(j) > -1$ and $ref\_p(ev\_p(j)) = NULL$, we assign $ref\_p(ev\_p(j))$ to the address of $lv\_p(j)$, i.e. $ref\_p(ev\_p(j)) := adr(lv\_p(j))$.

The additional memory required for equality vectors is $O(n * m')$. The additional time required for calculating the equality vectors is $O(m')$ at every node.

The initial approach renders global run time improvements of 12% to 15%. These result from an acceleration of the likelihood evaluation function between 19% and 22%, which in turn is achieved by a reduction in the number of floating point operations between 23% and 26% in the specific function.

It is important to note that the initial optimization is only applicable to the likelihood evaluation function, and *not* to the branch length optimization function. This limitation is due to the fact that the SEV calculated for the *virtual* root placed into the topology under evaluation, at either end of the branch being optimized, is very sparse, i.e. has few entries > -1. Therefore, the additional overhead induced by SEV calculation does not amortize well with the relatively small reduction in the number of floating point operations (2% - 7%). Note however, that the SEVs of the *real* nodes at either end of the specific branch do not need to be sparse, but depends on the number of tips in the respective subtrees.

We now show how to efficiently exploit the information provided by an SEV, in order to achieve a further significant reduction in the number of floating point operations by extending this mechanism to the branch length optimization function.

To make better use of the information provided by an SEV at an inner node p with children r and q, it is sufficient to analyze at a high level how a single entry i of the likelihood vector at p, $lv\_p(i)$, is calculated:

$$lv\_p(i) := f(g(lv\_q(i), z(p, q)), g(lv\_r(i), z(p, r)), \tag{2}$$

where $z(p, q)$ $(z(p, r))$ is the length of the branch from p to q (p to r, respectively). The function $g()$ is a computationally expensive function, that calculates the likelihood of the left (right) branch of p, depending on the branch length $z(p, q)$ $(z(p, r))$ and the value of $lv\_q(i)$ $(lv\_r(i)$, respectively). Whereas $f()$ performs some simple arithmetic operations for combining the

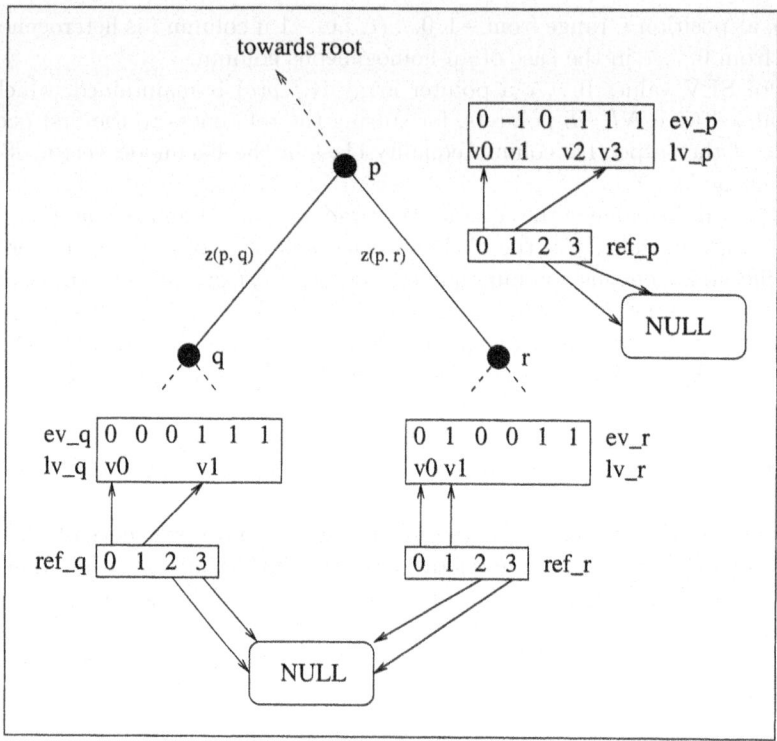

Fig. 2. Example likelihood-, equality- and reference-vector computation for the subtree at p

results of $g(lv\_q(i), z(p,q))$ and $g(lv\_r(i), z(p,r))$ into the value of $lv\_p(i)$. Note that $z(p,q)$ and $z(p,r)$ do not change with i.

If we have $ev\_q(i) > -1$ and $ev\_q(i) = ev\_q(j)$, $i < j$, we have $lv\_q(i) = lv\_q(j)$ and therefore $g(lv\_q(i), z(p,q)) = g(lv\_q(j), z(p,q))$ (the same equality holds for node r). Thus, for any node q we can avoid the recalculation of $g(lv\_q(i), z(p,q))$ for all $j > i$, where $ev\_q(j) = ev\_q(i) > -1$. We precalculate those values and store them in arrays $precalc\_q(c)$ and $precalc\_r(c)$ respectively, where c is the number of distinct character-value mappings found in the sequence alignment.

Our final optimization consists in the elimination of value assignments of type $lv\_q(i) := lv\_q(j)$, for $ev\_q(i) = ev\_q(j) > -1$, $i < j$ where i is the first entry for a specific homogeneous equality class $ev\_q(i) = 0, ..., c$ in ev_q. We need not assign those values due to the fact that $lv\_q(j)$ will never be accessed. Instead, since $ev\_q(j) = ev\_q(i) > -1$ and the value of $g\_q(j) = g\_q(i)$ has been precalculated and stored in $precalc\_q(ev\_p(i))$, we access $lv\_q(i)$ through its reference in $ref\_q(ev\_q(i))$.

During the main for-loop in the calculation of lv_p we have to consider 6 cases, depending on the values of ev_q and ev_r. For simplicity we will write $p\_q(i)$ instead of $precalc\_q(i)$ and $g\_q(i)$ instead of $g(lv\_q(i), z(p, q))$.

$$lv\_p(i) := \begin{cases} f(p\_q(ev\_q(i)), p\_r(ev\_r(i))) & if\ ev\_q(i) = ev\_r(i) > -1, \\ & ref\_p(ev\_r(i)) = NULL \\ skip & if\ ev\_q(i) = ev\_r(i) > -1, \\ & ref\_p(ev\_r(i)) \neq NULL \\ f(p\_q(ev\_q(i)), p\_r(ev\_r(i))) & if\ ev\_q(i) \neq ev\_r(i), \\ & ev\_q(i), ev\_r(i) > -1 \\ f(p\_q(ev\_q(i)), g\_r(i)) & if\ ev\_q(i) > -1, ev\_r(i) = -1 \\ f(g\_q(i), p\_r(ev\_r(i))) & if\ ev\_r(i) > -1, ev\_q(i) = -1 \\ f(g\_q(i), g\_r(i)) & if\ ev\_q(i) = -1, ev\_r(i) = -1 \end{cases} \quad (3)$$

A simple example for the optimized likelihood vector calculation and the respective data-types used is given in Fig. 2.

3 Implementation

We integrated subtree equality vectors into three existing phylogeny programs: **fastDNAml** [2], **parallel fastDNAml** [5] and **TrExML** [9]. We name the optimized versions **AxML**, **PAxML** and **ATrExML** respectively. About 300 lines of code have been added to the various programs, thus demonstrating the efficiency, simplicity and applicability of our approach.

A simple analysis of **fastDNAml** with the `gprof` tool shows that the tree likelihood function `newview()` and the branch length optimization function `makenewz()` consume over 95% of overall execution time. The basic ideas of this paper have been implemented in functions `newview()`, `makenewz()`, `sigma()` and `evaluate()`, since those functions access the likelihood-vectors of the nodes and are affected by the changes induced by skipping assignments of type $lv\_p(i) := lv\_p(j), i < j, ev\_p(j) = ev\_p(i) > -1$.

In each of those functions the main for-loop over the sequence length m' has been modified in order to correspond to formula (3). Furthermore, an additional loop for initializing $precalc\_q(c)$ and $precalc\_r(c)$ has been inserted.

The remaining modifications concern mainly initialization matters, and the definition of a few additional data-types for storing the $precalc()$ and $ref()$ array information.

For **PAxML** we designed a special version consisting of a single binary (`paxml`) instead of three distinct ones (`master, foreman, slave`), for reasons of portability, since the execution of multiple binaries is not supported by all MPI environments.

4 Adaptation to the SR8000-F1

The parallel architecture of **PAxML** consists of a simple master-worker model, with the master distributing the tree topologies to be evaluated in a simple short string representation to the workers, i.e. communication overhead is considered as a matter of low priority for initial adaptation.

Thus, we decided to perform initial tests in intra-node MPI-mode, in order to keep each worker module as compact as possible and to evaluate the scalability of our optimization to the specific processor architecture of the SR8000-F1.

The first tests rendered rather unfavorable results in terms of run time improvement of **PAxML** over **parallel fastDNAml**, compared with the results obtained on conventional processor architectures (see Sect. 5). The problem could however be quickly identified. The case analysis of formula (3) was originally implemented within the computationally expensive for-loops of functions `newview()`, `makenewz()`, `sigma()` and `evaluate()` as nested conditional statement and significantly perturbs the pipelining and prefetch mechanisms of Hitachi's hardware architecture. Therefore, we split up the for-loops within the functions mentioned above, and implemented a distinct for-loop for each case, without further conditional statements within the respective loops. This modification boosted program efficiency, both, in terms of floating point performance and run time reduction, although some additional code had to be inserted for precalculating the loop split.

E.g. the non-adapted **PAxML** code rendered 25.47% run time improvement compared with 34.71% for the adapted one for the 41 taxa mitochondrial rRNA test set executed with two workers (see Sect. 5).

5 Results

The amount of performance improvement strongly depends on the number and length of the input sequences, as well as on the quality of the alignment. We note that whenever more subtree column equalities are expected, performance improves more.

We initially present the results and tests performed on conventional architectures for demonstrating the soundness of our approach and report about first results on the Hitachi SR8000-F1 at the end of this section.

We tested the performance of **AxML**, **PAxML** and **ATrExML** with data sets from various sources and obtained global run time improvements between 35% and 47%. We compiled the sequential programs with `gcc -O3` and executed them under `Solaris` on a `Sun-Blade-1000`. For the parallel programs we used `gcc -O2` with the master and foreman components located on a `Sun-Blade-1000` and two workers, each running on a `Sun Ultra 5/10`.

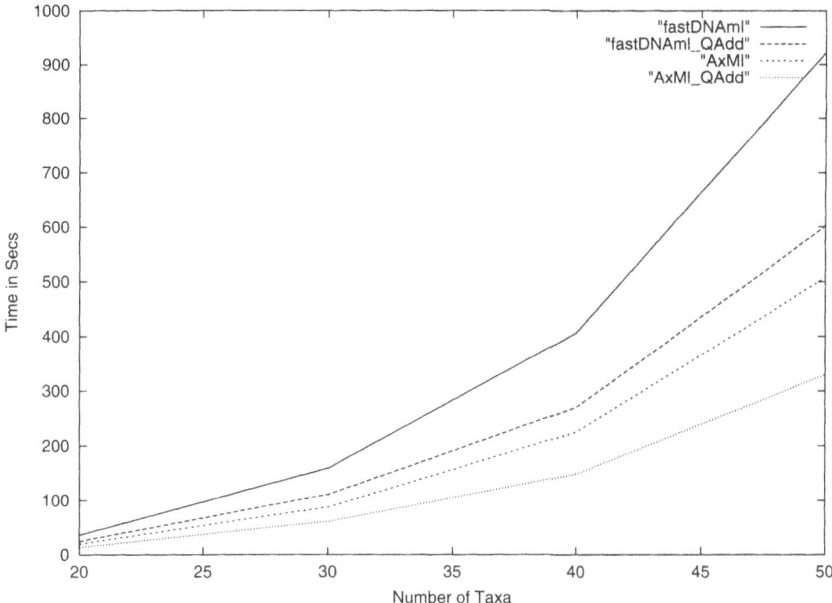

Fig. 3. Overall execution times of fastDNAml and AxML with the quickadd option enabled and disabled

For analyzing the global run time improvement of **AxML/PAxML** over **fastDNAml/parallel fastDNAml** tests with data-sets of 20, 30, 40 and 50 taxa with 303-386 distinct columns (also referred to as distinct data patterns) extracted from the alignment of 56 sequences delivered as test-set with `parallel fastDNAml`, as well as two alignments consisting of 161 mitochondrial rRNA sequences (418 distinct columns) from beetles and butterflies were used. We used different program options and data-sizes for demonstrating the scalability and generality of our method. In Table 1 we present the global run time improvement of **AxML/PAxML** over **fastDNAml/parallel fastD-NAml**, both with the quickadd (local branch length optimization) option enabled and disabled. For the mitochondrial rRNA test sets we executed tests with and without tree rearrangements (for details refer to the **fastDNAml** documentation).

Results for the 20 to 50 taxa sequential test runs are also depicted in Fig. 3. An important result is that **AxML** with the quickadd option disabled, still runs about 15% to 20% faster than **fastDNAml** with the quickadd option enabled, thus ensuring a higher tree quality. Furthermore, the algorithmic optimizations scale well to **PAxML**, even in the case of considerably small test sets for a parallel run as in the case of the 20 to 50 taxa test set. The results obtained from the 161 taxa tests demonstrate the scalability of our approach, both to large test sets, i.e. with a great number of taxa, as well as to the parallel algorithm. The good parallel performance improvement is due to the fact that the tree evaluation function is the core of the worker components, which perform the actual computation (for details refer to [5]).

Table 1. Global run time improvement (parallel) fastDNAml vs. (P)AxML

Data Set	Option	AxML	PAxML	Option	AxML	PAxML
20 taxa	Qadd	44.91%	36.74%	No Qadd	44.95%	35.17%
30 taxa	Qadd	44.81%	37.93%	No Qadd	44.77%	38.06%
40 taxa	Qadd	44.91%	37.30%	No Qadd	44.72%	37.58%
50 taxa	Qadd	45.09%	38.01%	No Qadd	44.97%	36.96%
161 taxa.1	rearr.	46.90%	39.32%	No rearr.	39.48%	37.81%
161 taxa.2	rearr.	46.98%	39.24%	No rearr.	39.53%	39.03%

For the performance analysis of **ATrExML** versus **TrExML**, presented in Table 2 we used the same data-sets as in the original **TrExML** publication. For details on the parameters and data used refer to [9].

Table 2. Global run time improvement TrExML vs. ATrExML

a	n	ATrExML	a	n	ATrExML	a	n	ATrExML
8	10	38.21%	9	10	38.24%	10	10	37.23%
8	11	38.67%	9	11	38.91%	10	11	38.39%
8	12	39.58%	9	12	39.91%	10	12	39.94%
8	13	40.02%	9	13	40.77%	10	13	41.08%
8	14	39.87%	9	14	40.19%	10	14	42.26%
8	15	40.68%	9	15	41.04%	10	15	43.00%
8	16	40.71%	9	16	41.10%	10	16	42.26%

Since **AxML** does not implement heuristics but only a purely algorithmic optimization in all tests **AxML** and **fastDNAml** rendered exactly the same results, a fact that can be verified by a simple `diff` on the output files.

For the initial tests performed on the SR8000-F1 we used the same 30, 40, 50 and 56 taxa test sets as in the previous tests and also a 41 taxa alignment (470 distinct columns) of mitochondrial rRNA sequences from beetles and butterflies, similar to the 161 taxa test set mentioned above. We compiled both, **parallel fastDNAml** and **PAxML** with `mpicc -O3 -model=F1`, measured Mflops/s/processor(worker) performance using PCL, and executed the test runs with 2 to 4 workers located on the same node in intra-node MPI-mode. The results including the respective program options are presented in Table 3. The measured Mflops/s/processor(worker) rate is already satisfying, although no special compiler options for further optimizing the code have been used so far, and varies less than 0.5% among the different workers, i.e. there is no load balancing problem.

Table 3. Global run time improvement PAxML vs. parallel fastDNAml and Mflops/s/proc. performance on the Hitachi SR8000-F1

Data Set	Workers	Options	**PAxML**	Mflops/s/proc.
30 taxa	2	rearr./Qadd	24.82%	123.77
40 taxa	2	rearr./Qadd	28.10%	128.06
50 taxa	2	rearr./Qadd	27.76%	128.62
56 taxa	2	rearr./Qadd	27.10%	128.55
41 taxa	2	rearr./Qadd	34.17%	119.74
41 taxa	4	rearr./Qadd	30.17%	106.51

Finally, for performing larger test runs, in terms of sequence length/number, and number of worker processes on the Hitachi SR8000-F1 with biologically significant data we extracted 3 alignments from the ARB small subunit ribosomal RNA database included in the most recent database release file (6spring2001.arb [7]) consisting of 150, 200 and 250 16S/18S rRNA sequences from organisms of the three kingdoms Eucarya, Bacteria and Archaea. The number of distinct columns in those alignments ranges from 2137 to 2330.

Table 4. Global run time improvement PAxML vs. parallel fastDNAml on the Hitachi SR8000-F1

Num. of Sequences	Num. of columns	Workers	Improvement
150	2137	14	26.57%
200	2253	14	28.52%
250	2330	14	28.40%

The large parallel tests were conducted with local and global rearrangements enabled and the quickadd option set. Table 1 indicates that there is no significant difference in run time improvement for runs with quickadd enabled and disabled.

In Table 4 we present the results obtained on the SR8000-F1, using the specially adapted program version with split up for-loops, for our test runs.

Those results are encouraging, since the algorithmic optimizations of **PAxML** lead to an already significant run time reduction on the Hitachi SR8000-F1.

6 Availability, Current and Future Work

The most recent distribution versions of **AxML**, **PAxML** and **ATrExML** are available at: wwwbode.in.tum.de/~stamatak/research.html.

Current work focuses on the evaluation of different Hitachi-specific parallelization and optimization concepts, such as pseudo-vectorization and further loop transformations, within the context of a more efficient adaptation of the program to the specific architecture. Analogous experiments to those described in Table 4 on a Linux cluster, as well as a comparison of Table 1 with Table 3 show, that the actual acceleration potential of **PAxML** is by far not fully exploited by the present program configuration on the Hitachi SR8000-F1. **PAxML** achieved a maximum run time improvement of 64.60% over **parallel fastDNAml** for the 250 sequence test set on a Linux cluster. For a detailed analysis of the hardware architecture impact on the efficiency of **PAxML** refer to [4].

After reiterating through this optimization process we will perform larger production runs using huge multiple alignments from the ARB database, in order to produce trees comprising 1000 taxa and more with **PAxML**.

Apart from the efforts made for appropriately adapting **PAxML** to the SR8000-F1, we are currently developing **GAxML** (Grid AxML), in cooperation with the MPI (Max Planck Institut) Potsdam.

GAxML is a "phylogenetic grid worm", i.e. an application that can be interrupted, checkpointed, migrated and then restarted on another supercomputer with free resources. Since interrupting, checkpointing and restarting **GAxML** is extremely simple and fast, in contrast to typical supercomputer applications, and due to the fact that the co-scheduling problem has not yet been satisfyingly resolved, **GAxML** is a promising approach for performing large scale phylogenetic tree computations on the grid.

Future work will cover the implementation and analysis of a new parallelization approach for the fast evaluation of a great number of input sequence permutations, which is required for improving the quality of the obtained tree, as well as the integration of empirical biological knowledge.

References

1. Höchstleistungsrechner in Bayern (HLRB): The Hitachi SR8000-F1:
   `http://www.lrz-muenchen.de/services/compute/hlrb`
2. Olsen, G.J., Matsuda, H., Hagstrom, R., and Overbeek, R. (1994): fastDNAml: A tool for construction of phylogenetic trees of DNA sequences using maximum likelihood. In: Comput. Appl. Biosci. 10: 41-48.
3. Stamatakis, A.P., Ludwig, T., Meier, H., and Wolf, M.J. (2002): AxML: A Fast Program for Sequential and Parallel Phylogenetic Tree Calculations Based on the Maximum Likelihood Method. In: Proceedings of the IEEE Computer Society Bioinformatics Conference, to be published. Stanford University, Palo Alto, California.
4. Stamatakis, A.P., Ludwig, T., Meier, H., and Wolf, M.J. (2002): Accelerating Parallel Maximum Likelihood-based Phylogenetic Tree Calculations using Subtree Equality Vectors. In: Proceedings of SC2002, to be published. Baltimore, Maryland.

5. Stewart, C.A., Hart, D. Berry D.K., Olsen G.J., Wernert, E., and Fischer, W. (2001): Parallel implementation and performance of fastDNAml - a program for maximum likelihood phylogenetic inference. In: Proceedings of SC2001. Denver, Colorado.

6. Stewart, C.A., Tan, T.W., Buchhorn, M., Hart, D., Berry, D., Zhang L., Wernert, E., Sakharkar, M., Fisher, W., and McMullen, D. (1999): Evolutionary biology and computational grids. In : IBM CASCON 1999 Computational Biology Workshop: Software Tools for Computational Biology.

7. The ARB project: `http://www.arb-home.de`

8. Top 500 supercomputer sites:
`http://www.top500.org/list/2002/06`

9. Wolf, M.J., Easteal, S., Kahn, M. McKay, B.D., and Jermiin, L.S. (2000): TrExML: A maximum likelihood program for extensive tree-space exploration. In: Bioinformatics 16:383-394.

Load Balancing for Spatial-Grid-Based Parallel Numeric Simulations on Clusters of SMPs – A Case Study from an Industrial CFD Simulation

Huaien Gao[1], Andreas Schmidt[1], Amitava Gupta[2], Peter Luksch[1], and Gerhard Kahl[3]

[1] Institut für Informatik, LRR-TUM
80290 Munich, Germany
{gao—Andreas.Schmidt—Peter.Luksch}@in.tum.de
[2] Institut für Informatik, LRR-TUM
80290 Munich, Germany
(on deputation from Jadavpur University, Calcutta, India)
Amitava.Gupta@in.tum.de
[3] MTU Aero Engines, Munich, Germany
Gerhard.Kahl@muc.mtu.de

Abstract. Load distribution is an essential factor to parallel efficiency of numerical simulations that are based on spatial grids, especially on clusters of symmetric multiprocessors (SMPs). This paper presents a method of mapping spatial grid nodes to processors that combines two load balancing methodologies, graph partitioning and graph matching, to achieve maximum parallel efficiency on SMP clusters. The method has been successfully applied to load distribution in a parallel Computational Fluid Dynamics (CFD) simulation code named TRACE. Test runs on the SR8000 prove the effectiveness of the method.

Keywords: Distributed Computing, Load Balancing, Graph Matching, Graph Partitioning, Homogeneous Cluster of SMPs.

1 Introduction

Scientific simulations in many domains, e.g., CFD, Computational Structural Mechanics (CSM), VLSI simulations, are based on the concept of a spatial grid. A natural way to execute this type of applications in parallel is to follow the *Single Program Multiple Data* (SPMD) approach, i.e., to distribute the spatial grid onto multiple processes, each of which is assigned a partition of the grid. A typical example of such a parallel CFD program is TRACE, which deals with aerodynamic simulation. The TRACE code solves the Reynolds - averaged Navier-Stokes equations in three dimensions for a compressible ideal gas. Turbulence is treated by an eddy-viscosity transport model according to Spalart and Allmaras, modified by Eulitz et al. [1]. The code uses a block

- structured grid topology and is parallelised using the MPI [2] standard. Further details about the numerical model are presented by Eulitz et al. [3] and Eulitz and Engel [4]. The spatial grids generated by suitable graphical pre-processors are assigned to parallel MPI pprocesses which perform computation and communication defined by the inter-dependencies amongst the grids. In this paper we present a methodology that achieves balanced distribution of these grids on a cluster of SMPs and present the results obtained through a trial run on the SR8000 at the *Leibniz Rechenzentrum* (LRZ) [5] in Munich, Germany. The methodoloy is general - in the sense that it can be applied to a genaral class of parallel applications running on a cluster of SMPs, using explicit parallelism like MPI and where the computation and communication costs of the modules can be determined and expressed in terms of time.

Load balancing is a key factor in achieving high parallel efficiency, especially on platforms with a large number of nodes and aims at minimising the *turnaround time* for a parallel application. For a parallel application, the *turnaround time* is defined as the maximum of all the times taken by the individual processors to complete the task. For a given problem, there also is an optimal number of processors for which the turnaround time is minimized. Knowing this number is important in order to maximize throughput. For instance, crash simulations usually involve parameter studies, where a large number of simulation jobs are executed with the same model, each with different parameter settings. Using more than the optimal number of processors for a single job would result in a waste of resources, because faster execution would be possible with less processors. With fewer processors per job, more jobs can run in parallel on a large cluster.

Approaches to task assignment in distributed systems can be classified into three broad categories, namely graph-theoretic([6], [7] as examples), mathematical programming ([8], [9] and heuristic [10]). Graph-theoretic algorithms view the task as a graph representing the inter-modular dependencies and apply graph partitioning methodologies to obtained equal partitions of the task with the inter-node communication or the volume of such communication minimized. The mathematical programming approach views *Load Balancing* as an optimization problem which is solved using mathematical programming techniques. Heuristic methods provide fast but often sub-optimal solutions within a finite time, where an optimal solution cannot be obtained within a finite time.

A concise tool for this purpose is Metis [11]. Using graph partitioning tools like Metis to obtain optimal task assignment on a cluster of SMPs, also has a problem. Metis, for example, cannot be applied directly for a cluster of SMPs, as the heterogenity in terms of difference in link speeds across two processors located in the same node and two processors located on two different nodes cannot be represented.

Shen and Tsai propose a method in [12], where the problem of *Load Balancing* by optimal task assignment is viewed as a graph matching problem. The task is represented as a *task graph* with each module represented by a vertex and the communication between these modules represented by edges. The weights associated with the vertices represent the computation cost associated with each module and the weight associated with each edge represents the commununication cost for the interaction between two adjacent vertices of the task graph. Both these costs are expressed in terms of time. The connectivity between the processors is viewed as a *processor graph* where each vertex represents a node and each edge represents a communication channel between two adjacent nodes. Each graph match corresponds to a specific task assignment. Assuming that each module can require different processing time on different nodes, and the communication links between the different nodes have different speeds, optimal task assignment for a task comprising of N modules on M processors would have to explore M^N possibilities, in the worst case. The work done by Shen and Tsai uses a heuristic approach based on A^* algorithm to reduce this search to obtain an optimal solution. The method can be used for any topology, a cluster of SMPs as an example, as it can tackle heterogenity.This method has a complexity that is exponential in the number of nodes. Hence it cannot be applied directly to real-world applications where the number of interacting modules is huge. In addition, SMP clusters have strongly connected *processor graphs*, which results in a huge number of searches.

This paper puts forward a methodology by which an optimal task assignment on a cluster of SMPs is obtained in two steps using a hybrid methodology comprising of graph partitioning followed by graph matching. Graph partitioning is used to assign a set of modules to each SMP node of the cluster. Subsequently, graph matching is used to assign these modules to individual processors(fewer in number compared to the original number of modules) of each SMP node. The basic algorithm presented in [10] is accordingly modified for this purpose.The results have been tested using TRACE with 64 grid blocks on the SR8000. Each grid block represents a module and is associated with a finite number of *cells* each of which represents a unit of computation and a finite number of physical variables each of which represents a unit of communication. The results establish this methodology as an efficient means of obtaining optimal task asignment on a cluster of SMPs like the SR8000.

2 A Test Case with Graph Matching and Graph Partitioning

To study the effectiveness of a graph matching based method, it was first applied to TRACE comprising of 10 modules defined by the set $\mathbf{T} = \{0,1,2,\ldots 9\}$ on a set of five identical nodes (each comprising of one processor) defined by the set $\mathbf{P} = \{ A,B,C,D,E \}$ such that the *turnaround time* is minimized. The computation time associated with these modules on any node is defined by the set $\mathbf{TA}_p$ = { 0.3, 0.4, 0.4, 0.4, 0.4, 0.4, 0.4, 0.4, 0.4, 0.3 } with the ith element in $\mathbf{TA}_p$ denoting the computation time in seconds, associated with the ith element in $\mathbf{T}$ on one processor of the SR8000. The inter module communication is defined by the matrix $\mathbf{C}$ below.

$$\mathbf{C} = \begin{matrix} 1 & 0 & 0 & 0 & 0 & 1 & 1 & 1 & 1 & 0 \\ 0 & 0 & 1 & 1 & 0 & 0 & 0 & 0 & 1 & 1 \\ 0 & 1 & 0 & 0 & 1 & 0 & 0 & 1 & 0 & 1 \\ 0 & 1 & 0 & 0 & 1 & 1 & 0 & 0 & 0 & 1 \\ 0 & 0 & 1 & 1 & 0 & 0 & 1 & 0 & 0 & 1 \\ 0 & 0 & 0 & 1 & 0 & 0 & 1 & 0 & 1 & 0 \\ 1 & 0 & 0 & 0 & 1 & 1 & 0 & 1 & 0 & 0 \\ 0 & 0 & 1 & 0 & 0 & 0 & 1 & 0 & 0 & 1 \\ 1 & 0 & 1 & 0 & 0 & 0 & 0 & 1 & 0 & 0 \\ 0 & 1 & 1 & 1 & 1 & 1 & 0 & 0 & 0 & 1 \end{matrix}$$

The communication time (in seconds) associated with each non-zero element of $\mathbf{C}$ is represented by the matrix $\mathbf{TA}_c$ defined as:

$$\mathbf{TA}_c = \begin{matrix} 0.000 & 0.000 & 0.000 & 0.000 & 0.000 & 0.004 & 0.004 & 0.004 & 0.004 & 0.000 \\ 0.000 & 0.000 & 0.008 & 0.004 & 0.000 & 0.000 & 0.000 & 0.000 & 0.004 & 0.004 \\ 0.000 & 0.008 & 0.000 & 0.000 & 0.004 & 0.000 & 0.000 & 0.004 & 0.000 & 0.004 \\ 0.000 & 0.004 & 0.000 & 0.000 & 0.008 & 0.004 & 0.000 & 0.000 & 0.000 & 0.004 \\ 0.000 & 0.000 & 0.004 & 0.008 & 0.000 & 0.000 & 0.004 & 0.000 & 0.000 & 0.004 \\ 0.004 & 0.000 & 0.000 & 0.004 & 0.000 & 0.000 & 0.008 & 0.000 & 0.004 & 0.000 \\ 0.004 & 0.000 & 0.000 & 0.000 & 0.004 & 0.008 & 0.000 & 0.004 & 0.000 & 0.000 \\ 0.004 & 0.000 & 0.004 & 0.000 & 0.000 & 0.004 & 0.000 & 0.000 & 0.008 & 0.000 \\ 0.004 & 0.004 & 0.000 & 0.000 & 0.000 & 0.004 & 0.000 & 0.008 & 0.000 & 0.000 \\ 0.000 & 0.004 & 0.004 & 0.004 & 0.004 & 0.000 & 0.000 & 0.000 & 0.000 & 0.000 \end{matrix}$$

The graph matching based load balancing algorithm is based on the work of Shen and Tsai described in [12] and uses the well known A^* algorithm to obtain an optimal task assignment. The A^* based serach algorithm starts with an initial partial mapping $0A$ meaning module 0 is mapped on node A. The algorithm then generates successive partial maps with successive tasks mapped on any of the nodes and applied the A^* based search on the state space so generated. The nth element in the state space was associated with a cost $f(n) = g(n) + h(n)$ where the cost functions $g(n)$ and $h(n)$ are defined in [12]. Typically $g(n)$ represents the minimum path cost from the start element to element n in the state-space and $h(n)$ is the lower-bound estimate, using any heurestic information available of the minimum path cost from the element n to the goal element in the state-space.

Since all nodes are identical, it is wise to start from an initial element with any of the tasks mapped on any of the processors, so as to reduce the number of elements generated in the state space. For a perfectly homogeneous cluster, comprising of M nodes, there would be M different optimal mappings, all with the same turn-around time. The optimal task assignment for the present case, found out by this method was *0D 1A 2B 3E 4B 5E 6D 7C 8C 9A* with an estimated turn-around time of 0.83 sec.

This same methodology was then repeated to obtain an optimal assignment for a version of TRACE comprising of 64 modules on 8 identical single processors belonging to 8 distinct nodes of the SR8000. The task is described in Appendix I. For the task described in Appendix I, vector **A** stores the adjacency information and communication weights associated with the edges in tuples representation. Each eadge between a vertex i and vertex j with a transfer of d bytes is represented by the tuple $\{i,j,d\}$. The vector **B** stores the computation weights associated with each module in terms of the number of units of computation. The number of *cells* and *physical variables* associated with a particular module or a grid block used by TRACE is obtained during the pre-processing stage and the actual weights in terms of computation and communication times associated with each module is computed using cluster specific information defined by COMP_WORTH and COMM_WORTH respectively in Appendix I with representative values for the SR8000. Graph matching based methodology failed to find an optimal task assignment even after generating as many as 90,000 elements in the state space. This is because of the *NP complete* nature of the problem. Accordingly, the graph matching algorithm tends to have an exponential complexity when the number of modules increases [12].

Following this, graph partitioning was used. For this purpose ready made graph partitioning routines provided by Metis was used. Metis produced the mapping presented in Fig. 1

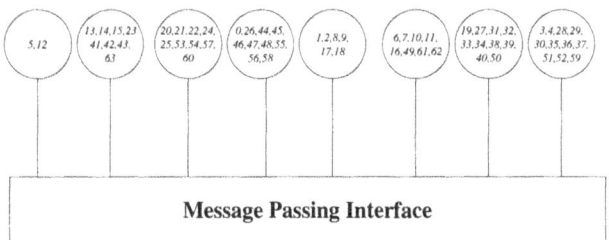

Fig. 1. An optimal task assignment using Metis for TRACE with 64 grid blocks

The load balancing using Metiis was based on the *Part_GraphVKway()* function[9] which aims at achieving load balancing by minimizing the volume of communication across the nodes.

It is to be noted that while processors belonging to the same SMP node can compute in parallel, they cannot communicate with processors which are not located on other SMP nodes at the same time, as a single SMP node has only one network interface. Thus, the minimum *turnaround* time for each SMP node can be obtained when the task assignment is such that computation is interleaved with communication. Metis does not provide a means of doing this.

Therefore, the methodology to obtain an optimal task assignment with N number of modules on a cluster of M number of P way SMPs essentially starts with the application of graph partitioning to obtain M equal partitions with the volume of communication minimised. Once this is done, it is possible to obtain M balanced partitions of the original task each of which comprises of smaller number of modules ($<< N$) and on each of these sub-tasks, graph matching can be applied to obtain an optimal task assignment on P processors for each node, which will produce an optimal *turnaround time* for the entire task.

3 Using the SMP advantage with Graph Matching

Taking the specific example of TRACE with 64 grid blocks discussed in the previous section, after having obtained the optimal task assignment on 8 nodes of the SR8000 the next step would be to treat each of these sub-graphs as a task comprising of a finite number of modules as sown in fig. 1 and obtain an optimal task assignment on two identical processors for each of the nodes. For graph matching, basically, the methodology proposed by Shen and Tsai described in [12] is followed with the following modifications to take care of the fact that computations and communications must be interleaved amongst the processors of a SMP node to obtain minimum node *turnaround* time.

As in [12], a partial mapping **S** is defined as a set of tuples, s_1, s_2, s_3,... s_n, each being of the form x,y , $x \in \mathbf{T}'$, $y = m(x)$, $y \in [0,1,2,...P\text{-}1]$. $\mathbf{T}'$ is the set of modules assigned to the node in question and $m(x)$ is the function which maps a module x on to a processor. The goal is the mapping in which all elements of $\mathbf{T}'$ are mapped on one of the P processors with the minimum turn-around time for the node in question. For each partial mapping, described in [12], the values of $g(n)$ and $h(n)$ are computed as follows:-

(i) Computation of $g(n)$

For a partial mapping **S**, let

TP_i be the time spent by processor i in computation;

TC_i be the time spent by processor i in communication;

$\mathbf{Y} \equiv [y_0, y_1, y_2, \ldots y_{p-1}]$ be the ordered set of processors, such that the computation time associated with a processor in **Y** is greater than or equal to the computation time associated with its preceeding element in **Y** ;

Now, if TP_j, TC_j and TA_j, denote the time spent in computation, the time spent in communication, and the *turnaround* time respectively by the processor corresponding to the jth element of **Y**, i.e. y_j, then the *turnaround* time TA_j, for this processor can be defined as follows:-

$$TA_j = TP_j + TC_j \quad \textit{if } j = 0;$$

else

$$TA_j = max((TP_j + TC_j), (TA_{j-1} + TC_j)) \tag{1}$$

$$g(n) = TA_{p-1} \tag{2}$$

Equation (2) takes into account the fact that in a SMP node comprising of P processors, the communication executed by the processors with modules on other nodes is essentially serialized (since each SMP node has only one network interface), while the computations proceed in parallel. Thus, the maximum time, a process can take to complete its computation and communication is represented by the second term of equation (1) assuming that the sum of its computation and communication times is less than this.

(ii) Computation of $h(n)$

$$h(n) = 0 \tag{3}$$

The result represented by equation (3) follows from the analysis presented in [12] and hence, this is a specific case, where the lower bound of $h(n)$ has a value zero. Experiments with various configurations have shown that this methodology is capable of finding an optimal distribution upto a maximum of about twenty modules per node.

For the present test, the assignment of modules on each individual node is shown in Fig. 1. If it is now assumed that each of these nodes comprises of 2 processors each, the details of the optimal distribution of a cluster of 8, 2 way SMPs is as shown in Table 1. This assignment is obtained using graph matching. The turnaround time for each node is also indicated against each distribution. In the table 1,the nodes are designated as *A,B,C,D,E,F,G &* *H*. The *turnaround* for each node corresponding to a distribution shown in Fig. 1 is also presented for ready comparison.

To understand the effectiveness of the assignment, it is now attempted to obtain an optimal task assignment on the processors associated with each node using Metis. Once again, the starting point is the assignment on 8 single processor nodes represented by Fig. 1. Table 2 shows the details of the distribution. For this distribution, each module is assumed to be represented by a vertex of a *task graph* with a weight which is equal to the sum of the computation time of the module plus the time spent by the module in communicating with other modules located in same or other nodes(for a single iteration). The idea is to find out partitions with equal *turnaround times*.

It is clearly seen from tables 1 and 2 that using the hybrid partitioning methodology, the *turnaround time* for most of the nodes is less than the corresponding *turnaround* times obtained by the application of Metis based

Table 1. Optimal Distribution on a cluster of 8, 2 way SMPs using Graph Matching

Node	Assignment of Blocks		Node Turnaround Time (secs.)	
	Processor 0	Processor 1	2 Way SMP	Single Processor (fig. 1)
A	5	12	3.01	3.48
B	13,42,43,63	14,15,23,41	1.77	3.34
C	20,21,25,54,57,60	22,24,53	1.93	3.67
D	0,26,47,55,56,58	44,45,46,48	1.94	3.63
E	1,17	2,8,9,18	2.04	3.90
F	6,7,10,11,62	16,49,61	1.86	3.50
G	19,32,33,39,40	27,31,34,38,50	1.84	3.49
H	3,4,28,30,51,52,59	29,35,36,37	1.92	3.62

Table 2. Optimal Distribution on a cluster of 8 2 way SMPs using Graph Partitioning

Node	Assignment of Blocks		Node Turnaround Time (secs.)	
	Processor 0	Processor 1	2 Way SMP	Single Processor (fig. 1)
A	12	5	3.01	3.48
B	14,15,23	13,41,42,43,63	1.98	3.34
C	20,21,24,53,54,57,60	22,25	2.04	3.67
D	26,46,47,48,55,56,58	0,44,45	1.99	3.63
E	1,8,9,18	2,17	2.04	3.90
F	16,49	6,7,10,11,61,62	1.98	3.50
G	27,31,32,33,50	19,34,38,39,40	1.85	3.49
H	3,28,29,30,51,52	4,35,36,37,59	1.99	3.62

approach. Inspite of this, the system *turnaround* time remains the same because of the fact that a single module viz. module 5 is associated with an overwhelmimgly large value.

4 Results and Discussion

To test proposed methodology, an optimal distribution of TRACE was obtained on the SR8000 cluster for the floowing cases:

(a) *64 grid blocks of TRACE assigned to 16 processors of SR8000 with 4 processors per node;*
(b) *64* grid blocks of TRACE assigned to 32 processors of SR8000 with 8 processors per node;

For each case, TRACE was run on the SR8000 cluster and the results are presented in Tables 3 and 4. The actual measured *turnaround* time for each case is presented, measured for 50 iterations with high precision timer routines. In each case, the corresponding *turnaround* time for a non-optimal assignment (original TRACE) is also presented. For the non-optimal assignment it is assumed that the blocks are assigned on the processors contiguously i.e. blocks 0,1,2,3 on processor 0 of the fist node, blocks 4,5,6,7 on procesor 1 of node 0 and so on for the case (a) cited above.

Table 3. Comparison of results: with and without optimal assignment of a cluster of 4, 4-way SMPs

Node	Assignment of Blocks				Actual Node Turnaround Time (secs.)	
	Proc. 0	Proc. 1	Proc. 2	Proc. 3	Optimal Distribution	Original TRACE
A	20,21,44, 45,60	0,46,47	23,25,26, 48,55	22,54,56	348.80	458.67
B	2	43,63	12,13,14, 15,24,41, 42	5	348.93	458.64
C	27,31,32, 33	3,28,29, 30,50,51, 52	4,34,35, 38	36,37,39, 40,53,57, 58,59	348.79	458.73
D	1,7,11, 61,62	6,10,49	8,9,18, 19	16,17	348.75	458.73

It is clearly seen that a *speedup of 1.31* or *31 p.c.* is obtained with optimal assignment for the case shown in Table 3.

Table 4. Comparison of results: with and without optimal assignment of a cluster of 4, 8-way SMPs

Node	Assignment of Blocks								Actual Node Turnaround Time (secs.)	
	Processor								Optimal Distri-bution	Original TRACE
	0	1	2	3	4	5	6	7		
A	0	20, 21			48, 60	45,46, 47,55, 56	23,25, 26,44, 54		349.00	408.41
B	2	22				43,63	12,13, 14,15, 24,41, 42	5	348.86	408.40
C	31, 33	27, 32	3,28, 30	29,50, 51,52	36,39	37,40, 53,57, 58,59	35,38	4, 34	348.95	408.41
D	1	7,61, 62	49	6,10, 11	8,9	18,19	16	17	349.22	408.44

Next, the number of processors is increased from 4 to 8 keeping the total number of nodes constant. This corresponds to case (b) cited above. The results (for 50 iterations) are presented in Table 4 below.

It is clearly seen that a *speedup speedup* comes down to 1.17 or 17 p.c. from 31 p.c. in the first case.

The results presented in Tables 3 and 4 indicate the following:-8

1. The hybrid methodology for task assignment is extremely effective and produces a balanced distribution of task on a cluster of SMPs.
2. Using optimal distribution, increasing the number of processors from 16 to 32 does not decrease the *turnaround* time further, as the block 5 has a large computation and communication time associated with it. Hence the speed up comes down in the case (b). This, in turn, indicates that a better scalability for the given problem may be obtained if some of the grid blocks could be split into *micro-blocks*. A general approach would be to split all grid blocks into *micro-blocks* and then obtain an optimal distribution in terms of these *micro-blocks*. This increases the number of *micro-blocks* per node to be tackled by the graph matching algorithm and an enhancement of the present algorithm based on a *method of bisection* strategy is targeted as a scope of further work.
3. The results are obtained considering a homogeneous cluster of SMPs where each node is assumed to be identical- comprising of same number of identical processors. In the actual scenario, jobs on a distributed computing environment are usually submitted using a batch processing

Table 5. Performance of TRACE with optimal distribution on a cluster of 8, 2way SMPs

Node	Assignment of Blocks		Actual Node Turnaround Time (secs.)
	Processor 0	Processor 1	
A	5	12	329.49
B	13,42,43,63	14,15,23,41	329.44
C	20,21,25,54,57,60	22,24,53	329.51
D	0,26,47,55,56,58	44,45,46,48	329.47
E	1,17	2,8,9,18	329.54
F	6,7,10,11,62	16,49,61	329.43
G	19,32,33,39,40	27,31,34,38,50	329.46
H	3,4,28,30,51,52,59	29,35,36,37	329.48

software, which takes an user defined number of processors and these processors may be distributed across the nodes in a non-uniform manner. Modification of the initial graph partitioning algorithm using the topology specific information is the next logical step.

Finally, to study the effect of increase in the number of nodes on the performance of TRACE with optimal distribution, the number of nodes is increased to 8 from 4 while keeping the number of processors constant as in case (a) above. The results(for 50 iterations) are presented in Table 5 below.

The results presented in Table 5 indicates that the latency for communication amongst the processors of a single node of the SR8000 is comparable to the inter-node communication latency. Infact, the node turnaround time reduces in the case represented by Table 5 compared to the case represented by Table 3, for every node, as the communication amongst the processors of a single node is also halved, with the number of processors per node reduced from 4 to 2.

5 Conclusion

The methodology presented in this paper combines two diverse load balancing strategies to obtain an optimal task assignment on a cluster of SMPs. It uses graph matching as a complement to graph partitioning to address the limitation of each. Graph partitioning tools suffer from the inability of being used directly to obtain an optimal task assignment on a cluster of SMPs as they cannot tackle heterogenity presented even by a homogeneous cluster of SMPs. The graph matching algorithms, on the other hand can tackle this homogenity, but cannot be used when the number of modules are large. The methodology of using graph matching on smaller size partitions produced by graph partitioning provess to be very effective. The methodolgy can be used for obtaining efficient assignments which optimize *turnaround times* for fairly large parallel applications on a homogeneous cluster of SMPs within a finite time.

Acknowledgement. This work is funded by the Bavarian Science Foundation within the scope of Competence Network for Technical, Scientific High Performance Computing in Bavaria, KONWIHR.

References

1. Eulitz, F., Engel, K. and Gebing, H.: *Application of a One-equation Eddy-viscosity Model to Unsteady Turbomachinery Flow*, Engineering Turbulence Modelling and Experiments 3, Rodi, W. and Bergeles, G (Eds.), Elsevier Science B.V., 1996.
2. Pacheco, P.: *Prallel Programming with MPI*, Morgan Kaufman, 1996, ISBN 1-55860-339-5
3. Eulitz, F., Engel, K. and Gebing, H.: *Numerical Investigation of the Clocking Effects in a Multistage Turbine*, ASME - Paper 96-GT-26, 1996
4. Eulitz, F. and Engel, K.: *Numerical Investigation of Wake Interaction in a Low Pressure Turbine*, ASME - Paper 98-GT-536, 1998.
5. LRZ: *System Description*,
   http://www.lrz-muenchen.de/services/compute/hlrb/system-en/
6. Stone, H. S., and Bokhari, S.H.: *Control of Distributed Processes*, Computer, vol. 11, pp. 97-106, July 1978
7. Chow, T. C. K., and Abraham, J. A.: *Load Balancing in Distributed Systems*, IEEE Transactions on Software Engineering, vol. SE-8, July 1982
8. Chu, W. W., Holloway, L.J., Lan, M.T., and Efe, K.: *Task Allocation in Distributed Data Processing*, Computer, vol. 13, pp. 57-69, Nov. 1980
9. Ma, P. R., et al: *A Task Allocation Model for Distributed Computing Systems*, IEEE Transactions on Computers, vol. C-31, pp. 41-47, Jan. 1982
10. Efe, K.: *Heuristic Models of Task Assignment Scheduling in Distributed Systems*, Computer, vol. 15, pp. 50-56, June 1982.
11. Metis: *Serial Graph Partitioning*,
    http://www-users.cs.umn.edu/ karypis/metis/metis/
12. Shen, Chien-Chung and Tsai, Wen-Hsiang: *A Graph Matching Approach to Optimal Task Assignment in Distributed Computing Systems using a Minimax Criterion*, IEEE Transactions on Computers, vol. C-34, No. 3, March 1985.
13. Eulitz, F., Engel, K. and Gebing, H.: *Numerical Investigation of the Clocking Effects in a Multistage Turbine*, ASME - Paper 96-GT-26, 1996
14. Eulitz, F. and Engel, K.: *Numerical Investigation of Wake Interaction in a Low Pressure Turbine*, ASME - Paper 98-GT-536, 1998.

6 Appendix

A = {

{0,5,278528},   {0,44,69632}, {0,22,139264},   {0,41,34816}, {0,63,104448},
{0,43,69632},   {0,47,69632}, {0,48,139264},   {0,45,69632},   {0,42,69632},
{0,46,69632},   {0,0,34816},   {0,0,34816},   {1,5,69632},   {1,8,69632},
{1,7,69632},   {1,6,43520},   {1,61,26112},   {1,17,69632},   {2,7,69632},
{2,12,69632},   {2,18,69632},   {2,17,69632},   {2,16,69632},   {3,8,69632},
{3,20,69632},   {3,29,69632},   {3,28,69632},   {3,27,69632},   {3,35,69632},
{4,18,69632},   {4,23,69632},   {4,28,69632},   {4,36,69632},   {4,35,69632},
{4,34,69632},   {5,0,278528},   {5,44,52224},   {5,41,13056},   {5,45,52224},
{5,63,39168},   {5,42,52224},   {5,1,69632},   {5,8,69632},   {5,7,69632},
{5,9,69632},   {5,12,69632},   {5,14,69632},   {5,20,69632},   {5,21,69632},
{5,13,69632},   {5,15,69632},   {5,41,13056},   {5,43,52224},   {5,63,39168},
{6,1,43520},   {6,10,32640},   {6,61,52224},   {6,49,43520},   {6,17,52224},
{7,1,69632},   {7,5,69632},   {7,9,69632},   {7,61,17408},   {7,2,69632},
{7,62,52224},   {8,1,69632},   {8,5,69632},   {8,9,69632},   {8,10,69632},
{8,3,69632},   {8,17,69632},   {9,5,69632},   {9,7,69632},   {9,8,69632},
{9,11,69632},   {9,18,69632},   {9,28,69632},   {10,6,32640},   {10,8,69632},
{10,11,52224}, {10,61,19584}, {10,27,52224}, {10,49,69632}, {10,17,52224},
{11,9,69632}, {11,10,52224}, {11,61,13056}, {11,19,52224}, {11,62,39168},
{11,31,52224}, {11,49,69632},   {12,2,69632},   {12,5,69632}, {12,14,69632},
{12,13,69632}, {12,41,17408}, {12,63,52224}, {13,12,69632}, {13,17,69632},
{13,15,69632}, {13,41,17408}, {13,63,52224},   {13,5,69632},   {14,5,69632},
{14,12,69632}, {14,18,69632}, {14,15,69632}, {14,23,69632}, {14,42,69632},
{15,13,69632}, {15,14,69632}, {15,17,69632}, {15,60,34816}, {15,43,69632},
{15,24,34816},   {15,5,69632},   {16,2,69632}, {16,62,52224}, {16,19,52224},
{16,17,52224}, {16,49,69632},   {17,2,69632}, {17,13,69632}, {17,15,69632},
{17,18,69632}, {17,16,52224}, {17,19,52224}, {17,35,69632}, {17,38,52224},
{17,49,139264},   {17,1,69632},   {17,8,69632},   {17,6,52224}, {17,10,52224},
{18,2,69632},   {18,9,69632}, {18,14,69632}, {18,17,69632}, {18,19,69632},
{18,4,69632}, {19,11,52224}, {19,16,52224}, {19,18,69632}, {19,17,52224},
{19,34,52224}, {19,49,69632},   {20,3,69632},   {20,5,69632}, {20,22,69632},
{20,21,69632}, {20,44,69632}, {20,24,34816}, {20,60,34816},   {21,5,69632},
{21,20,69632}, {21,28,69632}, {21,22,69632}, {21,23,69632}, {21,45,69632},
{22,0,139264}, {22,44,52224}, {22,48,52224}, {22,45,52224}, {22,20,69632},
{22,29,69632}, {22,21,69632},   {22,30,8704}, {22,25,69632}, {22,51,52224},
{22,52,8704}, {22,48,52224}, {22,54,17408}, {22,56,17408}, {22,57,17408},
{22,59,17408},   {23,4,69632}, {23,14,69632}, {23,21,69632}, {23,25,69632},
{23,60,34816}, {23,46,69632}, {23,24,34816}, {24,15,34816}, {24,23,34816},
{24,47,69632}, {24,60,69632}, {24,26,17408}, {24,54,17408}, {24,20,34816},
{25,22,69632}, {25,23,69632}, {25,36,69632}, {25,53,17408}, {25,48,69632},

{25,26,17408},   {25,55,17408}, {25,58,17408}, {26,24,17408}, {26,25,17408},
{26,48,17408},   {26,53,17408}, {26,54,17408}, {26,55,17408},   {27,3,69632},
{27,10,52224},   {27,32,52224}, {27,31,52224}, {27,49,69632}, {27,38,52224},
{28,3,69632},   {28,9,69632}, {28,21,69632},   {28,30,8704}, {28,31,69632},
{28,50,34816},   {28,4,69632}, {28,51,26112},   {29,3,69632}, {29,22,69632},
{29,30,17408},   {29,32,69632}, {29,50,34816}, {29,52,17408}, {29,37,69632},
{30,22,8704},   {30,28,8704}, {30,29,17408},   {30,50,8704}, {30,51,17408},
{30,52,8704},   {31,11,52224}, {31,27,52224}, {31,28,69632}, {31,33,52224},
{31,34,52224},   {31,49,69632}, {32,27,52224}, {32,29,69632}, {32,33,52224},
{32,49,69632},   {32,40,52224}, {33,31,52224}, {33,32,52224}, {33,50,69632},
{33,39,52224},   {33,49,69632}, {34,4,69632}, {34,19,52224}, {34,31,52224},
{34,39,52224},   {34,38,52224}, {34,49,69632},   {35,4,69632}, {35,17,69632},
{35,60,69632},   {35,37,69632}, {35,38,69632},   {35,3,69632},   {36,4,69632},
{36,25,69632},   {36,51,34816}, {36,37,69632}, {36,39,69632}, {36,50,34816},
{37,35,69632},   {37,36,69632}, {37,53,17408}, {37,40,69632}, {37,58,17408},
{37,57,17408},   {37,59,17408}, {37,29,69632}, {38,17,52224}, {38,34,52224},
{38,35,69632},   {38,40,52224}, {38,49,69632}, {38,27,52224}, {39,33,52224},
{39,34,52224},   {39,36,69632}, {39,40,52224}, {39,49,69632}, {40,37,69632},
{40,38,52224},   {40,39,52224}, {40,49,69632}, {40,32,52224},   {41,0,34816},
{41,63,104448},   {41,5,13056}, {41,12,17408}, {41,13,17408},   {41,5,13056},
{42,0,69632},   {42,5,52224}, {42,63,52224}, {42,46,52224}, {42,43,52224},
{42,14,69632},   {43,0,69632}, {43,63,52224}, {43,47,52224}, {43,42,52224},
{43,15,69632},   {43,5,52224},   {44,0,69632},   {44,5,52224}, {44,22,52224},
{44,45,52224},   {44,20,69632}, {44,47,52224},   {45,0,69632},   {45,5,52224},
{45,44,52224},   {45,22,52224}, {45,46,52224}, {45,21,69632},   {46,0,69632},
{46,42,52224},   {46,45,52224}, {46,48,52224}, {46,47,52224}, {46,23,69632},
{47,0,69632},   {47,43,52224}, {47,48,52224}, {47,46,52224}, {47,24,69632},
{47,44,52224},   {48,0,139264}, {48,47,52224}, {48,22,52224}, {48,46,52224},
{48,25,69632},   {48,26,17408}, {48,54,17408}, {48,55,17408}, {48,56,17408},
{48,22,52224},   {49,6,43520}, {49,10,69632}, {49,61,43520}, {49,27,69632},
{49,11,69632},   {49,31,69632}, {49,16,69632}, {49,19,69632}, {49,62,52224},
{49,34,69632}, {49,17,139264}, {49,38,69632}, {49,32,69632}, {49,33,69632},
{49,39,69632},   {49,40,69632}, {49,49,34816}, {49,49,34816}, {50,28,34816},
{50,29,34816},   {50,33,69632},   {50,30,8704}, {50,36,34816}, {50,51,52224},
{50,52,8704},   {51,22,52224}, {51,28,26112}, {51,36,34816}, {51,50,52224},
{51,30,17408},   {51,52,17408},   {52,22,8704}, {52,29,17408}, {52,51,17408},
{52,30,8704},   {52,50,8704}, {53,25,17408}, {53,37,17408}, {53,60,17408},
{53,26,17408},   {53,57,17408}, {53,58,17408}, {54,24,17408}, {54,48,17408},
{54,26,17408},   {54,57,17408}, {54,56,17408}, {54,22,17408}, {55,25,17408},
{55,26,17408},   {55,48,17408}, {55,58,17408}, {55,56,17408}, {56,48,17408},
{56,54,17408},   {56,55,17408}, {56,59,17408}, {56,22,17408}, {57,54,17408},
{57,60,17408},   {57,53,17408}, {57,59,17408}, {57,37,17408}, {57,22,17408},
{58,25,17408},   {58,37,17408}, {58,53,17408}, {58,55,17408}, {58,59,17408},

{59,56,17408}, {59,57,17408}, {59,58,17408}, {59,37,17408}, {59,22,17408},
{60,15,34816}, {60,23,34816}, {60,35,69632}, {60,53,17408}, {60,24,69632},
{60,57,17408}, {60,20,34816}, {61,1,26112}, {61,7,17408}, {61,10,19584},
{61,11,13056}, {61,6,52224}, {61,49,43520}, {61,62,52224}, {62,7,52224},
{62,16,52224}, {62,11,39168}, {62,61,52224}, {62,49,52224}, {63,0,104448},
{63,41,104448}, {63,43,52224}, {63,5,39168}, {63,42,52224}, {63,12,52224},
{63,13,52224}, {63,5,39168}

};

B = {

27744,	8000,	8000,	8000,	8000,	41472,	4480,	8000,
8000,	8000,	6400,	6400,	8000,	8000,	8000,	8000,
6400,	23040,	8000,	6400,	8000,	8000,	23040,	8000,
4800,	8000,	1728,	6400,	8000,	8000,	1152,	6400,
6400,	6400,	6400,	8000,	8000,	8000,	6400,	6400,
6400,	4608,	6400,	6400,	6400,	6400,	6400,	6400,
11520,	27744,	4800,	3840,	1152,	1728,	1728,	1728,
1728,	1728,	1728,	1728,	4800,	4480,	5120,	9216

};

COMP_WORTH = 0.0928;
COMM_WORTH = 0.00846;

The actual computation time of a module i equals B[i] * COMP_WORTH
msec. COMM_WORTH is the average time taken to transmit a single byte
in msec. computed for the application in question.

Scientific Progress in the Par-EXPDE-Project

Alexander Linke[1], Christoph Pflaum[1], and Ben Bergen[2]

[1] Institut für Angewandte Mathematik und Statistik der Universität Würzburg
Am Hubland, 97074 Würzburg, Germany
alinke@mathematik.uni-wuerzburg.de, pflaum@mathematik.uni-wuerzburg.de
[2] Lehrstuhl für Systemsimulation, Universität Erlangen-Nürnberg
Cauerstraße 6, 91058 Erlangen, Germany
Ben.Bergen@cs.fau.de

Abstract. Here we present the scientific progress of the Par-EXPDE project, the primary goal of which is to provide an efficient library for the numerical computation of PDEs on the Hitachi SR8000. It is intended that this library will significantly reduce the implementation effort required by users for the development of parallel adaptive multigrid methods on arbitrary 3D-domains by providing them with a high level interface that closely resembles mathematical language. This high level user interface is accomplished through the use of a special programming technique called expression templates that effectively hides the complexity of the underlying data structures while still allowing for an efficient implementation of computationally intensive codes. The low level data structures themselves use a technique of patchwise regular refinement of unstructured grid elements to allow for more aggressive compiler optimization and the use of stencil based smoothers that are better able to exploit the memory hierarchy and pseudo vectorization capabilities of the SR8000. Geometry and visualization capabilities are provided by the GRIDLIB project.

1 Introduction

Currently the use of super computers is an important tool for the numerical solution of PDEs. Unfortunately, for many scientists, developing software that not only runs efficiently but also exploits the special characteristics of modern computer architectures poses a difficult problem. Among other things, researchers are faced with issues concerning parallelism, optimization of algorithms, and the handling of complex data structures that arise from the approximation of general domains in 3D. To address these problems we introduce the concept of expression templates for partial differential equations which has been implemented in the EXPDE library[1]. This library provides an interface that closely resembles mathematical language making it accessible to a much broader range of research scientists who are not necessarily expert programmers. The library can be applied to a wide range of partial

---

[1] The original EXPDE library was developed at the University of Würzburg and the Lawrence Livermore National Laboratory. Parallelization is based on MPI.

differential equations. An important property of the expression templates concept is that complexities such as parallelization and optimization can be hidden. Therefore, the user is not confronted with the difficulties associated with developing these constructs, but still has an efficient tool for using them.

In the following sections we give a more in depth description of our approach and provide bench marks to demonstrate that our ideas actually lead to high MFLOPS rates on the SR8000 at LRZ in Munich. In Sect. 3, the parallelization properties of EXPDE on the SR8000 are compared to those on ASCI Pacific Blue at LLNL in Livermore and show nearly equivalent performance for equal numbers of processors. To improve the efficiency of EXPDE on Hitachi SR8000, we decided to change the low level data grid structure to one that is better suitable for optimization on SR8000. This version EXPDE-cuboid contains an optimal data structure for cuboids. The numerical results in Sect. 4 show that special features such as COMPAS[2] and PVP[3] can be applied through the use of expression templates. It is crucial that this be the case since these features are integral to our ability to develop efficient numerical algorithms on the SR8000. Our aim is to use EXPDE-cuboid as a basic code for solving PDE's on general domains in 3D. To this end we present, in Sect. 5, the concept of patch-wise regular refinement of unstructured grids which allows the use of contiguous memory arrays as opposed to indirect indexing. Another logical step in the development of EXPDE is to implement adaptive semi-unstructured grids which are important in case of singularities and other local structure in the solution of a PDE. In order to do this it is necessary to subdivide boundary cells in an adaptive manner which is described in more detail in Sect. 6.

The development of any library for the solution of PDEs must be done with a view toward certain specific applications. The applications that we are currently interested in are: computation of bio-electric fields, laser simulation, and flow simulation of ground water in waste deposits. In some of these areas we have formed cooperative partnerships with local companies. These relationships can be extremely helpful in keeping us up to date with important research interests in industry and in some cases help to guide our own research efforts. For example we are currently working with LASCAD in the area of laser simulation where it has become clear that adaptive semi-unstructured grids would be of great benefit. We are also cooperating with the company Infineon, where we are interested in the numerical solution of semiconductor lasers. This second topic is currently an important research field with many companies doing development, especially in the area of time-dependent behavior of semiconductor lasers where new research is needed to overcome difficulties in simulation practice. These simulations require very high resolution of the emerging resonator waves to resolve their important time-dependent behavior in the first few nano seconds, in which the resonator

---

[2] Co-Operative Micro Processors in single Address Space

[3] Pseudo Vector Processing

waves are not yet stationary. Clearly these types of problems require not only the power of super computers but also efficient, accessible tools to make use of them.

2 Efficiency Questions and Expression Templates

A general library such as EXPDE is only useful, if it is reasonably efficient. Applications such as laser simulation require an excessive amount of computing time and therefore should be performed as quickly as possible. Parallelization and optimization must be carried out by the library itself, in order not to limit their applicability for users.

The complexity of EXPDE requires the use of modern object-oriented languages. These languages are, in a sense, extensible by techniques such as dynamic memory allocation, operator overloading, type inheritance and genericity. Thereby, their syntax and semantics can be adapted to the mathematical problem. Usually one has to pay for this by a significant decline in efficiency. Especially crucial is the operator overloading technique. Indeed this problem can be avoided by using C++ and expression templates, a programming technique originally proposed by Todd Veldhuizen [5].

Expression templates use genericity and operator overloading to construct a syntax tree from a given mathematical expression. Different expressions are represented by different template types. Expression templates are then evaluated by a user-defined strategy. Thereby, one can prevent the creation of extra temporary variables and additionally enforce optimization and parallelization. Construction of the syntax tree and determination of a suitable evaluation strategy is performed at compile time. The resulting intermediate code even traverses further code generation process and low-level-optimization. Therefore, an optimizing compiler can generate highly efficient code. Todd Veldhuizen has shown in [5], that expression templates can compete well with hand optimized C and FORTRAN code provided that a suitable optimizing C++ compiler is available on a given platform. Although the benchmark results are necessarily dependent on the machine and compiler available, they show that for vector expressions 95-99.5% the speed of hand coded C can be achieved.

The use of expression templates itself is a complex programming technique, however this complexity can be hidden from the library user by object-oriented concepts such as information-hiding and data encapsulation. Therefore, the user of EXPDE will not have to learn to use expression templates.

3 Efficiency Comparison between EXPDE on the ASCI Blue Pacific and EXPDE on the Hitachi SR8000

A first MPI-parallelized version of EXPDE has been developed on the parallel machine ASCI Pacific Blue at the Lawrence Livermore National Laboratory. To compare the ASCI Blue Pacific and the Hitachi SR8000, we consider the computational time on both machines for solving a linear elasticity problem. Table 1 shows the computational time for one conjugate gradient step with a multigrid preconditioner. One can see that on both machines the computational time is nearly the same. However, in this comparison, important features of the Hitachi SR8000 are still unused in Par-EXPDE, especially element parallel processing and Pseudo Vector Processing (PVP). From previous experience in the gridlib project, it is clear that the computational time can significantly be reduced by using these concepts. The main reason, why element parallel processing and PVP were not be used in EXPDE, was that the optimizing C++ compiler CC could not compile member template constructions in the library EXPDE. Therefore, we had to compile EXPDE with the KAI compiler, which is not able to use COMPAS and is also not able to generate additional additive instructions for profiling. However, we expect that the new native optimizing C++ compiler by Hitachi will overcome this technical problem. Alternatively it is possible to hand tune the library, although this may be very work intensive.

Table 1. Computational time (in seconds) of one cg step with a multigrid preconditioner.

# processors	# grid points	# unknowns	time Blue	time SR8000
24	1.973.996	5.921.988	81 sec	99 sec
88	1.973.996	5.921.988	18 sec	23 sec
131	15.356.509	46.069.527	84 sec	94 sec

4 EXPDE-cuboid on the Hitachi SR8000

To improve the efficiency of EXPDE on Hitachi SR8000 we decided to reimplement all the parts of EXPDE, in which numerical calculations with floating point numbers are performed. Thereby, the old EXPDE-library serves as a framework for the new version. Large parts of the program are reused. But the transition to the Hitachi SR8000 obviously affects the data structure used in EXPDE. Soon became clear, that one-dimensional arrays are most suitable, as described in [6].

The first fruit of our efforts is now a EXPDE version on Hitachi SR8000, which works on cuboids. Currently, EXPDE-cuboid possesses the full functionality of EXPDE apart from multigrid operators. It uses COMPAS parallelization and PVP on a single node. MPI parallelization will be implemented later.

Although the specialization on cuboids seems to be very restrictive, this specialization is justified by several reasons. First, we want to gain experience with use of expression templates on Hitachi SR8000 by solving a simplified problem. Second, our group works on simulation of semiconductor lasers, where only cuboids are used up to now. Third, by combining cuboids, prisms and tetrahedra it is possible to describe very general geometrical structures. Therefore, EXPDE cuboid is the first of three specialized EXPDE libraries, which will be embedded by MPI in a library for very general geometrical domains. More about this follows in the next section.

We now present some results, which performance can be achieved by the EXPDE-cuboid library. Therefore, we investigate the performance measurements of a Poisson problems with Dirichlet boundary conditions. Linear equations are solved by the cg method. Originally, we wanted to present a complex problem from laser simulation, but unfortunately the CC-Compiler cannot compile programs with extensive use of template constructions because of compiler bugs. These bugs will be corrected in near future.

The example in Fig. 1 shows, that EXPDE-cuboid really supports the implementation of quite complex problems. The problem is described by a short program, whose code closely resembles to mathematical language.

We now describe, how an efficient implementation of EXPDE-cuboid can be implemented on Hitachi SR8000. In the above cg method the Laplace-Operator is evaluated in the interior of the cuboid. The source code, which evaluates

```
g = Laplace_FE(d);
```

within the library is depicted in Fig. 2.

All the difficulties in the innermost loop are hidden by the expression templates technique

```
a[ind] = a_.Give_interior(grid, ind, Nx, Ny, Nz, N, h_mesh);
```

where a_ represents the expression and Give_interior is an associated inlined member function, which returns the result of the expression with index ind. The compiler directives /*poption indep*/ and /*voption indep*/ enable parallel element processing and PVP.

Fig. 3 depicts the performance of the previous cg solver for Poisson's equation on one node of Hitachi SR8000. The total performance is up to 2.4 GFLOPS and the performance of the Laplace-operator up to 4.2 GFLOPS. Since the maximal peak performance is 12 GFLOPS these are quite encouraging results. Our aim is to obtain the same efficiency for our laser simulation project with the new native C++ Compiler on Hitachi SR8000.

```
#include "source/extpde.h"

int main() {
   int iteration=1000, i, N=20;
   double delta, delta_prime, beta, tau,
          eps=0.0000000001, normi;
   D3vector V_org(0.0,0.0,0.0);
   Grid grid(N,N,N,1.0/(double)(N-1),V_org);
   Variable f(& grid), g(& grid), d(& grid), r(& grid), u(& grid);

   u = Sin(X()) * Sinh(Y()) * Z() | boundary_points;
   u = 0.0 | interior_points;
   f = 0.0; // Right hand side
   f = f | interior_points; r = r | interior_points;
   d = d | interior_points; g = g | interior_points;
   r = Laplace_FE(u) - Helm_FE(f);
   d = -r;
   delta = product(r,r);
   for(i=1; i <= iteration && delta > eps; i++) {
     g = Laplace_FE(d);
     tau = delta / product(d,g);
     r = r + tau * g;
     u = u + tau * d;
     delta_prime = product(r,r);
     beta = delta_prime / delta;
     delta = delta_prime;
     d = beta*d - r;
   }
}
```

Fig. 1. Code for solving Poisson's equation with cg-method.

5 Data Structures for a High-Speed Par-EXPDE — Patch-Wise Regular Refinement

The use of unstructured grids for the solution of partial differential equations on complex geometries offers advantages in flexibility and problem resolution over purely structured grids. However, the associated cost of solving the resulting systems often prohibits their use on modern supercomputers such as the SR8000. This cost arises from the data structures used to represent generally sparse systems. Indirect indexing, which would be used in straight forward implementation of unstructured grids, makes it impossible to apply certain optimizations and in the case of the SR8000 it is not possible to use the pseudo vector characteristics of the nodes efficiently.

```
// interior points
/*poption indep*/
for(i=1; i < Nx-1; i++)
   for(j=1; j < Ny-1; j++) {
     /*voption indep*/
     for(k=1; k < Nz-1; k++) {
       ind = index(i, j, k);
       a[ind] = a_.Give_interior(grid, ind, Nx, Ny, Nz, N, h_mesh);
     }

}
```

Fig. 2. One part of the source code of EXPDE-cuboid.

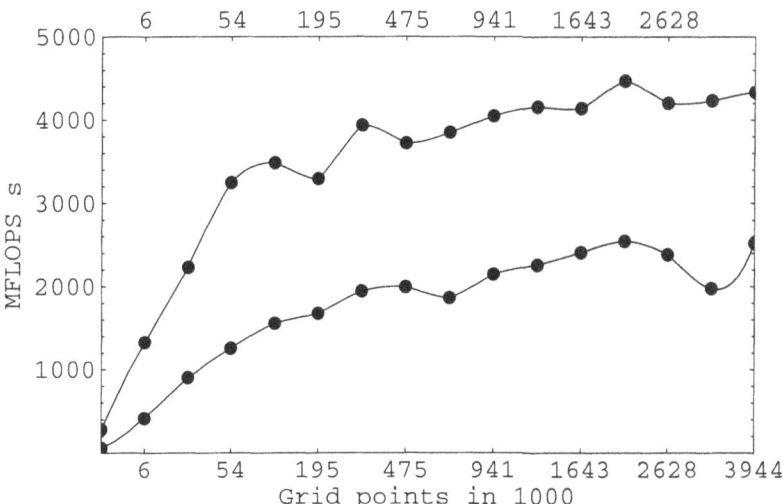

Fig. 3. Poisson problem with Dirichlet boundary conditions, solved by conjugate gradient method. The upper curve shows performance of the Laplace operator, while the lower curve shows total performance.

In order to overcome this difficulty and still allow the use of unstructured grids for geometry resolution we have introduced the notion of patch-wise regular refinement. The basic idea is as follows: given an unstructured grid that resolves only the geometry of the problem we apply regular refinement to each coarse grid element, it is then possible to treat the problem in a structured manner on a patch-wise basis (see also [1]). This strategy produces a grid hierarchy suitable for use with multigrid algorithms that uses stencil based smoothers in large parts of the domain and can therefore exploit both the memory characteristics and pseudo vectorization features of the SR8000.

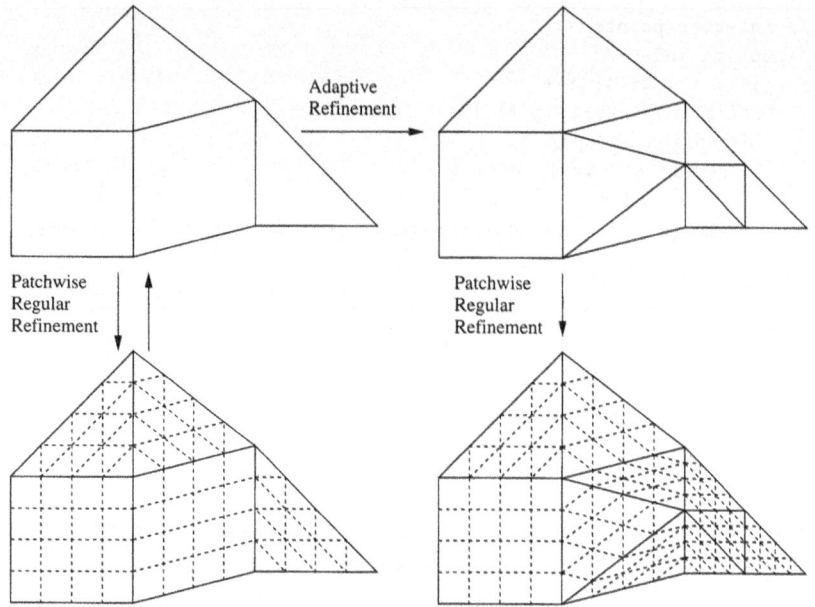

Fig. 4. Adaptive refinement

This technique also provides us with a straight forward method of implementing runtime adaptivity of the coarse grid elements. For an illustration of this idea consider Fig. 4. In this 2D-diagram, regular refinement is applied to the original coarse grid on the top left resulting in the lower left fine grid, during runtime it is determined that some adaptivity is required, the coarse grid is then adaptively refined resulting in the new coarse grid on the top right which is then regularly refined to produce the new fine grid on the lower right.

6 Subdivision of 3D-Boundary Cells for an Adaptive Version of Par-EXPDE

Adaptive discretization grids can reduce the computational time and storage requirement for solving a PDE, if the solution of the equation contains singularities or other local structures. Therefore, we are very interested in an efficient implementation of adaptive discretization grids.

The original EXPDE library uses uniform semi-unstructured grids to discretize general domains in 3D. In [2], the construction of these grids is described. The main idea of these grids is to use a large structured grid inside of the domain and a small unstructured grid near the boundary. To obtain adaptive semi-unstructured grids with the same property, the adaptive grid inside of the domain must be constructed in a different way than the one

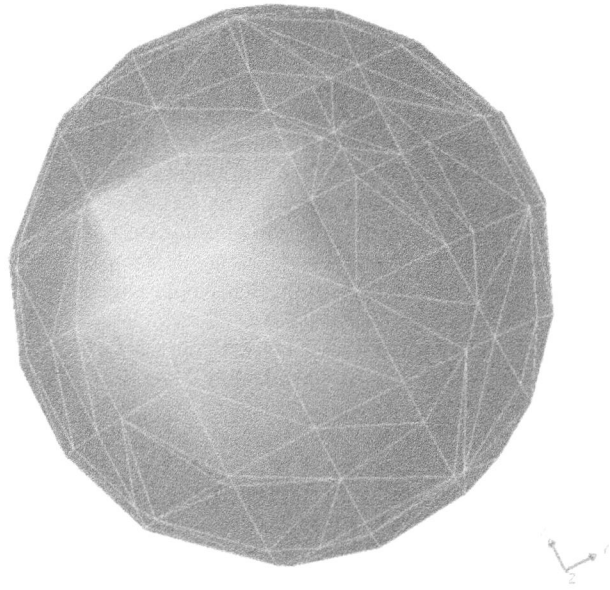

Fig. 5. Adaptive subdivision of a ball into tetrahedra. One boundary cell has half the mesh-size of the surrounding boundary cells. Now EXPDE can correctly divide these surrounding boundary cells into tetrahedra.

near the boundary. Inside of the domain a structured adaptive grid is realized by hanging nodes. Near the boundary an unstructured adaptive grid is constructed by an adaptive subdivision of boundary cells. This adaptive subdivision of boundary cells by tetrahedra has to be done in such a way that the following properties are satisfied:

– A maximal interior angle condition is satisfied.
– The resulting finite element space with linear functions is continuous.

To obtain a subdivision with the above properties one has to observe that there exist only 12 types of boundary cells (see [2]). Furthermore, certain adaptive boundary cells can be avoided by a suitable grid generation process. The resulting boundary cells can be subdivided similar to the uniform case (see [4]). Fig. 6 shows an example of an adaptive semi-unstructured grid.

We are currently working on the implementation of stiffness matrices and suitable finite element operators on semi-unstructured grids. Numerical results for solving a PDE on adaptive semi-unstructured grid will be presented in future.

References

1. Frank Hülsemann, Peter Kipfer, Ulrich Rüde, and Günther Greiner. gridlib: Flexible and efficient grid management for simulation and visualization.
   In *Computational Science-ICCS 2002*.
   Number 2331 in Lecture Notes in Computer Science, pages 652–661, 2002.
2. Pflaum, C: Semi-unstructured grids. Computing **67**, 141-166 (2001)
3. Pflaum, C: Expression templates for partial differential equation.
   Comput Visual Sci **4**, 1–8, (2001)
4. Pflaum, C: The maximum angle condition of semi-unstructured grids.
   In *Proceedings of the Conference: Finite Element Methods, Three-dimensional Problems (Jyväskylä, June 2000)*, Math. Sci. Appl., pages 229–242. GAKUTO Internat. Series, Tokio, 2001.
5. Veldhuizen, T: Expression Templates. C++ Report **7**(5), 26–31 (1995)
6. Super Technical Server Hitachi SR8000 Tuning Manual (C Language Version). Hitachi, Ltd. 2000.

gridlib — A Parallel, Object-Oriented Framework for Hierarchical-Hybrid Grid Structures in Technical Simulation and Scientific Visualization

Peter Kipfer[1], Frank Hülsemann[2], Stefan Meinlschmidt[1], Ben Bergen[2], Günther Greiner[1], and Ulrich Rüde[2]

[1] Computer Graphics Group
    University of Erlangen, Germany
    kipfer@cs.fau.de
[2] System Simulation Group
    University of Erlangen, Germany
    huelsemann@cs.fau.de

Abstract. This paper presents the *gridlib* software framework for integrated simulation and visualization. It provides modern object-oriented programming methods for supercomputer applications. While allowing the reuse of existing simulation codes, it enables writing new applications using advanced programming techniques. In this paper, we also present the concept of hierarchical hybrid grids as an efficient and flexible grid data structure for large scale computations. The resulting enormous amounts of data of nowadays applications can be visualized directly on the supercomputer as an integrated part of the whole simulation cycle, avoiding the typical bottleneck of supercomputers: external communication channels.

1 Introduction

The goal of the *gridlib* project is to develop a modern object-oriented software infrastructure for common grid-based numerical simulation problems on trans-teraflops machines. These supercomputers, like the Hitachi SR8000, and modern scalable algorithms allow numerical simulations to be performed at unprecedented grid resolutions. However, this also tremendously increases the sizes of the data sets, surpassing the capabilities of current pre- and post-processing tools by far. At the same time, pre- and post-processing has become more and more important. Current complex engineering solutions require the automatic generation of problem-specific, time-dependent, adaptive, hybrid 3D grids that can be partitioned for parallel simulation codes. Enormous amounts of data must be presented visually for easy interpretation.

The system hardware of current supercomputers also places non-trivial demands on the software architecture, in particular the gap between the low bandwidth of external communication channels and the available size of local data. This requires the execution of pre- and post-processing steps on the supercomputer, which is a significant problem due to missing generic software support. Other difficulties arise since only special data and software structures can be efficiently handled on the high performance architectures. A naive implementation may lead to unacceptable performance problems.

The *gridlib* project addresses these problems, acting as a middle-ware between existing software modules for pre- and post-processing and for implementing efficient solvers for complex simulation tasks. In this paper, we present an architectural overview of the *gridlib*, briefly describe its current functionality and give an outlook to further developments.

2 Overview

The *gridlib* architecture provides three major abstraction layers (see Fig. 1) [6, 7]. The lowest one is responsible for encapsulating the actual memory layout of data. Because the next layer entirely relys on this abstraction, the lowest layer can organize the storage freely. In particular, it can format its own memory layout to conform to the memory layout of other third party codes. We exploit this possibility for using a binary-only flow solver.

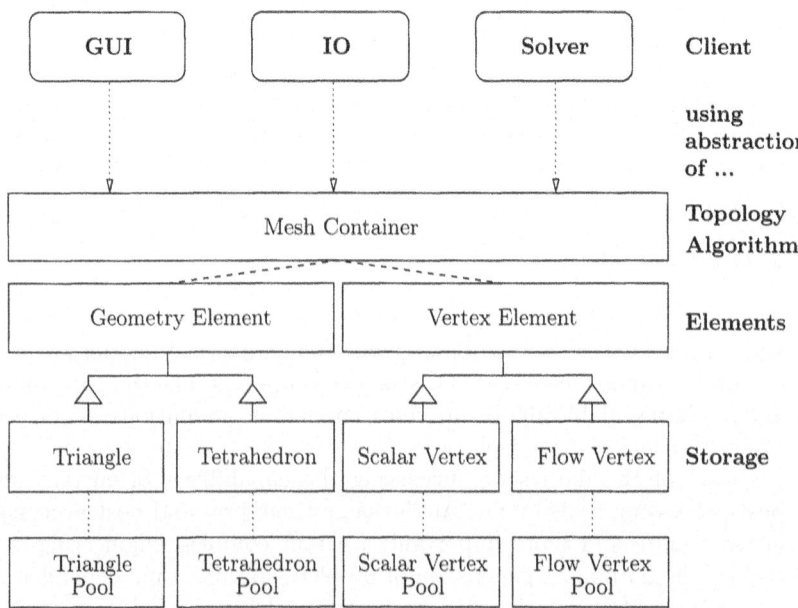

Fig. 1. The *gridlib* provides three major abstraction layers for both integration of binary-only codes and development of new object-oriented solvers.

The second abstraction layer is the main link to the object-oriented world. It provides interfaces for all primitive elements (triangles, quads, tetrahedra, hexahedra, prisms, pyramids, octahedra), edges and vertices as regular C++ classes. This sets the *gridlib* apart from other grid management libraries, as most of them do not allow the programmer to actually call methods on the objects.

The topmost layer provides the concept of a mesh container. It does not make any assumption on the mesh topology and implements abstract services, like neighborhood setup, subdivision functionality and management and content iterators.

The mesh container and the element abstraction layer provide powerful object-oriented programming support. For the library user, the *gridlib* further implements several clients that use the three-layer architecture for disk I/O, visualization and simulation. The performance of the interfaces for the data exchange between the grid management, the solver and the visualization and rendering subsystems has been evaluated by performing several simulations. However, please note that producing simulation results is not the main goal of the *gridlib* project.

For evaluation of the *gridlib* functionality, two solver codes have been ported to the SR8000 architecture to perform the full simulation cycle, consisting of grid management, solution and visualization. The solvers have been used for evaluating the functionality of the *gridlib* using two different approaches. The first solver presents the classic approach of a fluid flow simulation written in Fortran which can be integrated into the *gridlib*, even though it is available as object code only. Here the *gridlib* implements functionality for communication with the Fortran module by providing the data in exactly the right internal format that the Fortran program expects. The second solver is a generic C++ multigrid library that is currently under development and which uses the advanced *gridlib* functionality to obtain excellent runtime efficiency by exploiting the advanced architectural features of the SR8000.

2.1 Applications

The software architecture of the multigrid library is based on results of a DFG-funded project to develop highly efficient iterative solvers for hierarchical memory systems [1]. From the architectural viewpoint of the *gridlib*, this solver is a high-level client application. It is currently being used for several projects. This includes the computation of bioelectric fields [3] and the development of a simulation tool for the prognosis of the spreading and the degradation of contaminants in the saturated and vadose zone (funded by the Bavarian Environmental Protection Agency[1]).

---

[1] Bayerisches Landesamt für Umweltschutz

Finally, *gridlib* will be used with the solvers and software methodology developed within the ParEXPDE project [10]. The corresponding interface is currently under development.

3 Visualization and Rendering

During the first project year, the integrated visualization and rendering subsystem was ported to the SR8000 in order to provide post-processing tools on the supercomputer itself. Using the interfaces of high abstraction levels, it is completely independent of the numeric solver in use.

The visualization and rendering subsystem implements visualization methods for arbitrary planar slices through the unstructured grid [8], direct volume rendering by regular re-sampling, fast isosurface extraction [9, 12] and local exact particle tracing [5] (see Figs. 8 and 9). All methods use an abstract renderer for geometric primitives (triangles, quads, ...) for displaying the result (see Fig. 2). We have derived several concrete implementations from the abstract renderer that perform the actual image generation. There are pure software rasterizer classes as well as OpenGL based hardware accelerated rasterizer classes. This allows generating the visualization image on screen or in an offscreen rendering context using hardware acceleration if supported by the computers architecture.

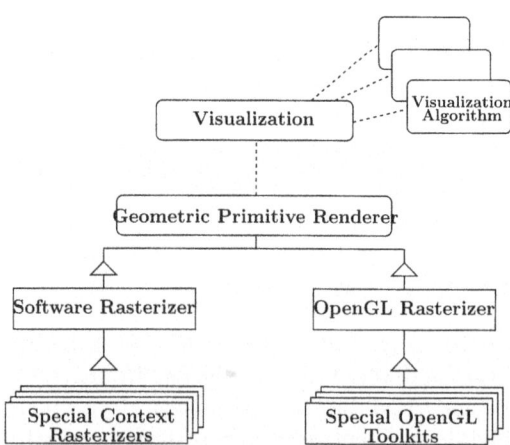

Fig. 2. The three-tier concept of the visualization and rendering subsystem.

The three-tier concept of the visualization and rendering system is very flexible and allows three basic usage scenarios:

– *remote rendering* on the supercomputer, using pure software algorithms for direct visualization on the high resolution simulation grid. Visualization and rendering parameters are passed to the subsystem along with simulation parameters at process startup time.

– *post-processed rendering* on the local desktop computer, using hardware accelerated algorithms to display (reduced) visualization geometry that is the result of a pre-visualization processing step on the supercomputer.

– *hybrid rendering*: By manipulating a reduced geometric representation of the simulation grid on the local desktop computer, the visualization parameters can be tuned interactively. The parameters are sent to the supercomputer for remote rendering. The result is sent back to the desktop where it is integrated into the local model.

The hybrid rendering approach allows especially easy handling of the grid, while maintaining very accurate visualization results. Furthermore, it allows simulation and visualization tasks to be run in parallel, display intermediate results and control grid management and numerical solvers on-the-fly. Thus, our approach allows computational steering, if supported by the solver.

So far, the emphasis has been on the implementation of the first two approaches, since they are most suitable for mainly batch-operated systems.

3.1 Parallel visualization and rendering

The *gridlib* supports parallel execution of visualization and rendering. Figure 3 shows the general data path for the application case "remote rendering". After distributing the simulation grid equally (in terms of number of cells) to all participating processing entities by a designated I/O process, a refined partition of the grid is computed in parallel. The refinement can be parameterized by geometric properties or arbitrary weights. In the general case, we map the grid nodes to the partition vertices. We also support the creation of the dual grid, where the cell centroids are mapped to the partition vertices. The visualization and rendering subsystem reuses the partition for executing the requested method on local data only. The distributed rendering contexts are finally merged into the final image by a synchronized method. The resulting image can be stored locally or it can be transferred back to the desktop computer [4].

Figure 4 demonstrates the achieved scalability of the rendering subsystem. We get almost linear speedup although the communication volume in the Z-buffer merging stage grows with the number of processors. Note that we minimize the amount of pixels to transport by projecting the bounding box of the triangles to render into screen space and by only selecting the projected rectangle for transport along with its screen space coordinates. The synchronization of the buffer merging stage is ensured by using blocking MPI calls. Although this seems to cause unnecessary delays, it is not a problem in practice, because the overall rendering time in most cases is typically 1 second. Dynamic redistribution of the triangles to rasterize incurs communication overhead that slows down the whole process considerably. The efficiency of

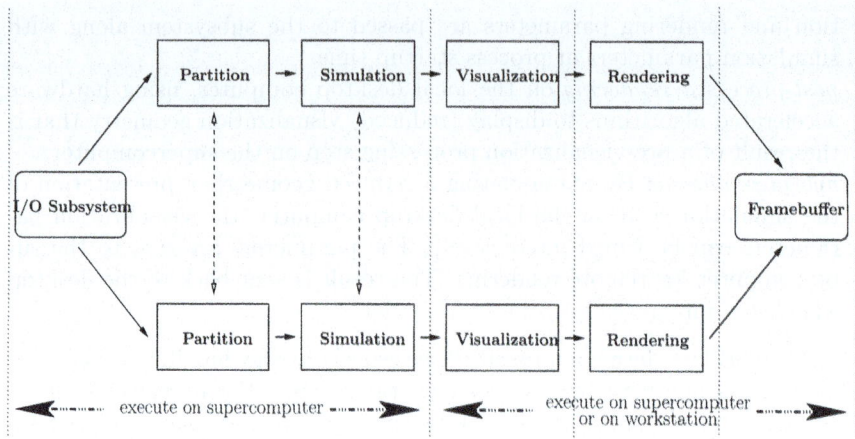

Fig. 3. The general data flow within the visualization and rendering subsystem.

our approach is shown in the left experiment of Fig. 5, where we compare two renderer instances running on the same node with two instances running on two separate nodes. The same was done in the right experiment using four instances on the same node and on four nodes respectively. The performance differences are within the normal measurement jitter using wall-clock time. The figures clearly show that there is no time penalty for distributing the rasterizers to different nodes: The overhead of the Z-buffer merging stage is not apparent.

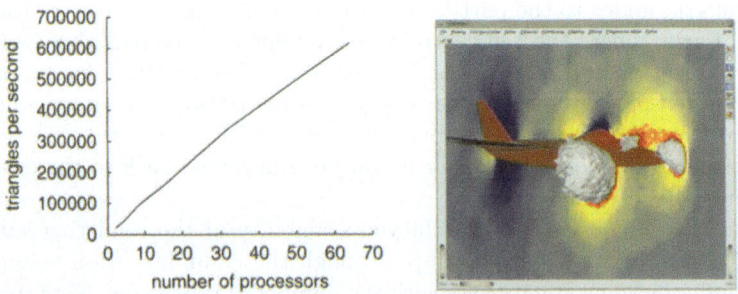

Fig. 4. Scaling of the rendering subsystem in terms of number of processors used and an example scene displaying density (color mapped slice) and an energy iso-surface.

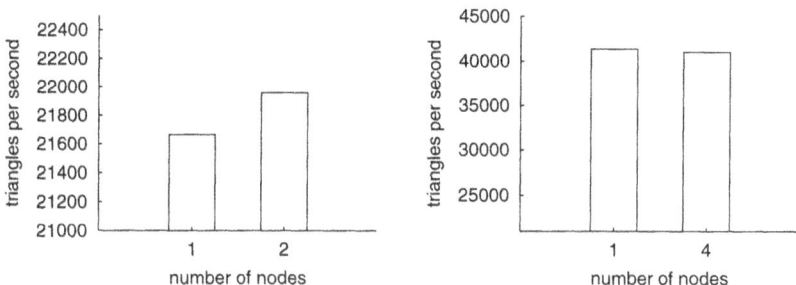

Fig. 5. Comparison of assigning N renderers to 1 node (left columns) or 1 renderer to N nodes (right columns). In the left experiment N=2, on the right N=4.

4 Hierarchical hybrid grids

Beyond the administration of a given grid, the *gridlib* project also aims to provide efficient and flexible grid data structures for large scale computations. Currently investigated are the so called *hierarchical hybrid grids*, which can be thought of as a hierarchy of block-structured grids generated from a potentially unstructured input grid. As first results illustrate, the hierarchical hybrid grid structure gives good performance as central data structure for the implementation of efficient numerical codes.

It is a fact that numerical simulations on completely unstructured grids do not match the floating point performance of those on structured grids.

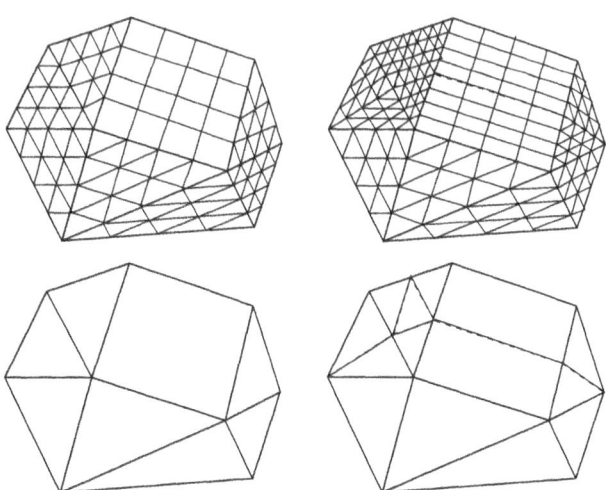

Fig. 6. The hierarchical hybrid grid structure is a multi-level regular subdivision performed on the cells of an unstructured geometry grid.

For the *gridlib* project, we have developed the concept of hierarchical hybrid grids which is able to compensate for the performance penalty of completely unstructured approaches, while maintaining their flexibility. It is based on the assumption that the accuracy demands placed on the simulation results necessitate the use of a much finer grid than the description of the problem domain. We therefore assign the distinct requirements placed on a simulation grid, namely geometric domain representation and resolution of simulation results, to different grid levels in our hierarchy.

The coarsest grid level is given by the input (geometry) grid. This grid may be completely unstructured. We then construct a hierarchy of simulation grids from the geometry grid by multiple steps of regular subdivision of the cells. Although the refined grids are still globally unstructured, we are now able to exploit the regularity within each cell using efficient solver algorithms.

We do not claim that the idea behind the generation of a block-structured grid from a given input grid is new. However, a combination of features does make the *gridlib* stand out. It has been designed from the start to cater to hybrid grids, it provides a hierarchy of refined grids with the aim to support geometric multigrid algorithms such that the operations on the refined grids can exploit the regularity for maximum efficiency. The top level user interface to the hierarchical hybrid grids is currently a work in progress but it is intended to provide building blocks for (multigrid) algorithms that hide the intricacies of the underlying data structures from the user.

In general, the nodes of a block structured hybrid grid can be classified into two groups: Internal nodes and nodes for which the neighborhood data has to be stored explicitly. While the first group consists of all nodes that have been created by the regular subdivision, the second group comprises the nodes of the geometry grid. After several subdivision steps, the first group will contain a much larger number of nodes than the second one. Therefore, in the context of node-based discretization, the vast majority of unknowns will lie in a regular environment. In what follows, we show that this approach outperforms completely unstructured grids by far.

As an example code we implement a Gauss-Seidel iteration which is frequently used as a smoother in multigrid methods. To highlight the difference between a general but unstructured approach and the exploitation of the grid regularity, we concentrate on the computations in the interior of a single cell of a potentially unstructured input grid. For ease of presentation, we assume in the following that the cell is a square which is regularly subdivided a number of times. If we view the resulting finest grid in the square as part of a globally unstructured grid, the straightforward implementation of one Gauss-Seidel step for the linear system $Au = f$ will read as follows:

```
for i from first vertex to last vertex:
  rhs = f(i)
  for j from 1 to number_of_neighbors(i)
    rhs = rhs - coeff(i,j) * u(neighbor(i,j))
  u(i) = rhs / coeff(i,i)
```

Here we assume that the matrix entries are stored in the `coeff` array.

In contrast, an implementation that takes advantage of the internal regular structure of the cell can access the neighbors of an unknown directly. It is well known that bilinear finite-element basis functions result in a 9-point stencil, which can be encoded in the following manner:

```
for i from first column to last column:
  for j from first row to last row:
    u(i,j) = (f(i,j)-c(i,j,1)*u(i-1,j-1)-c(i,j,2)*u(i-1,j)
                    -c(i,j,3)*u(i-1,j+1)-c(i,j,4)*u(i+1,j-1)
                    -c(i,j,5)*u(i+1,j)   -c(i,j,6)*u(i+1,j+1)
                    -c(i,j,7)*u(i,j+1)   -c(i,j,8)*u(i,j-1))/c(i,j,9)
```

Note that the unknowns inside the square are stored as a two-dimensional array and that the neighbors of an unknown are accessed directly by index arithmetic. Our implementation on the SR8000 replaces the division by appropriate pre-scaling of the matrix coefficients. It can then be parallelized in a straight forward manner: In order to avoid data dependencies, we introduce a red-black ordering of the *rows*. In the multigrid literature, this is referred to as a "zebra-smoother".

On the SR8000, we can implement the parallel version using Fortran and COMPAS:

```
*poption parallel
C       loop over rows
        do j = 2,NX,2

C               loop over columns
                do i = 2,NX
                    u(i,j) = c(1,i,j)-c(2,i,j)*u(i,j+1) -
        .                    c(3,i,j)*u(i+1,j+1)-c(4,i,j)*u(i+1,j) -
        .                    c(5,i,j)*u(i+1,j-1)-c(6,i,j)*u(i,j-1) -
        .                    c(7,i,j)*u(i-1,j-1)-c(8,i,j)*u(i-1,j) -
        .                    c(9,i,j)*u(i-1,j+1)

                end do
        end do

*poption parallel
C       loop over rows
        do j = 3,NX,2

C               loop over columns
                do i = 2,NX
                    u(i,j) = c(1,i,j)-c(2,i,j)*u(i,j+1) -
        .                    c(3,i,j)*u(i+1,j+1)-c(4,i,j)*u(i+1,j) -
        .                    c(5,i,j)*u(i+1,j-1)-c(6,i,j)*u(i,j-1) -
        .                    c(7,i,j)*u(i-1,j-1)-c(8,i,j)*u(i-1,j) -
        .                    c(9,i,j)*u(i-1,j+1)
```

```
        end do
    end do
```

We would like to point out that while one can exploit the grid structure further, this example shows that given an appropriate data structure, one can obtain a reasonably efficient, shared memory parallel program with comparatively little effort.

4.1 Results

The parallelization of the unstructured version is more involved because of its general character. On the SR8000, we obtain only 250 MFLOPS on a single node. Figure 7 compares the performance of the unstructured code (left column) to a unstructured version (middle column), that has additional knowledge about the type of refinement. Although this allows us to double the performance, it is still far from the theoretical peak performance of 12 GFLOPS for one node. The right column (5 GFLOPS) shows the performance of the structured version using the lexicographic processing within one line as shown above.

During the evaluation of the hierarchical hybrid approach, that was presented in Sect. 4, we experienced some unexpected behavior. Although the compiler was allowed to change the instruction sequence, the floating point performance of the SR8000 depends on the order in which the neighbors of the unknown appear in the statement. The piece of Fortran code in the previous section for example runs at 4 GFLOPS, while tuning the order of the instructions yields 5 GFLOPS as stated above.

As pointed out before, the computations inside a square illustrate most clearly what can be gained from exploiting explicit knowledge about the

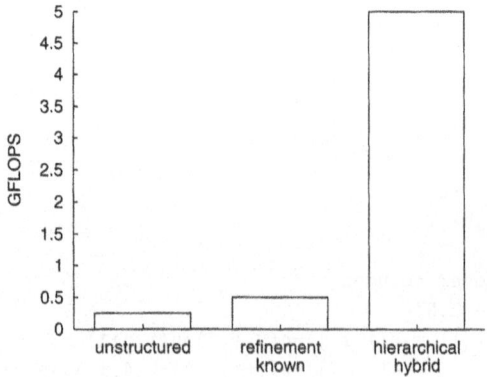

Fig. 7. Performing Gauss-Seidel iterations on a square patch of the geometry grid. The patch was subdivided ten times which yields in the order of 10^6 unknowns. All FLOP rates are given for one node of the SR8000 using COMPAS parallelization.

grid. Due to some inevitable synchronization between the regular regions, it would be unrealistic to expect the same factor of 20 when comparing whole program performance of completely unstructured and block structured implementations. However, as shown in [2], provided that the cells of the input grid can be refined sufficiently many times, the program performance will be dominated by the computations in the structured regions.

5 Conclusion

In this paper, we presented the *gridlib* architecture, a framework for integrated simulation and visualization. The *gridlib* introduces modern object-oriented programming paradigms to supercomputer applications for the highly dynamic programming tasks of grid management and visualization.

The parallel visualization and rendering subsystem successfully distributes the visualization task to all participating processors, making visualization algorithms possible that work on the refined hierarchical hybrid computational grid. The benefit is two-fold: no artifacts of grid reduction as introduced by common methods that create a visualization grid, and high speed thanks to parallel execution. Furthermore, we have shown that integrated simulation and visualization is possible and, given the bandwidth limitation of external communication channels, is inevitable for current numerical simulation applications. The *gridlib* therefore addresses the significant problem of missing tools for pre- and post-processing steps on the supercomputer using modern object-oriented software.

We have presented the concept of hierarchical hybrid grids which allows efficient implementations that satisfy both geometric accuracy because of their flexible unstructured nature and at the same time are able to maintain high numerical performance due to their regular subdivision scheme. Our experiments have shown that the program performance will be dominated by the computations in the structured regions. The hierarchical hybrid approach therefore is able to deliver competitive performance to Fortran code and integrates nicely with advanced object-oriented programming concepts.

Acknowledgement. The authors wish to thank Dr. Brenner from the fluid dynamics group of Erlangen University for helpful discussions, U. Labsik from the Computer Graphics Group at Erlangen University for his input concerning geometric modeling and mesh adaptivity and M. Kowarschik from the System Simulation Group, also at Erlangen University, for his insights in exploiting grid regularity for iterative methods. The *gridlib* project is funded by a KONWIHR grant of the Bavarian High Performance Computing Initiative, which also provided the access to the Hitachi SR8000.

Fig. 8. Two example visualizations: multi-modal visualization of a biomedical simulation on the left and direct volume rendering on the right.

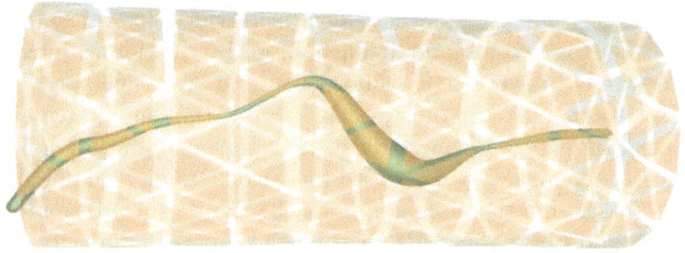

Fig. 9. Example: local exact particle tracing in unstructured grids.

References

1. C.C. Douglas, G. Haase, J. Hu, M. Kowarschik, U. Rüde, and C. Weiß. Portable Memory Hierarchy Techniques For PDE Solvers: Part II. *Siam News*, 33(6), July 2000.
2. F. Hülsemann, P. Kipfer, U. Rüde, and G. Greiner. *gridlib*: Flexible and efficient grid management for simulation and visualization. In *Lecture Notes in Computer Science, Computational Science*, 2002.
3. C. R. Johnson, M. Mohr, U. Rüde, A. Samsonov, and K. Zyp. Multilevel Methods for Inverse Bioelectric Field Problems. In T.J Barth, T. Chan, and R. Haimes, editors, *Multiscale and Multiresolution Methods*, volume 20 of *Lecture Notes in Computational Science and Engineering*, pages 331–346. Springer, 2001.
4. P. Kipfer and G. Greiner. Parallel rendering within the integrating simulation and visualization framework "*gridlib*". *VMV Conference Proceedings, Stuttgart*, 2001.
5. P. Kipfer, F. Reck, and G. Greiner. Local exact particle tracing on unstructured grids. *Computer Graphics Forum*, 2002. submitted.

6. Peter Kipfer. *gridlib*: System design. Technical Report 4/00, Computer Graphics Group, University of Erlangen-Nürnberg, 2000.

7. Peter Kipfer. *gridlib*: Numerical methods. Technical Report 2/01, Computer Graphics Group, University of Erlangen-Nürnberg, 2001.

8. U. Labsik, P. Kipfer, and G. Greiner. Visualizing the structure and quality properties of tetrahedral meshes. Technical Report 2/00, Computer Graphics Group, University of Erlangen-Nürnberg, 2000.

9. U. Labsik, P. Kipfer, S. Meinlschmidt, and G. Greiner. Progressive isosurface extraction from tetrahedral meshes. *Pacific Graphics Conference Proceedings, Tokio*, 2001.

10. A. Link, C. Pflaum, and B. Bergen. Scientific Progress in the Par-expde Project. Technical Report Preprint No. 246, Universität Würzburg, 2000.

11. M. Sabanca, G. Brenner, and N. Alemdaroglu. Improvements to compressible Euler methods for low-Mach number flows. *Int. Journal Numerical Methods in Fluids*, 34:167–185, 2000.

12. M. Schrumpf. Beschleunigte Isoflächenberechnung auf unstrukturierten Gittern. *Studienarbeit*, 2001. Computer Graphics Group, University of Erlangen-Nürnberg.

Peter Lipps, "Instruction systems," in *The use of Ingot 2000 Compiler Code Through Breadth of Computer Architects,* 2001.

Peter Lipps, Andrea Lunardini, and H. ... Rahnema, *Journal of Computer Graphics Group,* University of Informatics in Theory, 2001.

Andreas R. Lipps, and H. Lunardini, "Evaluating the execution intermediate Operation of Instructed codes," *Technical Report 1460, Computer Operation Group University of Chicago,* Information Group, 2000.

S. Lewis, I. Brighton, S. Rothschild, and D. Cooper, "Programmable Compiler Instructed Data-Flow and modeling," *Contract Report, Computer Processing,* 2004.

Annual Indian Science and technology — proof of action in the IEEE H. I. Lipps, *Journal of Informatics in Information Systems,* 2004.

The manufacturer's authorised representative in the EU is Springer
Nature Customer Service Centre GmbH, Europaplatz 3, 69115 Heidelberg,
Germany. If you have any concerns regarding our products, please
contact ProductSafety@springernature.com

Printed and bound by CPI Group (UK) Ltd, Croydon, CR0 4YY

28/04/2026

02098453-0009